D1626018

Atlas of
Human Tumor Cell Lines

Atlas of
Human Tumor Cell Lines

Edited by

Robert J. Hay
Cell Culture Department
American Type Culture Collection
Rockville, Maryland

Jae-Gahb Park
Department of Surgery
Seoul National University Hospital
Seoul, Korea

Adi Gazdar
Simmons Cancer Center
Southwestern Medical School
Dallas, Texas

Academic Press, Inc.
A Division of Harcourt Brace & Company

San Diego   New York   Boston   London   Sydney   Tokyo   Toronto

Front cover photograph: Scanning electron micrograph of KATO-III gastric carcinoma cells. The cultured cells are covered with numerous microvilli, and the typical signet ring-shaped cells have cave-like depressions on the cell wall. Courtesy of T. Suzuki.

This book is printed on acid-free paper. ∞

Copyright © 1994 by ACADEMIC PRESS, INC.
All Rights Reserved.
No part of this publication may be reproduced or transmitted in any form or by any means, electronic or mechanical, including photocopy, recording, or any information storage and retrieval system, without permission in writing from the publisher.

Academic Press, Inc.
525 B Street, Suite 1900, San Diego, California 92101-4495

United Kingdom Edition published by
Academic Press Limited
24–28 Oval Road, London NW1 7DX

Library of Congress Cataloging-in-Publication Data

Atlas of human tumor cell lines / edited by Robert J. Hay, Jae-Gahb Park, Adi Gazdar.
     p.   cm.
   Includes bibliographical references and index.
   ISBN 0-12-333530-2
   1. Cancer cells--Atlases.   2. Human cell culture--Atlases.
 I. Hay, Robert.   II. Park, Jae-Gahb.   III. Gazdar, Adi F.
   [DNLM: 1. Tumor Cells, Cultured--atlases.   2. Tumor Stem Cells-
-atlases.   3. Cell Line--atlases.   4. Neoplasms--pathology-
atlases.]
   RC267.A84   1994
   616.99'2'0072--dc20   93-25709
                                                               CIP

PRINTED IN THE UNITED STATES OF AMERICA
94   95   96   97   98   99   EB   9   8   7   6   5   4   3   2   1

Contents

Preface *xi*

1 Quality Control and Characterization of Cell Lines
Robert J. Hay

    I. Introduction   1
    II. Seed Stock Concept   2
    III. Microbial Contamination   3
    IV. Cellular Cross-Contamination   8
    V. Origin and Function   13
    VI. Conclusions   14
       References   14

2 Human Glioma Cell Lines
Monica Nistér   Bengt Westermark

    I. Introduction   17
    II. Cell Lines Derived from Glioblastoma Multiforme and Anaplastic Astrocytoma   18
    III. Cell Lines Derived from Oligodendroglioma   31
    IV. Cell Lines Derived from Primitive Neuroectodermal Tumor/Medulloblastoma   32
    V. Concluding Remarks   33
       References   34

3 **Tumor Cell Lines of the Peripheral
   Nervous System**
Mark A. Israel   Carol J. Thiele

   I. Neuroblastoma   45
   II. Peripheral Neuroepithelioma   61
   References   70

4 **Head and Neck Tumor Cell Lines**
Thomas E. Carey

   I. Introduction   79
   II. Methods of Establishment and Maintenance   81
   III. Morphology   96
   IV. Other Characteristics   108
   V. Discussion   113
   VI. Future Prospects   115
   References   117

5 **Cell Culture of Lung Cancers**
Adi Gazdar

   I. Cellular Origins of Lung Cancers   121
   II. Major Forms of Lung Cancer   123
   III. Defined Media for the Culture of Lung Tumors   124
   IV. Establishment of Small-Cell Lung Carcinoma Cell Lines   125
   V. Other Neuroendocrine Tumors   134
   VI. Cell Lines Derived from Non-Small-Cell Lung Carcinoma   136
   VII. Mesotheliomas   146
   VIII. Conclusions   146
   References   146

6 **Cell Lines from Human Breast**
Albert Leibovitz

   I. Introduction   161
   II. Methods of Establishment and Maintenance   162
   III. Morphology   162

IV. Discussion   180
V. Future Prospects   181
   References   181

7 Hepatocellular Carcinomas
Masahiro Miyazaki   Masayoshi Namba

I. Introduction   185
II. Undifferentiated Hepatocellular Carcinoma Cell Line HLE   186
III. Differentiated Hepatocellular Carcinoma Cell Lines   189
IV. Differentiated Hepatoblastoma Cell Lines   199
V. Conclusions   205
   References   205

8 Hematopoietic Cell Lines
Hans G. Drexler   Suzanne M. Gignac   Jun Minowada

I. Introduction   213
II. Culture of Hematopoietic Cells   214
III. Leukemia–Lymphoma Cell Lines   217
IV. Characterization of Leukemia Cell Lines   219
V. Growth Factor-Dependent Leukemia Cell Lines   242
VI. Hodgkin's Disease-Derived Cell Lines   243
VII. Future Prospects   244
VIII. Lists of Leukemia Cell Lines   249
   References   249

9 Human Sarcoma Cells in Culture
Richard B. Womer   Albert E. Wilson

I. Introduction   251
II. Methods of Establishment and Maintenance   252
III. Specific Cell Line Characteristics   255
   References   265

10 Cell Lines from Esophageal Tumors
Tetsuro Nishihira   Masafumi Katayama
Yuh Hashimoto   Takashi Akaishi

I. Introduction   269
II. Methods of Establishment and Maintenance   270
III. Morphology   273
IV. Other Characteristics   277
V. Discussion   279
VI. Future Prospects   282
   References   282

11 Gastric Tumor Cell Lines
M. Sekiguchi   T. Suzuki

I. Introduction   287
II. Methods of Establishment and Maintenance   288
III. Morphology   292
IV. Other Characteristics   309
V. Future Prospectives   312
   References   314

12 Colorectal Cancer Cell Lines
Jae-Gahb Park   Han-Kwang Yang   Robert J. Hay
Adi Gazdar

I. Introduction   317
II. Establishment of Cell Lines   320
III. Culture Characteristics   324
IV. Morphological Characteristics   325
V. Other Characteristics   334
   References   338

13 Cell Lines from Urinary Bladder Tumors
Sonny L. Johansson   Bertil Unsgaard   Carol O'Toole

I. Background   342
II. Patient History and History of Original Tumor   342

III. Histological Examination of the Original Tumors and Developed
Cell Cultures   343
IV. Establishment and Maintenance of Bladder Cancer Lines   349
V. Mycoplasma Testing   349
VI. Cryopreservation of Cell Lines   349
VII. Characterization of Lines   350
VIII. Summary   355
References   356

14 The Female Reproductive System: Cell Lines from Tumors of the Human Ovary and Uterus
Y.-C. Hung   S. Tabibzadeh   P. G. Satyaswaroop

I. Introduction   359
II. Methods of Establishment and Maintenance of Cell Lines   360
III. Morphological Aspects   363
IV. Growth and Other Characteristics   377
V. Future Prospects   384
References   384

15 The Male Reproductive System: Prostatic Cell Lines
Donna M. Peehl

I. Introduction   387
II. Methods of Establishment and Maintenance   388
III. Morphology   391
IV. Other Characteristics   396
V. Discussion and Future Prospects   405
References   407

16 Melanocyte and Melanoma Cell Lines
Tibor Györfi   Meenhard Herlyn

I. Introduction   413
II. Establishment and Maintenance   414
III. Morphology of Cultured Melanocytes and
Melanoma Cells   416

IV. Growth Characteristics   416
V. Growth Factor Production by Melanoma Cells   419
VI. Chromosomal Abnormalities in Melanoma Cells   420
VII. Antigen Expression by Melanocytic Cells   421
VIII. Invasion and Metastasis of Melanoma Cells   422
IX. Conclusions   424
X. Origin of Cell Lines   425
    References   426

17 Exocrine Pancreatic Tumor Cell Lines
Richard S. Metzgar

I. Introduction   429
II. Methods of Establishment and Maintenance   432
III. Morphology   432
IV. Other Characteristics   436
V. Induction of Differentiation   436
VI. Discussion   439
VII. Future Prospects   440
    References   440

18 Cell Lines from Human Germ-Cell Tumors
Peter W. Andrews   Ivan Damjanov

I. Introduction   443
II. Establishment and Maintenance of Cell Lines   445
III. Morphology   457
IV. Other Characteristics   458
V. Discussion and Future Prospects   463
VI. Reported Cell Lines Derived from Human Germ-Cell
    Tumors   466
    References   470

Index   477

Preface

The utility of human tumor cell lines in culture for a diverse range of research programs involving modern cellular and molecular biology, genetics, and oncology is recognized today much more extensively than ever before. Numerous cell lines of interest have been developed and are generally available. Data on those lines that are both useful and popular accumulate at a breathtaking pace, and the task of remaining up-to-date has become increasingly difficult. This atlas brings together in a single volume much of the most relevant information relating to human lines from major tissue categories, including all of the high-cancer-risk organs. It should be of use not only to those in a broad range of disciplines working with many different human tumor cell lines in culture, but also to those searching for continuous human lines with specific characteristics.

We recognized at the outset that the morphological definition of a cell line is only one of its characteristics and that this alone is a notoriously flawed tool for precise identification. In fact, many of the contributing authors point this out and provide and refer to additional features of specific types of cell lines that can be used for more positive identification and for further experimentation. Particulars on cell banking and characterization are included, along with representative data on the absolute identification of human cell lines by DNA fingerprinting. Most chapters detail methods for establishment and maintenance, morphology, identification of surface and intracellular antigens, and secreted products. If pertinent, lists of the available lines and future prospects for their use in research are included. Phase-contrast and electron photomicrographs are generally supplemented with additional related data on, for example, cell line derivation, growth properties in culture, antigenic traits, karyology, and related critical genetic attributes such as chromosomal abnormalities and locations of oncogenes, if known—all with appropriate reference to recent publications.

With the exception of several textbooks on cell culture methods, which generally do not dwell on morphology and other detailed descriptive characteristics of cell lines, only one major volume describing derivation and properties of many human tumor cell lines has been published [Fogh, J. (ed.) (1975).

"Human Tumor Cells in Vitro." Plenum, New York]. Since publication of that volume almost two decades ago, there has been a tremendous increase in both the number and the availability of new lines with unique characteristics. The *Atlas of Human Tumor Cell Lines* will fill this gap in information and provide timely and useful summaries for scientists working in the fields of cell biology and cancer research.

Robert J. Hay
Jae-Gahb Park
Adi Gazdar

Quality Control and Characterization of Cell Lines

Robert J. Hay
Cell Culture Department, American Type Culture Collection
Rockville, Maryland 20852

I. Introduction 1

II. Seed Stock Concept 2

III. Microbial Contamination 3
A. Bacteria and Fungi 4
B. Mycoplasma Infection 4
C. Viruses 5

IV. Cellular Cross-Contamination 8
A. Species Verification 9
B. Intraspecies Cross-Contamination 9

V. Origin and Function 13
VI. Conclusions 14
References 14

I. Introduction

The scientific literature documents over 250 instances of cross-contamination in cell culture systems (Nelson-Rees, 1978; Nelson-Rees et al., 1981; Hukku et al., 1984), and many more have certainly gone unreported. The novice technician or student using cell culture techniques soon is made painfully aware of the potential for bacterial and fungal infection. Generally, however, one must be alerted to the more insidious problems of animal cell cross-contaminations, the presence of mycoplasma, and especially the potential for latent or otherwise inconspicuous viral infection. The financial losses in research and production efforts resulting from the use of contaminated cell lines is incalculable but certainly equivalent to many millions of dollars. Accordingly, frequent reiteration of the details of cell culture contaminations and of precautionary steps to avoid and detect such problems clearly is warranted.

This chapter includes comments on quality control methods applied to authenticate cell lines, that is, to insure absence of microbial, viral, and cellular contamination, as well as potential tests to verify the identity of human cell lines for which previous data are readily available. The approach suggested has been developed during establishment of a national cell repository. Specific rationales for applying the tests indicated are included in this chapter or are discussed in more detail elsewhere (Hay, 1992). All these tests provide information supplementary to the primarily morphological data customarily presented in an atlas, such as this one, on human tumor cell lines.

Most established cell lines have been characterized by the originator and collaborators well beyond the steps essential for quality control. Specific details are provided in each subsequent chapter of this book and include, for example, phase contrast and ultrastructural morphologies; detailed cytogenetic analysis; definition of proto-oncogene, oncogene, or oncogene product presence, nature, and location; detailed evaluation of intermediate filament proteins; and demonstration of tissue-specific antigens or production of other specific products. These characterizations obviously increase the value of each cell line for research and, perhaps for production work. However, cell banking organizations need not attempt to repeat all these tests before distributing the stock cultures. Decisions must be made to establish the most acceptable authentication steps, consistent with maintaining the lowest possible cost, to provide a high quality cell stock. Authentication can be considered the act of confirming or verifying the identity of a specific line, whereas characterization is the definition of the many traits of the cell line, some of which may be unique and also may serve later to identify or authenticate that line specifically. Essential steps for quality control will vary with the type of cell bank constructed; such minimal descriptive data frequently will be supplemented with a much broader characterization base for each particular cell line.

II. Seed Stock Concept

Definitions of public repository seed stocks may vary from those used for specific applications such as production of vaccines or other biologicals. A scheme illustrating the steps involved in developing the seed stocks is presented in Fig. 1.

Generally, starter cultures or ampules are obtained from the originator and progeny are propagated according to instructions to yield the first "token" freeze. Cultures derived from such token material then are tested for bacterial, fungal, and mycoplasmal contamination. The species of each cell line is verified. These quality control steps are the minimum ones that must be performed before eventual release of a line. If these tests confirm that further efforts are warranted, the material is expanded to produce the seed and distribution stocks. Note that, under ideal conditions, additional major quality control and characterization efforts are applied to cell populations from seed stock ampules. Test results refer to specific numbered stocks. The distribu-

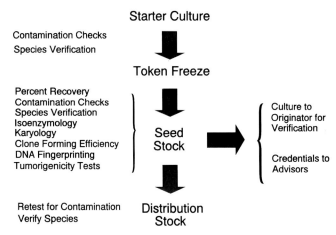

Fig. 1. A quality control scheme suggested for the addition (accessioning) of a cell line to a collection. Representative tests for token, seed, and distribution stocks are indicated (left), as are steps for review by originating investigators and advisors (right). Additional characterizations such as general or specific screens for viruses, functional tests, and definitions of particular inclusions or ultrastructure may be applied when critical, or with special support. Immunological tests are required, especially to identify specific protein inclusions or other products such as immunoglobulins released by hybridomas.

tion stock consists of ampules that are distributed on request to investigators. The reference seed stock, on the other hand, is retained to generate further distribution stocks as the initial distribution stock becomes depleted. The degree of characterization applied to master cell banks or master working cell banks in production facilities is generally more rigorous, although the seed stock, like the master cell bank, is used as a reservoir to replenish depleted distribution lots over the years. By adherence to this principle, problems associated with genetic instability, cell line selection, senescence, or transformation can be avoided.

III. Microbial Contamination

Microbial contamination in cell culture systems remains a serious problem. Cryptic contaminants, even of readily isolatable bacteria and fungi, are missed by many laboratories. The American Type Culture Collection (ATCC) still receives cultures, even for the patent depository, that contain yeast, filamentous fungi, and/or mycoplasma contaminants.

A. Bacteria and Fungi

Microscopic examination is only sufficient for detection of gross contaminations; even some of these cannot be detected readily by simple observations. Therefore, an extensive series of culture tests also is required to provide reasonable assurance that a cell line stock or medium is free of fungi and bacteria (Hay, 1992).

B. Mycoplasma Infection

Contamination of cell cultures by mycoplasma can be a much more insidious problem than that created by growth of bacteria or fungi. Although the presence of some mycoplasma species may be apparent because of the degenerative effects induced, other mycoplasmas metabolize and proliferate actively in the culture without producing any overt morphological change in the contaminated cell line. Thus, cell culture studies relating to metabolism, surface receptors, virus–host interactions, and so forth are certainly suspect in interpretation, if not negated entirely, when conducted with cell lines that harbor mycoplasma. The seriousness of these problems can be documented through published data from testing services and cell culture repositories. The high incidence of mycoplasma contamination from human operators is supported by the fact that *Mycoplasma orale* and others of human origin (*Mycoplasma hominus, Mycoplasma salivarium,* and *Mycoplasma fermentans*) are among those most frequently isolated. In the study by Del Giudice and Gardella (1984), of the 34,697 lines tested, 3955 (11%) were positive; 36% of these isolates were mycoplasmas of human origin (Table I). A high incidence of isolation of *Mycoplasma hyorhinis* was noted, that may result from use of contaminated sera or by culture-to-culture spread in laboratories working with infected biologicals. After a more recent study, Uphoff *et al.* (1992) reported that 84 (33%) of 253 cell lines submitted for their developing cell repository in Germany were infected with mycoplasma. A comparative analysis of the sensitivity of six different detection methods was conducted,

Table I

Mycoplasma Isolation, 1966–1982[a]

Number of specimens	34,697
Number of positives[b]	3,955
Mycoplasmas of human origin (% of total)	36
Mycoplasmas of bovine origin (% of total)	31

[a] Summarized from Del Giudice and Gardella (1984).
[b] About 14% of these were infected with more than one species of mycoplasma.

with the recommendation that a combination of two or three be selected for any routine testing regimen.

Nine general methods are available to detect mycoplasma. The direct culture test and the indirect test employing a bis-benzimidazole fluorochrome stain (Hoechst 33258) for DNA are used routinely in many laboratories to check incoming cell lines and all working cell stocks. Hay *et al.* (1989) reviewed detection and elimination methods. Advanced techniques involving the polymerase chain reaction (PCR) provide the most sensitive methods for detection (Spaepen *et al.*, 1992; Uemori *et al.*, 1992).

Four general recommendations can be offered to avoid mycoplasma infection. The implementation of an effective regime to monitor cell lines for mycoplasma is one critical step. Quarantining all new untested lines and using mechanical pipetting aids are others. Most experts also strongly suggest that the use of antibiotics be eliminated when possible. Antibiotic-free systems permit overgrowth by bacteria and fungi to provide ready indication whenever a lapse in aseptic technique occurs. When the initial tissue is used, for example, a human tumor sample, antibiotics may be employed intially, but after the primary population has grown out and been cryopreserved, reconstituted cells should be propagated further in antibiotic-free medium.

C. Viruses

Of the various tests to detect adventitious agents associated with cultured cells, those detecting endogenous and contaminant viruses are the most problematic. Representative viruses possibly present in human and other cell lines are presented in Table II. Development of an overt and characteristic cytopathogenic effect (CPE) certainly will provide an early indication of viral contamination. However, the absence of CPE definitely does not indicate that the culture is virus free. In fact, persistent latent infections may exist in cell lines and remain undetected until the appropriate immunological, cytological, ultrastructural, and/or biochemical tests are applied. Additional host systems or manipulations, for example, treatment with halogenated nucleosides, may be required for virus activation and isolation.

Screening techniques may be applied as an expedient compromise to monitor for readily detectable viruses associated with cell lines. Egg inoculations, in addition to selected co-cultivations and hemadsorption tests, can be included, as can routine examinations for CPE using phase contrast microscopy. Similar general tests are recommended by government agencies when cell lines are to be used for biological production work. Detailed protocols have been provided elsewhere (Hay, 1992; Lubiniecki and May, 1985).

Despite such screens, latent viruses and viruses that do not produce overt CPE or hemadsorption will escape detection. Some of these viruses could be potentially dangerous for the cell culture technician. For example, Hantaan

Table II

Representative Viruses of Special Concern in Cell
Production Work[a]

Host	Virus
Human	Human immunodeficiency viruses
	Human T-cell leukemia viruses
	Other endogenous retroviruses
	Hepatitis viruses
	Human herpes virus-6
	Cytomegalovirus
	Human papilloma virus
	Epstein–Barr virus
Other	Hantavirus
	Lymphocytic choriomeningitis virus
	Ectromelia virus
	Murine hepatitis virus
	Simian viruses
	Sendai virus
	Avian leukosis virus
	Bovine viral diarrhea virus

[a] See Hayflick and Hennessen (1989) for more detail.

virus, the causative agent of Korean hemorrhagic fever, replicates in rat, certain human tumor, and other cell lines. Outbreaks of the disease in individuals exposed to infected colonies of laboratory rats have been reported separately in five countries. An incident of transmission during passage of a cell line was confirmed in Belgium. As a result of these findings, several rat and rat/mouse hybrid cell lines expanded in this laboratory were screened using an indirect immunofluorescent antibody assay (LeDuc et al., 1985) and were found to be negative.

Concern over laboratory transmission of the human immunodeficiency viruses (HIV) also should be expressed. Several cases of probable infection during processing in United States laboratories have been described, one presumed to be caused by parenteral exposure and another by work with highly concentrated preparations (Weiss et al., 1988). In this latter circumstance, strict adherence to Biosafety Level 3 containment practices is essential. More detailed discussion of safety precautions for work with human tissues and cell lines is presented by Grizzle and Polt (1988) and by Caputo (1988).

Other viruses that may present general problems in cell culture work include ectromelia virus, bovine viral diarrhea virus (BVDV), and Epstein–Barr virus (EBV). Ectromelia virus, a member of the orthopoxvirus genus, is a natural pathogen of mice. Murine products such as ascitic fluids may transmit the virus; in the past, such infections have led to serious outbreaks of mousepox in animal colonies. Accordingly, screening murine products and lines for presence of this agent is prudent. The African Green monkey cell line BS-C-1 (ATCC.CCL 26) can be used to produce ectromelia stocks and to estimate virus infectivity effectively. With reference to murine cell lines, a study conducted at the National Institutes of Health (NIH) and the ATCC (Buller *et al.,* 1987) established that ectromelia virus replicated in all murine lymphoma lines and in a small proportion of the hybridomas tested. Further, certain hybridoma lines passed in ectromelia virus-infected mice yielded ectromelia virus infectivity on return to cell culture. Note that ectromelia-infected cultures were processed only in the special P4 facility available at the NIH. A screen of all murine lines in the ATCC collection involving two serial passages of the test line with BS-C-1 cells was completed in 1987. No infection with the virus was detected on observation for the characteristic CPE.

The frequent presence of BVDV in bovine sera has been recognized for years as a special problem for those engaged in production of human and animal viral vaccines. Infection of susceptible cell lines by noncytopathic BVDV strains can be unapparent. The presence of BVDV can be verified by immunofluorescence or by immunoperoxidase labeling. Screens performed in the past using the immunofluorescence method indicated that many lots of bovine sera were contaminated (Nuttal *et al.,* 1977); lines from a variety of other species (e.g., pig, cat, rabbit, and others) may be infected chronically (Levings and Wessman, 1991; Bolin *et al.,* 1992).

Accordingly, researchers might anticipate that many cell lines from these susceptible species will be infected with BVDV unless gamma irradiation (Erikson *et al.,* 1989), very stringent selection of bovine sera, or nonbovine sera were used during isolation and all subsequent propagation. Importantly, for the purpose of this volume, no reports of BVDV contamination of human or primate cell lines have been confirmed to date.

EBV, a member of the herpes virus group associated with Burkitt's lymphoma and infectious mononucleosis, is used extensively now to immortalize human B lymphoblasts to produce continuous lines. The nuclear and/or viral capsid antigens for EBV are found in most human B lymphoblastoid lines, although not all these lines give rise to biologically active virus. A common source of EBV for immortalization is the supernatant medium supporting growth of B95-8, an EBV-transformed marmoset cell line (ATCC.CRL 1612) (Miller and Lipman, 1973; Hay *et al.,* 1992). A second virus resembling a Type D retrovirus also is produced by some stocks of the line (Popovic *et al.,* 1982; Tumilowicz *et al.,* 1984). Since this primate cell type releases one virus known to be associated with human lymphoblasts as well as a second retrovirus, both the marmoset line and the human lymphoblastoid lines produced by transformation with B95-8 supernatants should be handled using appropriate precautions. At a minimum, class 2 containment facilities and procedures would be recommended.

IV. Cellular Cross-Contamination

Wherever cells are grown in culture, serious risk exists of the inadvertent addition of and subsequent overgrowth by cells of another individual or species. We hope this volume will assist in the rapid identification of potential problems by providing photomicrographs that illustrate the typical morphological characteristics of many human cell lines. As stated in the preface and in many chapters of this book, however, one cannot rely on morphological criteria alone to recognize cell lines or problems. Data have been collected over the years by groups offering identification services for cell culture laboratories in the United States and elsewhere.

In one study, 466 lines from 62 different laboratories were examined. Of these, 75 (16%) were found to be identified incorrectly. A total of 43 lines (9%) were not of the species expected, and 32 lines (7%) were either incorrect mixtures of two or more lines or were not the individual line stated (Nelson-Rees, 1978). More recently, Hukku *et al.* (1984) examined 275 lines over a period of 18 months. Results of their analyses are summarized in Table III. A total of 96 lines (35%) were not as indicated by the donor laboratories. For purported human lines, 36% were not as expected; 25% were a different species and 11% a different human individual.

To minimize the risk of cellular cross-contamination, culture technicians require a laminar flow hood for ideal operation. These individuals must be instructed periodically to work only with one cell line at any given time, use one reservoir of medium for each line, and avoid introducing pipets that have been used to dispense or mix cells into any medium reservoir. Technicians must be reminded repeatedly to label each cell culture legibly with designations, passages, and dates. Labels of different colors can be used to distinguish one cell line from another further during expansion. Technicians also must be instructed to allow at least 5 min hood clearance time, with ultra-

Table III
Summary of Cell Line Cross-Contamination[a]

Reported cell species	Cultures received	Interspecies	Intraspecies	Percentage of total
Human	160	40 (25%)	18 (11%)	36
Mouse	27	2	—	8
Cat	24	1	—	4
Others	64	35	—	54
Total	275	78	18	35

[a] Reprinted with permission from B. Hukku *et al.* (1984).
[b] Data from an 18-month survey.

violet lights and blower on, between cell lines when working on more than one line during a particular work period. The inner surfaces of each hood should be swabbed with 70% ethanol between such uses.

The studies outlined earlier illustrate the severity of the problem of cellular cross-contamination and provide a strong rationale for vigilance in careful handling, characterization, and authentication of cell lines.

A. Species Verification

Species of origin can be determined for cell lines by a variety of immunological tests, by isoenzymology, and/or by cytogenetics (Nelson-Rees, 1978; Hukku et al., 1984; Hay, 1992). The indirect fluorescence antibody staining technique is used in many laboratories to verify the species of a cell line (see Hay, 1992, for details). Isoenzyme analyses performed on homogenates of cell lines from over 25 species have demonstrated clearly the utility of these biochemical characteristics for species verification (O'Brien et al., 1977). By determining the mobilities of three isoenzyme systems—glucose 6-phosphate dehydrogenase (G6PD), lactic acid dehydrogenase (LDH), and nucleoside phosphorylase (NP)—using vertical starch gel electrophoresis, the species of origin of cell lines can be identified with a high degree of certainty. Alternatively, a standardized kit employing agarose gels and stabilized reagents may be obtained for this purpose.

Karyologic techniques long have been used informatively to monitor for interspecies contamination among cell lines. In many instances, the chromosomal constitutions are so dramatically different that even cursory microscopic observations are adequate. In others, for example, in comparisons among cell lines from closely related primates, careful evaluation of banded preparations is required (Nelson-Rees, 1978; Hukku et al., 1984). Cytogenetics has the advantage of detecting even very minor contaminants, on the order 1% or less in some circumstances. However, this procedure is time consuming and interpretation may require a high degree of skill. The karyotype is constructed by cutting chromosomes from a photomicrograph and arranging them according to arm length, position of centromere, presence of secondary constrictions, and so forth. Consult the "Atlas of Mammalian Chromosomes" (Hsu and Benirschke, 1967–1975) for examples of conventionally stained preparations from over 550 species.

B. Intraspecies Cross-Contamination

With the dramatic increase in numbers of cell lines being developed, especially from human tissues, the risk of intraspecies cross-contamination rises proportionately. The problem is especially acute in laboratories in which work is in progress with the many different cell lines of human and murine origin that are available today.

Methods for verifying cell line species employing enzyme mobility studies have been mentioned. Using similar technology with different enzyme systems, one also can screen for intraspecies cellular cross-contamination. Cell lines from various individuals of the same species often show different co-dominant alleles for a given enzyme locus, the products of which are polymorphic and electrophoretically resolvable. In most cases, the phenotype for these allelic isoenzymes (allozymes) is extremely stable. Consequently, when allozyme phenotypes are determined over a suitable spectrum of loci, they can be used effectively to provide an allozyme genetic signature for each line under study (O'Brien et al., 1977; Hukku et al., 1984).

The application of recombinant DNA technology and cloned DNA probes to identify and quantitate allelic polymorphisms provides additional powerful means for cell line identification. These polymorphisms can be recognized as extremely useful markers, even if they are not expressed through transcription and translation to yield structural or enzymatically active proteins.

Hybridization probes to regions of the human genome that are highly variable have been produced and are already in use for DNA fingerprinting applications including cell line individualization (Gilbert et al., 1990). Representative probes of demonstrated utility for cell line identification are listed in Table IV. Fingerprints derived from human cell lines using multilocus probes such as the 33.6 can be interpreted best using scanning devices since the patterns are complicated (see Gilbert et al., 1990, for examples).

Alternatively, or in addition, one may use single locus probes or, more appropriately, a mixture of several single locus probes to effect individual cell line identification (Reid et al., 1990; Honma et al., 1992). With single locus probes, one can not only differentiate easily among cell lines, but also detect cross-contamination. Reid and associates (1990) showed that 13 certified cell lines could be identified using pYNH24 and, through mixing experiments, found that cellular contamination could be detected at 5–10%. Fig-

Table IV

Examples of Probes Used for DNA Fingerprinting of
Human Cell Lines

Type	Designation	Originator/Reference
Multilocus	33.6	Jeffreys et al. (1985)
Single locus	pYNH24	Nakamura et al. (1987)
	ChdTC-15	Honma et al. (1991)
	ChdTC-114	Honma et al. (1991)
	TM-18	Honma et al. (1991)
	p79-2-23	Chimera et al. (1989)
	pCMM86	Nakamura et al. (1988)

ures 2 and 3 are autoradiograms showing typical results with pYNH24 and two other single locus probes. Each pair of lines from the same donor (Fig. 2) exhibited an identical pattern of mobility and hybridization, as expected. Lines from different individuals are distinguished from one another easily after their DNA is separated and hybridized with mixtures of labeled single locus probes (Fig. 3). Honma et al. (1991) also used pYNH24 and a cocktail of three other single locus probes (ChdTC-15, ChdTC-114, and TM-18) to identify 14 human cell lines. The profiles are reportedly more stable and more easily interpreted than those generated by the multilocus probe.

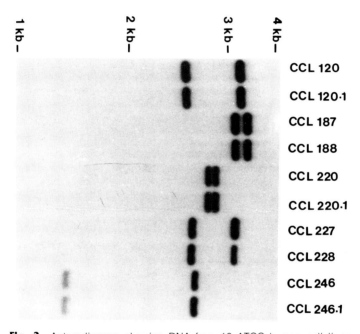

Fig. 2. Autoradiogram showing DNA from 10 ATCC human cell lines after hydrolysis with the restriction endonuclease HinfI, electrophoresis, Southern blotting, and hybridization with P³²-labeled PYNH24, a single locus probe. Results with DNA from five pairs of cell lines from five separate tissue donors are shown. CCL 120 and 120.1 are human B- and T-lymphoblastic cell lines, respectively, both from an 11.5 year-old white male with acute lymphoblastic leukemia. CCL 187 and 188 are colon cancer lines from a 58-year-old female. CCL 187 was developed without trypsin treatment whereas CCL 188 is a trypsinized variant. CCL 220 and CCL 220.1 are also colon carcinoma lines, from a 55-year-old female. CCL 220.1 developed as a variant with karyology somewhat different from CCL 220. CCL 227 and 228 were derived from a metastasis and from the primary adenocarcinoma, respectively, of the colon of a 51-year-old white male. CCL 246 and 246.1 were isolated from a 59-year-old white male with erythroleukemia. The two lines have different morphological and cytological staining properties. [See Reid et al. (1990) and Hay et al. (1992) for additional details. Photo provided by Y. Reid.]

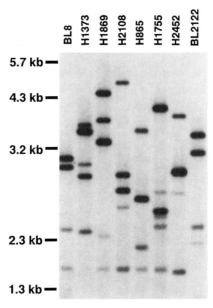

Fig. 3. Autoradiogram similar to that in Fig. 2 but showing DNA hydrolyzed with Hinf1 and hybridized with the three ^{32}P-labled, single locus probes pYNH24, p79-2-23, and pCMM86. The DNA was extracted from eight different human B-lymphoblastic (BL prefix) and human lung tumor (H prefix) cell lines. (Photo provided by Y. Reid.)

The polymerase chain reaction (PCR) provides yet another related means for specifically verifying cell line identities, even when very little cellular material is available. Thus, for example, to compare DNA from a small tissue sample with that of a cell line derived from it, PCR amplification of the tissue DNA would be the method of choice (Reid and Luo, 1993). Data shown in Fig. 4 illustrate the power of this technique to differentiate among seven distinct cell lines.

The best method of intraspecies cell line individualization involves karyotype analysis after treatment with trypsin and the Giemsa stain (Giemsa or G-banding). The banding patterns made apparent by this technique are characteristic of each chromosome pair and permit recognition, by an experienced cytogeneticist, of even comparatively minor inversions, deletions, or translocations. Many lines retain multiple marker chromosomes, readily recognizable by this method, that identify the cells specifically and positively (Chen et al., 1987). If readily recognized marker chromosomes are present, contaminations at less than 1% can be recognized with careful scrutiny. This technique even permits discrimination among lines from the same individual that are not identified as different by DNA fingerprinting, at least with probes currently readily available. For example, Chen (1988) showed distinct differ-

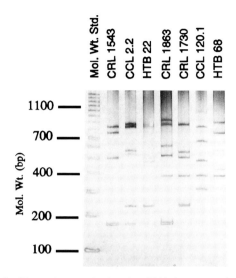

Fig. 4. Photomicrograph showing DNA fragments formed by the polymerase chain reaction (PCR) and used for human cell line identification. Extracted DNA was amplified by standard PCR in three separate mixtures with primer sequences ApoB, D1S80, and D17S5. Fragments were then pooled, separated by gel electrophoresis, and silver stained (Reid and Luo, 1993). (Photo provided by Y. Reid.)

ences among HeLa (ATCC.CCL 2), HeLa S3 (ATCC-CCL 2.2), and the HeLa derivative Hep-2 (ATCC.CCL 23) cell lines. These lines have similar marker chromosomes, indicating a common source, but each also has markers unique in type and copy number. In contrast, Gilbert *et al.* (1990) noted no distinct differences among the nine HeLa derivatives examined using the minisatellite probe 33.6 after *Hae* III digestion. Metabolic differences among the HeLa derivatives (Nelson-Rees *et al.*, 1980) ultimately should be traced to genetic differences among the lines, as reflected by the unique cytogenetic profiles. On this basis, the importance of documenting the precise cytogenetics of cell lines used for production purposes is very clear.

At this stage, DNA fingerprinting alone will not suffice for the most thorough characterization of given cell lines. The technique will be extremely valuable for authentication, however, especially when standard methods are in place and DNA fingerprint data on most common human cell lines are more readily available.

V. Origin and Function

The markers used for verification of the source tissues for cell lines are probably as numerous as the types of metazoan cells. Major methods of

demonstration include analysis of fine structure, immunological tests for cytoskeletal and tissue-specific proteins, and, of course, any of an extremely
broad range of biochemical tests for specific functional traits of tissue cells.
For examples, see Hay (1992).

Ultrastructural features such as desmosomes or Weibel–Palade bodies
identify epithelia and endothelia, respectively. The nature of intermediate
filament proteins, now demonstrated using monoclonal antibodies, permits
differentiation among epithelial subtypes, mesenchymal, and neurological
cells. Tissue- and tumor-specific antigens can be used when reagents are
reliable and available. In addition, tissue-specific biochemical reactions or
syntheses may be used for absolute identification if these features are retained by the cell line in question. One excellent example is the cell line
NCI-H820 (ATCC.HTB 181) isolated from a metastatic lesion of a human papillary lung adenocarcinoma. Cells of this line reportedly retain multilamellar
bodies suggestive of Type 2 pneumocytes and express the three surfactant-
associated proteins SP-A (constitutively), SP-B, and SP-C (after dexamethasone stimulation; Gazdar et al., 1990).

VI. Conclusions

The overall utility of any bank of cultured cell lines depends on the degree
of characterization of the holdings that has been performed by the originators, the banking agency, and other individuals within the scientific commu-
nity. Ready availability at reasonable cost of both the lines and such data as
well as the ability to track distribution of the biologicals are additional critical
considerations. Documenting verification of species and identity of each cell
line, when possible, is considered essential. Freedom from bacterial, fungal,
and mycoplasmal infection must be assured. However, from the cell banking
perspective, applying all possible characterizations to every seed or master
cell stock developed is neither essential nor practical. At ATCC, for example,
screens for particular viruses have been applied when specific program support is available for such testing. Similarly, definition of ultrastructural, tumorigenicity, and functional traits, for example, is performed with appropriate
external support and adequate rationale. The central responsibility is to produce reference stocks, authenticated and well characterized for multiple purposes, and to return to those preparations over the years for development of
working stocks for distribution or other specific applications. Each
replacement distribution stock requires re-authentication prior to distribution
to intended users.

References

Bolin, S., Ridpath, J., Black, J., Macy, M., and Roblin, R. (1992). Survey of ATCC cell lines for
   Bovine Viral Diarrhea Virus (BVD). World Federation for Culture Collections, p. 114. Abstract
   presented at The Seventh International Congress, Beijing, China.

Buller, R. M. L., Corman-Weinblatt, A., Hamburger, A. W., and Wallace, G. D. (1987). Observations on the replication of ectromelia virus in mouse-derived cell lines: Implications for epidemiology mousepox. *Lab. Anim. Sci.* **37,** 28–32.

Caputo, J. (1988). Biosafety procedures in cell culture. *J. Tissue Culture Meth.* **11,** 223–228.

Chen, T. R. (1988). Re-evaluation of HeLa, HeLa S3 and Hep-2 karyotypes. *Cytogenet. Cell Genet.* **48,** 19–24.

Chen, T. R., Drabkowksi, D., Hay, R. J., Macy, M. L., and Peterson, W., Jr. (1987). WiDr is a derivative of another colon adenocarcinoma cell line, HT-29. *Cancer Genet. Cytogenet.* **27,** 125–134.

Chimera, J. A., Harris, C. R., and Litt, M. (1989). Population genetics of the highly polymorphic locus D16S7 and its use in paternity evaluation. *Am. J. Hum. Genet.* **45,** 926–931.

Del Giudice, R. A., and Gardella, R. S. (1984). Mycoplasma infection of cell culture: Effects, incidence and detection. *In Vitro Monogr.* **5,** 104–115.

Erikson, G. A., Landgraf, J. G., Wessman, S. J., Koski, T. A., and Moss, L. M. (1989). Detection and elimination of adventitious agents in continuous cell lines. *Dev. Biol. Standard.* **70,** 59–66.

Gazdar, A., Linnoila, R. I., Kurita, Y., Oie, H. K., Mulshine, J. L., Clark, J. C., and Whitsett, J. A. (1990). Peripheral airway cell differentiation in human lung cancer cell lines. *Cancer Res.* **50,** 5481–5487.

Gilbert, D. A., Reid, Y. A., Gail, M. H., Pee, D., White, C., Hay, R. J., and O'Brien, S. J. (1990). Application of DNA fingerprints for cell-line individualization. *Am. J. Hum. Genet.* **47,** 499–514.

Grizzle, W. E., and Polt, S. H. (1988). Guidelines to avoid personnel contamination by infective agents in research laboratories that use human tissues. *J. Tissue Culture Meth.* **11,** 191–200.

Hay, R. J. (1992). Cell line preservation and characterization. *In* "Animal Cell Culture, A Practical Approach" (R. I. Freshney, ed.), 2d Ed., pp. 95–148. IRL Press, Washington, D. C.

Hay, R. J., Macy, M. L., and Chen, T. R. (1989). Mycoplasma infection of cultured cells. *Nature (London)* **339,** 487–488.

Hay, R. J., Caputo, J., Chen, T. R., Macy, M. L., McClintock, P., and Reid, Y. A. (1992). "Catalogue of Cell Lines and Hybridomas," *7th Ed.* American Type Culture Collection, Rockville, Maryland.

Hayflick, L., and Hennessen, W. (eds.) (1989). Continuous cell lines as substrates for biologicals. *Dev. Biol. Standard.* **70.**

Honma, M., Kataoka, E., Ohnishi, K., Ohno, T., Takeuchi, M., Nomura, N., and Mizusawa, H. (1992). A new DNA profiling system for cell line identification for use in cell banks in Japan. *In Vitro Cell Dev. Biol.* **28A,** 24–28.

Hsu, T. C., and Benirschke, K. (1967–1975). "An Atlas of Mammalian Chromosomes" (9 vols.). Springer-Verlag, New York.

Hukku, B., Halton, D. M., Mally, M., and Peterson, W. D., Jr. (1984). Cell characterization by use of multiple genetic markers. *In* "Eukaryotic Cell Cultures" (R. T. Acton and J. D. Lynn, eds.). pp. 13–31. Plenum Publishing, New York.

Jeffreys, A. J., Wilson, L., and Thein, S. L. (1985). Hypervariable minisatellite regions in human DNA. *Nature (London)* **314,** 67—73.

LeDuc, J. W., Smith, G. A., Macy, M. L., and Hay, R. J. (1985). Certified cell lines of rat origin appear free of infection with hantavirus. *J. Infect. Dis.* **152,** 1081–1082.

Levings, R. L., and Wessman, S. J. (1991). Bovine viral diarrhea virus contamination of nutrient serum, cell cultures and viral vaccines. *Dev. Biol. Standard.* **75,** 177–181.

Lubiniecki, A. S., and May, L. H. (1985). Cell bank characterization for recombinant DNA mammalian cell lines. *Dev. Biol. Standard.* **60,** 141—146.

Miller, G., and Lipman, M. (1973). Release of infectious Epstein-Barr virus by transformed marmoset leukocytes. *Proc. Natl. Acad. Sci. U.S.A.* **90,** 190–194.

Nakamura, Y., Leppert, M., O'Connell, P., Wolff, R., Holm, T., Culver, M., Martin, C., Fujimoto, E., Hoff, M., Kumlin, E., and White, R. (1987). Variable number of tandem repeat (VNTR) markers for human gene mapping. *Science* **235,** 1616–1622.

Nakamura, Y., Martin, C., Myers, R., Ballard, L., Leppert, M., O'Connell, P., Lathrop, G. M., Lalouel, J.-M., and White, R. (1988). Isolation and mapping of a polymorphic DNA sequence (pCMM86) on chromosome 17q [D17S74]. *Nucleic Acids Res.* **16,** 5223.

Nelson-Rees, W. A. (1978). The identification and monitoring of cell line specificity. *Progr. Clin. Biol. Res.* **20,** 25–79.

Nelson-Rees, W. A., Hunter, L., Darlington, G. J., and O'Brien, S. J. (1980). Characteristics of HeLa strains: Permanent vs. variable features. *Cytogenet. Cell Genet.* **27,** 216–231.

Nelson-Rees, W. A., Daniels, D. W., and Flandermeyer, R. R. (1981). Cross-contamination of cell lines. *Science* **212,** 446–452.

Nuttal, P.A., Luther, P. D., and Stott, E. J. (1977). Viral contamination of bovine foetal serum and cell cultures. *Nature (London)* **266,** 835–837.

O'Brien, S. J., Kleiner, G., Olson, R., and Shannon, J. E. (1977). Enzyme polymorphisms as genetic signatures in human cell culture. *Science* **195,** 1345–1348.

Popovic, M., Kalyanaraman, V. S., Reitz, M. S., and Sarngadharan, M. G. (1982). Identification of the RPMI 8226 retrovirus and its dissemination as a significant contaminant of some widely used human and marmoset cell lines. *Int. J. Cancer* **30,** 93–99.

Reid, Y. A., and Luo, X. (1993). The use of PCR-amplified hypervariable regions for the identification and characterization of human cell lines. *In Vitro Cell Dev. Biol.* **29A,** 120A.

Reid, Y. A., Gilbert, D. A., and O'Brien, S. J. (1990). The use of DNA hypervariable probes for human cell line identification. *Am. Type Culture Coll. Newsl.* **10 (4),** 1–3.

Spaepen, M., Angulo, A. F., Marynen, P., and Cassiman, J. J. (1992). Detection of bacterial and mycoplasma contamination in cell cultures by polymerase chain reaction. *FEMS Microbiol. Lett.* **99,** 89–94.

Tumilowicz, J. J., Gallick, G. E., East, J. L., Pathak, S., Trentin, J. J., and Arlinghaus, R. B. (1984). Presence of retrovirus in the B95-8 Epstein–Barr virus-producing cell line from different sources. *In Vitro* **20,** 486–492.

Uemori, T., Asada, K., Kato, I., and Harasawa, R. (1992). Amplification of the 16S-23S spacer region in rRNA operons of mycoplasmas by the polymerase chain reaction. *Syst. Appl. Microbiol.* **15,** 181–186.

Weiss, S. H., Goedert, J. J., Gartner, S., Popovic, M., Waters, D., Markham, P., Veronese, F. M., Gail, M. H., Barkley, W. E., Gibbons, J., Gill, F. A., Leuther, M., Shaw, G. M., Gallo, R. C., and Blattner, W. A. (1988). Risk of human immunodeficiency virus (HIV-1) infection among laboratory workers. *Science* **239,** 68–71.

Human Glioma Cell Lines

Monica Nistér and Bengt Westermark
Department of Pathology,
University of Uppsala
University Hospital
S-751 85 Uppsala, Sweden

I. Introduction 17

II. Cell Lines Derived from Glioblastoma Multiforme and Anaplastic Astrocytoma 18
A. Phenotypic Properties 18
B. Chromosomal Abnormalities 23
C. Tumor Suppressor Genes 26
D. Studies of Oncogenes and Their Products 27
E. Growth Factors 29

F. Spheroid Cultures 31

III. Cell Lines Derived from Oligodendroglioma 31

IV. Cell Lines Derived from Primitive Neuroectodermal Tumor/Medulloblastoma 32

V. Concluding Remarks 33
References 34

I. Introduction

Primary tumors of the central nervous system (CNS) are of neuroepithelial origin. The classification of these tumors is based primarily on morphology (Kernohan et al., 1949; Ringertz, 1950; Zülch, 1979; Russell and Rubinstein, 1989; Kleihues et al., 1993). The most frequent tumors are neuroglial in nature, that is, astrocytomas, anaplastic astrocytomas, glioblastoma multiforme, oligodendrogliomas, and ependymomas. Others are thought to originate from neuronal cells and from primitive neuroepithelial cells (e.g., primitive neuroectodermal tumors, PNETs) that appear mostly in cerebellum; in this location, the tumors are designated medulloblastomas. Rare "tumors of specialized tissues of central neuroepithelial origin" also occur (Russell and Rubinstein, 1989). In this chapter, we discuss only those CNS tumors from which permanent cell lines have been established in culture.

Early attempts to culture human tumor cells as permanent lines were rather disappointing. Although primary cultures of tumor cells could be obtained

17

relatively easily, the most common long-term result was the deterioration of the tumor cell population and overgrowth of nonneoplastic stromal cells (mainly fibroblasts). The pioneering studies by Manuelidis (1965,1969) and Pontén and Macintyre (1968) showed that human glioblastoma is an exception to the rule, since permanent cell lines could be established from high grade gliomas at a relatively high frequency. These results have been confirmed by a number of reports. Currently, several laboratories have established their own collection of glioma cell lines (for review, see Collins, 1983). However, despite the improvement in tissue culture techniques and the design of new culture medium formulas, brain tumors other than high grade gliomas and medulloblastomas are still difficult to grow as permanent cell lines.

Although malignant gliomas as a group frequently form established cell lines, the "take frequency" is well below 100%. In a study of a large series of brain tumor biopsies, some 20% of the high grade gliomas could be grown as permanent lines (Westermark et al., 1973). Inspection of the primary cultures showed that this event was not random; thus, a clear and consistent correlation existed between cell morphology and long-term growth capacity. In the successful cases, the preponderant cell type was spindle shaped (Fig. 1a) whereas, in the unsuccessful cases, the primary cultures consisted mainly of astrocyte-like cells with cytoplasmic projections that progressively increased in size and number (Fig. 1b). In the latter type of culture, mitotic figures were seen only during the first days of explantation. Thus, the explanted astrocyte-like glioma cells seemed to differentiate rather than multiply. These early studies, which have been confirmed repeatedly in our laboratory in recent years, suggest that the ability to grow as a permanent cell line is an inherent property of the individual tumor.

II. Cell Lines Derived from Glioblastoma Multiforme and Anaplastic Astrocytoma

A. Phenotypic Properties

Permanent glioma lines can be categorized according to cell morphology. Thus, cell lines have been described as resembling astrocytes, epithelial cells, or fibroblasts (Bullard et al., 1981; Nistér et al., 1988a). Many cell surface markers and antigenic determinants have been studied and found to be expressed heterogeneously. However, no "markers" have been as firmly correlated with cell morphology as the astrocyte-specific intermediate filament glial fibrillary acidic protein (GFAP) and the extracellular matrix protein fibronectin (FN). Table I lists several glioma cell lines in which GFAP and FN expression

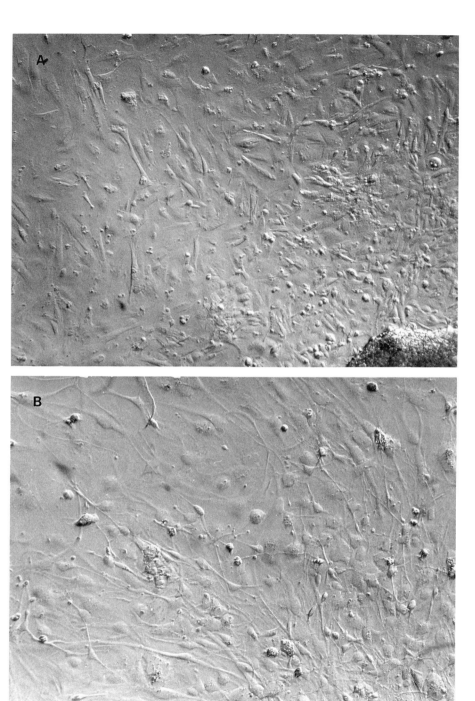

Fig. 1. Phase-contrast microscopy of primary glioma cultures. (A) U-343 MG. Spindle-shaped cells that proliferated and established as a permanent cell line *in vitro*. (B) U-348 MG. Astrocyte-like cells with long cytoplasmic projections. These cells failed to establish in culture.

Table I

GFAP/FN Expression in Human Glioma Cell Lines[a]

Cell line	GFAP[b]	FN[b]	Morphology[c]	References
U-251 MG	+	−/+	pleo/astro	Vaheri et al. (1976); Bigner et al. (1981); Jones et al. (1982); Rutka et al. (1987); LaRocca et al. (1989); Nishiyama et al. (1989)
U-251 MG sp	+	−/+	fascicular	Bigner et al. (1981); Jones et al. (1982); LaRocca et al. (1989); Bongcam-Rudloff et al. (1991)
U-251 MG AgCl 1	+		astro	B. Westermark (unpublished observations)
U-343 MGa	+	−	polygonal	Vaheri et al. (1976)
U-343 Mga Cl 2:6	+	−	polygonal	Bongcam-Rudloff et al. (1991)
U-343 Mga 31L	+	−	pleo/astro	Bongcam-Rudloff et al. (1991)
U-373 MG	+	−	pleo/astro	LaRocca et al. (1989); Bongcam-Rudloff et al. (1991)
U-1231 MG	+	−	epithelioid	Bongcam-Rudloff et al. (1991)
SNB-19	+		pleo/glial	LaRocca et al. (1989); Saxena and Ali (1992)
IPSB-18	+	−	fusiform/polygonal	Knott et al. (1990)
NCE-G112	+		astro	Westphal et al. (1990)
D-263 MG	+		epithelioid	Bigner et al. (1986a)
U-87 MG	−		epithelioid	Nistér et al. (1988a); LaRocca et al. (1989)
U-105 MG	−	+	bipolar/fibro	Vaheri et al. (1976); Bigner et al. (1981)
U-118 MG	−	+	bipolar/fibro	Vaheri et al. (1976); Bigner et al. (1981); Jones et al. (1982)
U-138 MG	−	+	bipolar/fibro	Bigner et al. (1981)
U-178 MG	−	+	bipolar/fibro	Bongcam-Rudloff et al. (1991)
U-343 MG	−	+	fibro/fascicular	Bigner et al. (1981); Bongcam-Rudloff et al. (1991)
U-372 MG	−		bipolar/fibro	Nistér et al. (1988a); B. Westermark (unpublished observations)
U-399 MG	−		bipolar/fibro	Nistér et al. (1988a); B. Westermark (unpublished observations)
U-410 MG	−	+	bipolar/fibro	Bigner et al. (1981)

Cell line	GFAP	FN	Morphology	Reference
U-489 MG	–		bipolar/fibro	Nistér et al. (1988a); B. Westermark (unpublished observations)
U-539 MG	–		epithelioid	Nistér et al. (1988a); B. Westermark (unpublished observations)
U-563 MG		–	polygonal	Bongcam-Rudloff et al. (1991)
U-706 T	–		epithelioid	LaRocca et al. (1989)
U-1240 MG	–		fascicular	Bongcam-Rudloff et al. (1991)
U-1242 MG	–	+	bipolar/fibro	Bongcam-Rudloff et al. (1991)
U-1796 MG	–		bipolar/fibro	Nistér et al. (1988a); LaRocca et al. (1989)
D-18 MG	–	(+)	glial	Bigner et al. (1981)
D-32 MG	–	+	epithelioid	Bigner et al. (1981)
D-37 MG	–	(+)	fibro	Bigner et al. (1981)
D-54 MG	–	–	glial	Bigner et al. (1981); Jones et al. (1982)
D-65 MG	–	(+)	epithelioid	Bigner et al. (1981)
D-245 MG	–		spindle	Bigner et al. (1986a)
A-172 MG	–	(+)	glial	Bigner et al. (1981)
NCE-G62	–	+	fibro	Westphal et al. (1990)
SNB-56	–		bipolar	LaRocca et al. (1989); Saxena and Ali (1992)
SNB-78	–		bipolar	LaRocca et al. (1989)
SF-126	–	+	bipolar/fibro	Rutka et al. (1987); LaRocca et al. (1989)
SF-188	–	+	epithelioid	Rutka et al. (1987); LaRocca et al. (1989)
SF-210	–	+	epithelioid	Rutka et al. (1987); LaRocca et al. (1989)
SF-268	–	–	fascicular	Rutka et al. (1987)
SF-295	–	–	bipolar/fibro	Rutka et al. (1987); LaRocca et al. (1989)

[a] A selection of established human glioma cell lines was listed, all of which have been passaged so many times in vitro that the GFAP/FN (glial fibrillary acidic protein/fibronectin) expression pattern should be relatively stable.

[b] +, Positive; (+), positive but in low concentration; –, negative; no listing, no information available.

[c] Pleo, pleomorphic; astro, astrocytoid; fibro, fibroblastic.

have been described. The GFAP$^+$ cell lines are composed of astrocytoid, pleomorphic, or small polygonal cells whereas fibroblast-like or bipolar cells are invariably GFAP$^-$. Although GFAP$^+$ cells in culture are, no doubt, of glial origin, whether permanent glioma lines of the GFAP$^-$/FN$^+$ phenotype are genuine representatives of gliomas *in vivo* is unkown. The majority of established glioma lines is, in fact, GFAP$^-$/FN$^+$. FN expression normally is seen in cells of mesenchymal origin. The GFAP$^-$/FN$^+$ cell lines have been proposed to originate from mesenchymal elements that typically are localized around or within thick-walled capillary vessels in the glioma tissue (Paetau *et al.*, 1980; Kennedy *et al.*, 1987; McKeever *et al.*, 1987). Whether such areas constitute part of the neoplastic process has even been questioned. However, GFAP$^-$/ FN$^+$ cells obviously are not derived from nonneoplastic mesenchymal cells, since they give rise to permanent tumorigenic cell lines with abnormal karyotypes, as described by several investigators. Bigner *et al.* (1986a) karyotyped original biopsies, followed the karyotype of corresponding cell lines during establishment, and found that GFAP$^-$ cell lines retained the characteristic abnormalities found in tumor biopsies.

The GFAP$^-$/FN$^+$ cells are likely to represent a subpopulation of glioma cells present in the original tumor that is selected for during the process of *in vitro* culturing. The most probable reason for the selection is a higher *in vitro* growth potential of cells with the GFAP$^-$/FN$^+$ phenotype (McKeever *et al.*, 1987). Lumsden (1971) and Westermark *et al.* (1973) had noted that the "mesenchymal" elements constituted the fast growing elements of the culture. The existence of an *in vitro* selection force favoring FN$^+$ cells has been confirmed by other authors (McKeever and Chronwall, 1985; Paetau, 1988). Shapiro *et al.* (1981) noted that astrocyte-like and squamous-like cells were often near diploid, whereas fibroblast-like cells were mostly hyperploid and fast growing, indicating a relationship between karyotype, morphology, and growth properties.

The nature of the CNS cell(s) that is(are) the origin of the neoplastic process and the question of a mono- or polyclonal origin of glioblastoma multiforme have been of interest to researchers for decades. Some information has been gained from the study of GFAP/FN expression in glioma cell lines. Several observations are in favor of the possibility that the GFAP$^-$/FN$^+$ and GFAP$^+$/ FN$^-$ cells coexisting within one glioma have a common origin. Although generally GFAP and FN expression seem to be mutually exclusive, some cell lines such as U-251 MG sp and a subclone of U-251 MG (Bigner *et al.*, 1981; Jones *et al.*, 1982; Rutka *et al.*, 1987) contain a subpopulation of cells that is GFAP$^+$/ FN$^+$. McKeever *et al.* (1987) also found single dual-positive explants during early culture. Such dual-positive cells might represent a transitional stage between the two phenotypes. The strongest evidence for a common origin has been obtained from the study of clonal variants of single tumor cases (Nistér *et al.*, 1987; Westphal *et al.*, 1988). Two cell lines, U-343 MG and U-343 MGa, derived from one glioblastoma multiforme biopsy (Westermark, 1973; Westermark et al., 1973; Pontén and Westermark, 1978) were found to be GFAP$^-$/FN$^+$ and GFAP$^+$/FN$^-$ respectively. Many clonal variants of U-343 MGa were analyzed further and all were GFAP$^+$ (Nistér et al., 1986). Eight of the GFAP$^+$ U-343 MGa clones are shown in Fig. 2. The morphologies of the U-343 MG/MGa cells, derived from one single biopsy encompass most of the described morphologies of glioma cell lines. Many of the U-343 MGa clones and the U-343 MG cells also have been karyotyped (Mark et al., 1974; Nistér et al., 1987). Despite their phenotypical heterogeneity, these clones have several chromosome markers in common, for example, the unique 1p−q+;inv(1)(p13q43), implying that these cells have a common origin.

Mathiesen et al. (1989) found that, when GFAP$^-$ glioma cells were injected into the anterior chamber of the rat eye, an initial proliferation, induced GFAP expression, and a morphological change to a more astrocytic morphology occurred, indicating that inducing a differentiated phenotype in glioma cells is possible. In conjunction with the process of selection, changes in specific regulatory components in the extracellular milieu or in the spatial composition of the explants might be responsible for the failure to maintain GFAP$^+$ phenotypes in vitro. Additional evidence for a shift between the two phenotypes is provided by the finding that transformation of GFAP$^+$ astrocyte progenitor cells by SV 40 large T antigen results in a GFAP$^-$/FN$^+$ phenotype (Geller and Dubois-Dalcq, 1988). Several authors have suggested that loss of GFAP expression represents a sign of anaplasia in glioma in vivo; during serial passage in vitro, a more anaplastic phenotype is likely to be selected.

Malignant gliomas in vivo also express high levels of the intermediate filament protein nestin, characteristic of neuroepithelial stem cells (Lendahl et al., 1990). Interestingly, Tohyama et al. (1992) described the coexpression of nestin and GFAP in the glioma lines U-373 MG and U-251 MG. Thus, glioma cell lines seem to have an uncoordinated expression of intermediate filaments, perhaps reflecting a perturbed differentiation program.

In addition to FN, glioma cells express other extracellular matrix proteins such as laminin and type IV and type I collagen (McKeever et al., 1989), suggesting that these cells have both epithelial and mesenchymal characteristics and, in this respect, resemble leptomeningeal rather than glial cells (McKeever et al., 1989). Also, established cell lines from gliosarcomas have such characteristics (Rutka et al., 1986; McKeever et al., 1987). These results may be interpreted as an ability for the GFAP$^-$/FN$^+$ cells to differentiate further as mesenchymal/leptomeningeal elements.

B. Chromosomal Abnormalities

The pioneering work by Mark (1971), performed before the chromosomal banding technique was in use, elucidated most of the chromosomal abnormalities of human malignant glioma biopsies. A genotypic heterogeneity among gliomas was detected. Diploid-near-diploid stem lines were found in approxi-

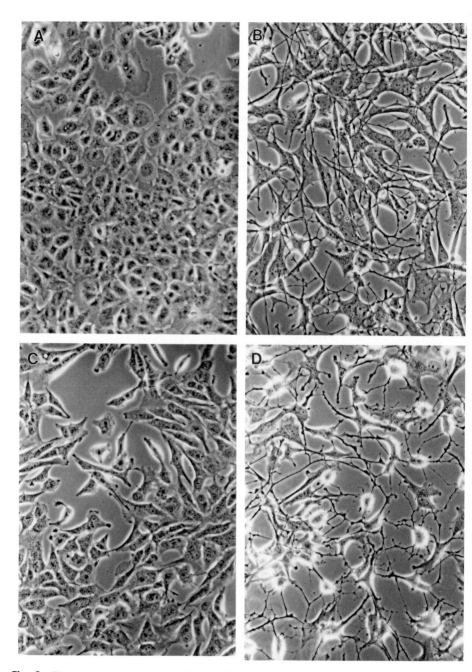

Fig. 2. Phase-contrast microscopy of eight U-343 MGa glioma clones. All clones are GFAP$^+$, but extensive phenotypic variation still exists among them. The spectrum covers most of the described glioma cell morphologies, but does not include bipolar/fibroblastic cells. (A) 7L flat polyhedral epithelioid cells. (B) 35L astrocytoid cells. (C) 26L middle-sized polygonal cells. (D) 21L astrocytoid cells. (E) 34 L astrocytoid cells with short projections. (F) 30L astrocytoid cells with long projections. (G) 35H fascicular cells. (H) Cl2:13 small polygonal cells (cf. Nistér *et al.*, 1986).

24

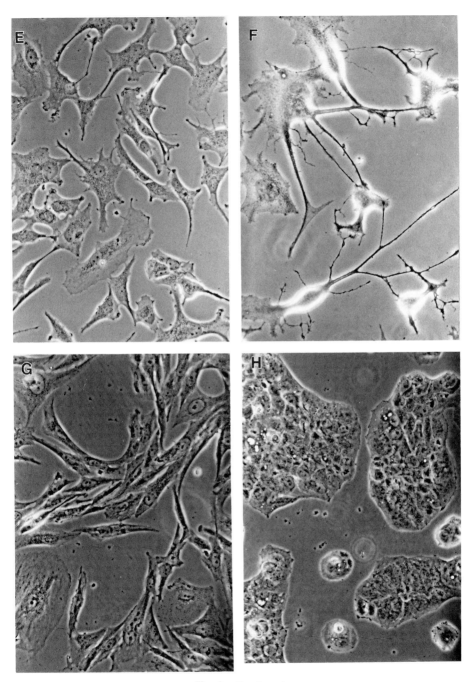

Fig. 2. *Continued*

25

mately 75% of the cases; among the remaining 25%, near-triploids were more common than near-tetraploids. Among the near-diploids, the stem lines had chromosome numbers of 47 or 44–45. Marker chromosomes and deviations were found in chromosomal groups C and D in most cases (75%), and double minute chromosomes (DMs) were seen in several cases. Later studies using banding techniques have confirmed these findings. Bigner *et al.* (1984,1986b) described three groups of primary gross deviations: (1) loss of a gonosome, (2) gain of chromosome 7 and loss of chromosome 10, and (3) loss of chromosome 22. DMs were seen in half the cases. Structural abnormalities of 9p and 19q were most consistent, but abnormalities on chromosomes 1, 6p, 11, 13q, and 15q were found with some frequency also (Bigner *et al.*, 1988a).

Most low-grade astrocytomas have a homogeneous karyotype with a near-diploid chromosome number and numerical deviations of chromosomes 7, 10, and 22 or a gonosome. This karyotype is similar to that of high-grade gliomas, but the frequency of structural deviations is higher in the high-grade cases (Rey *et al.*, 1987; Griffin *et al.*, 1992). Clonal and regional variations of the karyotype are more frequent findings in high-grade tumors (Shapiro, 1981; Shapiro and Shapiro, 1985; Shapiro, 1986). Glioblastoma multiforme also has a more complex karyotype than anaplastic astrocytomas (Bigner *et al.*, 1988a), as well as loss of heterozygocity in a higher number of chromosomal regions, as defined by restriction fragment length polymorphism (RFLP) analysis (Fults *et al.*, 1990). Thus, a chromosomal progression seems to occur during the development of malignant gliomas *in vivo*.

Glioma cells with relatively homogeneous near-diploid karyotypes are stable through short-term serial passage and fail to establish in culture (Mark, 1971), whereas, anaplastic astrocytomas and glioblastoma multiforme, with their more complex karyotypes, frequently establish as cell lines *in vitro*. The chromosomal progression described *in vivo* continues during the establishment of the glioma cell lines (Bigner and Mark, 1984; Bigner *et al.*, 1987a). The cell lines generally double their near-diploid stem line, subsequently gain and lose individual chromosomes, and become near-tetraploid. During this process, the lines retain original marker chromosomes but also gain new markers; most often, DMs are lost. Direct transplantation and passage of original biopsies in athymic mice, however, results in a karyotype that is more similar to the original biopsies, often without doubling of the stem line and with retention of the DMs (Bigner *et al.*, 1985,1989). The advanced changes in karyotype that glioma cells undergo in tissue culture is a warning signal, since these changes imply that the cells are subject to profound adaptation to the culture conditions. Attempts to extrapolate from *in vitro* experiments to the *in vivo* situation must, therefore, be made with caution.

C. Tumor Suppressor Genes

Molecular genetic analysis of human gliomas using RFLPs has revealed a high frequency of nonrandom loss of genetic material on several chromosomes, particularly chromosomes 9, 10, and 17 (Bigner *et al.*, 1988a; James *et al.*, 1988,1989; El-Azozouzi *et al.*, 1989; Fujimoto *et al.*, 1989; Fults *et al.*, 1989, 1990). These findings suggest that the corresponding chromosomal regions in the normal genome contain loci for tumor suppressor genes, the loss of which contributes to neoplastic transformation and tumorigenesis. The lost material on chromosome 9 has been shown to contain the interferon gene cluster on 9p (Miyakoshi *et al.*, 1990), but these sequences probably do not constitute the critical information that is lost in this area (Olopade *et al.*, 1992). Several laboratories currently are involved in the identification of tumor suppressor genes on chromosome 10. The loss of genetic material on chromosome 17 coincides with the deletion of the *p53* tumor suppressor gene; ample evidence suggests that deletion or mutational inactivation of *p53* is a critical event in the genesis of malignant astrocyte-derived tumors (Chung *et al.*, 1991; Frankel *et al.*, 1992; Fults *et al.*, 1992a). The progression of human malignant gliomas *in vivo* is accompanied by *p53* mutations (Sidransky *et al.*, 1992).

Generally studies of glioma cell lines to date (Godbout *et al.*, 1992) have contributed very little to the understanding of the role of deletion, rearrangement, and mutation of tumor suppressor genes in the development of human gliomas. We may anticipate, however, that well-characterized glioma cell lines will be more important in the future, when the functional characteristics of identified tumor suppressor genes must be elucidated. Experiments to date have shown that introduction of a normal coding sequence for *p53* into human glioma cells leads to growth arrest (Mercer *et al.*, 1990), implying a possible function of the *p53*-encoded protein in the regulation of the cell cycle. Envisioning future experiments in which the consequences of reconstituting lost tumor suppressor genes for growth behavior will be studied is easy; for such experiments, established glioma lines will be the systems of choice.

D. Studies of Oncogenes and Their Products

As mentioned earlier, cytogenetic studies of human malignant glioma have revealed a frequent occurrence of DMs. The pioneering work by Robert Schimke showed that DMs harbor amplified gene sequences (Haber and Schimke, 1981; for review, see Schimke, 1984); in human malignant glioma, most tumors with DMs have an amplified c-*erb*B/EGFR gene (Libermann *et al.*, 1985; Bigner *et al.*, 1987b). Although formal proof is lacking, researchers believe that, in these cases, the amplified epidermal growth factor receptor (EGFR) gene is contained in the DMs. Most established human glioblastoma cell lines express EGF receptors (Nistér *et al.*, 1988a); however, the level of expression is, at most, increased only moderately and no amplification of the EGFR gene has been reported (Nistér *et al.*, 1988a). Two primary explanations are possible for the discrepancy between *in vivo* and *in vitro* findings. First, the establishment of permanent cell lines is a highly selective process; only a fraction of glioma biopsies gives rise to continuous cell lines and, in the successful cases, the emerging cell line will be founded only by a small population in the explant. The presence of an amplified EGFR gene may be directly or indirectly inversely correlated to the ability of the cells to proliferate *in vitro*; cells without EGFR amplification may not be selected. Second, because of their lack of centromeres, DMs do not segregate as chromosomes at mitosis and are lost rapidly on serial passage in the absence of selective pressure. The transfer of cells from *in vivo* to *in vitro* conditions may nullify the importance of the amplified EGF receptor for growth, and thereby abolish the selective pressure that permits the maintenance of the DMs that harbor the amplified gene sequences. Results of studies by Bigner (Bigner *et al.*, 1990a) favor the latter notion. Clearly these studies show that cell lines can be established from gliomas with EGFR amplification, and that the amplified sequences are lost rapidly *in vitro*. EGFR gene amplification is retained on serial transplantation of glioma biopsies in athymic (nu/nu) mice (Humphrey *et al.*, 1988; Bigner *et al.*, 1990a). This observation fits with the notion that the selective pressure on the amplified EGFR gene *in vivo* is different from that *in vitro*.

Sporadic cases of gliomas with amplified c-*myc* (Trent *et al.*, 1986), N-*myc*, *gli* (Bigner *et al.*, 1987b; Kinzler *et al.*, 1987; Wong *et al.*, 1987), and platelet-derived growth factor (PDGF) α-receptor genes have been reported (Fleming *et al.*, 1992a; Kumabe *et al.*, 1992) but, like the amplified EGFR gene, these amplifications (except for that of the c-*myc* gene) do not seem to be represented in glioma cell lines.

Analysis of c-*ros* mRNA in a large series of human tumor cell lines showed a surprisingly selective expression in glioma lines (Birchmeier *et al.*, 1987). Further, in one of the lines (U-118 MG), the c-*ros* gene seemed to have undergone rearrangement (Sharma *et al.*, 1989). c-*ros* encodes a putative receptor molecule with protein tyrosine kinase activity and is the human homolog of the *sevenless* gene of *Drosophila melanogaster* (Birchmeier *et al.*, 1990). Given the finding that Sevenless has a specific function in one particular cell type in the developing eye of the fruit fly, the selective expression of c-*ros* mRNA in glioma cell lines is certainly an intriguing finding. Even more intriguing is the finding that c-*ros* mRNA does not seem to be expressed in gliomas *in vivo* (Wu and Chikaraishi, 1990) or in the mammalian nervous system during development. Rather, evidence suggests that Ros has a critical function in the embryonal development of the kidney (Sonnenberg *et al.*, 1991).

Searches for other abnormalities in structure or expression of oncogenes and proto-oncogenes in human glioma cell lines have yielded a relatively incoherent picture. In the glioblastoma cell line A-172, rearrangement of c-*abl* and a loss of both germ-line alleles is seen (Heisterkamp *et al.*, 1990). The rearranged gene encodes a truncated mRNA that lacks the coding sequence for a functioning Abl protein, lacking the tyrosine kinase domain. The significance of this finding is unknown. Studies of four glioma cell lines (HeRo, A-172, T-406, T-508) showed a moderate amplification of c-*myb* in two cases (HeRo and T-406) but no relative increase in the corresponding mRNA level (Welter *et al.*, 1990). A large survey of c-*myc*, c-*raf*-1, c-*myb*, and c-*erb*B2 mRNA expression in 20 glioma cell lines of the SF, SNB, and U series showed a relatively uniform pattern: most genes were expressed at a moderate level in all cell lines (LaRocca *et al.*, 1989). All cell lines but one were shown to express c-*myc* mRNA, whereas none expressed N-*myc* mRNA. With respect to the *ras* family of proto-oncogenes, a study of five glioblastoma cell lines of the Hu series (Gerosa *et al.*, 1989) showed a relative overexpression of N-*ras* but not of Ha-*ras* or Ki-*ras*.

E. Growth Factors

Knowledge of the factors regulating the growth and differentiation of glial cell progenitors has emerged. The proliferation of neuroepithelial stem cells in culture can be stimulated by EGF (Reynolds and Weiss, 1992). Glial progenitor cells derived from rat optic nerves express PDGF α-receptors (Richardson *et al.*, 1988; Hart *et al.*, 1989) and proliferate without any signs of differentiation or senescence when given a combination of PDGF and basic fibroblast growth factor (bFGF) (Noble *et al.*, 1988; Bögler *et al.*, 1990).

Receptors for EGF, bFGF, and PDGF are also present on glioma cells. An unexpected finding is that the pattern of expression of PDGF receptors in glioma cell lines is related to their morphology (Nistér *et al.*, 1988a; Bongcam-Rudloff *et al.*, 1991). PDGF α-receptors are found only on cell lines that are GFAP$^+$, some of which have an astrocytic morphology, whereas fibroblast-like FN$^+$ glioma cells express both PDGF β- and α-receptors. This type of correlation has not been described for other growth factor receptors and implies that differential PDGF receptor expression in glioma cells relates to their differentiation stage or their lineage derivation.

The growth factors involved in glioma cell proliferation have been studied extensively. Compelling evidence for an autocrine mechanism in the development of glioma is provided by studies on simian sarcoma virus (SSV). First, the v-*sis* oncogene of SSV encodes a PDGF-like growth factor (Doolittle *et al.*, 1983; Waterfield *et al.*, 1983). Studies on cultured cells have shown that transformation by this oncogene is mediated by an autocrine activation of PDGF receptors (Johnsson *et al.*, 1985). Second, intracerebral injection of SSV into newborn marmosets induces the development of glial tumors (Deinhardt, 1980), the histology of which is indistinguishable from that of human glioma. PDGF-B/c-*sis* mRNA is expressed in glioma cells (Eva *et al.*, 1982) and, since glioma cells secrete PDGF (Nistér *et al.*, 1984,1988b) and express the cognate receptor (Harsh *et al.*, 1990; Nistér *et al.*, 1991), the stage is set for an autocrine stimulation. Glioma cells also coexpress transforming growth factor α (TGF α) and EGF receptors (Nistér *et al.*, 1988a), as well as acidic fibroblast growth factor (aFGF) and bFGF and their receptors (Libermann *et al.*, 1987; Gross *et al.*, 1990; Morrison *et al.*, 1990; Takahashi *et al.*, 1991; Murphy *et al.*, 1992). Glioma cells thus may depend on several autocrine factors for their growth.

Suramin, which displaces several growth factors from their receptors, inhibits the growth of the vast majority of glioma cells (LaRocca et al., 1990; Fleming et al., 1992b). In a few cases, the growth of glioma cells has been inhibited by interfering with a defined autocrine loop. U-87 MG cells can be inhibited by antibodies against bFGF (Takahashi et al., 1991) or antisense oligonucleotides against the same factor (Murphy et al., 1992). A-172 cells, which express PDGF-B (Pantazis et al., 1985) and PDGF β-receptors, can be inhibited specifically by antibodies against PDGF (Vassbotn et al., in press).

Another family of growth factors with potential autocrine effects in gliomas is the insulin/insulin-like growth factor (IGF) family, since insulin receptors (Grunberger et al., 1986) as well as IGF I and IGF II receptors (Gammeltoft et al., 1988) were found on glioblastoma cells, and gliomas can express IGF I and IGF II (Sandberg et al., 1988; Antoniades et al., 1992).

Transforming growth factor β (TGFβ) is a cytokine with pleiotropic effects on a multitude of cells. Primarily $TGF\beta_2$ has been isolated from glioma cell cultures (dc Martin et al., 1987; Bodmer et al., 1989), but $TGF\beta_1$ and $TGF\beta_3$ (Constam et al., 1992; Olofsson et al., 1992) are products of glioma cell lines also. TGFβ is produced as a biologically inactive large latent complex in which the growth factor is coupled covalently to a binding protein. Glioma cells, however, can make an as yet unidentified activating substance (Huber et al., 1991). Since glioma cells appear to have receptors for TGFβ, this cytokine also may have an autocrine activity. However, a clear picture has not yet emerged from these studies; some glioma cells are inhibited, whereas others are stimulated by TGFβ (Helseth et al., 1988). Interestingly, cells derived from high-grade gliomas are growth stimulated whereas low-grade tumors are inhibited (Jennings et al., 1991).

One characteristic and necessary feature of malignant gliomas in vivo is neovascularization, probably induced by the tumor cells. Glioma cell lines can be used for the study of this process since they elaborate growth factors that are well-known angiogenic factors, for example, bFGF, aFGF, and TGFα. Also, PDGF-B is mitogenic for capillary brain endothelial cells (Smits et al., 1989). In addition, glioma cells in culture produce a vascular permeability factor (VPF) (Criscuolo et al., 1988).

The CNS generally is regarded as an immunologically privileged tissue, so cells of the immune system are not normally present in brain tissue. However, astrocytes and glioma cells have an immunomodulating capacity, and glioma cells can secrete, as well as react to, different cytokines. One important function of glioma cells may be antigen presentation, since Class II major histocompatibility complex (MHC) antigens are present on some glioma cells, and the expression can be induced by interferon γ (IFNγ) (Takiguchi et al., 1985; Basta et al., 1988). Glioblastoma cells secrete interleukin 8 (IL-8), the effect of which is attraction of T lymphocytes (Kasahara et al., 1991; van Meir et al., 1992). The $TGF\beta_2$ produced by glioma cells originally was identified as a

T-cell suppressor factor (Fontana *et al.*, 1984; de Martin *et al.*, 1987; Kuppner *et al.*, 1988). Glioma cell-produced IL-6 (Yasukawa *et al.*, 1987; Leppert *et al.*, 1989) may modulate the activity of immunoglobulin-producing cells and natural killer (NK) cells; the expression of this cytokine in glioma cells can be induced by IL-1β and TNFα (Isshiki *et al.*, 1990; van Meir *et al.*, 1990).

F. Spheroid Cultures

Several human glioma cell lines have been grown successfully as free-floating spheroids in suspension culture. A particular advantage for this type of culture system is that it mimics three-dimensional tumor growth *in vivo*. The population doubling times of the spheroid cultures are somewhat longer than those of conventional monolayer cultures. The slower growth rate is likely to be related to the uneven distribution of the proliferative pool of cells. Autoradiographic studies of the distribution of [^{3}H]thymidine incorporation have shown that only cells at the periphery of the spheroid proliferate rapidly, whereas cells more centrally located near the center necrosis are growth arrested. Such a gradient of proliferating cells resembles tumors *in vivo*. The multicellular spheroids have been particularly useful in *in vitro* studies of nutrient gradients, oxygen tension, pH gradients, tumor tissue penetration of cytostatic drugs and monoclonal antibodies, release of catabolic products, and dependence of cell properties on location in a three-dimensional structure (for review, see Carlsson *et al.*, 1983; Carlsson and Acker, 1988; Carlsson, 1992). Spheroids also have been used successfully for the study of glioma cell invasion (de Ridder *et al.*, 1987; Lund-Johansen *et al.*, 1990).

III. Cell Lines Derived from Oligodendroglioma

Two oligodendroglioma cell lines, HOG (Post and Dawson, 1992) and TC 620 (Manuelidis *et al.*, 1977; Merrill and Matsushima, 1988), have been established; both express the oligodendrocyte specific marker galactocerebroside (GalC) on their cell surface. The TC 620 cells are described as bipolar and epithelioid with short processes (Kashima *et al.*, 1993). The two lines resemble immature oligodendrocytes because they only express myelin basic protein (MBP) transcripts of sizes that are seen in early brain development. Moreover, these cells do not express the myelin proteolipid protein that is a marker for mature oligodendrocytes. These cells do not express GFAP or neurofilament, but specifically express cytokeratin K7 mRNA and protein (Kashima *et al.*, 1993). Another cell line, ONS-21 (Okamoto *et al.*, 1990), derived from an anaplastic (grade III) oligodendroglioma, expresses MBP, GalC, and S100 and is GFAP$^-$

IV. Cell Lines Derived from Primitive Neuroectodermal Tumor/Medulloblastoma

The origin of PNET/medulloblastoma is thought to be a bipotential cell capable of differentiating into either glial or neuronal cells (Molenaar et al., 1989), but the potential for glial differentiation has been debated. Valtz et al. (1991) used SV40 large T antigen to immortalize rat cerebellar cells. The resulting ST15A cell line has the characteristics of a nestin-positive primitive neuroectodermal cell line that can differentiate into neuronal and glial cells, and is suggested to represent the target cell for PNET/medulloblastoma development.

Medulloblastoma is the most frequent tumor in the group of PNETs of the CNS. This tumor occurs preferentially in the cerebellum of young children. The classification of these tumors as a subgroup of PNETs was adopted (Rorke, 1983) because primitive neuroepithelial cells anywhere in the CNS are purported to be the targets of transformation, resulting in similar tumors, an event that seems to be most frequent in the cerebellum. Microscopically, medulloblastomas consist mainly of rounded or slightly elongated small cells with a high nucleus:cytoplasm ratio.

As a consequence of careful adjustment of cell culture media and other in vitro conditions, quite a few medulloblastoma cell lines have been established (Friedman et al., 1985,1988; Trojanowski et al., 1987; He et al., 1991; for review, see Friedman et al., 1991). The medulloblastoma lines also have been karyotyped and, as are the tumor biopsies (Bigner et al., 1988b; Griffin et al., 1988; Biegel et al., 1989), are mostly near-diploid and have the characteristic marker i(17q) (Bigner et al., 1990b). DMs have been found in some cases, which also had an amplified c-myc gene. Thus D-341 Med, D-384 Med, D-425 Med, and the xenograft of D-382 Med have an amplified c-myc gene (Bigner et al., 1990b), whereas no amplification of N-myc, EGFR, or gli occurs in the D-Med cell lines. Another medulloblastoma cell line, MTS, has a 150-fold amplification of the N-myc gene (Wasson et al., 1990). Altogether, medulloblastoma cell lines have a higher frequency of c-myc amplification than the original tumor biopsies, suggesting that c-myc is important to their establishment as cell lines in vitro (Wasson et al., 1990).

Several of the medulloblastoma cell lines show a tendency for neuronal differentiation. The lines D-283 Med, D-341 Med, D-384 Med, D-425 Med, and D-458 Med all grow in suspension and form spheroidal structures spontaneously. Nestin is expressed in all five cell lines (Tohyama et al., 1992). These lines are characterized by expression of synaptophysin and, mostly, all three forms of neurofilaments, but are devoid of GFAP and Class I and II MHC antigens. The cells also have a ganglioside expression profile that resembles that of other tumors of neuroectodermal derivation, rather than that of gliomas. The medulloblastoma cells fail to adhere to fibronectin, laminin, and collagen, but readily express the adhesion molecules neural cell adhesion molecule (NCAM) and L1 (Wikstrand *et al.*, 1991). In this respect, these cell lines also have the characteristics of neuroblastic cells.

The cell line Daoy (D-324 Med) differs from the majority of established medulloblastoma cell lines by growing adherently, having a tetraploid karyotype, and not expressing any of the neuronal markers. Daoy could, therefore, represent a medulloblastoma with glial differentiation, not be representative of medulloblastoma, or even represent a very immature stem cell before the expression of nestin (Tohyama *et al.*, 1992). This line has been shown to have a 5-fold amplification of the *erb*B1 gene (Wasson *et al.*, 1990) and a *p53* mutation in codon 242 (Saylors *et al.*, 1991).

Tamura *et al.* (1989) described two cell lines from medulloblastoma cases, ONS-76 and ONS-81, that expressed neurofilament but not GFAP and S100 antigens. These lines grew adherently and, with addition of IFNγ, expressed Class II MHC antigens. In the latter respect, these lines resembled cells of glial rather than of neuronal differentiation. The authors interpreted this finding as an indication of the origin of medulloblastomas as a bipotential cell that also differentiates into glial cells.

Another cell line described originally as a medulloblastoma (TE 671) (McAllister *et al.*, 1977) actually was a cross-contamination with the rhabdomyosarcoma cell line RD, as judged by genetic finger print analysis (Stratton *et al.*, 1989).

Fults *et al.* (1992b) described the establishment of the cell line PFSK from a PNET of the cerebral hemisphere of a child. Three sublines evolved. PFSK-1 had "multinucleated cells with broad, flat morphology," the karyotype was hypotetraploid with many chromosomal abnormalities. The PFSK-2 cells grew in clumps and were poorly adherent to the plastic; this line had a pseudodiploid karyotype with specific translocations t(1;11), t(3;10), and t(17;22). The only chromosomal abnormality of the PFSK-2C cells was trisomy of chromosome 8. The PFSK cell line had the characteristic of a primitive neuroepithelial stem cell, since it expressed high levels of nestin but no neurofilament protein, GFAP, or GalC.

V. Concluding Remarks

Established cell lines are available from anaplastic astrocytomas, glioblastomas, oligodendrogliomas, and PNET/medulloblastomas. A striking heterogeneity exists, especially among the astrocyte-derived cell lines, and the selective and adaptive processes that occur during establishment *in vitro* make each cell line only partly representative of gliomas *in vivo*. Therefore researchers should remain aware that the established lines can be used only as model systems for studying specific subjects *in vitro*. In this respect, the abundance of established glioma cell lines and the variability among them is a great advantage. These lines constitute an important source of material for researchers in many fields.

References

Antoniades, H. N., Galanopoulos, T., Neville-Golden, J., and Maxwell, M. (1992). Expression of insulin-like growth factors I and II and their receptor mRNAs in primary human astrocytomas and meningiomas; In vivo studies using in situ hybridization and immunocytochemistry. *Int. J. Cancer* **50**, 215–222.

Basta, P. V., Sherman, P. A., and Ting, J. P.-Y. (1988). Detailed delineation of an interferon-γ-responsive element important in human HLA-DRA gene expression in a glioblastoma multiforme line. *Proc. Natl. Acad. Sci. U.S.A.* **85**, 8618–8622.

Biegel, J. A., Packer, R. J., Rorke, L. B., and Emanuel, B. S. (1989). Non-random chromosomal changes in CNS primitive neuroectodermal tumors/medulloblastoma. Proceedings of the Third International Workshop on Chromosomes in Solid Tumors. *Cancer Genet. Cytogenet.* **41**, 228.

Bigner, D. D., Bigner, S. H., Pontén, J., Westermark, B., Mahaley, M. S. Jr., Ruoslahti, E., Herschman, H., Eng, L. F., and Wikstrand, C. J. (1981). Heterogeneity of genotypic and phenotypic characteristics of fifteen permanent cell lines derived from human gliomas. *J. Neuropathol. Exp. Neurol.* **XL**, 201–229.

Bigner, S. H., and Mark, J. (1984). Chromosomes and chromosomal progression of human gliomas in vivo, in vitro and in athymic nude mice. *Prog. Exp. Tumor Res.* **27**, 67–82.

Bigner, S. H., Mark, J., Mahaley, M. S., and Bigner, D. D. (1984). Patterns of the early gross chromosomal changes in malignant human gliomas. *Hereditas* **101**, 103–113.

Bigner, S. H., Mark, J., Schold, S. C., Jr., Eng, L. F., and Bigner, D. D. (1985). A serially transplantable human giant cell glioblastoma that maintains a near-haploid stem line. *Cancer Genet. Cytogenet.* **18**, 141–154.

Bigner, S. H., Friedman, H. S., Biegel, J. A., Wikstrand, C. J., Mark, J., Gebhardt, R., Eng, L. F., and Bigner, D. D. (1986a). Specific chromosomal abnormalities characterize four established cell lines derived from malignant human gliomas. *Acta Neuropathol. (Berlin)* **72**, 86–97.

Bigner, S. H., Mark, J., Bullard, D. E., Mahaley, M. S., Jr., and Bigner, D. D. (1986b). Chromosomal evolution in malignant human gliomas starts with specific and usually numerical deviations. *Cancer Genet. Cytogenet.* **22**, 121–135.

Bigner, S. H., Mark, J., and Bigner, D. D. (1987a). Chromosomal progression of malignant human gliomas from biopsy to establishment as permanent lines in vitro. *Cancer Genet. Cytogenet.* **24**, 163–176.

Bigner, S. H., Wong, A. J., Mark, J., Muhlbaier, L. H., Kinzler, K. W., Vogelstein, B., and Bigner, D. D. (1987b). Relationship between gene amplification and chromosomal deviations in malignant human gliomas. *Cancer Genet. Cytogenet.* **29**, 165–170.

Bigner, S. H., Mark, J., Burger, P. C., Mahaley, M. S., Jr., Bullard, D. E., Muhlbaier, L. H., and Bigner, D. D. (1988a). Specific chromosomal abnormalities in malignant human gliomas. *Cancer Res.* **48**, 405–411.

Bigner, S. H., Mark, J., Friedman, H. S., Biegel, J. A., and Bigner, D. D. (1988b). Structural chromosomal abnormalities in human medulloblastoma. *Cancer Genet. Cytogenet.* **30**, 91–101.

Bigner, S. H., Schold, S. C., Friedman, H. S., Mark, J., and Bigner, D. D. (1989). Chromosomal composition of malignant human gliomas through serial subcutaneous transplantation in athymic mice. *Cancer Genet. Cytogenet.* **40**, 111–120.

Bigner, S. H., Humphrey P. A., Wong, A. J., Vogelstein, B., Mark, J., Friedman, H. S., and Bigner,

D. D. (1990a). Characterization of the epidermal growth factor receptor in human glioma cell lines and xenografts. *Cancer Res.* **50**, 8017–8022.

Bigner, S. H., Friedman, H. S., Vogelstein, B., Oakes, W. J., and Bigner, D. D. (1990b). Amplification of the c-myc gene in human medulloblastoma cell lines and xenografts. *Cancer Res.* **50**, 2347–2350.

Birchmeier, C., Sharma, S., and Wigler, M. (1987). Expression and rearrangement of the ROS1 gene in human glioblastoma cells. *Proc. Natl. Acad. Sci. U.S.A.* **84**, 9270–9274.

Birchmeier, C., O'Neill, K., Riggs, M., and Wigler, M. (1990). Characterization of ROS1 cDNA from a human glioblastoma cell line. *Proc. Natl. Acad. Sci. U.S.A.* **87**, 4799–4803.

Bodmer, S., Strommer, K., Frei, K., Siepl, C., de Tribolet, N., Heid, I., and Fontana, A. (1989). Immunosuppression and transforming growth factor-β in glioblastoma. Preferential production of transforming growth factor-β2. *J. Immunol.* **143**, 3222–3229.

Bögler, O., Wren, D., Barnett, S. C., Land, H., and Noble, M. (1990). Cooperation between two growth factors promotes extended self-renewal and inhibits differentiation of oligodendrocyte-type-2 astrocyte (O-2A) progenitor cells. *Proc. Natl. Acad. Sci. U.S.A.* **87**, 6368–6372.

Bongcam-Rudloff, E., Nistér, M., Betsholtz, C., Wang, J.-L., Stenman, G., Huebner, K., Croce, C. M., and Westermark, B. (1991). Human glial fibrillary acidic protein: Complementary DNA cloning, chromosome localization, and messenger RNA expression in human glioma cell lines of various phenotypes. *Cancer Res.* **51**, 1553–1560.

Bullard, D. E., Bigner, S. H., and Bigner, D. D. (1981). The morphologic response of cell lines derived from human gliomas to dibutyryl adenosine 3':5' cyclic monophosphate. *J. Neuropathol. Exp. Neurol.* **40**, 230–246.

Carlsson, J. (1992). Tumor spheroids in studies of immunotherapy. In "Spheroid Culture in Cancer Research" (R. Bjerkvig, ed.), pp. 277–306. CRC Press, Boca Raton, Florida.

Carlsson, J., and Acker, H. (1988). Relations between pH, oxygen partial pressure and growth in cultured cell spheroids. *Int. J. Cancer* **42**, 715–720.

Carlsson, J., Nilsson, K., Westermark, B., Pontén, J., Sundström, C., Larsson, E., Bergh, J., Påhlman, S., Busch, C., and Collins, V. P. (1983). Formation and growth of multicellular spheroids of human origin. *Int. J. Cancer* **31**, 523–533.

Chung, R., Whaley, J., Kley, N., Anderson, K., Louis, D., Menon, A., Hettlich, C., Freiman, R., Hedley-Whyte, E. T., Martuza, R., Jenkins, R., Yandell, D., and Seizinger, B. R. (1991). TP53 gene mutations and 17p deletions in human astrocytomas. *Genes Chrom. Cancer* **3**, 323–331.

Collins, V. P. (1983). Cultured human glial and glioma cells. *Int. Rev. Exp. Pathol.* **24**, 135–202.

Constam, D. B., Philipp, J., Malipiero, U. V., ten Dijke, P., Schachner, M., and Fontana, A. (1992). Differential expression of transforming growth factor-β1, -β2, and -β3 by glioblastoma cells, astrocytes, and microglia. *J. Immunol.* **148**, 1404–1410.

Criscuolo, G. R., Merrill, M. J., and Oldfield, E. H. (1988). Further characterization of malignant glioma-derived vascular permeability factor. *J. Neurosurg.* **69**, 254–262.

Deinhardt, E. (1980). Biology of primate retroviruses. In "Viral Oncology" (G. Klein, ed.), pp. 357–398. Raven Press, New York.

de Martin, R., Haendler, B., Hofer-Warbinek, R., Gaugitsch, H., Wrann, M., Schlüsener, H., Seifert, J. M., Bodmer, S., Fontana, A., and Hofer, E. (1987). Complementary DNA for human glioblastoma-derived T cell suppressor factor, a novel member of the transforming growth factor-β gene family. *EMBO J.* **6**, 3673–3677.

de Ridder, L. I., Laerum, O. D., Mørk, S. J., and Bigner, D. D. (1987). Invasiveness of human glioma cell lines in vitro: Relation to tumorigenicity in athymic mice. *Acta Neuropathol. (Berlin)* **72**, 207–213.

Doolittle, R. F., Hunkapiller, M. W., Hood, L. E., Devare, S. G., Robbins, K. C., Aaronson, S. A., and Antoniades, H. N. (1983). Simian sarcoma virus onc gene, v-sis, is derived from the gene (or genes) encoding a platelet-derived growth factor. *Science* **221**, 275–277.

El-Azozouzi, M., Chung, R. Y., Farmer, G. E., Martuza, R. L., Black, P. M., Rouleau, G. A., Hettlich, C., Hedley-Whyte, E. T., Zervas, N. T., Panagopoulos, K., Nakamura, Y., Gusella, J. F., and Seizinger, B. R. (1989). Loss of distinct regions on the short arm of chromosome 17 associated with tumorigenesis of human astrocytomas. *Proc. Natl. Acad. Sci. U.S.A.* **86,** 7186–7190.

Eva, A., Robbins, K. C., Andersen, P. R., Srinivasan, A., Tronick, S. R., Reddy, E. P., Ellmore, N. W., Galen, A. T., Lantenberger, J. A., Papas, T. S., Westin, E. H., Wong-Staal, F., Gallo, R. C., and Aaronson, S. A. (1982). Cellular genes analogous to retroviral *onc* genes are transcribed in human tumour cells. *Nature (London)* **295,** 116–119.

Fleming, T. P., Saxena, A., Clark, W. C., Robertson, J. T., Oldfield, E. H., Aaronson, S. A., and Ali, I. U. (1992a). Amplification and/or overexpression of platelet-derived growth factor receptors and epidermal growth factor receptor in human glial tumors. *Cancer Res.* **52,** 4550–4553.

Fleming, T. P., Matsui, T., Heidaran, M. A., Molloy, C. J., Artrip, J., and Aaronson, S. A. (1992b). Demonstration of an activated platelet-derived growth factor autocrine pathway and its role in human tumor cell proliferation *in vitro. Oncogene* **7,** 1355–1359.

Fontana, A., Hengarter, H., de Tribolet, N., and Weber, E. (1984). Glioblastoma cells release interleukin-1 and factors inhibiting interleukin-2-mediated effects. *J. Immunol.* **132,** 1837–1844.

Frankel, R. H., Bayona, W., Koslow, M., and Newcomb, E. W. (1992). p53 mutations in human malignant gliomas: Comparison of loss of heterozygosity with mutation frequency. *Cancer Res.* **52,** 1427–1433.

Friedman, H. S., Burger, P. C., Bigner, S. H., Trojanowski, J. Q., Wikstrand, C. J., Halperin, E. C., and Bigner, D. D. (1985). Establishment and characterization of the human medulloblastoma cell line and transplantable xenograft D283. *Med. J. Neuropathol. Exp. Neurol.* **44,** 592–605.

Friedman, H. S., Burger, P. C., Bigner, S. H., Trojanowski, J. Q., Brodeur, G. M., He, X., Wikstrand, C. J., Kurtzberg, J., Berens, M. E., Halperin, E. C., and Bigner, D. D. (1988). Phenotypic and genotypic analysis of a human medulloblastoma cell line and transplantable xenograft (D341 Med) demonstrating amplification of c-*myc. Am. J. Pathol.* **130,** 472–484.

Friedman, H. S., Oakes, W. J., Bigner, S. H., Wikstrand, C. J., and Bigner, D. D. (1991). Review, Medulloblastoma: Tumor biological and clinical perspectives. *J. Neurooncol.* **11,** 1–15.

Fujimoto, M., Fults, D. W., Thomas, G. A., Nakamura, Y., Heilbrun, M. P., White, R., Story, J. L., Naylor, S. L., Kagan-Hallet, K. S., and Sheridan, P. J. (1989). Loss of heterozygosity on chromosome 10 in human glioblastoma multiforme. *Genomics* **4,** 210–214.

Fults, D., Tippets, R. H., Thomas, G. A., Nakamura, Y., and White, R. (1989). Loss of heterozygosity for loci on chromosome 17p in human malignant astrocytoma. *Cancer Res.* **49,** 6572–6577.

Fults, D., Pedone, C. A., Thomas, G. A., and White, R. (1990). Allelotype of human malignant astrocytoma. *Cancer Res.* **50,** 5784–5789.

Fults, D., Brockmeyer, D., Tullous, M. W., Pedone, C. A., and Cawthon, R. M. (1992a). p53 mutation and loss of heterozygosity on chromosomes 17 and 10 during human astrocytoma progression. *Cancer Res.* **52,** 674–679.

Fults, D., Pedone, C. A., Morse, H. G., Rose, J. W., and McKay, R. D. G. (1992b). Establishment and characterization of a human primitive neuroectodermal tumor cell line from the cerebral hemisphere. *J. Neuropathol. Exp. Neurol.* **51,** 272–280.

Gammeltoft, S., Ballotti, R., Kowalski, A., Westermark, B., and Van Obberghen, E. (1988). Expression of two types of receptor for insulin-like growth factors in human malignant glioma. *Cancer Res.* **48,** 1233–1237.

Geller, H. M., and Dubois-Dalcq, M. (1988). Antigenic and functional characterization of a rat central nervous system-derived cell line immortalized by a retroviral vector. *J. Cell Biol.* **107,** 1977–1986.

Gerosa, M. A., Talarico, D., Fognani, C., Raimondi, E., Colombatti, M., Tridente, G., De Carli, L., and Della Valle, G. (1989). Overexpression of N-*ras* oncogene and epidermal growth factor receptor gene in human glioblastomas. *J. Natl. Cancer Inst.* **81,** 63–67.

Godbout, R., Miyakoshi, J., Dobler, K. D., Andison, R., Matsuo, K., Allalunis-Turner, M. J., Takebe, H., and Day, R. S., III (1992). Lack of expression of tumor-suppressor genes in human malignant glioma cell lines. *Oncogene* **7,** 1879–1884.

Griffin, C. A., Hawkins, A. L., Packer, R. J., Rorke, L. B., and Emanuel, B. S. (1988). Chromosome abnormalities in pediatric brain tumors. *Cancer Res.* **48,** 175–180.

Griffin, C. A., Long, P. P., Carson, B. S., and Brem, H. (1992). Chromosome abnormalities in low-grade central nervous system tumors. *Cancer Genet. Cytogenet.* **60,** 67–73.

Gross, J. L., Morrison, R. S., Eidsvoog, K., Herblin, W. F., Kornblith, P. L., and Dexter, D. L. (1990). Basic fibroblast growth factor: A potential autocrine regulator of human glioma cell growth. *J. Neurosci. Res.* **27,** 689–696.

Grunberger, G., Lowe, W. L., Jr., MeElduff, A., and Glick, R. P. (1986). Insulin receptor of human cerebral gliomas. Structure and function. *J. Clin. Invest.* **77,** 997–1005.

Haber, D. A., and Schimke, R. T. (1981). Unstable amplification of an altered dihydrofolate reductase gene associated with double minute chromosomes. *Cell* **26,** 355–362.

Harsh, G. R., Keating, M. T., Escobedo, J. A., and Williams, L. T. (1990). Platelet derived growth factor (PDGF) autocrine components in human tumor cell lines. *J. Neurooncol.* **8,** 1–12.

Hart, I. K., Richardson, W. D., Heldin, C.-H., Westermark, B., and Raff, M. (1989). PDGF receptors on cells of the oligodendrocyte-type-2 astrocyte (O-2A) cell lineage. *Development* **105,** 595–603.

He, X., Wikstrand, C. J., Friedman, H. S., Bigner, S. H., Pleasure, S., Trojanowski, J. Q., and Bigner, D. D. (1991). Differentiation characteristics of newly established medulloblastoma cell lines (D384 Med, D425 Med, and D458 Med) and their transplantable xenografts. *Lab. Invest.* **64,** 833–843.

Heisterkamp, N., Morris, C., Sender, L., Knoppel, E., Uribe, L., Cui, M.-Y., and Groffen, J. (1990). Rearrangement of the human ABL oncogene in a glioblastoma. *Cancer Res.* **50,** 3429–3434.

Helseth, E., Unsgaard, G., Dalen, A., and Vik, R. (1988). The effect of type beta transforming growth factor on proliferation of clonogenic cells from human gliomas. *Acta Neurochir. Suppl.* **43,** 118–120.

Huber, D., Fontana, A., and Bodmer, S. (1991). Activation of human platelet-derived latent transforming growth factor-$\beta 1$ by human glioblastoma cells. Comparison with proteolytic and glycosidic enzymes. *Biochem. J.* **277,** 165–173.

Humphrey, P. A., Wong, A. J., Vogelstein, B., Friedman, H. S., Werner, M. H., Bigner, D. D., and Bigner, S. H. (1988). Amplification and expression of the epidermal growth factor receptor gene in human glioma xenografts. *Cancer Res.* **48,** 2231–2238.

Isshiki, H., Akira, S., Tanabe, O., Nakajima, T., Shimamoto, T., Hirano, T., and Kishimoto, T. (1990). Constitutive and interleukin-1 (IL-1)-inducible factors interact with the IL-1 responsive element in the IL-6 gene. *Mol. Cell. Biol.* **10,** 2757–2764.

James, C. D., Carlbom, E., Dumanski, J. P., Hansen, M., Nordenskjöld, M., Collins, V. P., and Cavenee, W. K. (1988). Clonal genomic alterations in glioma malignancy stages. *Cancer Res.* **48,** 5546–5551.

James, C. D., Carlbom, E., Nordenskjöld, M., Collins, V. P., and Cavenee, W. K. (1989). Mitotic recombination of chromosome 17 in astrocytomas. *Proc. Natl. Acad. Sci.* **86,** 2858–2862.

Jennings, M. T., Maciunas, R. J., Carver, R., Bascom, C. C., Juneau, P., Misulis, K., and Moses, H. L. (1991). TGF$\beta 1$ and TGF$\beta 2$ are potential growth regulators for low-grade and malignant gliomas *in vitro:* Evidence in support of an autocrine hypothesis. *Int. J. Cancer* **49,** 129–139.

Johnsson, A., Betsholtz, C., Heldin, C.-H., and Westermark, B. (1985). Antibodies against platelet-derived growth factor inhibit acute transformation by simian sarcoma virus. *Nature (London)* **317,** 438–440.

Jones, T. R., Ruoslahti, E., Schold, S. C., and Bigner, D. D. (1982). Fibronectin and glial fibrillary acidic protein expression in normal human brain and anaplastic human gliomas. *Cancer Res.* **42,** 168–177.

Kasahara, T., Mukaida, N., Yamashita, K., Yagisawa, H., Akahoshi, T., and Matsushima, K. (1991). IL-1 and TNF-α induction of IL-8 and monocyte chemotactic and activating factor (MCAF) mRNA expression in a human astrocytoma cell line. *Immunology* **74,** 60–67.

Kashima, T., Tiu, S. N., Merrill, J. E., Vinters, H. V., Dawson, G., and Campagnoni, A. T. (1993).

Expression of oligodendrocyte-associated genes in cell lines derived from human gliomas and neuroblastomas. *Cancer. Res.* **53,** 170–175.

Kennedy, P. G. E., Watkins, B. A., Thomas, D. G. T., and Noble, M. D. (1987). Antigenic expression by cells derived from human gliomas does not correlate with morphological classification. *Neuropathol. Appl. Neurobiol.* **13,** 327–347.

Kernohan, J. W., Mabon, R. F., Svien, H. J., and Adson, A. W. (1949). Symposium on a new and simplified concept of gliomas (a simplified classification of gliomas). *Proc. Mayo Clin.* **24,** 71.

Kinzler, K. W., Bigner, S. H., Bigner, D. D., Trent, J. M., Law, M. L., O'Brien, S. J., Wong, A. J., and Vogelstein, B. (1987). Identification of an amplified, highly expressed gene in a human glioma. *Science* **236,** 70–73.

Kleihues, P., Burger, P. C., and Scheithauer, B. W. (1993). Histological typing of tumors of the central nervous system. *In* "International Histological Classification of Tumors." World Health Organization, Geneva.

Knott, J. C. A., Edwards, A. J., Gullan, R. W., Clarke, T. M., and Pilkington, G. J. (1990). A human glioma cell line retaining expression of GFAP and gangliosides, recognized by A2B5 and LB1 antibodies, after prolonged passage. *Neuropathol. Appl. Neurobiol.* **16,** 489–500.

Kumabe, T., Sohma, Y., Kayama, T., Yoshimoto, T., and Yamamoto, T. (1992). Amplification of a platelet-derived growth factor receptor gene lacking an exon coding for a portion of the extracellular region in a primary brain tumor of glial origin. *Oncogene* **7,** 627–633.

Kuppner, M. C., Hamou, M.-F., Bodmer, S., Fontana, A., and de Tribolet, N. (1988). The glioblastoma-derived T-cell suppressor factor/transforming growth factor beta$_2$ inhibits the generation of lymphokine-activated killer (LAK) cells. *Int. J. Cancer* **42,** 562–567.

LaRocca, R. V., Rosenblum, M., Westermark, B., and Israel, M. A. (1989). Patterns of protooncogene expression in human glioma cell lines. *J. Neurosci. Res.* **24,** 97–106.

LaRocca, R. V., Stein, C. A., and Myers, C. E. (1990). Suramin: Prototype of a new generation of antitumor compounds. *Cancer Cells* **2,** 106–115.

Lendahl, U., Zimmerman, L. B., and McKay, R. D. G. (1990). CNS stem cells express a new class of intermediate filament protein. *Cell* **60,** 585–595.

Leppert, D., Frei, K., Gallo, P., Yasargil, M. G., Hess, K., Baumgartner, G., and Fontana, A. (1989). Brain tumors: Detection of B-cell stimulatory factor-2/interleukin-6 in the absence of oligoclonal bands of immunoglobulins. *J. Neuroimmunol.* **24,** 259–264.

Libermann, T. A., Nusbaum, H. R., Razon, N., Kris, R., Lax, I., Soreq, H., Whittle, N., Waterfield, M. D., Ullrich, A., and Schlessinger, J. (1985). Amplification enhanced expression, and possible rearrangement of the EGF receptor gene in primary human brain tumours of glial origin. *Nature (London)* **313,** 144–147.

Libermann, T. A., Friesel, R., Jaye, M., Lyall, R. M., Westermark, B., Drohan, W., Schmidt, A., Maciag, T., and Schlessinger, J. (1987). An angiogenic growth factor is expressed in human malignant glioma cells. *EMBO J.* **6,** 1627–1632.

Lumsden, C. E. (1971). The study by tissue culture of tumors of the nervous system. *In* "Pathology of Tumors of the Nervous System" (C. S. Russel and L. J. Rubinstein, eds.), 3d Ed., pp. 334–420. Edward Arnold, London.

Lund-Johansen, M., Bjerkvig, R., Humphrey, P. A., Bigner, S. H., Bigner, D. D., and Laerum, O.-D. (1990). Effect of epidermal growth factor on glioma cell growth, migration, and invasion *in vitro*. *Cancer Res.* **50,** 6039–6044.

McAllister, R. M., Isaacs, H., Rongey, R., Peer, M., Au, W., Soukup, S. W., and Gardner, M. B. (1977). Establishment of a human medulloblastoma cell line. *Int. J. Cancer* **20,** 206–212.

McKeever, P. E., and Chronwall, B. M. (1985). Early switch in glial protein and fibronectin markers on cells during the culture of human gliomas. *Ann. N.Y. Acad. Sci.* **435,** 457–459.

McKeever, P. E., Smith, B. H., Taren, J. A., Wahl, R. L., Kornblith, P. L., and Chronwall, B. M. (1987). Products of cells cultured from gliomas. VI. Immunofluorescent, morphometric, and ultrastructural characterization of two different cell types growing from explants of gliomas. *Am. J. Pathol.* **127,** 358–372.

McKeever, P. E., Fligiel, S. E. G., Varani, J., Castle, R. L., and Hood, T. W. (1989). Products of cells cultured from gliomas. VII. Extracellular matrix proteins of gliomas which contain glial fibrillary acidic protein. *Lab. Invest.* **60,** 286–295.

Manuelidis, E. E. (1965). Long-term lines of tissue cultures of intracranial tumors. *J. Neurosurg.* **22,** 368–373.

Manuelidis, E. E. (1969). Experiments with tissue culture and heterologous transplantation of tumors. *Ann. N.Y. Acad. Sci.* **159(2),** 409–443.

Manuelidis, L., Yu, R. K., and Manuelidis, E. E. (1977). Ganglioside content and pattern in human gliomas in culture: Correlation of morphological changes with altered gangliosides. *Acta Neuropathol. (Berlin)* **38,** 129–135.

Mark, J. (1971). Chromosomal characteristics of neurogenic tumours in adults. *Hereditas* **68,** 61–100.

Mark, J., Pontén, J., and Westermark, B. (1974). Cytogenetical studies with G-band technique of established cell lines of human malignant gliomas. *Hereditas* **78,** 304–308.

Mathiesen, T., Björklund, H., Collins, V. P., Granholm, L., and Olson, L. (1989). Induction of GFAP production in human glioma lines grafted into the anterior chamber of the rat eye. *Neurosci. Lett.* **97,** 291–297.

Mercer, W. E., Shields, M. T., Amin, M., Sauve, G. J., Appella, E., Romano, J. W., and Ullrich, S. J. (1990). Negative growth regulation in a glioblastoma tumor cell line that conditionally expresses human wild-type p53. *Proc. Natl. Acad. Sci. U.S.A.* **87,** 6166–6170.

Merrill, J. E., and Matsushima, K. (1988). Production of and response to interleukin 1 by cloned human oligodendroglioma cell lines. *J. Biol. Regul. Homeo. Agents* **2,** 77–86.

Miyakoshi, J., Dobler, K. D., Allalunis-Turner, J., McKean, J. D. S., Petruk, K., Allen, P. B. R., Aronyk, K. N., Weir, B., Huyser-Wierenga, D., Fulton, D., Urtasun, R. C., and Day, R. S. III. (1990). Absence of IFNA and IFNB genes from human malignant glioma cell lines and lack of correlation with cellular sensitivity to interferons. *Cancer Res.* **50,** 278–283.

Molenaar, W. M., Jansson, D. S., Gould, V. E., Rorke, L. B., Franke, W. W., Lee, V. M.-Y., Packer, R. J., and Trojanowski, J. Q. (1989). Molecular markers of primitive neuroectodermal tumors and other pediatric central nervous system tumors. Monoclonal antibodies to neuronal and glial antigens distinguish subsets of primitive neuroectodermal tumors. *Lab. Invest.* **61,** 635–643.

Morrison, R. S., Gross, J. L., Herblin, W. F., Reilly, T. M., LaSala, P. A., Alterman, R. L., Moskal, J. R., Kornblith, P. L., and Dexter, D. L. (1990). Basic fibroblast growth factor-like activity and receptors are expressed in a human glioma cell line. *Cancer Res.* **50,** 2524–2529.

Murphy, P. R., Sato, Y., and Knee, R. S. (1992). Phosphorothioate antisense oligonucleotides against basic fibroblast growth factor inhibit anchorage-dependent and anchorage-independent growth of a malignant glioblastoma cell line. *Mol. Endocrinol.* **6,** 877–884.

Nishiyama, A., Onda, K., Washiyama, K., Kumanishi, T., Kuwano, R., Sakimura, K., and Takahashi, Y. (1989). Differential expression of glial fibrillary acidic protein in human glioma cell lines. *Acta Neuropathol.* **78,** 9–15.

Nistér, M., Heldin, C.-H., Wasteson, Å., and Westermark, B. (1984). A glioma-derived analog to platelet-derived growth factor: Demonstration of receptor competing activity and immunological cross-reactivity. *Proc. Natl. Acad. Sci. U.S.A.* **81,** 926–930.

Nistér, M., Heldin, C.-H., and Westermark, B. (1986). Clonal variation in the production of platelet-derived growth factor-like protein and expression of corresponding receptors in a human malignant glioma. *Cancer Res.* **46,** 332–340.

Nistér, M., Wedell, B., Betsholtz, C., Bywater, M., Pettersson, M., Westermark, B., and Mark, J. (1987). Evidence for progressional changes in the human malignant glioma line U-343 MGa: Analysis of karyotype and expression of genes encoding the subunit chains of platelet-derived growth factor. *Cancer Res.* **47,** 4953–4960.

Nistér, M., Libermann, T. A., Betsholtz, C., Pettersson, M., Claesson-Welsh, L., Heldin, C.-H., Schlessinger, J., and Westermark, B. (1988a). Expression of messenger RNAs for platelet-derived growth factor and transforming growth factor-α and their receptors in human malignant glioma cell lines. *Cancer Res.* **48**, 3910–3918.

Nistér, M., Hammacher, A., Mellström, K., Siegbahn, A., Rönnstrand, L., Westermark, B., and Heldin, C.-H. (1988b). A glioma-derived PDGF A chain homodimer has different functional activities from a PDGF AB heterodimer purified from human platelets. *Cell* **52**, 791–799.

Nistér, M., Claesson-Welsh, L., Eriksson, A., Heldin, C.-H., and Westermark, B. (1991). Differential expression of platelet-derived growth factor receptors in human malignant glioma cell lines. *J. Biol. Chem.* **266**, 16755–16763.

Noble, M., Murray, K., Stroobant, P., Waterfield, M. D., and Riddle, P. (1988). Platelet-derived growth factor promotes division and motility and inhibits premature differentiation of the oligodendrocyte/type-2 astrocyte progenitor cell. *Nature (London)* **333**, 560–562.

Okamoto, Y., Minamoto, S., Shimizu, K., Mogami, H., and Taniguchi, T. (1990). Interleukin 2 receptor β chain expressed in an oligodendroglioma line binds interleukin 2 and delivers growth signal. *Proc. Natl. Acad. Sci. U.S.A.* **87**, 6584–6588.

Olofsson, A., Miyazono, K., Kanzaki, T., Colosetti, P., Engström, U., and Heldin, C.-H. (1992). Transforming growth factor-β1, -β2, and -β3 secreted by a human glioblastoma cell line. Identification of small and different forms of large latent complexes. *J. Biol. Chem.* **267**, 19482–19488.

Olopade, O. I., Jenkins, R. B., Ransom, D. T., Malik, K., Pomykala, H., Nobori, T., Cowan, J. M., Rowley, J. D., and Diaz, M. O. (1992). Molecular analysis of deletions of the short arm of chromosome 9 in human gliomas. *Cancer Res.* **52**, 2523–2529.

Paetau, A. (1988). Glial fibrillary acidic protein, vimentin and fibronectin in primary cultures of human glioma and fetal brain. *Acta Neuropathol. (Berlin)* **75**, 448–455.

Paetau, A., Mellström, K., Vaheri, A., and Haltia, M. (1980). Distribution of a major connective tissue protein, fibronectin, in normal and neoplastic human nervous tissue. *Acta Neuropathol. (Berlin)* **51**, 47–51.

Pantazis, P., Pelicci, P. G., Dalla-Favera, R., and Antoniades, H. N. (1985). Synthesis and secretion of proteins resembling platelet-derived growth factor by human glioblastoma and fibrosarcoma cells in culture. *Proc. Natl. Acad. Sci. U.S.A.* **82**, 2404–2408.

Pontén, J., and Macintyre, E. H. (1968). Long term culture of normal and neoplastic human glia. *Acta Pathol. Microbiol. Scand.* **74**, 465–486.

Pontén, J., and Westermark, B. (1978). Properties of human malignant glioma cells in vitro. *Med. Biol.* **56**, 184–193.

Post, G. R., and Dawson, G. (1992). Regulation of carbachol- and histamine-induced inositol phospholipid hydrolysis in a human oligodendroglioma. *Glia* **5**, 122–130.

Rey, J. A., Bello, M. J., de Campos, J. M., Kusak, M. E., Ramos, C., and Benitez, J. (1987). Chromosomal patterns in human malignant astrocytomas. *Cancer Genet. Cytogenet.* **29**, 201–221.

Reynolds, B. A., and Weiss, S. (1992). Generation of neurons and astrocytes from isolated cells of the adult mammalian central nervous system. *Science* **255**, 1707–1710.

Richardson, W. D., Pringle, N., Mosley, M. J., Westermark, B., and Dubois-Dalcq, M. (1988). A role for platelet-derived growth factor in normal gliogenesis in the central nervous system. *Cell* **53**, 309–319.

Ringertz, N. (1950). "Grading" of gliomas. *Acta Pathol. Microbiol. Scand.* **27**, 51–64.

Rorke, L. B. (1983). The cerebellar medulloblastoma and its relationship to primitive neuroectodermal tumors. *J. Neuropathol. Exp. Neurol.* **42**, 1–15.

Russell, D. S., and Rubinstein, L. J. (1989). "Pathology of Tumours of the Nervous System," 5th Ed. Edward Arnold, London.

Rutka, J. T., Giblin, J. R., Høifødt, H. K., Dougherty, D. V., Bell, C. W., McCulloch, J. R., Davis, R. L., Wilson, C. B., and Rosenblum, M. L. (1986). Establishment and characterization of a cell line from a human gliosarcoma. *Cancer Res.* **46**, 5893–5902.

Rutka, J. T., Giblin, J. R., Dougherty, D. Y., Liu, H. C., McCulloch, J. R., Bell, C. W., Stern, R. S.,

Wilson, C. B., and Rosenblum, M. L. (1987). Establishment and characterization of five cell lines derived from malignant gliomas. *Acta Neuropathol. (Berlin)* **75,** 92–103.

Sandberg, A.-C., Engberg, C., Lake, M., von Holst, H., and Sara, V. R. (1988). The expression of insulin-like growth factor I and insulin-like growth factor II genes in the human fetal and adult brain and in glioma. *Neurosci. Lett.* **93,** 114–119.

Saxena, A., and Ali, I. U. (1992). Increased expression of genes from growth factor signaling pathways in glioblastoma cell lines. *Oncogene* **7,** 243–247.

Saylors, R. L., III, Sidransky, D., Friedman, H. S., Bigner, S. H., Bigner, D. D., Vogelstein, B., Brodeur, G. M. (1991). Infrequent p53 gene mutations in medulloblastomas. *Cancer Res.* **51,** 4721–4723.

Schimke, R. T. (1984). Gene amplification in cultured animal cells. *Cell* **37,** 705–713.

Shapiro, J. R. (1986). Biology of gliomas: Heterogeneity, oncogenes, growth factors. *Sem. Oncol.* **13,** 4–15.

Shapiro, J. R., and Shapiro, W. R. (1985). The subpopulations and isolated cell types of freshly resected high-grade human gliomas: Their influence on the tumor's evolution in vivo and behavior and therapy in vitro. *Cancer Metab. Rev.* **4,** 107–124.

Shapiro, J. R., Yung, W.-K. A., and Shapiro, W. R. (1981). Isolation, karyotype and clonal growth of heterogeneous subpopulations of human malignant gliomas. *Cancer Res.* **41,** 2349–2359.

Sharma, S., Birchmeier, C., Nikawa, J., O'Neill, K., Rodgers, L., and Wigler, M. (1989). Characterization of the *ros*1-gene products expressed in human glioblastoma cell lines. *Oncogene Res.* **5,** 91–100.

Sidransky, D., Mikkelsen, T., Schwechheimer, K., Rosenblum, M. L., Cavanee, W., and Vogelstein, B. (1992). Clonal expansion of p53 mutant cells is associated with brain tumour progression. *Nature (London)* **355,** 846–847.

Smits, A., Hermansson, M., Nistér, M., Karnushina, I., Heldin, C.-H., Westermark, B., and Funa, K. (1989). Rat brain capillary endothelial cells express functional PDGF B-type receptors. *Growth Factors* **2,** 1–8.

Sonnenberg, S., Gödecke, A., Walter, B., Bladt, F., and Birchmeier, C. (1991). Transient and locally restricted expression of the *ros1* protooncogene during mouse development. *EMBO J.* **10,** 3693–3702.

Stratton, M. R., Darling, J., Pilkington, G. J., Lantos, P. L., Reeves, B. R., and Cooper, C. S. (1989). Characterization of the human cell line TE671. *Carcinogenesis* **10,** 899–905.

Takahashi, J. A., Fukumoto, M., Kozai, Y., Ito, N., Oda, Y., Kikuchi, H., and Hatanaka, M. (1991). Inhibition of cell growth and tumorigenesis of human glioblastoma cells by a neutralizing antibody against human basic fibroblast growth factor. *FEBS Lett.* **288,** 65–71.

Takiguchi, M., Ting, J. P.-Y., Buessow, S. C., Boyer, C., Gillespie, Y., and Frelinger, J. A. (1985). Response of glioma cells to interferon-γ: Increase in class II RNA, protein and mixed lymphocyte reaction-stimulating activity. *Eur. J. Immunol.* **15,** 809–814.

Tamura, K., Shimizu, K., Yamada, M., Okamoto, Y., Matsui, Y., Park, K. C., Mabuchi, E., Moriuchi, S., and Mogami, H. (1989). Expression of major histocompatibility complex on human medulloblastoma cells with neuronal differentiation. *Cancer Res.* **49,** 5380–5384.

Tohyama, T., Lee, V. M.-Y., Rorke, L. B., Marvin, M., McKay, R. D. G., and Trojanowski, J. (1992). Nestin expression in embryonic human neuroepithelium and in human neuroepithelial tumor cells. *Lab. Invest.* **66,** 303–313.

Trent, J., Meltzer, P., Rosenblum, M., Harsh, G., Kinzler, K., Mashal, R., Feinberg, A., and Vogelstein, B. (1986). Evidence for rearrangement, amplification, and expression of c-*myc* in a human glioblastoma. *Proc. Natl. Acad. Sci. U.S.A.* **83,** 470–473.

Trojanowski, J. Q., Friedman, H. S., Burger, P. C., and Bigner, D. D. (1987). A rapidly dividing human medulloblastoma cell line (D283 MED) expresses all three neurofilament subunits. *Am. J. Pathol.* **126,** 358–363.

Vaheri, A., Ruoslahti, E., Westermark, B., and Pontén, J. (1976). A common cell-type specific surface antigen in cultured human glial cells and fibroblasts: Loss in malignant cells. *J. Exp. Med.* **143,** 64–72.

Valtz, N. L. M., Hayes, T. E., Norregaard, T., Liu, S., and McKay, R. D. G. (1991). An embryonic origin for medulloblastoma. *New Biologist* **3**, 364–371.

Van Meir, E., Sawamura, Y., Diserens, A.-C., Hamou, M.-F., and de Tribolet, N. (1990). Human glioblastoma cells release interleukin 6 in vivo and in vitro. *Cancer Res.* **50**, 6683–6688.

Van Meir, E., Ceska, M., Effenberger, F., Walz, A., Grouzmann, E., Desbaillets, I., Frei, K., Fontana, A., and de Tribolet, N. (1992). Interleukin-8 is produced in neoplastic and infectious diseases of the human central nervous system. *Cancer Res.* **52**, 4297–4305.

Vassbotn, F. S., Östman, A., Langeland, N., Holmsen, H., Westermark, B., Heldin, C.-H., and Nistér, M. (1993). *J. Cell. Physiol.,* in press.

Wasson, J. C., Saylors, R. L., III, Zeltzer, P., Friedman, H. S., Bigner, S. H., Burger, P. C., Bigner, D. D., Look, A. T., Douglass, E. C., and Brodeur, G. M. (1990). Oncogene amplification in pediatric brain tumors. *Cancer Res.* **50**, 2987–2990.

Waterfield, M. D., Scrace, G. T., Whittle, N., Stroobant, P., Johnsson, A., Wasteson, Å., Westermark, B., Heldin, C.-H., Huang, J. S., and Deuel, T. F. (1983). Platelet-derived growth factor is structurally related to the putative transforming protein p28[sis] of simian sarcoma virus. *Nature (London)* **304**, 35–39.

Welter, C., Henn, W., Theisinger, B., Fischer, H., Zang, K. D., and Blin, N. (1990). The cellular *myb* oncogene is amplified, rearranged and activated in human glioblastoma cell lines. *Cancer Lett.* **52**, 57–62.

Westermark, B. (1973). The deficient density-dependent growth control of human malignant glioma cells and virus-transformed glia-like cells in culture. *Int. J. Cancer* **12**, 438–451.

Westermark, B., Pontén, J., and Hugosson, R. (1973). Determinants for the establishment of permanent tissue culture lines from human gliomas. *Acta Pathol. Microbiol. Scand. A* **81**, 791–805.

Westphal, M., Hansel, M., Müller, D., Laas, R., Kunzmann, R., Rohde, E., König, A., Hölzel, F., and Herrmann, H.-D. (1988). Biological and karyotypic characterization of a new cell line derived from human gliosarcoma. *Cancer Res.* **48**, 731–740.

Westphal, M., Nausch, H., and Herrmann, H.-D. (1990). Antigenic staining patterns of human glioma cultures: primary cultures, long-term cultures and cell lines. *J. Neurocytol.* **19**, 466–477.

Wikstrand, C. J., Friedman, H. S., and Bigner, D. D. (1991). Medulloblastoma cell-substrate interaction in vitro. *Invasion Metastasis* **11**, 310–324.

Wong, A. J., Bigner, S. H., Bigner, D. D., Kinzler, K. W., Hamilton, S. R., and Vogelstein, B. (1987). Increased expression of the epidermal growth factor receptor gene in malignant gliomas is invariably associated with gene amplification. *Proc. Natl. Acad. Sci. U.S.A.* **84**, 6899–6903.

Wu, J. K., and Chikaraishi, D. M. (1990). Differential expression of *ros* oncogene in primary human astrocytomas and astrocytoma cell lines. *Cancer Res.* **50**, 3032–3035.

Yasukawa, K., Hirano, T., Watanabe, Y., Muratani, K., Matsuda, T., Nakai, S., and Kishimoto, T. (1987). Structure and expression of human B-cell stimulatory factor-2 (BSF-2/IL-6) gene. *EMBO J.* **6**, 2939–2945.

Zülch, K. J. (1979). Histological typing of tumours of the central nervous system. *In* "International Histological Classification of Tumours No. 21." World Health Organization, Geneva.

Tumor Cell Lines of the Peripheral Nervous System

3

Mark A. Israel
The Preuss Laboratory
Brain Tumor Research Center
Department of Neurological Surgery, University of California at San Francisco
San Francisco, California 94080

Carol J. Thiele
Pediatric Branch
Cell and Molecular Biology Section
National Cancer Institute
Bethesda, Maryland 20892

I. Neuroblastoma 45
   A. Introduction 45
   B. Neuroblastoma Tumor Cell Lines 47

II. Peripheral Neuroepithelioma 61

   A. Introduction 61
   B. Neuroepithelioma Tumor Cell
      Lines 64

References 70

The peripheral nervous system (PNS) arises from the embryonic neural crest and includes the cranial and spinal sensory ganglia, the autonomic ganglia, the adrenal medulla, and a variety of paraendocrine cells that are found in organs throughout the body. Tumors of the human PNS occur in the various cell types that make up these structures, including the neurons themselves and the supportive tissues, especially glial cells such as Schwann cells, which form the nerve sheath of peripheral neurons and fibroblasts.

Although cell lines derived from human schwannomas and neurofibrosarcomas have been reported occasionally, none have been characterized extensively or widely used experimentally. No human tumor cell lines have been established to date from pheochromocytomas or other neuroendocrine cell types other than carcinoid and small-cell lung cancers, which generally are not considered to be of PNS origin. Thus, the focus of this chapter is on tumors arising in neuronal and ganglionic tissues.

Tumors do not arise in fully differentiated, mature neurons. Presumably, these cells are incapable of proliferating; hence, the origins of known neuronal tumors are in nervous system precursor cells. Since nervous system maturation

Table I

Selected Neuronal Tumor Cell Lines of the Peripheral Nervous System

Neuroblastoma cell lines	Reference	Neuroepithelioma cell lines	Reference
CHP-126	Schlesinger et al. (1976)	A4573	Whang-Peng et al. (1986)
CHP-134B	Schlesinger et al. (1976)	CHP-100	Schlesinger et al. (1976)
CHP-212	Schlesinger et al. (1976)	LAP-35	Bagnara et al. (1990)
CHP-234	Schlesinger et al. (1976)	NUB-20	Yeger et al. (1990)
CHP-382	Schlesinger et al. (1976)	TC-32	Whang-Peng et al. (1986)
CHP-404	Schlesinger et al. (1976)	N1000	Whang-Peng et al. (1986)
GI-CA-N	Donti et al. (1988)	N1008	Whang-Peng et al. (1986)
GI-LI-N	Cornaglia et al. (1992)	N1016	Whang-Peng et al. (1986)
GI-ME-N	Ponzoni et al. (1988)	SK-PN-LO	Helson and Helson (1985)
GOTO	Sekiguichi et al. (1979)	SK-PN-LI	Helson and Helson (1985)
IGR-N-835	Bettan et al. (1989)	SK-PN-DW	Helson and Helson (1985)
IMR-32	Tumilowicz et al. (1970)	SK-N-MC	Biedler et al. (1983)
LA-N-1	Seeger et al. (1977)	TC-106	Whang-Peng et al. (1986)
LA-N-5	Seeger et al. (1977)	TC-268	Cavazzana et al. (1988)
MHH-NB11	Pietsch et al. (1988)		
NB-69	Gilbert et al. (1982)		
NB1-G	Carachi et al. (1987)		
NBL-W	Foley et al. (1991)		
NGP	Brodeur et al. (1977)		
NGP-2	Brodeur et al. (1977)		
NLF	Brodeur et al. (1977)		
NMB	Brodeur et al. (1977)		
RN-GA	Scarpa et al. (1989)		
SK-N-AS	Helson and Helson (1985)		
SK-N-BE(2)	Biedler and Spengler (1976)		
SK-N-DZ	Helson and Helson (1985)		
SK-N-FI	Helson and Helson (1985)		
SK-N-LE	Helson and Helson (1985)		
SK-N-SH	Biedler et al. (1983)		
SMS-KAN	Reynolds et al. (1986)		
SMS-KANR	Reynolds et al. (1986)		
SMS-KCN	Reynolds et al. (1986)		
SMS-KCNR	Reynolds et al. (1986)		
VA-N-BR	Helson and Helson (1985)		

is completed at an early age, most such tumors arise during early childhood (see subsequent discussion). The two known tumor types from which cell lines have been derived are neuroblastoma and neuroepithelioma. Such tumors grow easily as short-term cultures using standard techniques. Permanent cell lines can be established from 10–20% of such tumors. Scientists agree that cell lines can be established more frequently from bone marrow metastases of such tumors. Also, the use of irradiated feeder layers and the orthotopic transfer to immunosuppressed mice are thought to offer good opportunities for increasing the frequency with which cell lines can be established. Table I lists some widely used cell lines that have been derived from these tumors.

I. Neuroblastoma

A. Introduction

1. Clinical Features

Neuroblastoma (NB) occurs most commonly in the abdomen where tumors typically arise in the adrenal medulla and other sites of known sympathetic nervous system tissue. NB originating in the thoracic cavity, usually in close association with a dorsal root ganglion, represents about 15% of cases (Hayes and Smith, 1988). Low stage, anatomically limited disease is seen most commonly in young children. Overall, metastatic disease is present at presentation in 60–70% of patients. At presentation, children with disseminated NB are often cachexic, pale, and in substantial pain from marrow and bony metastases. This presentation contrasts sharply with the less symptomatic presentation that is observed in children with thoracic NB (McLatchie and Young, 1980). These contrasting clinical features are important because they suggest that NB of the abdomen, which presumably arises in a chromaffin precursor cell (see subsequent discussion), and thoracic NB, which presumably arises in precursors of the ganglion cells in the spinal cord, are distinctly different biological entities.

Analysis of survival in patients with NB has resulted in the identification of numerous prognostic factors (Bagnara *et al.,* 1990; Brodeur, 1990; Triche, 1990). Certain clinical characteristics such as Evans clinical stage, age, and Shimada pathological classification are thought to be independent prognostic variables. Other prognostic variables that have been identified include clinical characteristics (site at presentation, lymph node involvement), biochemical characteristics (catecholamine secretion, serum ferritin levels), and genetic characteristics of the tumor (cellular DNA content, N-*myc* amplification).

2. Pathological Features

Histological examination of NB specimens usually reveals small round or slightly oval-shaped cells with hyperchromatic nuclei and stippled chromatin.

A hallmark of the light microscopic appearance of NB is the presence of Homer–Wright rosettes characterized by tumor cells clustered around a central mesh of cell processes (the neuropile). Although sheets of monomorphic tumor cells interrupted by these abortive attempts at histological differentiation usually dominate the light microscopic appearance of this tumor, sometimes necrosis and calcification are seen. Undifferentiated NB occasionally can be a diagnostic dilemma, mimicking other small round cell tumors of childhood. Ultrastructural examination of tumor tissue by electron microscopy always demonstrates the presence of cytoplasmic neurosecretory, dense core granules and neural processes, providing important diagnostic information for tumors without other evidence of neuroblastic differentiation (Triche and Askin, 1983).

Various enzymes and proteins of neural origin are expressed in NB and aid in its diagnosis. Antibodies against neuron-specific enolase always test positive in tissue from these tumors (Triche and Askin, 1983). Neuron-specific enolase also can be found in the serum of patients with advanced disease and may be of prognostic value (Viallard et al., 1988). Neuropeptide Y also has been detected in NB cells (Cohen et al., 1990) and has been suggested as a marker that may be of clinical value (Kogner et al., 1990). NB cells synthesize sympathomimetic catecholamines which, in addition to their precursors and metabolites, can be measured in blood or urine. These measurements frequently corroborate the histopathological diagnosis and may be useful adjuncts with which to follow tumor response as well (Hayes and Smith, 1988).

3. Genetic Features

Using modern banding techniques, nearly 80% of NB studied to date have shown some chromosomal abnormality (Brodeur et al., 1977,1980). The most common cytogenetic abnormality recognized in NB tumor specimens is a deletion or rearrangement of the short arm of chromosome 1. However, the chromosomal rearrangements in NB that are of greatest importance to date have been homogeneously staining regions (HSRs) and double minute chromosomes (DMs) (Biedler and Spengler, 1976). These cytogenetic structures are now widely recognized to contain regions of gene amplification. The N-myc gene, a gene with considerable homology to the cellular proto-oncogene c-myc, has been shown to be amplified in HSRs and DMs found in NB-derived cell lines and tumor specimens (Montgomery et al., 1983; Schwab et al., 1983; Kohl et al., 1984). N-myc amplification now is believed to be associated with a shorter time to relapse and can be detected in the tumors of approximately one-third of patients with advanced stage disease (Brodeur et al., 1984). Tumor cell N-myc amplification, when it occurs, seems to be an independent prognostic indicator (Seeger and Brodeur, 1985) and can be detected in cell lines derived from NB (Table I).

4. Treatment

Treatment for NB primarily reflects the clinical stage of the tumor at presentation. Lower stage disease, Stage I and Stage II, is limited in its localization, whereas Stages III and IV represent more extensive local and metastatic disease, respectively. The mainstay of therapy for infants without bone metastases and Stage I and II tumors is surgical excision. With this treatment alone, survival is excellent (Matthay *et al.,* 1989). Chemotherapy is the mainstay of therapy for Stage III and IV disease and is effective in inducing complete remissions in 30–40% of patients (Finkelstein *et al.,* 1979; Shafford *et al.,* 1984; Sawaguchi *et al.,* 1990). Currently, the efficacy of very intensive therapy requiring bone marrow transplantation and other extensive supportive approaches is being examined.

B. Neuroblastoma Tumor Cell Lines

As indicated in Table I, a large number of NB tumor cell lines has been reported rather extensively in the scientific literature (Tumilowicz *et al.,* 1970; Schlesinger *et al.,* 1976; Brodeur *et al.,* 1977; Seeger *et al.,* 1977; Sekiguichi *et al.,* 1979; Gilbert *et al.,* 1982; Helson and Helson, 1985, 1992; Reynolds *et al.,* 1986; Carachi *et al.,* 1987; Cavazzana *et al.,* 1988; Donti *et al.,* 1988; Pietsch *et al.,* 1988; Ponzoni *et al.,* 1988; Bettan *et al.,* 1989; Scarpa *et al.,* 1989; Cornaglia *et al.,* 1990; Foley *et al.,* 1991; Gazitt *et al.,* 1992a,b). Cell lines have been isolated, both from primary tumor sites and from metastatic sites such as the bone marrow. Interestingly, systematically identifying differences between cell lines with such divergent origins has not been possible, although some differences have been cited (Sugimoto *et al.,* 1986). As for other tumor types, the cloning efficiency of primary tumor tissue is low; that any particular cell line is representative of the tumor in general is unlikely. Isolating multiple tumor cell lines from patients at different times during the course of their illness has been possible (Reynolds *et al.,* 1986; Kees *et al.,* 1992). In some cases, documenting differences in the cell lines has been possible (Rosen *et al.,* 1986), although whether these differences reflect authentic differences *in vivo* at the time the tumor cell line was isolated is not known. The safest assumption is that these differences reflect tumor heterogeneity and occasionally represent changes associated with the pathological progression of the tumor. Of course, the implication of this interpretation is that any given cell line may be representative of a particular cell type present in a tumor at any given time, and that any given tumor cell line is unlikely to represent the heterogeneity that characterizes most individual tumors.

1. Morphology

The morphological characteristics of NB cell lines vary somewhat, but are distinct from those of most other tumor cell types and are characterized by evidence of neuronal rather than obvious chromaffin cell characteristics. These characteristics include short, although rarely well developed, neuritic outgrowths that possess varicosities visible even by light microscopy (Fig. 1). Immunofluorescent studies indicate the presence of catecholamines within these varicosities, reminiscent of similar structures that occur in neurons of the PNS. NB cells typically have one outgrowth or only a few per cell; these lack extensive arborization or terminal sprouting unless induced to differentiate (see subsequent discussion).

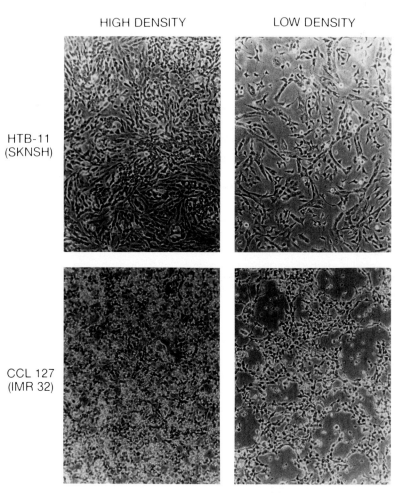

Fig. 1. Neuroblastoma cell lines. Light microscopic analysis (58X) of the human neuroblastoma cell lines ATCC HTB-11 (SK-N-SH) and ATCC CCL 127 (IMR-32) at high and low density. Cells are loosely substrate adherent and, at low density, short neuritic processes can be seen. Note the heterogeneous morphology of cells in the HTB-11 cell line.

The ultrastructural features of NB cell lines also have been examined (Barnes *et al.*, 1981; Kees *et al.*, 1992). These cells generally do not show features typical of mature neuronal differentiation, such as neurites with microtubules and neurofilaments. Instead, these cells contain abundant free ribosomes, Golgi complexes, and scant granular endoplasmic reticulum, characteristic of early embryonic autonomic neurons. Also, identifying dense core granules in these cell lines is routinely possible, although these structures are typically present in rather small numbers per cell. As noted in the literature, however, the appearance of differentiated features of the NB cells does not follow the pattern observed for normal neurons, suggesting that the expression of some neuronal morphological properties is not regulated correctly in these cells (Barnes *et al.*, 1981).

Cultures of most NB tumor cell lines have a rather homogeneous cell morphology, similar to that just described. Some NB cell lines possess a variety of morphological variants; this observation has been studied extensively (Barnes *et al.*, 1981; Ross *et al.*, 1983; Sadee *et al.*, 1987). These morphological variants, which arise at a low rate in the cell line, can be enriched for greatly by differential cloning and cautious expansion (Fig. 2). Such cells can revert to other morphologies at a very low rate (Ross *et al.*, 1983). These variants were described first by Biedler and colleagues. The morphological variants derived from the SK-N-SH cell line remain the best characterized and most widely studied. The presence of such morphological variants has been recognized to occur in many different NB cell lines, and clonal morphological variants of the LAN-1 and SMS-KCN cell lines have been described (Ciccarone *et al.*, 1989).

These cellular subclones appear neuroblastic, epithelial-like, or of intermediate morphological appearance. Their common origin could be documented by the presence of shared marker chromosomes (Ross *et al.*, 1983). Evidence suggests a coordinated expression of morphological and biochemical phenotypic characteristics in these distinct cell subclones. Only neuroblastic subclones express tyrosine hydroxylase and dopamine-beta-hydroxylase, enzymes unique to catecholaminergic neurons. Epithelial-like cells lack these enzymatic activities. The epithelial morphology of these clonal derivatives is similar to that observed following the treatment of some NB cell lines with retinoic acid (RA) or dibutyryl cyclic AMP (Tsokos *et al.*, 1987). In this case, the changes include the induction of a basal lamina and the enzyme cyclic nucleotidyl phosphohydrolase. Among neural crest-derived lineages, these findings are most typical of Schwann cells, further increasing the variety of lineages mimicked by NB tumor cell lines (Table II).

2. Biochemical and Biological Characteristics

Several different characteristics further suggest the origin of NB tumor cell lines in neuronal or neuroendocrine precursors of the PNS (Tables II and III). These features include their biochemical characteristics and the expression of a large number of different nervous system markers. Biochemically, these cell

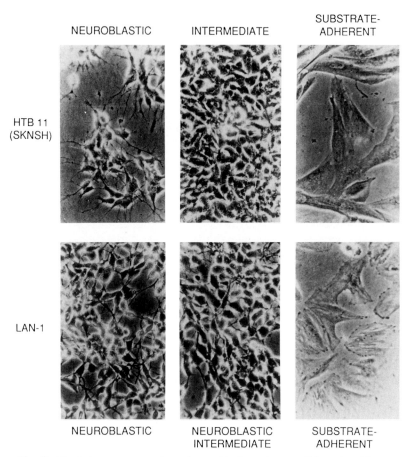

Fig. 2. The heterogeneous nature of neuroblastoma tumor cell lines is evident, since clonal populations of each morphological type can be isolated. The neuroblastoma cell line HTB-11 (SK-N-SH) has a single copy of N-*myc* and subclones have a neuroblastic (N), intermediate (I), or substrate-adherent or epithelial (S) morphology (*top*). The neuroblastoma cell line LAN-1 has multiple copies of N-*myc;* clonal sublines having N, I, and S morphology types have been isolated.

lines typically express enzymes and proteins known to be important for the biosynthesis and storage of catecholamines. Catecholamines are the major neurotransmitter substances of the sympathetic ganglion cells in the sympathetic nervous system and of the chromaffin cells of the adrenal medulla and other abdominal ganglia, the major sites of origin for NB tumors. These biosynthetic enzymes include tyrosine hydroxylase and dopamine beta-hydroxylase, enzymes unique to catecholamine neurons (Ross *et al.,* 1983). These characteristics are consistent with the observation that patients with NB

Table II

Characteristics of Cell Lines from Peripheral Nervous System Tumors[a]

	Neuroblastoma		
	N-*myc* copy (1 copy)	N-*myc* copy (>1 copy)	Neuroepithelioma
Histopathology			
neural granules	+	+	$-(\pm)^b$
neuropil	+	+	$-(\pm)^b$
rosettes	+	+	$-(\pm)^b$
Immunocytochemistry			
Class I MHC	±	−	+
HSAN 1.2	+	+	−
NCAM	+	±	+
NSE	+	+	−
LEU 7	+	+	$-(\pm)^b$
NF	+	+	$-(\pm)^b$
CGA	+	+	±
β-2μ	±	−	+
Genetic alterations			
N-*myc* amplifications	−	+	−
t(11;22)	−	−	+
LOH 11q	−	−	NA[c]
LOH 14	−	−	NA
1p deletion	+	+	−
Neurotransmitter biosynthetic enzymes			
CAT	−	−	+
DBH	+	+	−
TH	+	+	−
Oncogene expression			
N-*myc*	−	+	−
c-*myc*	+	−	+
c-*myb*	+	+	−
ets-1	+	+	−
dbl	−	−	+
rdc-1	−	−	+
Growth factors			
IGF I	−	−	+
IGF II	+	+	−
Differentiation			
RA	−	+	−
cAMP	−	+	−
ARA-C (cell death)	+	+	+
NGF	−	−	−
Combination (RA, cAMP, no serum)	NT	NT	+

(*continues*)

Table II

Continued

| | Neuroblastoma | | |
	N-*myc* copy (1 copy)	N-*myc* copy (>1 copy)	Neuroepithelioma
Clinical parameters			
Patient <5 years of age	+	+	−
Patient >5 years (adolescence)	−	−	+
Race			
Caucasian	+	+	+
Black	+	+	−
Asian	+	+	−

[a] Abbreviations: NE, neuroepithelioma; ES, Ewing's sarcoma; MHC, major histocompatibility antigen; NCAM, neural cell adhesion molecule; NSE, neuron-specific enolase; NF, neurofilament protein; CGA, chromogranin A; β-2μ, β_2-microglobulin; CAT, choline acetyltransferase; DBH, dopamine-beta-hydroxylase; TH, tyrosine hydroxylase; IGF, insulin-like growth factor; RA, retinoic acid; ARA-C, cytosine arabinoside; NGF, nerve growth factor.
[b] Seen in NE and occasionally in ES.

Table III

Proto-oncogene Expression in Cell Lines of Peripheral Nervous System Tumors

| | Neuroblastoma | | Neuroepithelioma t(11;22)[c] |
	NB[+a]	NB[−b]	
Proto-oncogene expression			
N-*myc*	+[d]	±[e]	±
myc	−[f]	+	+
myb	+	+	+
mil/raf	+	+	+
ets-1	+	±	±
sis	−	−	−
src	+	+	+
rdc-1	−	−	+
Differentiation gene expression			
cholecystokinin	−	−	+
Na$^+$/K$^+$ ATPase	+	+	±

[a] NB$^+$, neuroblastoma containing amplified N-*myc* gene.
[b] NB$^-$, neuroblastoma containing a single copy of N-*myc* gene.
[c] t(11;22), tumor containing reciprocal translocation involving chromosomes 11 and 22.
[d] +, expressed.
[e] ±, variably expressed.
[f] −, not expressed.

tend to have high circulating levels of catecholamines and excrete the breakdown products of these substances and the products of their intermediary metabolism in their urine. Enzymes of these biosynthetic pathways are not detectable in all NB cell lines. In some cell lines, tyrosinase has been detected (Ross and Biedler, 1985); this enzyme is important in the synthesis of melanin, consistent with the neural crest origin of melanocytes. Other evidence of melanocytic and schwannian differentiation in NB cell lines has been presented (Ross and Biedler, 1985; Tsokos et al., 1987; Tsunamoto et al., 1988). Melanocytic NB and melanocytic schwannoma are tumors that are widely recognized, although rare.

NB cell lines have been examined extensively for the expression of various cellular and cell surface markers that are typical of neuronal (Cooper et al., 1990b) and neuroendocrine or chromaffin (Perez-Polo et al., 1979; Cooper et al., 1990b) cell types. Although overlap among these markers occurs, they can be distinct; clearly some cell lines derived from NB express none of the markers commonly associated with chromaffin tissue. Typical of chromaffin cells are very high levels of expression of catecholaminergic biosynthetic enzymes and chromogranin (Fig. 3). Neuronal cells of the sympathetic nervous system also can express these markers, but invariably at much lower levels.

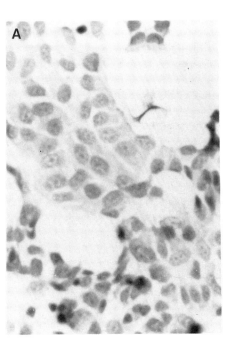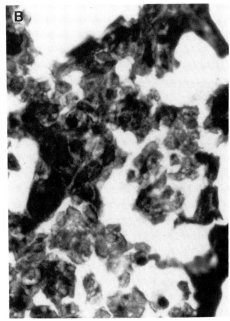

Fig. 3. Immunoperoxidase staining for chromogranin A in a neuroblastoma cell line. The chromogranin A gene product can be detected in (A) normal and (B) malignant chromaffin tissues by standard avidin–biotin peroxidase-linked immunohistochemical techniques.

Synaptophysin (Perez-Polo *et al.*, 1979), neuron-specific enolase (Kimhi *et al.*, 1976), and neuropeptide Y (Cohen *et al.*, 1990; Kogner *et al.*, 1990) are neuronal markers found in NB tumor cell lines. Although these proteins also are expressed in chromaffin tissue, they generally are found at levels much lower than those observed in neuronal tissues.

Studies of marker gene expression in NB tumor cell lines suggest that this tumor arises in close association with a disturbance of the differentiation of primitive neural crest cells (Ross *et al.*, 1983). The neural crest gives rise to at least three distinct lineages that are present in the PNS: chromaffin, ganglionic, and sustentacular or glial-like cells (Le-Douarin and Smith, 1988; Cooper *et al.*, 1990a). These lineages can be marked immunocytochemically, and have been studied systematically in NB tumor specimens only recently. As alluded to earlier, NB tumors can exhibit varying degrees of morphological differentiation along a ganglionic pathway, although most NB tumor cell lines that have been examined have evidence of chromaffin differentiation (Cooper *et al.*, 1990a,b).

Interestingly, the SK-N-SH cell line, which is widely used for the study of NB, is the only NB cell line known to have been isolated from a tumor arising in the thorax. This cell line expresses ganglionic markers (Ross *et al.*, 1983; Sadee *et al.*, 1987) and is biologically quite distinct from most other NB cell lines (Ross *et al.*, 1983). These differences seem likely to reflect a distinct cell of origin for thoracic tumors, perhaps a thoracic dorsal root ganglion cell, a finding compatible with their distinctive biology and clinical behavior.

NB cell lines typically do not express or only weakly express the Class I major histocompatibility (MHC) antigens and may not express beta 2-microglobulin on their membrane (Donner *et al.*, 1985), although expression of beta 2-microglobulin mRNA is readily detectable in many NB cell lines and is developmentally regulated in the human adrenal medulla, in which it is expressed at high levels in adult tissues (Cooper *et al.*, 1990b). Once again, heterogeneity of expression for Class I MHC antigens is reflected in the variability of its expression in NB tumor cell lines (Fig. 4).

Many other markers that may be developmentally regulated in the PNS have been examined also in NB tumor cell lines, including the extracellular matrix proteins laminin, fibronectin, and Type IV collagen (Tsokos *et al.*, 1987), which are expressed, and the S100 protein, which may not be (Pietsch *et al.*, 1988). Others genes, the expression of which might not be anticipated in NB cell lines, include alpha-smooth muscle actin and desmin (Sugimoto *et al.*, 1991). Markers such as MC25 (Rettig *et al.*, 1985) and HSAN 1.2 (Smith and Reynolds, 1987) have been selected specifically for their frequent and apparently quite specific reactivity with NB cell lines.

Treatment of NB cells *in vitro* with RA and a variety of other nonpolar planar molecules can mimic differentiation and has been an important model system for studies of the role of disturbed maturation in the pathogenesis of this tumor (Sidell, 1982; Thiele *et al.*, 1985). Changes in cell morphology, biochemical activities, and gene expression that are thought to be indicative of differentia-

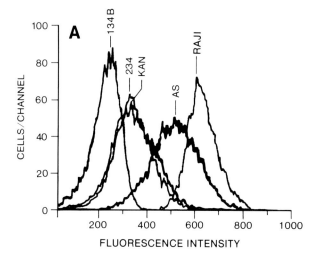

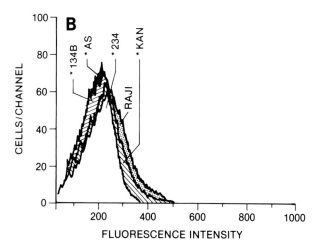

Fig. 4. Class I major histocompatibility complex (MHC) antigen expression by human neuroblastoma cell lines. Flow cytometric analysis of Class I MHC antigen expression (A) compared with isotype controls (B) in different neuroblastoma cells, compared with a lymphoid cell line (Raji). The SK-N-AS (AS) is the most positive cell line, whereas CHP 134B (134B) is essentially negative. The other lines examined in this experiment are CHP 234 (234) and SMS-KAN (KAN). Reprinted with permission of editors J. Immunol. (Feltner et al., 1989).

tion of neural crest-derived cells have been documented in these systems. The variety of agents that can contribute to the morphological and biochemical maturation of NB cell lines is summarized in Table III. These findings have led to the use of NB cell lines in many different studies examining the mechanisms of tissue-specific gene regulation and nervous system maturation.

Morphological changes in the appearance of NB cell lines after treatment with RA, cyclic AMP, or nerve growth factor (NGF) have been studied most extensively. In cell lines that are responsive to these agents, the most remarkable changes are those that mimic neuronal differentiation, including the sprouting of numerous thin neuritic processes, extensive terminal arborization, apparent synapse formation with distant cells, and readily observable axonal swellings in which neurotransmitters are thought to be stored (Prasad, 1975; Sidell, 1982). Biochemical evidence of differentiation along a neuronal lineage also has been documented extensively, including changes in the expression of genes encoding neuron specific proteins (Thiele *et al.,* 1988) and changes in neurofilaments (Chen *et al.,* 1990; Ponzoni *et al.,* 1992) and neurotransmitter biosynthetic enzyme activity (Sidell, 1982). Depending on the cell line examined and the specific inducing agent used, researchers have been able to demonstrate evidence for differentiation along schwannian (Tsokos *et al.,* 1987; Tsunamoto *et al.,* 1988), melanocytic (Ross and Biedler, 1985; Tsokos *et al.,* 1987), and chromaffin (Cooper *et al.,* 1988) cellular pathways (Table III). In some cases, combinations of differentiating agents have been found to have an apparently synergistic effect (Ponzoni *et al.,* 1992), although such studies often are complicated by the absence of reproducible measures of differentiation that can be quantified accurately.

NGF is a pleiotropic mediator of many different biological activities in cells of the developing PNS. Therefore, the fact that most NB cell lines do not respond more vigorously to treatment with NGF is enigmatic (Reynolds and Perez, 1989; Azar *et al.,* 1990). Although some cell lines respond with evidence of morphological differentiation or with altered growth, surprisingly SK-N-SH does not extend neurites in response to NGF; however, this cell line exhibits extensive morphological changes in response to other differentiating agents including RA. In contrast, a clone of this line, SY-5Y, extends neurites in response to NGF (Reynolds and Perez, 1989) dibutyryl cyclic AMP, RA, and phorbol esters and its proliferation is inhibited (Fig. 5). These findings raise the possibility that defects in the response of neural crest precursors to neurotrophins contribute to the pathogenesis of this tumor although, to date, documenting specific alterations in pathways known to be of physiological importance has not been possible.

Cellular differentiation and proliferation are linked closely since fully differentiated cells invariably have a diminished growth potential. Tumor cells typically appear undifferentiated; a central feature of their malignant potential is their unregulated growth. The growth potential of NB cells, both in culture and *in vivo* when examined as xenografts, generally is diminished after treatment with agents that cause morphological differentiation (Tsokos *et al.,* 1987; Reynolds and Perez, 1989; Cornaglia *et al.,* 1992; Ponzoni *et al.,* 1992). This effect has been studied most extensively in RA-treated NB cells. RA treatment of some NB cell lines causes an accumulation of cells in the G_1 portion of the cell cycle within 24 hr of treatment. Down-regulation of N-*myc* gene expression

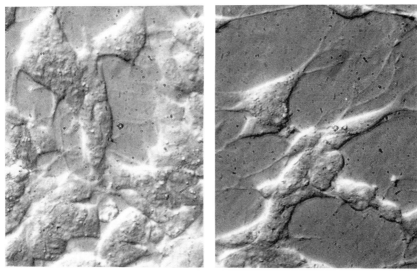

| 6 days control | 6 days 1µM retinoic acid |

6 days 16µM PMA

Fig. 5. Morphological differentiation in the SY5Y neuroblastoma cell line. The neuroblastic subclone of HTB-11 (SK-N-SH) was treated for 6 days with 1 μM retinoic acid or 16 μM phorbol 12-myristate 13-acetate (PMA). Growth inhibition occurs with both agents and extension of neuritic processes can be seen in the retinoic acid- and PMA-treated cells compared with control cultures.

occurs within hours of RA treatment and may be necessary for the implementation of a differentiation program (Thiele *et al.,* 1985). After 4 days of treatment, 70% of cells have begun to extend neuritic processes; by 8 days, a dramatic network of neurites exists that biochemically, electrophysiologically, and ultrastructurally is similar to normal neural processes (Fig. 6). Agents that raise intracellular cAMP levels have been used to induce differentiation of NB cells.

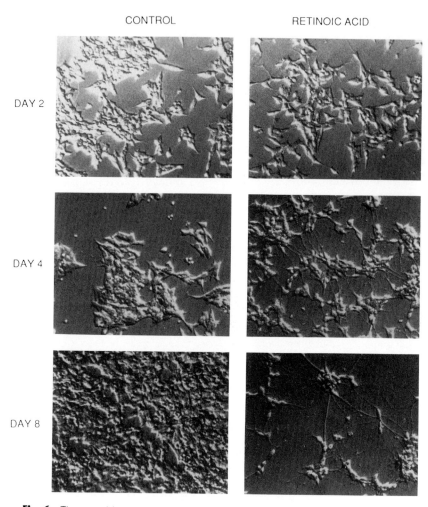

Fig. 6. The neuroblastoma cell line SMS-KCNR was treated for the indicated periods of time with 5 μM retinoic acid (RA) or solvent control. At 2 days of treatment, RA-treated cells are more substrate adherent than cells in control cultures, although little evidence exists of neuronal differentiation. By 4 days of culture in RA, over 75% of the cells are extending short neuritic processes. By 8 days, a dramatic decrease in cell number occurs in the RA-treated cells, and an extensive network of neuritic processes is seen. (Reproduced from Thiele *et al.,* 1988, with permission.)

Although cell growth is decreased by these agents, cell cycle analyses reveal no significant change in the distribution of cells in the cell cycle (Gaetano et al., 1992). The growth regulatory mechanisms by which NB tumor cell lines proliferate have been studied widely. Insulin-like growth factor II (IGF-II) is now known to be a particularly important stimulatory ligand for the in vivo and in vitro growth of NB tumor cells (El-Badry et al., 1989,1991). This mitogen may mediate the autocrine stimulation of NB tumors in some cases, but seems to function more commonly as a paracrine mediator of cell growth. In NB tissue specimens, IGF-II expression can be detected in infiltrating tumor and adjacent normal tissues, including normal adrenal cortical tissue. This finding is especially interesting because IGF-II expression occurs in the adrenal cortex during embryonic life but is not detectable after birth, raising the possibility that tumor tissue induces the production of this growth factor. IGF-I may promote neuronal differentiation of these cells (Pahlman et al., 1991).

3. Genetic Characteristics

The cytogenetic characteristics of NB tumors have been studied extensively (Brodeur et al., 1977; Biedler et al., 1980; Franke et al., 1986; Reynolds et al., 1986; Ritke et al., 1989; Scarpa et al., 1989; Suzuki et al., 1989; Brodeur, 1990; Cornaglia et al., 1990; Donner, 1991; Nojima et al., 1991; Petkovic and Cepulic, 1991; Rudolph et al., 1991; Srivatsan et al., 1991; Karnes et al., 1992). Many of the observed changes have been documented in cell lines as well (Potluri et al., 1987; Nojima et al., 1991; Rudolph et al., 1991). Using modern banding techniques, nearly 90% of NB studied to date have shown some chromosomal abnormality (Brodeur et al., 1977,1980). The most common cytogenetic abnormality recognized in NB tumor specimens and tumor cell lines is a deletion or rearrangement of the short arm of chromosome 1 (Biedler and Spengler, 1976; Brodeur et al., 1981; Gilbert et al., 1982). Other chromosomal alterations found in tumor tissue, including chromosomal loss, gain, and rearrangements involving chromosomes 10, 14, 17, and 19, have not been observed in cell lines regularly to date.

The chromosomal rearrangements in NB that are of greatest importance have been HSRs and DMs, which were detected first in human NB tumor cell lines (Biedler and Spengler, 1976; Biedler et al., 1980) and contain regions of gene amplification. The N-myc gene, a gene with considerable homology to the cellular proto-oncogene c-myc, has been shown to be amplified in HSRs and DMs found in NB-derived cell lines and tumor specimens (Montgomery et al., 1983; Schwab et al., 1983; Kohl et al., 1984). Retrospective analysis of hundreds of NB tumor tissues and patient outcome indicates that N-myc amplification, as well as chromosome Ip deletions (Brodeur, 1990), are associated with a shorter time to relapse (Seeger and Brodeur, 1985). These genetic changes can be detected in the tumors of approximately one-third of patients with advanced stage disease (Brodeur et al., 1984; Seeger and Brodeur, 1985).

N-*myc* amplification, when it occurs, seems to be an independent prognostic indicator (Seeger and Brodeur, 1985). DMs and HSRs (Biedler and Spengler, 1976; Rudolph *et al.*, 1991) have been identified in NB tumor cell lines, although they are not present in all NB tumors (Bettan *et al.*, 1989). Several NB cell lines without evidence for the amplification of the N-*myc* gene have been described as well (Cohn *et al.*, 1990).

The molecular biology of NB also has been studied widely, although a comprehensive picture of the pathogenesis of this disease has not yet emerged. Several different proto-oncogenes including N-*myc*, c-*myc*, c-*myb*, *ets*-1, *dbl*, *src*, and *rdc*-1 have been recognized to be expressed highly in NB cell lines (Bolen *et al.*, 1985; Thiele *et al.*, 1987; McKeon *et al.*, 1988; Collum *et al.*, 1989) (Table III). Approximately 50% of NB tumors have amplified N-*myc* and most of these are higher clinical stage tumors (Brodeur *et al.*, 1984). N-*myc* is likely to be highly expressed in some cell lines in which it is not amplified, although this concept has not been examined systematically. An important activity of N-*myc* is its ability to block RA-induced NB differentiation (see subsequent discussion) (Thiele *et al.*, 1985; Thiele and Israel, 1988).

The use of NB cell lines to study the differentiation of ganglionic precursors has revealed numerous changes that occur at the level of gene expression. These studies have provided some insights into the cellular mechanisms that contribute to the regulation of cellular maturation in this lineage (Rupniak *et al.*, 1984; Thiele *et al.*, 1985; Tsokos *et al.*, 1987; Bettan *et al.*, 1989; Collum *et al.*, 1989; Feltner *et al.*, 1989; Parodi *et al.*, 1989; Chen *et al.*, 1990; Weiler *et al.*, 1990; Busse *et al.*, 1991b; Reed *et al.*, 1991; Table IV). When NB tumor cell lines are induced to differentiate, the expression of several different proto-oncogenes has been documented to change (Thiele *et al.*, 1988; Feltner *et al.*, 1989). N-*myc* seems to play an important role in the regulation of NB tumor cell maturation (Thiele *et al.*, 1985; Thiele and Israel, 1988). Within 20 min of the treatment of the NB tumor cell line SMS-KCNR with RA a sharp decline occurs in the steady-state levels of N-*myc* mRNA, although the morphological changes associated with differentiation become evident only much later (see Fig. 6; Thiele and Israel, 1988). Transfection of this cell line with an N-*myc* construct that constitutively expresses N-*myc* from a promoter not regulated by RA blocks the ability of RA to induce differentiation (Thiele *et al.*, 1985).

4. Models for the Development of New Therapies

The growth characteristics of NB cell lines, both *in vivo* and *in vitro*, vary considerably. Only a few cell lines have been grown widely as xenografts (Bernal *et al.*, 1983; Danon and Kaminsky, 1985; Pietsch *et al.*, 1988; Cornaglia *et al.*, 1992), although the examination of *in vivo* growth potential has emerged as an important characteristic of newly isolated cell lines (Bettan *et al.*, 1989; Cornaglia *et al.*, 1990). The capacity of some NB cell lines to grow as xenografts in immunosuppressed animals has led to a number of studies in which animals bearing cell line-induced tumors have contributed to the preclinical evaluation of novel diagnostic as well as therapeutic strategies. Diagnosti-

Table IV
Biological Response of Neuroblastoma Cell Lines

Biological response modifier	Cell phenotype[a]	Reference
Ara-C	N	Ponzoni *et al.* (1988)
cAMP	N	Prasad (1975)
DMSO[b]	N	Kimhi *et al.* (1976)
Interferon γ	N	Cornaglia *et al.* (1992)
Herbamycin-A	N	Preis *et al.* (1988)
Mitomycin C	N	Goldstein and Plurad (1980)
Nerve growth factor	N	Perez-Polo *et al.* (1979)
Retinoic acid	N	Sidell (1982)
	N, S	Reynolds and Maples (1985)
	N, S, M	Tsokos *et al.* (1987)
Tumor necrosis factor	N	Ponzoni *et al.* (1988)
TPA[c]	N	Weiler *et al.* (1990)

[a] N, Neuronal phenotype; S, schwannian phenotype; M, melanocytic phenotype.
[b] DMSO, Dimethylsulfoxide.
[c] TPA, 12-O-Tetradecanoyl phorbol-13-acetate.

cally, the use of antiganglioside (Cheung *et al.*, 1987) as well as other monoclonal antibodies has been evaluated in the pursuit of approving imaging technologies (Drebin *et al.*, 1986; Etoh *et al.*, 1988). Attempts to characterize the *in vivo* expression of the multidrug resistance transport protein, MDR, also have been pursued using xenograft models of NB (Mehta *et al.*, 1992).

Therapeutically, NB xenograft models have been used to examine the effects of chemotherapeutic regimens on tumor growth (Busse *et al.*, 1991a; Mehta *et al.*, 1992). More extensive studies have pursued the use of differentiating agents and biological response modifiers in these models (Busse *et al.*, 1991b). Of the agents that have been examined to date (Danon and Kaminsky, 1985; Busse *et al.*, 1990), RA has attracted sufficient attention to be considered for clinical trials.

II. Peripheral Neuroepithelioma

A. Introduction

1. Clinical Features

Peripheral neuroepithelioma (PN) first was recognized as a distinct pathological entity in 1918 by Stout, who characterized a tumor with neuronal characteristics arising in a peripheral nerve (Stout, 1918). This tumor probably occurs more commonly than generally is appreciated, and has been described variously as adult NB, peripheral NB, PNET of the chest wall (Askin's Tumor), and other classifications (Askin *et al.*, 1979; Llombart-Bosch *et al.*, 1988b; Williams *et al.*, 1988). Several studies have determined, however, that tumors such as these carry an easily recognized and apparently invariant chromosomal translocation, rcp(11;22)(q24;q12) (Whang-Peng *et al.*, 1984,1986; Israel *et al.*, 1985). This finding, in addition to a variety of recently recognized biological differences (Donner *et al.,* 1985; Thiele *et al.*, 1987), strongly suggests that this tumor originates in a different cell of the PNS than that in which the histologically indistinguishable tumor, undifferentiated NB, arises (Thiele *et al.*, 1987). Ewing's sarcoma, a round cell tumor of unknown histogenesis, also is characterized by a rcp(11;22)(q24;q12) translocation (Turc-Carel *et al.*, 1983); several other characteristics suggest it is closely related or identical to PN (Williams *et al.*, 1988; Zucman *et al.*, 1993).

PN occurs in patients ranging in age from infant to adult. This tumor most commonly presents in the second decade of life (Miser *et al.*, 1985; Marina *et al.*, 1989), albeit with an apparent slight female predominance. Patients with PN of the thorax present with a variety of clinical syndromes, including pleural-based masses with or without pleural effusion and rib erosion. Occasionally, the first evaluation reveals a paraspinal mass with or without signs of spinal cord compression. Urine catecholamine excretion is absent in patients with PN, further distinguishing this disease from NB.

Current data suggest that PN tends to recur locally, although the disease can be metastatic at presentation or recurrence. When metastases occur, they are found most commonly in the lung and bones (Miser *et al.*, 1985). Bone marrow involvement, at diagnosis and in relapse, appears to be more frequent than previously recognized.

2. Pathological Features

Histopathologically, PN is indistinguishable from primitive NB and frequently has been referred to as "peripheral NB" or "metastatic NB with a regressed primary tumor" because of its occurrence outside known sites of sympathetic nervous system tissue (Seemayer *et al.*, 1975; Biedler *et al.*, 1978; Askin *et al.*, 1979; Triche and Askin, 1983; Gonzalez-Crussi *et al.*, 1984; Triche *et al.*, 1989). Histologically, this tumor appears as an undifferentiated, small, blue, round cell tumor with varying degrees of neuronal differentiation. Ewing's sarcoma, another undifferentiated tumor that typically occurs in the first decades of life, also is characterized by a t(11;22) translocation and is probably identical to or very closely related to PN (Zucman *et al.*, 1993), although this possibility has been questioned by some investigators (Rettig *et al.*, 1992). PN can be periodic acid–Schiff (PAS) positive, another common feature of Ewing' sarcoma. Although not always present, an important feature of the light microscopic appearance of this tumor is the presence of rare rosettes. Ultrastructural analysis invariably reveals evidence of neuronal differentiation, including neurites, neurofilaments, neurotubules, and neurosecretory granules (Ross *et al.*, 1983; Triche and Askin, 1983; Linnoila *et al.*, 1986).

PN expresses neuron-specific enolase and is usually negative for S100 protein. Biochemical evaluation of PN cells has demonstrated high levels of choline acetyltransferase, an enzyme important for the synthesis of acetylcholine (McKeon et al., 1988). Acetylcholine largely is confined to postganglionic parasympathetic neurons outside the central nervous system (CNS), suggesting that these cells are the origins of PN. This finding contrasts to the finding that NB cells have high levels of catecholamine biosynthetic enzymes, but cholinergic biosynthetic enzymes such as choline acetyltransferase are not elevated (Thiele et al., 1987). Cholecystokinin also has been identified as a marker capable of distinguishing NB from Ewing's sarcoma or PN (Friedman et al., 1992). The absence of morphological evidence of ganglionic differentiation and the lack of detectable neuronal markers in Ewing's sarcoma tumors are important characteristics that may distinguish this tumor from PN. Although the immunocytochemical profile of Ewing's sarcoma may be different from that characterizing PN (Donner et al., 1985), ongoing work invariably reveals evidence of commitment to a neuronal differentiation pathway in Ewing's sarcoma tissues as well.

3. Genetic Features

The identification of a t(11;22)(q24;q12) chromosomal rearrangement in PN was an important observation in clarifying the relationship of this tumor to other neuroblastic malignancies, especially childhood NB (Whang-Peng et al., 1984,1986; Fig. 7). This translocation also was the first evidence of a close relationship between this tumor and Ewing's sarcoma. t(11;22)(q24;q12) now has been reported in several other malignancies with evidence of neuronal differentiation as well (Griffen, et al., 1986). These tumors ultimately also may be peripheral neuroepitheliomas, arising in previously undescribed sites.

Interestingly, many peripheral neuroepitheliomas that have been examined for cytogenetic rearrangements also contain more than the expected two copies of chromosome 8. The significance of this event is unknown, although the tumors do express high steady-state levels of mRNA encoded by the c-*myc* gene that is located on chromosome 8 (Thiele et al., 1987).

4. Treatment

Because this rare tumor has been identified as an entity separate from "adult NB" only in the last decade, limited information is available about its optimal treatment. The average survival of patients presenting with metastatic disease is approximately 8 months, even with a multimodal approach to therapy consisting of surgery, radiation, and chemotherapy (Miser et al., 1985). Most often a combination of vincristine, actinomycin D, doxorubicin, and cyclophosphamide has been used (Miser et al., 1985; Marina et al., 1989). Other agents known to be active in this disease are VM-26, cisplatin, and VP-16. Local irradiation at doses of 45 to 60 Gy generally are given to treat primary unresectable tumors. Because of the poor prognosis of this tumor and the presence of a genetic alteration indistinguishable from that seen in Ewing's sarcoma, treat-

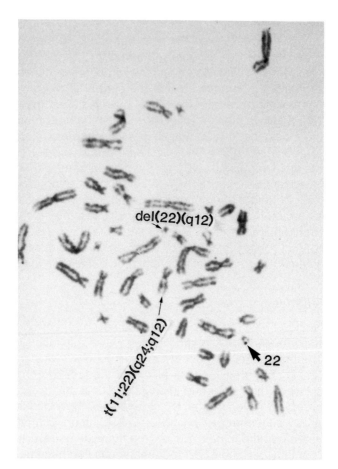

Fig. 7. *In situ* hybridization of ^{3}H-labeled c-*sis* probe to Wright's stained metaphase spreads from the neuroepithelioma cell line TC-32. The large arrow indicates a silver grain at the 22q13 region on the normal chromosome 22. The two small arrows indicate the reciprocally translocated chromosome 11 and 22, der(22).

ment of these patients in a manner reflective of the experience gathered in the management of Ewing's sarcoma patients holds great hope for the identification of more effective treatments.

B. Neuroepithelioma Tumor Cell Lines

1. Morphology

As indicated in Table I, a rather large number of tumor cell lines has been isolated from PN tumors (Schlesinger *et al.*, 1976; Biedler *et al.*, 1983; Helson and Helson, 1985; Whang-Peng *et al.*, 1986; Potluri *et al.*, 1987; Bagnara *et al.*,

1990; Llombart-Bosch *et al.*, 1990; Yeger *et al.*, 1990). Cell lines from tumors pathologically classified as Ewing's sarcoma also are listed in Table I because the differences between these tumors, if any, have not been defined precisely (see subsequent discussion). Ewing's sarcoma and PN are unlikely to be distinguished easily from institution to institution, and have not yet been differentiated by cytogenetic or molecular criteria. To date, no differences between tumor cell lines isolated from these two different tumor types have been observed routinely. Some cell lines originally designated as derived from PN may present more obvious morphological evidence of neuronal differentiation; however, this feature probably is not a reproducible distinguishing characteristic.

Morphologically, the neuronal features of cell lines from these tumors are subtle and limited to rather abortive, short outgrowths that rarely mimic the neuritic extensions seen in NB tumor cell lines (Fig. 8). Ultrastructurally, both PN and Ewing's sarcoma cell lines have been observed to show evidence of neuronal differentiation (Cavazzana *et al.*, 1987; Llombart-Bosch *et al.*, 1990; Navarro *et al.*, 1990; Hasegawa *et al.*, 1991). Although Ewing's sarcoma may be a tumor of somewhat less mature neuronal precursors than PN, no direct evidence for this model exists (Rettig *et al.*, 1992). Esthesioneuroblastoma or olfactory neuroblastoma also should be included in the Ewing's sarcoma family of tumors, when these tumors are of neuronal rather than glial origin. The esthesioneuroblastoma cell line TC-268 contains a t(11;22) translocation; light microscopic analysis indicates an epithelial morphology with short neuritic processes (Cavazzana *et al.*, 1988; Fig. 9). This cell line also expresses the high levels of Class I MHC and c-*myc* that are characteristic of PN cell lines.

2. Biochemical and Biological Characteristics

The biochemical features of PN-derived cell lines are indistinguishable from those of Ewing's sarcoma-derived cell lines and constitute strong evidence for the origin of these tumors in cells of the PNS (Tables II and III). In all cell lines from these tumors that have been examined to date, evidence for the expression of choline acetyltransferase has been identified (McKeon *et al.*, 1988). This enzyme is the rate limiting enzyme for the biosynthesis of acetylcholine, a neurotransmitter used outside the CNS by only a few neuronal cell types. One such cell type is the postganglionic parasympathetic neuron. This neuron has a cell body in the periphery, uses acetylcholine as a neurotransmitter, and enervates all blood vessels. These characteristics are compatible with its being the cell of origin of PN, which occurs at sites throughout the body. Such cells do not make catecholamines, explaining the absence of catecholamine precursors in the urine of PN patients (McKeon *et al.*, 1988).

PN and Ewing's sarcoma cell lines express the cholecystokinin gene, another neurotransmitter found only in cells of neuronal origin. Few cell types express completely processed immunoreactive cholecystokinin. However, a radioimmunoassay that detects precursors of cholecystokinin can be used to demonstrate the synthesis of cholecystokinin precursor-like peptides in all the

HIGH DENSITY   LOW DENSITY

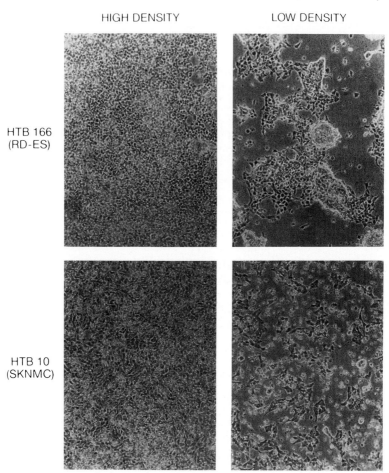

HTB 166
(RD-ES)

HTB 10
(SKNMC)

Fig. 8. Ewing's sarcoma/neuroepithelioma cell lines. Light microscopy reveals that cell lines grow as loosely adherent monolayers. Evidence of neuritic processes is less apparent in the Ewing's sarcoma and neuroepithelioma cell lines than in neuroblastoma cell lines, even at low density.

Ewing's sarcoma and PN cell lines that have been examined (Friedman *et al.*, 1992).

Markers of neuronal differentiation in addition to the neurotransmitters just mentioned also have been examined in PN tumor cell lines (Donner *et al.*, 1985). Although these cell lines are invariably reactive to antibodies against neuron-specific enolase, their reactivity to other gene products known to be expressed in many different types of ganglionic cells is somewhat less predictable. A newly derived monoclonal antibody that recognizes a cell-surface glycoprotein encoded by the *mic2* gene has been characterized and found to be reactive with cell lines derived from PN and Ewing's sarcoma tumor pa-

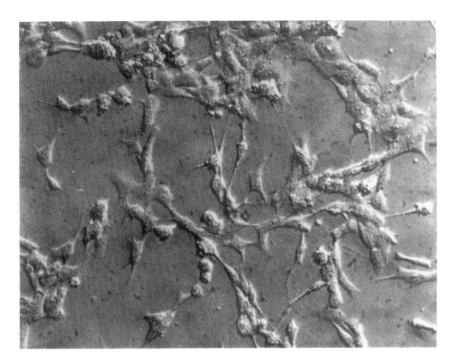

Fig. 9. Esthesioneuroblastoma cell line. The esthesioneuroblastoma or olfactory neuroblastoma cell line TC-268 is a loosely substrate adherent cell line with some evidence of neuritic processes. This line is classified more precisely as a member of the Ewing's sarcoma family of tumors and contains a t(11;22) translocation.

tients, but not with NB-derived tumor cell lines (Fellinger *et al.*, 1991,1992). PN and Ewing's sarcoma cell lines also can be distinguished from most NB tumor cell lines because they show strong binding of antibodies to Class I MHC antigens and to beta 2-microglobulin. PN cells also show positive binding to an anti-human fibronectin antibody.

In contrast to the experimental success that has been achieved with NB cell lines for the study of neuronal differentiation *in vitro*, differentiating PN cell lines in culture has been difficult. Inducing the expression of morphological evidence for neural differentiation in Ewing's sarcoma-derived cell lines is possible (Cavazzana *et al.*, 1987; Noguera *et al.*, 1992). Treatment with cAMP or 12-O-tetradecanoyl phorbol-13-acetate (TPA) results in cells developing elongated processes with varicosities that are visible by phase contrast microscopy. Neurofilaments, microtubules, and uraniffin-positive dense core granules also could be detected by electron microscopy. Neural markers such as neurofilament protein also were detected readily after treatment. Increased neurite extension and synthesis of acetylcholinesterase have been detected in PN and Ewing's sarcoma cell lines cultured in serum-free medium in the presence of cAMP and RA. The growth of a number of these cell lines can be inhibited by agents such as cAMP, cytosine arabinoside, tumor necrosis factor α (TNF α), and interferon γ (IFN γ) without evidence of morphological differentiation (C. J. Thiele, unpublished data). NGF does not induce morphological differentiation of PN cell lines and apparently cannot enhance the expression of neuronal characteristics in these cell lines (Chen et al., 1990). This result is somewhat unanticipated; learning the physiological mediators of differentiation in this lineage will be of interest so their effectiveness in promoting tumor cell differentiation might be examined.

Few studies have been done to examine the biochemical pathways that mediate the growth of PN cell lines. In contrast to the importance of IGF-II in promoting the proliferation of NB-derived cell lines, some evidence exists that IGF-I plays a role in the growth of PN tumor cells (Yee et al., 1990). Most PN cell lines express mRNA encoding IGF-I. Secretion of biologically active IGF-I and an IGF binding protein, IGFBP-2, in one cell line, CHP100, has been demonstrated. This cell line also expresses the type I IGF receptor mRNA. Blockage of this receptor by a monoclonal antibody that is known to block the binding of IGF-I inhibited serum-free growth, indicating that IGF-I is an autocrine growth factor for this cell line.

3. Genetic Characteristics

Both PN- and Ewing's sarcoma-derived cell lines typically carry a t(11;22)(q24;q12) chromosomal rearrangement, although variations of this rearrangement have been observed rarely (Whang-Peng et al., 1984; Israel et al., 1985; Potluri et al., 1987; Bagnara et al., 1990; Llombart-Bosch et al., 1990; Miozzo et al., 1990; Yeger et al., 1990). Although these tumors originally were thought to be distinct, the finding of a cytogenetically indistinguishable breakpoint in these two tumor types suggests that they may be pathologically similar entities. This translocation breakpoint has been cloned; 95% of the tumors have translocations that lie within a 7-kb region on chromosome 22 and a 40-kb region on chromosome 11 (Zucman et al., 1993). The translocation results in a chimeric gene involving a previously unknown gene on chromosome 22 termed wes and the human homolog of fli-1, a member of the ets-1 DNA-binding transcription factor family, on chromosome 11 (Fig. 10). The ews gene encodes a protein that has limited homology with a portion of the large subunit of eukaryotic RNA polymerase II protein and has an RNA binding domain. The ews gene is expressed constitutively in a wide variety of tissues, in sharp contrast to fli-1, which has a more restricted pattern of expression. The resultant chimeric protein replaces the RNA binding domain of the ews product with the ets-like DNA binding domain of the fli-1 protein. The chimeric protein may lead to aberrant regulation of fli-1 target genes or interfere with normal ews gene function.

The expression of a number of different oncogenes has been characterized in PN cell lines (Table IV) (Thiele et al., 1987; Mckeon et al., 1988; Sacchi et al.,

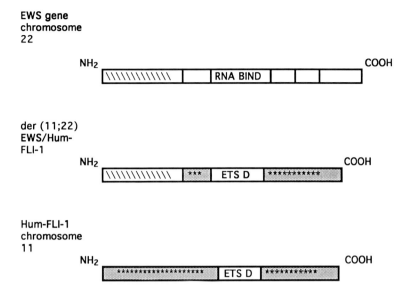

Fig. 10. Schematic diagram of the normal gene on chromosome 11 (*fli-1*) and the normal gene on chromosome 22 (*ews-1*) involved in the t(11;22) translocation in the Ewing's sarcoma family of tumors. The hybrid gene represented in the middle is the most common translocated product detected in these tumors.

1991). Of particular interest is the high level of c-*myc* expression, which may be associated with the increased copy number of chromosome 8, that frequently is observed in these cell lines. The pattern of oncogene expression in these cell lines is distinguishable from the pattern of oncogene expression observed in NB tumor cell lines (Thiele *et al.*, 1987). In contrast, the pattern of oncogene expression in PN cell lines is indistinguishable from that observed in cell lines derived from Ewing's sarcoma (McKeon *et al.*, 1988).

One putative oncogene that has an apparently limited range of expression, but is expressed in all PN- and Ewing's sarcoma-derived cell lines that have been examined, is *rdc*-1 (Collum *et al.*, 1989). *rdc*-1 maps to chromosome 13 and seems to have a *myc*-like transforming activity when assayed *in vitro*. *rdc*-1 is expressed in a limited set of normal tissues, most prominently in developing human retina, but also in brain and spinal cord. The only significant difference in oncogene expression in PN and Ewing's sarcoma cell lines to date indicates that the *dbl* gene is expressed in Ewing's sarcoma but not in NB or PN (Vecchio *et al.*, 1989).

4. Models for the Development of New Therapies

Cell lines from Ewing's sarcoma and peripheral PN have been grown as xenografts in immunosuppressed mice (Floersheim *et al.*, 1980,1982,1986; Llombart-Bosch *et al.*, 1988a,1989,1990). These tumors retain biochemical, immunohistochemical, and ultrastructural evidence of their neuronal differentiation when grown *in vivo*. Histological study reveals a small round cell pattern with minimal or no signs of morphological differentiation (Llombart-Bosch *et al.*, 1989). Immunohistochemically, neural markers such as HNK-1, neuron-specific enolase, and neurofilament protein are detected in the xenografts (Llombart-Bosch *et al.*, 1990). The translocation t(11;22)(q24;q12) found in PN-derived cell lines is retained in the xenografted tumor tissue (Llombart-Bosch *et al.*, 1990).

These xenografts have been used for a series of studies seeking to characterize and evaluate tumor-specific antineoplastic therapy. These lines have been used to establish models in which the value of various chemotherapies (Floersheim *et al.*, 1980,1982,1986), as well as new approaches of therapeutic potential including biological response modifiers such as interferon (Strander, 1987) and monoclonal antibodies (Cheung *et al.*, 1987) has been examined. In a particularly innovative undertaking, a PN xenograft model has been used to examine the *in vivo* efficacy of continuous subcutaneous perfusion of unmodified phosphodiester oligodeoxynucleotides (Whitesell *et al.*, 1991a). N-*myc*, an oncogene that has been associated closely with the regulation of growth and differentiation of NB tumor cell lines, is expressed at very low levels in some PN cell lines and, as such, may be an efficacious target for antisense therapy. After the study of the effect of N-*myc* antisense on the growth and differentiation of PN *in vitro* (Whitesell *et al.*, 1991b), the *in vivo* effects of antisense inhibition of N-*myc* were examined in athymic mice carrying a PN xenograft. Specific effects of the antisense molecules included inhibition of N-*myc*-encoded protein expression, moderate growth inhibition, loss of secretogranin 1 expression, and morphological alterations.

Acknowledgments

The authors thank June Beidler and Barbara Spengler for their assistance in producing figures that demonstrate the morphological variation of neuroblastoma cells in culture; the ATCC and Robert Hay for assistance in obtaining photomicrographs used in this manuscript, and Lucy de la Calzada and Norma Shipp for their assistance in preparing the manuscript.

References

Askin, K. F., Rosai, J., and Sibley, R. (1979). Malignant small cell tumor of the thoracopulmonary region in childhood. A distinctive clinicopathologic entity of uncertain histogenesis. *Cancer* **43**, 2438–2451.

Azar, C. G., Scavarda, N. J ., Reynolds, C. P., and Brodeur, G. M. (1990). Multiple defects of the nerve growth factor receptor in human neuroblastomas. *Cell Growth Diff.* **1**, 421–428.

Bagnara, G. P., Serra, M., Giovannini, M., Badiali, M., Stella, M., Montaldi, A., Granchi, D., Paolucci, P., Rocchi, P., Pession, A. (1990). Establishment and characterization of a primitive neuroectodermal tumor of bone continuous cell line (LAP-35). *Int. J. Cell Cloning* **8**, 409–424.

Barnes, E. N., Biedler, J. L., Spengler, B. A., and Lyser, K. M. (1981). The fine structure of continuous human neuroblastoma lines SK-N-SH, SK-N-BE(2), and SK-N-MC. *In Vitro* **17,** 619–631.

Bernal, S., Thompson, R., Gilbert, F., and Baylin, S. B. (1983). *In vitro* and *in vivo* growth characteristics of two different cell populations in an established line of human neuroblastoma. *Cancer Res.* **43,** 1256–1260.

Bettan, R. L., Bayle, C., Teyssier, J. R., and Benard, J. (1989). Stability of phenotypic and genotypic traits during the establishment of a human neuroblastoma cell line, IGR-N-835. *Int. J. Cancer* **44,** 460–466.

Biedler, J. L., and Spengler, B. A. (1976). A novel chromosome abnormality on human neuroblastoma and anti-folate resistant Chinese hamster cell lines in culture. *J. Natl. Cancer Inst.* **57,** 683.

Biedler, J. L, Roffler, T. S., Schachner, M., and Freedman, L. S. (1978). Multiple neurotransmitter synthesis by human neuroblastoma cell lines and clones. *Cancer Res.* **38,** 3751–3757.

Biedler, J., Ross, R., Sharske, S., and Spengler, B. (1980). Human neuroblastoma cytogenetics. Search for significance of homogeneously staining regions in double minute chromosomes. *In* "Advances in Neuroblastoma Research (A. Evans, ed.), pp. 81–96. Raven Press, New York.

Biedler, J. L., Helson, L., and Spengler, G. A. (1983). Morphology and growth, tumorigenicity and cytogenetics of human neuroblastoma cells in continuous culture. *Cancer Res.* **33,** 2643–2652.

Bolen, J. B., Rosen, N., and Israel, M. A. (1985). Increased pp60c-src tyrosyl kinase activity in human neuroblastomasis associated with amino-terminal tyrosine phosphorylation of the src gene product. *Proc. Natl. Acad. Sci. U.S.A.* **82,** 7275–7279.

Brodeur, G. M. (1990). Neuroblastoma: Clinical significance of genetic abnormalities. *Cancer Surv.* **9,** 673–688.

Brodeur, G., Sekhon, G., and Goldstein, M. (1977). Chromosomal aberrations in human neuroblastoma. *Cancer* **40,** 2256–2263.

Brodeur, G., Green, A., and Hayes, F. (1980). Cytogenetic studies of primary human neuroblastoma. *In* "Advances in Neuroblastoma Research" (A. Evans, ed.). Raven Press, New York.

Brodeur, G. M., Green, A. A., Hayes, F. A., Williams, K. J., Williams, D. L., and Tsiatis, A. A. (1981). Cytogenetic features of human neuroblastomas and cell Lines. *Cancer Res.* **41,** 4678.

Brodeur, G., Seeger, R., and Schwab, M. (1984). Amplification of N-*myc* in untreated human neuroblastomas correlates with advanced disease stage. *Science* **224,** 1121–1124.

Busse, E., Baum, R. P., Hor, G., and Kornhuber, B. (1990). *In vitro* and *in vivo* effect of thyroid hormones on the growth of neuroblastoma cells. II. The effect of thyroxine *in vivo*. *Nuklearmed.* **29,** 125–128.

Busse, E., Bartsch, O., and Kornhuber, B. (1991a). Non-myelotoxic antitumour effects of L-dopa, buthionine sulphoximine and tamoxifen on neuroblastoma cells *in vitro* and *in vivo*. *J. Cancer Res. Clin. Oncol.* **117,** 449–453.

Busse, E., Bartsch, O., and Kornhuber, B. (1991b). Research on the differentiation of human and murine neuroblastoma cells. *Oncology* **48,** 196–201.

Carachi, R., Raza, T., Robertson, D., Wheldon, T. W., Wilson, L., Livingstone, A., Van, H. V., Spowart, G., Middleton, P., Gosden, J. R., *et al.* (1987). Biological properties of a tumour cell line (NB1-G) derived from human neuroblastoma. *Br. J. Cancer* **55,** 407–411.

Cavazzana, A. O., Miser, J. S., Jefferson, J., and Triche, T. J. (1987). Experimental evidence for a neural origin of Ewing's sarcoma of bone. *Am. J. Pathol.* **127,** 507–518.

Cavazzana, A. O., Navarro, S., Noguera, Reynolds, C. P., and Triche, T. (1988). Olfactory neuroblastoma is not a neuroblastoma but is related to primitive neuroectodermal tumor (PNET). *Prog. Clin. Biol. Res.* **271,** 463–473.

Chen, J., Chattopadhyay, B., Venkatakrishnan, G., and Ross, A. H. (1990). Nerve growth factor-induced differentiation of human neuroblastoma and neuroepithelioma cell lines. *Cell Growth Diff.* **1,** 79–85.

Cheung, N. K., Neely, J. E., Landmeier, B., Nelson, D., and Miraldi, F. (1987). Targeting of ganglioside GD 2 monoclonal antibody to neuroblastoma. *J. Nucl. Med.* **28,** 1577–1583.

Ciccarone, V., Spengler, B. A., Meyers, M. B., Biedler, J. L., and Ross, R. A. (1989). Phenotypic diversification in human neuroblastoma cells: Expression of distinct neural crest lineages. *Cancer Res.* **49,** 219–225.

Cohen, P. S., Cooper, M. J., Helman, L. J., Thiele, C. J., Seeger, R. C., and Israel, M. A. (1990). Neuropeptide Y expression in the developing adrenal gland and in childhood neuroblastoma tumors. *Cancer Res.* **50,** 6055–6061.

Cohn, S. L., Salwen, H., Quasney, M. W., Ikegaki, N., Cowan, J. M., Herst, C. V., Kennett, R. H., Rosen, S. T., DiGiuseppe, J. A., and Brodeur, G. M. (1990). Prolonged N-*myc* protein half-life in a neuroblastoma cell line lacking N-*myc* amplification. *Oncogene* **5,** 1821–1827.

Collum, R. G., DePinho, R., Mellis, S., Thiele, C. J., Israel, M. A., and Alt, F. W. (1989). A novel gene expressed specifically in neuroepitheliomas and related tumors. *Cancer Cells* **7,** 113–116.

Cooper, M., Helman, L., Evans, A., Swamy, S., O'Connor, D. T., Helson, L. and Israel, M. A. (1988). Chromagranin A expression in childhood peripheral neuroectodermal tumors. *Prog. Clin. Biol. Res.* **271,** 175–184.

Cooper, M. J., Hutchins, G. M., and Israel, M. A. (1990a). Histogenesis of the human adrenal medulla: An evaluation of gene expression in chromaffin and non-chromaffin lineages. *Am. J. Pathol.* **137,** 605–615.

Cooper, M. J., Hutchins, G. M., Cohen, P. S., Mennie, R. J., and Israel, M. A. (1990b). Human neuroblastoma tumor cell lines correspond to the arrested differentiation of chromaffin adrenal medullary neuroblasts. *Cell Growth Diff.* **2,** 149–159.

Cornaglia, F. P., Ponzoni, M., Montaldo, P., Mariottini, G. L., Donti, E., Di, M. D., and Tonini, G. P. (1990). A new human high tumorigenic neuroblastoma cell line with undetectable expression of N-myc. *Pediatr. Res.* **27,** 1–6.

Cornaglia, F. P., Mariottini, G. L., and Ponzoni, M. (1992). Gamma-interferon and retinoic acid synergize in inhibiting the growth of human neuroblastoma cells in nude mice. *Cancer Lett.* 215–220.

Danon, Y. L., and Kaminsky, E. (1985). Dimethyl sulfoxide-induced differentiation does not alter tumorigenicity of neuroblastoma cells. *J. Neurooncol.* **3,** 43–51.

Donner, L. R. (1991). Cytogenetics and molecular biology of small round-cell tumors and related neoplasms. Current status. *Cancer Genet. Cytogenet.* **54,** 1–10.

Donner, L., Triche, T. J., Israel, M. A., Seeger, R. C., and Reynolds, C. P. (1985). A panel of monoclonal antibodies which discriminate neuroblastoma from Ewing's sarcoma, rhabdomyosarcoma, neuroepithelioma, and hematopoietic malignancies. *Prog. Clin. Biol. Res.* **175,** 347–366.

Donti, E., Longo, L., Tonini, G. P., Verdona, G., Melodia, A., Lanino, E., and Cornaglia, F. P. (1988). Cytogenetic and molecular study of two human neuroblastoma cell lines. *Cancer Genet. Cytogenet.* **30,** 225–231.

Drebin, J. A., Link, V. C., Weinberg, R. A., and Greene, M. I. (1986). Inhibition of tumor growth by a monoclonal antibody reactive with an oncogene-encoded tumor antigen. *Proc. Natl. Acad. Sci. U.S.A.* **83,** 9129–9133.

El-Badry, O. M., Romanus, J. A., Helman, L. J., Cooper, M. J., Rechler, M. M., and Israel, M. A. (1989). Autonomous growth of a human neuroblastoma cell line is mediated by insulin-like growth factor II. *J. Clin. Invest.* **84,** 829–839.

El-Badry, O. M., Helman, L. J., Chatten, J., Steinberg, S. M., Evans, A. E., and Israel, M. A. (1991). Insulin-like growth factor II-mediated proliferation of human neuroblastoma. *J. Clin. Invest.* **87,** 648–657.

Etoh, T., Takahashi, H., Maie, M., Ohnuma, N., and Tanabe, M. (1988). Tumor imaging by antineuroblastoma monoclonal antibody and its application to treatment. *Cancer* **62,** 1282–1286.

Fellinger, E. J., Garin, C. P., Triche, T. J., Huvos, A. G., and Rettig, W. J. (1991). Immunohistochemical analysis of Ewing's sarcoma cell surface antigen p30/32 MIC2. *Am. J. Pathol.* **139,** 317–325.

Fellinger, E. J., Garin, C. P., Glasser, D. B., Huvos, A. G., and Rettig, W. J. (1992). Comparison of cell surface antigen HBA71 (p30/32MIC2), neuron-specific enolase, and vimentin in the immunohistochemical analysis of Ewing's sarcoma of bone. *Am. J. Surg. Pathol.* **16,** 746–755.

Feltner, D. E., Cooper, M., Weber, J., Israel, M. A., and Thiele, C. J. (1989). Expression of class I histocompatibility antigens in neuroectodermal tumors is independent of the expression of a transfected neuroblastoma myc gene. *J. Immunol.* **143,** 4292–4299.

Finkelstein, J., Kemperer, M., and Evans, A. (1979). Multiagent chemotherapy for children with metastatic neuroblastoma: A report from the children's cancer study group. *Med. Pediatr. Oncol.* **6,** 179–188.

Floersheim, G. L., Nassenstein, D., and Torhorst, J. (1980). Growth of human tumors in mice after short-term immunosuppression with procarbazine, cyclophosphamide, and antilymphocyte serum. *Transplantation* **30,** 275–280.

Floersheim, G. L., Looser, R., Grundmann, H. P., and Meyer, J. C. (1982). Inhibition of growth of human cancers by extracts from *Trichophyton verrucosum*. *Lancet* **1,** 708–710.

Floersheim, G. L., Bieri, A., and Chiodetti, N. (1986). Xenografts in pharmacologically immunosuppressed mice. *Int. J. Cancer* **37,** 9–14.

Foley, J., Cohn, S. L., Salwen, H. R., Chagnovich, D., Cowan, J., Mason, K. L., and Parysek, L. M. (1991). Differential expression of N-*myc* in phenotypically distinct subclones of a human neuroblastoma cell line. *Cancer Res.* **51,** 23–31.

Franke, F., Rudolph, B., and Lampert, F. (1986). Translocation in two stage II neuroblastomas. *Cancer Genet. Cytogenet.* **20,** 129–135.

Friedman, J. M., Vitale, M., Maimon, J., Israel, M. A., Horowitz, M. E., and Schneider, B. S. (1992). Expression of the cholecystokinin gene in pediatric tumors. *Proc. Natl. Acad. Sci. U.S.A.* **89,** 5819–5823.

Gaetano, C., Matsumoto, K., and Thiele, C. J. (1992). *In vitro* activation of distinct molecular and cellular phenotypes after induction of differentiation in a human neuroblastoma cell line. *Cancer Res.* **52,** 4402–4407.

Gazitt, Y., He, Y. J., Chang, L., Koza, S., Fisk, D., and Graham, P. J. (1992a). Expression of N-*myc*, c-*myc*, and MDR-1 proteins in newly established neuroblastoma cell lines: A study by immunofluorescence staining and flow cytometry. *Cancer Res.* **52,** 2957–2965.

Gazitt, Y., He, Y. J., and Graham, P. J. (1992b). A novel methodology for the establishment of neuroblastoma cell lines from metastatic marrow. Expression of surface markers, neurofilaments, MDR-1 and myc proteins. *J. Immunol. Methods* **148,** 171–178.

Gilbert, F., Balaban, G., Moorhead, P., Bianchi, D., and Schlessinger, H. (1982). Abnormalities of chromosome 1p in human neuroblastoma tumors and cell lines. *Cancer Genet. Cytogenet.* **7,** 33.

Goldstein, M. N., and Plurad, S. (1980). Transformation into ganglion cells with Mitomycin-C. *Results Prol. Cell Diff.* **11,** 259–264.

Gonzalez-Crussi, F., Wolfson, S., and Misugi, K. (1984). Peripheral neuroectodermal tumors of the chest wall in childhood. *Cancer* **54,** 2519–2527.

Griffin, C. A. A., McKeon, C., Israel, M. A., Gegonne, A., Chysdael, J., Stehelin, D., Douglass, E., C., Green, A. E., and Emanuel, B. S. (1986). Comparison of constitutional and tumor associated 11;22 translocations: Nonidentical breakpoints on chromosomes 11 and 22. *Proc. Natl. Acad. Sci. U.S.A.* **83,** 6122–6126.

Hasegawa, T., Hirose, T., Kudo, E., Hizawa, K., Yamawaki, S., and Ishii, S. (1991). Atypical primitive neuroectodermal tumors. Comparative light and electron microscopic and immunohistochemical studies on peripheral neuroepitheliomas and Ewing's sarcomas. *Acta Pathol. Jpn.* **41,** 444–454.

Hayes, F., and Smith, E. (1988). Neuroblastoma. *In* "Principles and Practice of Pediatric Oncology" (O. Pizzo and D. G. Poplack, eds.), pp. 607–622. Lippincott, Philadelphia.

Helson, L., and Helson, C. (1985). Human neuroblastoma cells and 13 cis-retinoic acid. *J. Neuro-Oncol.* **3,** 39–41.

Helson, L., and Helson, C. (1992). Establishment of a new cell line, VA-N-BR, from a primitive neuroblastoma tumor of the abdomen. *Anticancer Res.* **4,** 67–72.

Israel, M. A., Thiele, C., Whang, P. J., Kao, S. C. S., Triche, T. J., and Miser, J. (1985). Peripheral neuroepithelioma: Genetic analysis of tumor derived cell lines. *Prog. Clin. Biol. Res.* **175,** 161–170.

Karnes, P. S., Tran, T. N., Cui, M. Y., Raffel, C., Gilles, F. H., Barranger, J. A., and Ying, K. L. (1992). Cytogenetic analysis of 39 pediatric central nervous system tumors. *Cancer Genet. Cytogenet.* **59,** 12–19.

Kees, U. R., Ford, J., Dawson, V. M., Ranford, P. R., and Armstrong, J. A. (1992). Three neuroblastoma cell lines established from consecutive samples of one patient which show distinct morphologic features, MYCN amplification, and surface marker expression. *Cancer Genet. Cytogenet.* **59,** 119–127.

Kimhi, Y., Palfrey, C., and Spector, I. (1976). Maturation of neuroblastoma cells in the presence of dimethylsulfoxide. *Proc. Natl. Acad. Sci. U.S.A.* **73,** 462–466.

Kogner, P., Bjork, O., and Theodorsson, E. (1990). Neuropeptide Y as a marker in pediatric neuroblastoma. *Pediatr. Pathol.* **10,** 207–216.

Kohl, N. E., Gee, C. E., and Alt, F. W. (1984). Activated expression of the N-*myc* gene in human neuroblastomas and related tumors. *Science* **226,** 1335–1337.

Le-Douarin, N. M., and Smith, J. (1988). Development of the peripheral nervous system from the neural crest. *Ann. Rev. Cell Biol.* **4,** 375–404.

Linnoila, R. I., Tsokos, M., Triche, T. J., Marangos, P. J., and Chandra, R. S. (1986). Evidence for neural origin and pas-positive variants of the malignant small cell tumor of the thoracopulmonary region ("Askin tumor"). *Am. J. Surg. Pathol.* **10,** 124–133.

Lipinski, M., Braham, K., and Philips, T. (1987). Neuroectodermal-associated antigens on Ewing's sarcoma cell lines. *Cancer Res.* **47,** 183–187.

Llombart-Bosch, A., Carda, C., Boix, J., Pellin, A., and Peydro, O. A. (1988a). Value of nude mice xenografts in the expression of cell heterogeneity of human sarcomas of bone and soft tissue. *Pathol. Res. Pract.* **183,** 683–692.

Llombart-Bosch, A., Lacombe, M. J., Peydro, O. A., Perez, B. M., and Contesso, G. (1988b). Malignant peripheral neuroectodermal tumours of bone other than Askin's neoplasm: Characterization of 14 new cases with immunohistochemistry and electron microscopy. *Virchows Arch. A Pathol. Anat. Histopathol.* **412,** 421–430.

Llombart-Bosch, A., Carda, C., Peydro, O. A., Noguera, R., Boix, J., and Pellin, A. (1989). Pigmented esthesioneuroblastoma showing dual differentiation following transplantation in nude mice. An immunohistochemical, electron microscopical, and cytogenetic analysis. *Virchows Arch. A Pathol. Anat. Histopathol.* **414,** 199–208.

Llombart-Bosch, A., Carda, C., Peydro, O. A., Noguera, R., Perez, B. M., Pellin, A., and Boix, J. (1990). Soft tissue Ewing's sarcoma. Characterization in established cultures and xenografts with evidence of a neuroectodermic phenotype. *Cancer* **66,** 2589–2601.

McKeon, C., Thiele, C. J., Ross, R. A., Kwan, M., Triche, T. J., and Miser, J. S. (1988). Indistinguishable patterns of protooncogene expression in two distinct but closely related tumors: Ewing's sarcoma and neuroepithelioma. *Cancer Res.* **48,** 4307–4311.

McLatchie, G., and Young, D. G. (1980). Presenting features of thoracic neuroblastoma. *Arch. Dis. Child.* **55,** 958–962.

Marina, N. M., Etcubanas, E., Parham, D. M., Bowman, L. C., and Green, A. (1989). Peripheral primitive neuroectodermal tumor (peripheral neuroepithelioma) in children. A review of the St. Jude experience and controversies in diagnosis and management. *Cancer* **64,** 1952–1960.

Matthay, K. K., Sather, H. N., Seeger, R. C., Haase, G. M., and Hammond, G. D. (1989). Excellent outcome of stage II neuroblastoma is independent of residual disease and radiation therapy. *J. Clin. Oncol.* **17,** 236–244.

Mehta, B. M., Rosa, E., Fissekis, J. D., Bading, J. R., Biedler, J. L., and Larson, S. M. (1992). *In vivo* identification of tumor multidrug resistance with tritium-3-colchicine. *J. Nuc. Med.* **33,** 1373–1377.

Miozzo, M., Sozzi, G., Calderone, C., Pilotti, S., Lombardi, L., Pierotti, M. A., and Della, P. G. (1990). T(11;22) in three cases of peripheral neuroepithelioma. *Genes Chrom. Cancer* **2,** 163–165.

Miser, J., Steis, R., and Longo, D. (1985). Treatment of newly diagnosed high risk sarcoma and primitive neuroectodermal tumors (PNET) in children and young adults. *Proc. Am. Soc. Clin. Oncol.* **4,** 240.

Montgomery, K., Biedler, J. L., and Spengler, B. A. (1983). Specific DNA sequence amplification in human neuroblastoma cells. *Proc. Natl. Acad. Sci. U.S.A.* **80,** 5724–5728.

Navarro, S., Gonzalez, D. M., Ferrandez, I. A., Triche, T. J., and Llombart-Bosch, A. (1990). Scanning electron microscopic evidence for neural differentiation in Ewing's sarcoma cell lines. *Virchows Arch. A Pathol. Anat. Histopathol.* **416,** 383–391.

Noguera, R., Triche, T. J., Navarro, S., Tsokos, M., and Llombart-Bosch, A. (1992). Dynamic model of differentiation in Ewing's sarcoma cells. Comparative analysis of morphologic, immunocyto-chemical, and oncogene expression parameters. *Lab. Invest.* **66,** 143–151.

Nojima, T., Abe, S., Furuta, Y., Nagashima, K., Alam, A. F., Takada, N., Sasaki, F., and Hata, Y. (1991). Morphological and cytogenetic characterization and N-*myc* oncogene analysis of a newly established neuroblastoma cell line. *Acta Pathol. Jpn.* **41,** 507–515.

Pahlman, S., Meyerson, G., Lindgren, E., Schalling, M., and Johansson, I. (1991). Insulin-like growth factor I shifts from promoting cell division to potentiating maturation during neuronal differentiation. *Proc. Natl. Acad. Sci. U.S.A.* **88,** 9994–9998.

Parodi, M. T., Cornaglia, F. P., and Ponzoni, M. (1989). Effects of gamma-interferon on the growth, morphology, and membrane and cytoskeletal proteins expression of LAN-1 cells. *Exp. Cell Res.* **185,** 327–341.

Perez-Polo, J., Werrbach-Perez, K., and Tiffany-Castiglioni, E. (1979). A human clonal cell line model of differentiating neurons. *Dev. Biol.* **71,** 341–355.

Petkovic, I., and Cepulic, M. (1991). Cytogenetic analysis of primary neuroblastoma with del(1), del(14), hsr, and dmin chromosomes. *Cancer Genet. Cytogenet.* **55,** 231–234.

Pietsch, T., Gottert, E., Meese, E., Blin, N., Feickert, H. J., Riehm, H., and Kovacs, G. (1988). Characterization of a continuous cell line (MHH-NB-11) derived from advanced neuroblastoma. *Anticancer Res.* **8,** 1329–1333.

Ponzoni, M., Melodia, A., and Cirillo, C. (1988). Effect of cytosine arabinoside on the growth and phenotypic expression of G-ME-N: A new human neuroblastoma cell line. *Prog. Clin. Biol. Res.* **271,** 437–448.

Ponzoni, M., Casalaro, A., Lanciotti, M., Montaldo, P. G., and Cornaglia, F. P. (1992). The combination of gamma-interferon and tumor necrosis factor causes a rapid and extensive differentiation of human neuroblastoma cells. *Cancer Res.* **52,** 931–939.

Potluri, V. R., Gilbert, F., Helsen, C., and Helson, L. (1987). Primitive neuroectodermal tumor cell lines: Chromosomal analysis of five cases. *Cancer Genet. Cytogenet.* **24,** 75–86.

Prasad, K. N. (1975). Differentiation of neuroblastoma cells in culture. *Biol. Rev.* **50,** 129–265.

Preis, P., Saya, H., and Nadasdi, L. (1988). Neuronal differentiation of human neuroblastoma cells by retinoic acid plus herbamycin A. *Cancer Res.* **48,** 6530–6534.

Reed, J. C., Meister, L., Tanaka, S., Cuddy, M., Yum, S., Geyer, C., and Pleasure, D. (1991). Differential expression of BCL 2 protooncogene in neuroblastoma and other human tumor cell lines of neural origin. *Cancer Res.* **51,** 6529–6538.

Rettig, W. J., Chesa, P. G., Jennings, M. T., Spengler, B. A., Melamed, M. R., Oettgen, H. F., Biedler, J. L., and Old, L. J. (1985). Cell surface antigen of human neuroblastomas in related to nuclear antigen of normal cells. *Proc. Natl. Acad. Sci. U.S.A.* **82,** 6894–6898.

Rettig, W. J., Garin, C. P., and Huvos, A. G. (1992). Ewing's sarcoma: New approaches to histogenesis and molecular plasticity. *Lab. Invest.* **66,** 133–137.

Reynolds, C. P., and Maples, J. (1985). Modulation of cell surface antigens accompanies morpho-logical differentiation of human neuroblastoma cell lines. *Prog. Clin. Biol. Res.* **175,** 13–37.

Reynolds, C. P., and Perez, P. J. R. (1989). Nerve growth factor induces neurite outgrowth in a clone derived from an NGF-insensitive human neuroblastoma cell line. *Int. J. Dev. Neurosci.* **7**, 125–132.

Reynolds, C. P., Biedler, J. L., Spengler, B. A., Reynolds, D. A., Ross, R. A., Frenkel, E. P., and Smith, R. G. (1986). Characterization of human neuroblastoma cell lines established before and after therapy. *J. Natl. Cancer Inst.* **76**, 375–387.

Ritke, M. K., Shah, R., Valentine, M., Douglass, E. C., and Tereba, A. (1989). Molecular analysis of chromosome 1 abnormalities in neuroblastoma. *Cytogenet. Cell Genet.* **50**, 84–90.

Rosen, N., Reynolds, C. P., Thiele, C. J., Biedler, J. L., and Israel, M. A. (1986). Increased N-*myc* expression following progressive growth of human neuroblastoma. *Cancer Res.* **46**, 4139–4142.

Ross, R., and Biedler, J. (1985). Presence and regulation of tyrosinase activity in human neuroblastoma variants *in vitro. Cancer Res.* **45**, 1628–1632.

Ross, R. A., Spengler, B. A., and Biedler, J. L. (1983). Coordinate morphological and biochemical interconversion of human neuroblastoma cells. *J. Natl. Cancer Inst.* **71**, 741–747.

Rudolph, G., Schilbach, S. K., Handgretinger, R., Kaiser, P., and Hameister, H. (1991). Cytogenetic and molecular characterization of a newly established neuroblastoma cell line LS. *Hum. Genet.* **86**, 562–566.

Rupniak, H. T., Rein, G., Powell, J. F., Ryder, T. A., Carson, S., Povey, S., and Hill, B. T. (1984). Characteristics of a new human neuroblastoma cell line which differentiates in response to cyclic adenosine 3′ : 5′-monophosphate. *Cancer Res.* **44**, 2600–2607.

Sacchi, N., Wendtner, C. M., and Thiele, C. J. (1991). Single-cell detection of ETS-1 transcripts in human neuroectodermal cells. *Oncogene* 6, 2149–2154.

Sadee, W., Yu, V. C., Richards, M. L., Pries, P. N., Schwab, M. R., Brodsky, F. M., and Biedler, J. L. (1987). Expression of neurotransmitter receptors and myc protooncogenes in subclones of a human neuroblastoma cell line. *Cancer Res.* **47**, 5207–5212.

Sawaguchi, S., Kaneko, M., Uchino, J., Takeda, T., Iwafuchi, M., and Matsuyama, S. (1990). Treatment of advanced neuroblastoma with emphasis on intensive induction chemotherapy. A report from the Study Group of Japan. *Cancer* **66**, 1879–1887.

Scarpa, S., Dominici, C., Grammatico, P., Del, P. G., Raschella, G., Castello, M., Forni, G., and Modesti, A. (1989). Establishment and characterization of a human neuroblastoma cell line. *Int. J. Cancer* **43**, 645–651.

Schlesinger, H. R., Gerson, J. M., Moorhead, P. S., Maguire, H., and Hummeler K. (1976). Establishment and characterization of human neuroblastoma tumor cell lines. *Cancer Res.* **36**, 3094–3100.

Schwab, M., Alitalo, K., and Lempnauer, K. (1983). Amplified DNA with limited homology to myc cellular oncogene is shared by human neuroblastoma cell lines and a neuroblastoma tumor. *Nature (London)* **305**, 245–248.

Seeger, R., and Brodeur, G. S. (1985). Association of multiple copies of the N-myc oncogene with rapid progression of neuroblastomas. *N. Engl. J. Med.* **313**, 1111–1116.

Seeger, R. C., Rayner, R. C., Banerjee, A., Chung, H., Laug, W. E., Neustein, H. B., and Benedict, W. F. (1977). Morphology, growth, chromosomal pattern, and fibrinolytic activity of activity of two new human neuroblastoma cell lines. *Cancer Res.* **37**, 1365–1371.

Seemayer, T., Thelmo, W., and Bolande, R. (1975). Peripheral neuroectodermal tumors. *Perspect. Pediatr. Pathol.* **2**, 151.

Sekiguchi, M., Oota, T., Sakakibara, K., Inui, N., and Fujii, G. (1979). Establishment and characterization of a human neuroblastoma cell line in tissue culture. *Jpn. J. Exp. Med.* **49**, 67–83.

Shafford, E., Rogers, D., and Pritchard, J. (1984). Advanced neuroblastoma: Improved response rate using a multiagent regimen (OPEC) including sequential cisplatin and VM26. *J. Clin. Oncol.* **2**, 742–747.

Sidell, N. (1982). Retinoic acid-induced growth inhibition and morphologic differentiation of human neuroblastoma cells *in vitro. J. Natl. Cancer Inst.* **68**, 589–593.

Smith, R. G., and Reynolds, C. P. (1987). Monoclonal antibody recognizing a human neuroblastoma-associated antigen. *Diagn. Clin. Immunol.* **5**, 209–220.

Srivatsan, E. S., Murali, V., and Seeger, R. C. (1991). Loss of heterozygosity for alleles on chromosomes 11q and 14q in neuroblastoma. *Prog. Clin. Biol. Res.* **366,** 91–98.
Stout, A. P. (1918). A tumor of the ulnar nerve. *Proc. N.Y. Pathol. Soc.* **18,** 2.
Strander, H. (1987). Interferon use for sarcomas. *In* "The Interferon System" (S. Baron, ed.), pp. 517–525. University of Texas Press, Austin.
Sugimoto, T., Sawada, T., Matsumura, T., Kemshead, J. T., Ishii, T., Horii, Y., Morioka, H., Morita, M., and Reynolds, C. P. (1986). Identical expression of cell surface membrane antigens on two parent and eighteen cloned cell lines derived from two different neuroblastoma metastases of the same patient. *Cancer Res.* **46,** 4765–4769.
Sugimoto, T., Ueyama, H., Hosoi, H., Inazawa, J., Kato, T., Kemshead, J. T., Reynolds, C. P., Gown, A. M., Mine, H., and Sawada, T. (1991). Alpha-smooth-muscle actin and desmin expressions in human neuroblastoma cell lines. *Int. J. Cancer* **48,** 277–283.
Suzuki, T., Yokota, J., Mugishima, H., Okabe, I., Ookuni, M., Sugimura, T., and Terada, M. (1989). Frequent loss of heterozygosity on chromosome 14q in neuroblastoma. *Cancer Res.* **49,** 1095–1098.
Thiele, C. J., and Israel, M. A. (1988). Regulation of N-*myc* expression is a critical event controlling the ability of human neuroblasts to differentiate. *Exp. Cell Biol.* **56,** 321–333.
Thiele, C. J., Reynolds, P. C., and Israel, M. A. (1985). Decreased expression of N-*myc* precedes retinoic acid induced phenotypic differentiation of human neuroblastoma. *Nature (London)* **313,** 404–406.
Thiele, C. J., McKeon, C., Triche, T. J., Ross, R. A., Reynolds, C. P., and Israel, M. A. (1987). Differential protooncogene expression characterizes histopathologically indistinguishable tumors of the peripheral nervous system. *J. Clin. Invest.* **80,** 804–811.
Thiele, C. J., Deutsch, L. A., and Israel, M. A. (1988). The expression of multiple proto-oncogenes is differentially regulated during retinoic acid induced maturation of human neuroblastoma cell lines. *Oncogene* **3,** 281–288.
Triche, T. J. (1990). Neuroblastoma and other childhood neural tumors: A review. *Pediatr. Pathol.* **10,** 175–193.
Triche, T., and Askin, F. (1983). Neuroblastoma and the differential diagnosis of small, round, blue cell tumors. *Hum. Pathol.* **14,** 569–594.
Triche, T., and Cavazzana, A. O. (1989). Principles in pediatric pathology. *In* "Principles and Practice of Pediatric Oncology" (P. A. Pizzo and D. E. Poplack, eds.), pp. 93–125. Lippincott, Philadelphia.
Tsokos, M., Scarpa, S., Ross, R. A., and Triche, T. J. (1987). Differentiation of human neuroblastoma recapitulates neural crest development. Study of morphology, neurotransmitter enzymes, and extracellular matrix proteins. *Am. J. Pathol.* **128,** 484–496.
Tsunamoto, K., Todo, S., and Imashuku, S. (1988). Induction of S100 protein by 5-bromo-2'-deoxyuridine in human neuroblastoma cell lines. *Cancer Res.* **48,** 70–174.
Tumilowicz, J. J., Nichols, W. W., Cholon, J. J., and Greene, A. E. (1970). Definition of a continuous human cell line derived from neuroblastoma. *Cancer Res.* **30,** 2110–2118.
Turc-Carel, C., Philip, I., and Berger, M. (1983). Chromosomal translocations in Ewing's sarcoma. *N. Engl. J. Med.* **309,** 309–497.
Vecchio, G., Cavazzana, A. O., Triche, T. J., Ron, D., Reynolds, C. P., and Eva, A. (1989). Expression of the DBL proto-oncogene in Ewing's sarcomas. *Oncogene* **4,** 897–900.
Viallard, J. L., Tiget, F., Hartmann, O., Lemerle, J., Demeocq, F., Malpuech, G., and Dastugue, B. (1988). Serum neuron-specific/nonneuronal enolase ratio in the diagnosis of neuroblastomas. *Cancer* **62,** 2546–2553.
Weiler, R., Meyerson, G., Fischer, C. R., Laslop, A., Pahlman, S., Floor, E., and Winkler, H. (1990). Divergent changes of chromogranin A/secretogranin II levels in differentiating human neuroblastoma cells. *FEBS Lett.* **265,** 27–29.
Whang-Peng, J., Triche, J. J., Knutsen, T., Miser, J., Douglass, E. C., and Israel, M. A. (1984). Chromosome translocation in peripheral neuroepithelioma. *N. Engl. J. Med.* **311,** 584–585.
Whang-Peng, J., Triche, T. J., Knutsen, T., Miser, J., Kao, S. S., Tsai, S., and Israel, M. A. (1986).

Cytogenetic characterization of selected small round cell tumors of childhood. *Cancer Genet. Cytogenet.* **21,** 185–208.

Whitesell, L., Rosolen, A., and Neckers, L. M. (1991a). *In vivo* modulation of N-*myc* expression by continuous perfusion with an antisense oligonucleotide. *Antisense Res. Dev.* **1,** 343–350.

Whitesell, L., Rosolen, A., and Neckers, L. M. (1991b). Antisense suppression of N-*myc* expression inhibits transdifferentiation of neuroectoderm tumor cell lines. *Prog. Clin. Biol. Res.* **366,** 45–54.

Williams, C. J., Krikorian, J. C., Green, M. R., and Raghavan, D. (1988). Peripheral nueroepithelioma. *In* "Textbook of Uncommon Cancer" (C. J. Williams, J. C. Krikorian, M. R. Green, and D. Raghavan, eds.), pp. 683–690. John Wiley & Sons, New York.

Yee, D., Favoni, R. E., Lebovic, G. S., Lombana, F., Powell, D. R., Reynolds, C. P., and Rosen, N. (1990). Insulin-like growth factor I expression by tumors of neuroectodermal origin with the t(11;22) chromosomal translocation. A potential autocrine growth factor. *J. Clin. Invest.* **86,** 1806–1814.

Yeger, H., Mor, O., Pawlin, G., Kaplinsky, C., and Shiloh, Y. (1990). Importance of phenotypic and molecular characterization for identification of a neuroepithelioma tumor cell line, NUB-20. *Cancer Res.* **50,** 2794–2802.

Zucman, J., Delattre, O., Desmaze, C., Plougastel, B., Joubert, I., Melot, T., Peter, M., De Jong, P., Rouleau, G., Aurias, A., and Thomas, G. (1992). Cloning and characterization of the Ewing's sarcoma and peripheral neuroepithelioma t(11;22) translocation breakpoints. *Genes Chrom. Cancer* **5,** 271–277.

Head and Neck Tumor Cell Lines

Thomas E. Carey

Laboratory of Head and Neck Cancer Biology
Department of Otolaryngology/Head and Neck Surgery
University of Michigan Cancer Center
Ann Arbor, Michigan 48109-0506

I. Introduction 79
  A. Head and Neck Cancer 79
  B. Etiology 80

II. Methods of Establishment and Maintenance 81
  A. Historical Perspective 81
  B. Cell Line Establishment and Maintenance Methodology 86
  C. Cell Line Nomenclature 96

III. Morphology 96
  A. Squamous Carcinomas: UM-SCC Series 96

B. SCC Cell Lines from the American Type Culture Collection Repository 105
C. Other Tumors of the Head and Neck Region 106

IV. Other Characteristics 108
  A. High and Low Malignant Types 108
  B. Antigen Phenotype: Blood Group, Pemphigus, Pemphigoid, and the $\alpha 6 \beta 4$ Integrin 108

V. Discussion 113

VI. Future Prospects 115
References 117

I. Introduction

A. Head and Neck Cancer

Cancer of the head and neck affects over 40,000 patients each year in the United States and causes 20,000 deaths in the same population. Head and neck cancers typically arise from the mucosal surfaces of the upper aerodigestive tract, including the oral and nasal cavities, larynx, and pharynx. Squamous cell carcinomas (SCCs) account for 85–90% of tumors that typically are classified as head and neck cancer. The majority of the remaining head and neck neoplasms arise from the major and minor salivary glands and the thyroid. The types and frequency of head and neck neoplasms vary around the world. Nasopharyngeal carcinoma (NPC), a type of poorly differentiated

79

squamous carcinoma that arises from the mucosa overlying the lymphoid-rich tissues of the nasopharynx, is associated with the Epstein–Barr virus and is relatively rare in the general population of the United States and other western countries, with an incidence of about 1/100,000. However, NPC is comparatively frequent in people from certain parts of China, with incidence rates as high as 10–14/100,000 in some Chinese populations, even if the individuals live outside of China. Burkitt's lymphoma, a tumor that often presents in the head and neck, also is associated with the Epstein-Barr virus. However, this tumor, which is endemic to central Africa, is rare in western populations. Because this tumor is derived from transformed lymphocytes of the B-cell series, it is classified by cell type with the leukemias and other lymphomas. Basal cell carcinomas, epidermal squamous cell carcinomas, and malignant melanomas are not uncommon on the sun-exposed areas of the head and neck but, because they are not unique to the region, they generally are not classified as head and neck cancers. Similarly, sarcomas of the bone or soft tissues also may arise in the head and neck, but usually are classified by cell type rather than by anatomic location. Cell lines from these tumors will be addressed in separate chapters. The term "head and neck cancer" also does not include brain or ocular tumors. Brain tumor cell lines will be discussed in the chapter on tumors of the central nervous system. Uncommon tumors unique to the head and neck region include odontogenic tumors that develop in the jaws, acoustic neuromas that arise from the auditory nerve, glomus tumors or chemodectomas that arise from the vascular sensory organs in the carotid body in the neck or the jugular bulb in the temporal bone, and esthesioneuroblastomas or olfactory neuroblastomas, rare neurogenic tumors that arise from the olfactory epithelium. To my knowledge, no human tumor cell lines have been established from acoustic neuromas, glomus tumors, esthesioneuroblastomas, or odontogenic tumors. With the exception of a few cell lines from thyroid and salivary gland tumors, most cell lines established from head and neck tumors are squamous carcinomas.

B. Etiology

Head and neck cancer in the United States is predominantly a disease of 60- to 70-year-old men with a long history of smoking and heavy alcohol consumption. However, the increase in tobacco use among women over the past four decades that has contributed to the epidemic of lung cancer among women also has contributed to an increased incidence of head and neck cancer among women.

Although tobacco smoking and alcohol are clearly the most important etiologic factors in head and neck cancer, other causes are implicated in some cases. Chewing tobacco is an etiologic factor for cancer of the oral cavity and buccal mucosa. Workplace exposure to heavy metals or to sawdust from hardwood furniture manufacturing has been linked to cancer of the nasal sinuses. Exposure to sunlight is a risk factor for squamous carcinoma of the lips and skin of the head and neck. Viruses also have been implicated. The Epstein–Barr virus is an etiologic factor in both Burkitt's lymphoma and nasopharyngeal carcinoma. The role of the papillomavirus family in head and neck cancer etiology is still unclear, but studies demonstrating human papillomavirus (HPV) genetic material incorporated into the DNA of head and neck tumors suggest that this virus is involved in the development of a minority of head and neck cancers. Head and neck squamous cancers are among the cancers that arise in immunosuppressed individuals such as transplant recipients, suggesting that some head and neck cancers may be under immunological control initially and become clinically detectable only when the immune system is suppressed. Preliminary evidence from our laboratory suggests that papillomaviruses may be factors in head and neck tumors that arise in immunosuppressed individuals.

The etiology of most thyroid and salivary gland tumors is unknown. These tumors do not show the same sex and age distribution as the squamous carcinomas, nor do they demonstrate the same association with carcinogen exposure as the typical head and neck squamous cancer. These tumors are more likely to arise in younger patients and show a more equal distribution between men and women. In mice, salivary gland tumors arise after neonatal infection with polyomavirus. Whether an analog of the murine polyomavirus plays a role in human salivary gland tumors has not been determined.

II. Methods of Establishment and Maintenance

A. Historical Perspective

Prior to 1980, very few head and neck tumor cell lines existed, in part because the majority of head and neck tumors are squamous cell carcinomas, a tumor type that has been viewed as somewhat difficult to cultivate, and in part because of an absence of research on head and neck cancer. When we began our studies of head and neck cancer in the late 1970s, the literature contained only isolated reports of individual cell lines established from head and neck cancer (see Carey, 1985, for review); few systematic efforts had been made to cultivate tumors from this region.

By the early 1980s, reports from Easty et al. (1981) in England, from Rheinwald and Beckett (1981) in Boston, and from our group (Krause et al., 1981) showed that the cultivation of head and neck squamous carcinomas was feasible in a reasonable proportion of cases. As additional groups have focused on head and neck cancer, many reports of head and neck squamous carcinoma cell lines have been published (see Table I). Our own laboratory bank now includes 91 head and neck squamous carcinoma cell lines from 82 patients. These cell lines have been used in our laboratory and elsewhere for

Table I

Reported Head and Neck Tumor Cell Lines

Cell lines	Patient Age/sex	TNM class[a]	Primary site[b]	Specimen site[c] (prior therapy[d])	Method of culture[e]	Reference
Squamous carcinoma						
Hep 2 (HeLa?)[f]	57/M	NS	Larynx	Recurrence	Rat xenograft	Moore et al. (1955)
Hep 3 (HeLa?)	62/M	NS	Buccal cavity	LN metastasis	Rat xenograft	Moore et al. (1955)
KB (HeLa?)	54/M	NS	Buccal cavity	Primary tumor	Explant culture	Eagle (1955)
RPMI 2650	?	NS	Nasal Septum	Pleural fluid (RT/S)	Cell pellet from fluid	Moore and Sandberg (1964)
FaDu	56/M	NS	Hypopharynx	Primary tumor (none)	Explant on rat collagen	Rangan (1972)
Detroit 562	?/F	NS	Pharynx	Pleural fluid (?)	Cell pellet from fluid	Peterson et al. (1971)
A-253	54/M	NS	Salivary gland	Primary tumor? (?)	Explant	Giard et al. (1973)
SW579	59/M	NS	Thyroid SCC	Primary tumor? (?)	Explant ?	Liebovitz (1973)
T3M-1	33/M	NS	Lower gingiva	Pleural effusion (?)	Nude mouse	Okabe et al. (1978)
HLaC 78	?/M?	NS	Larynx	Neck met (?)	Explant	Zenner et al. (1979)
HSmC 78	?/M?	NS	Submand. gland	Neck met (?)	Explant	
MC	66/F	NS	Maxillary sinus	Primary tumor (?)	Trypsin digest	Nakashima et al. (1980)
SCC-4	55/M	NS	Tongue	Primary (RT/MTX)	Trypsin collagenase digest or explant on fibroblast feeder layer	Rheinwald and Beckett (1981)
SCC-9	25/M	NS	Tongue	Primary (none)		
SCC-12	60/M	NS	Facial skin	Primary ? (?)		
SCC-13	56/F	NS	Facial skin	Primary ? (RT)	(SCC-13 required feeder layer)	
SCC-15	55/M	NS	Tongue	Primary (none)		
SCC-25	70/M	NS	Tongue	Primary (none)		
HN-1	51/M	T2N1M0	Tongue	Primary ? (CT/RT/S)	Explant	Easty et al. (1981)
HN-2	49/M	T3N0M0	Larynx	Rec ? (CT/S/RND)	Explant	
HN-3	63/M	T3N0M0	Tongue	Rec ? (CT/RT/S/ND)	Explant	
HN-4	57/M	T2N0M0	Larynx	Rec ? (CT/RT/S)	Explant	
HN-5	73/M	T2N0M0	Tongue	Rec ? (CT/RT/S)	Explant	
HN-6	54/M	T2N0M0	Tongue	Rec ? (CT/RT/S)	Explant	
HN-7	56/M	T2N0M0	Tongue	Rec ? (CT/RT/S)	Explant	
HN-8	56/M	T2N0M0	Tongue	Rec ? (RT/S)	Explant	
HN-9	67/F	T2N0M0	Tongue	Rec ? (RT/S)	Explant	
HN-10	57/M	T2N0M0	Larynx	Rec ? (RT/S)	Explant	
UM-SCC-1	73/M	T2N0M0	FOM	Rec (S/RT)	Explant	Krause et al. (1981)

UM-SCC-3	73/F	T1N0M0	Nasal columella	LN metastasis (S)	Explant	
UM-SCC-4	47/F	T3N2aM0	Tonsillar pillar	BOT (CX)	Explant	
UM-SCC-5	59/M	T2N1M0	Larynx	Larynx (CX)	Explant	
UM-SCC-6	37/M	T2N0M0	BOT	Primary BOT (none)	Explant	
UM-SCC-7	64/M	NS	Alveolus	Primary (none)	Explant	
UM-SCC-8	76/F	T2N1M0	Alveolus	Rec (RT)	Explant	
UM-SCC-9	71/F	T2N0M0	Ant. Tongue	LN metastasis (RT)	Explant	
UM-SCC-10A	57/M	T3N1M0	Larynx	Primary (none)	Explant	Zenner et al. (1983)
H LaC 79	?/M?	NS	Larynx	Neck met (?)	Explant	Carey et al. (1983)
UM-SCC-10B	58/M	(see 10A)	Larynx	LN metastasis (S)	Explant	
UM-SCC-11A	65/M	T2N2aM0	Larynx	Primary (none)	Explant	
UM-SCC-11B	65/M	(see 11A)	Larynx	Primary (CX)	Explant	
UM-SCC-12	72/M	T2N1M0	Larynx	Primary (RT)	Explant	
UM-SCC-13	60/M	T3N0M0	Larynx	Rec esophagus (RT)	Explant	
UM-SCC-14A	58/F	T1N0M0	FOM	Rec FOM (S/RT)	Explant	
UM-SCC-14B	59/F	(see 14A)		Rec FOM (S/RT/S)	Explant	
UM-SCC-15[g]	70/M	T4N1M0	Hypopharynx	Primary (none)	Explant	
UM-SCC-16	61/F	T2N0M0	Larynx	Primary (none)	Explant	
UM-SCC-17A	47/F	T1N0M0	Larynx	Primary (RT)	Explant	
UM-SCC-17B	47/F	(see 17A)		Soft tissue-neck (RT)	Explant	
UM-SCC-18	68/M	T3N1M0	BOT	Rec pharynx (S/RT)	Explant	
UM-SCC-19	67/M	T2N1M0	BOT	Primary (none)	Explant	
UM-SCC-20	67/M	T2N1M0	Larynx	Neck met (RT/S)	Explant	
UM-SCC-21	65/M	NS	Skin	Ethmoid sinus (S/RT)	Explant	
UM-SCC-22A	58/F	T2N1M0	Hypopharynx	Primary (none)	Explant	
UM-SCC-22B	58/F	(see 22A)		Neck met (none)	Explant	
UM-SCC-23	36/F	T2N0M0	Larynx	Primary (none)	Explant	
TR126	58/F	NS	Tongue	Rec (S/RT)	Explant	Rupniak et al. (1985)
TR131	57/M	NS	Larynx	Rec (RT/S)	Explant	
TR138	38/M	NS	Larynx	Rec (RT/S)	Explant	
TR146	67/F	NS	Buccal mucosa	Rec (RT/S)	Explant	
SCC-35	?	T4N0	Pyriform sinus	Rec (RT)	Minced tumor fragments co-cultivated with irradiated 3T3 fibroblasts	Weichselbaum et al. (1986)
SCC-49	?	T2N0	Tonsil	Rec (RT)		
SCC-61	?	T4×N2bx	Tongue	Primary ? (?)		
SCC-71	?	T2N0	Soft palate	Primary (none)		

Table I

Continued

Cell lines	Patient Age/sex	TNM class[a]	Primary site[b]	Specimen site[c] (prior therapy[d])	Method of culture[e]	Reference
SCC-73	?	T4N0	RMT	Primary ? (?)		
SCC-76	?	T4N0	Maxillary antrum	Persistent (RT)		
SQ-9G	?	T3N1	Tonsil	Rec (RT)		
SQ-20B	?	T2N0	Larynx	Rec (RT)		
SQ-29	?	T3N1	RMT	Persistent (CX/S)		
SQ-31	?	T2N0	Pyriform sinus	Rec (RT)		
SQ-38	?	T3N0	RMT	Primary ? (?)		
SQ-39	?	T3N2a	RMT	Persistent (CX/S)		
SQ-43	?	T4N2	Supraglottis	Rec (RT)		
SQ-50	?	T4N2	Supraglottis	Rec (RT)		
TYS	81/F	NS	Oral mucosa	Primary (?)	Explant	Yanagawa et al. (1986)
183	54/M	T3N0M0	Tonsil	Primary ? (?)	Explant	Sacks et al. (1988)
1483	66/M	T2N1M0	RMT	Primary ? (?)	Explant	
MDA 886Ln	64/M	T3N3aM0	Larynx	LN met (?)	Explant	Sacks et al. (1989)
PCI-01	65/M	?	Larynx	Rec larynx (S)	Explant	Heo et al. (1989)
PCI-02	51/M	T3N0M0	FOM	Primary (none)	Explant	
PCI-03	50/M	T3N0M0	RMT	Primary (none)	Explant	
PCI-04A	51/M	T3N0M0	Larynx	Primary (none)	Explant	
PCI-04B	?	(see 04A)		LN Met (none)	Explant	
PCI-05	63/M	T3N1M0	Hypopharynx	Primary (none)	Explant	
PCI-06A	81/M	T3N3M0	Tonsil	Primary (none)	Explant	
PCI-06B	?	(see 06A)		Rec LN (S/RT)	Explant	
PCI-07	65/M	T4N2M0	BOT	Primary (none)	Explant	
PCI-08	54/M	T3N0M0	Pyriform sinus	LN met (none)	Explant	
PCI-09A	56/M	T4N3M0	BOT	Primary (none)	Explant	
PCI-09B	?	(see 09A)		LN met (none)	Explant	
PCI-10	61/M	T3N1M0	BOT	Primary (none)	Explant	
PCI-11	83/M	T4N0M0	Pyriform sinus	Primary (none)	Explant	
PCI-12	65/M	?	Hypopharynx	Rec LN (RT)	Explant	

PCI-13	50/M	T3N1M0	RMT	Primary (none)	Explant	
PCI-14	37/M	?	Larynx	Rec LN (S)	Explant	
PCI-15	69/M	T3N1M0?	Pyriform sinus	Primary (none)	Explant	
PCI-16	61/M	T2N1M0	Epiglottis	LN Met (none)	Explant	
PCI-17	61/F	T2N0M0	Epiglottis	Primary (none)	Explant	
PCI-18	46/M	?	Larynx	Primary (RT)	Explant	
HC-2,3,4,7,9	64/F	T3N0M0	Maxillary sinus	Rec. (RT/S)	Nude mouse	Komiyama et al. (1989)
UM-SCC-35	51/M	T4N1M0	Tonsillar fossa	Primary (none)	Explant	Grenman et al. (1991)
UM-SCC-38	60/M	T2N2aM0	Tonsillar pillar	Primary (none)	Explant	
UM-SCC-69	35/M	T4N0M0	Hard palate	Primary (CX)	Explant	
UT-SCC-1A	75/F	T3N0M0	Gingiva	Primary (RT)	Explant	
UT-SCC-2	60/M	T4N1M0	FOM	Primary (none)	Explant	
Non-squamous tumors						
TT	77/F	NS	Thyroid (Med Ca)	Primary tumor (?)	Explant ?	Leong (1981)
UT-MUC-1	74/F	NS	Palate (Muco-Ep)	Rec-Met (RT/S)	Explant	Grenman et al. (1992)

[a] Clinical staging classification based on Tumor size, lymph Node, involvement, and presence of distant Metastasis; NS, Not staged or staging not given.

[b] Submand., submandibular; FOM, floor of mouth; BOT, base of tongue; Ant. tongue, anterior portion of tongue; Med Ca, medullary carcinoma of the thyroid gland; Muco Ep, mucoepidermoid carcinoma of the salivary gland; RMT, retromolar trigone; Supraglottis, supraglottic larynx.

[c] LN, Lymph node; pleural fluid, malignant effusion from tumor metastasis; neck met, metastasis to soft tissues of the neck; rec, recurrence; persistent, tumor did not completely respond to the original therapy.

[d] None, the patient received no therapy prior to the in vitro culture; S, surgery; RT, radiation therapy; CX, chemotherapy; MTX, methotrexate; RND, radical neck dissection, ND, neck dissection.

[e] Rat xenograft: tumors were transferred from the patient to cortisone treated rats; then at a later time the tumors were placed in culture. Explant: in vitro cultures were started from small pieces of tumor. Nude mouse: tumors were transferred from the patient to athymic/nude mice, then from the nude mouse tumor to in vitro culture.

[f] Several laboratories have shown that these cell lines have HeLa characteristics including marker chromosomes and the type A glucose-6-phosphate dehydrogenase enzyme that is rare in Caucasians but is expressed in the HeLa cell line that came from an African-American woman. A cross-contamination with HeLa is thought to have occurred during the development or propagation of these cell lines. A ? indicates uncertainty or unknown data.

[g] The UM-SCC-15 cell culture undergoes spontaneous lysis at passage level 6–8 and thus does not represent a true established cell line.

85

studies of tumor formation in nude mice (Baker, 1985), of membrane antigens (Carey et al., 1983; Kimmel and Carey, 1986; Kimmel et al., 1986, Vlock et al., 1989), of synthesis of extracellular matrix proteins (Frenette et al., 1988), of radiation resistance (Grenman et al., 1991), of consistent chromosomal abnormalities (Carey et al., 1989; Bradford et al., 1991; Van Dyke et al., 1993), and of the role of the $\alpha6\beta4$ integrin in the biology of squamous cell carcinoma (Van Waes et al., 1991; Van Waes and Carey, 1992). In this chapter, we use examples of these and a few other widely used head and neck cancer cell lines to illustrate the morphological characteristics of this tumor type.

B. Cell Line Establishment and Maintenance Methodology

Most successful long-term cultures from head and neck squamous cell carcinomas have been established from tumor explants (see Table I). In these cases, the cultures are initiated from minced tumor fragments placed directly in tissue culture flasks. The other major method for initiating head and neck squamous carcinoma cell lines was described by Rheinwald and Beckett (1981) and employs dissaggregated tumor cells plated on irradiated 3T3 mouse-fibroblast feeder-cell monolayers. The relative success rate of each method seems to be equivalent, based on the largest reported series (Easty et al., 1981; Krause et al., 1981; Rheinwald and Beckett, 1981). However, whether one method works preferentially for a different class of tumor cells has never been determined. In other words, is it possible to grow tumor cells on a feeder layer that otherwise would not have been cultured successfully using the explant method? We have not compared the two approaches because we were concerned that the tumor cells might take up foreign DNA or viral RNA from the feeder cells that might alter subsequent studies of the tumor line. An alternative to in vitro cell culture for developing tumor lines has been to culture head and neck tumors in immunosuppressed animals (Fogh et al., 1980). For the most part, we have avoided this approach to establishing tumors for the same reasons that we do not use feeder cells. However, in vivo cultivation in athymic mice of human tumors from cell cultures is very useful to confirm that the cells cultured in vitro faithfully recreate the histological features of the patient's original tumor (Baker, 1985; Carey, 1985).

1. Explant Cultures

a. Tissue processing. Tumor specimens are transported to the laboratory as soon after surgical removal as possible and are immersed immediately in PSA, a triple antibiotic solution consisting of cold buffered balanced salt solution (BSS) containing penicillin (100 U/ml), streptomycin (100 μg/ml), and amphotericin B (10 μg/ml), and placed at 4°C until the tissue can be processed. A glucose-containing salt solution such as Earle's BSS (and antibiotics purchased from Gibco, Grand Island, New York or Sigma, St. Louis, Missouri) should be used to prepare the PSA solution. The rationale for this type of solution is to minimize damage to the tumor cells and, simultaneously, maximally suppress the growth of microorganisms. Healthy cultured tumor cells suspended in saline alone exhibit loss of viability, whether at room temperature or at 4°C; therefore, normal saline is not recommended to transport, store, or wash specimens. Similarly, a rich growth medium such as minimal essential medium (MEM) with fetal bovine serum (FBS) should be avoided for washing and storing the tumors because, if microorganisms are present, suppressing their growth in a nutrient-rich environment is very difficult.

Usually the specimens are processed within 1 hr of arrival in the laboratory, but if the specimens are placed immediately in cold PSA solution and stored in the refrigerator, successful cultures can be obtained even after overnight storage. At the time of processing, the samples are washed vigorously in three changes of PSA to remove surface contaminants. Then the tumor pieces are trimmed of fat and connective tissue and the tumor is minced carefully into 1- to 2-mm^3 cubes using two scalpel blades drawn in opposite directions through the tissue. This method of cutting is preferable to mincing with scissors since it minimizes crushing injury, so undamaged cells in the tissue are at a minimum migration distance from the freshly cut surface. The cubes of tissue are transferred into tissue culture flasks, allowed to adhere, and covered with a drop of Eagle's or Dulbecco's modified Eagle's MEM supplemented with 1% nonessential amino acids, 2 mM L-glutamine, and 10–20% FBS. The cultures should be left undisturbed for several days. Then the cultures can be examined under the microscope and a small amount of fresh medium may be added. Tumor cell outgrowth usually appears within 4–7 days; medium changes consisting of 1–2 ml are performed at 2- to 5-day intervals, depending on the growth responses of individual cultures.

  b. Removing fibroblasts.   Fibroblast outgrowth usually begins to appear during the second week of culture, and should be monitored carefully so promising epithelial cell colonies are not undermined. To remove isolated fibroblast colonies, we occasionally employ a sterile cell scraper, but in most cases selective trypsinization can be used to remove fibroblasts without disturbing the epithelial cells. Avoid being overzealous in removing fibroblasts! Partial removal of fibroblasts on several different days is preferable to losing tumor cells from very small islands or to dislodging the tissue fragments from which the tumor cells are proliferating and migrating. We use 0.1% recrystallized porcine trypsin with an activity of 13,000–20,000 BAEE units per ml (Sigma) and 0.02% ethylenediamine tetraacetic acid (EDTA) prepared in Puck's saline A solution. To remove fibroblasts, the flask is rinsed with calcium- and magnesium-free phosphate buffered saline (PBS); then the trypsin–EDTA solution is added and the areas of interest are monitored under the microscope. When the fibroblasts are rolling up but before the epithelial cells begin to detach, the trypsin action is stopped by the addition of fresh serum-containing medium, taking care to avoid dislodging the tumor tissue frag-

ments. The trypsin–medium mixture is transferred from the flask into a centrifuge tube and the flask is washed gently with more medium. This rinse medium
is combined with the solution already in the centrifuge tube and the flask is
refed with frsh medium. The cells in the tube are spun down, resuspended in
fresh medium, and placed in a new flask. These fibroblast subcultures provide
a source of "normal" cells from the patient but also may allow the early isolation
of some tumor cells capable of subculture; therefore, the cells in these "fibroblast" flasks should always be scrutinized closely for small tumor islands, as
illustrated in Fig. 1. Figure 1a shows tumor cells in a primary culture of
UM-SCC-29. This cell line was established in 1983 from a carcinoma of the
alveolar ridge in a 67-year-old male who had used chewing tobacco and
moonshine whiskey for many years. When the flask containing these cells was
treated by selective trypsinization to remove the fibroblasts in other parts of the
flask, some tumor cells also were removed; these are shown as a small island
surrounded by fibroblasts in the passage 1 flask (Fig. 1b). Subsequent selective trypsin treatment resulted in a pure culture of tumor cells. The early culture
shown in Fig. 1c is from Passage 2 of UM-SCC-6 derived from a squamous
carcinoma of the base of tongue in a 32-year-old male. This passage 1 flask
contains a mix of fibroblasts and small islands of tumor cells. The photograph
in Fig. 1d shows the same culture after selective trypsinization.

    c. Subculture. Tumor cell cultures are usually ready for subculture when
large patches of cells filling at least half the original flask are present and when
mitotic cells are numerous throughout the tumor cell islands during the first or
second day after feeding. Usually, fibroblasts are removed first; then a subculture is attempted. An example of the primary culture of UM-SCC-18 from a
squamous carcinoma of the pyriform sinus in a 72-year-old male is shown in
Fig. 1e. Several mitotic cells are present in this field, and the culture was ready
for passage. Note large spreading cells with low nucleus : cytoplasm ratios
and the evidence of terminally differentiating cells being sloughed off into the
supernatant. Nevertheless, this culture persisted *in vitro* and became a permanent cell line. The appearance of UM-SCC-18 at passage 12, as shown in Fig.
1f, is changed very little from that in the primary culture.

    In early cultures, attempting to remove all cells for subculture is not advisa-
ble. Recovering only those cells that come off easily from a primary tumor
culture is better than harvesting the entire culture because the plating efficiency of the removed cells may be poor. If so, the entire culture could be lost.

Fig. 1. Photomicrographs of early cultures from head and neck tumors. (a) Primary culture of
UM-SCC-29. High power (320×) view shows a colony of squamous carcinoma cells growing out
from a tumor explant. (b) Passage 1 flask from the same UM-SCC-29 culture containing mostly
fibroblasts that were removed from the primary culture by selective trypsinization. Note the small
island of epithelial cells (arrows). (c) Passage 2 subculture from UM-SCC-6. Note the islands of
epithelial cells completely surrounded by fibroblasts.

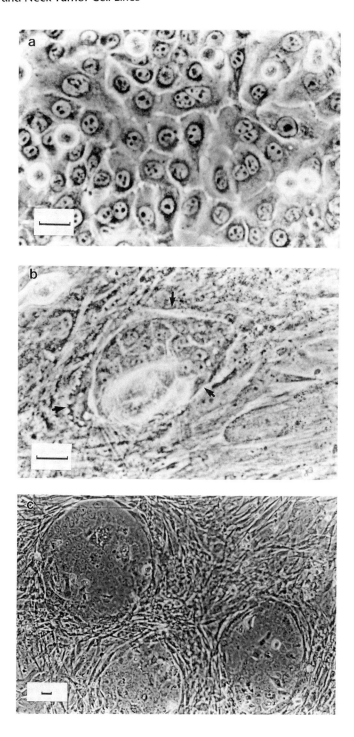

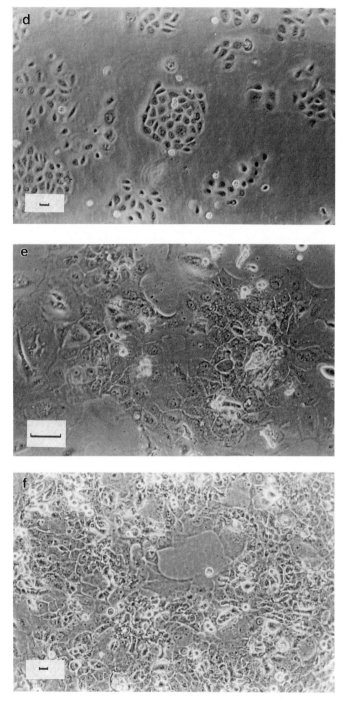

Fig. 1. *Continued* (d) The flask shown in c, 1 day after selective trypsinization to remove fibroblasts. (e) Primary culture of UM-SCC-18 prior to subculture. (f) UM-SCC-18 at passage 12. Note the similarity in morphology to the primary culture. Bars, 10 μm.

Further, the long exposure to trypsin from attempting to remove cells that are attached firmly to each other and to the flask may result in damage to the cells that come off early in the procedure. After partial trypsinization, the cells remaining in the flask, as well as those removed, should be washed in serum-containing medium and refed with fresh medium. Partial harvests can be repeated several times; thus, the chances of losing the whole culture are diminished.

Careful records of cell lineages should be maintained on all cultures, usually by designating each primary culture flask with a letter A, B, and so on. Then, at subculture, the new flask is designated passage 1 (P-1), with an A, B, and C (and so on) depending on the flask from which the cells were taken. As soon as sufficient cells are available at any early passage, some cells should be frozen in liquid nitrogen for later reference or to recover the culture in case an untoward event such as microbial contamination or incubator failure destroys the cells in active culture. Representative passages should be frozen at periodic intervals through all early pssages and at 5- or 10-passage intervals thereafter. Cells can be frozen in either 9 parts DMEM containing 20% serum and 1 part dimethylsulfoxide (DMSO) or in 9 parts serum and 1 part DMSO. DMSO is a membrane-active agent that prevents ice crystal formation and damage to the cell during freezing and thawing. DMSO can be toxic to cells at warm temperatures. Having the freezing medium at 4°C at the time the cells are added for freezing is important. The cryovials should be placed immediately in the biological freezer adjusted for a freezing rate of $-1°C$ per min. At the time of thawing, similar care should be taken to insure that, after quickly thawing the cells by holding the vial under running tap water at 33–37°C, the DMSO is diluted immediately with a 10 : 1 volume of cold culture medium and the cells are removed immediately from the DMSO by centrifugation and decantation of the DMSO-containing medium.

2. Feeder Layer Cultures

The feeder layer method of cultivating squamous carcinomas was adapted by Rheinwald and Beckett (1981) from a protocol originally developed by Rheinwald and Green (1975) for studying normal skin keratinocytes *in vitro*. For this method, the tumor tissue is treated as described earlier, through the step at which it is minced into 2-mm fragments. After this stage, the tissue fragments are minced further with scissors (*sic*) into pieces less than 1 mm in size. These fragments then are placed either directly into culture flasks containing feeder cells for explant culture or into in a stir flask containing a mixture of 0.2% trypsin and 0.2% collagenase for disaggregation at 37°C. The cells in the enzyme mix are withdrawn at 30-min intervals, washed and suspended in Dulbecco's modified Eagle's medium supplemented with 20% FBS and 0.4 μg/ml hydrocortisone, and plated on mitomycin C-treated or lethally irradiated mouse 3T3 fibroblast feeder cells. For passage, the feeder cells and human fibroblasts are removed by incubation in 0.02% EDTA for 30 sec and vigorous pipetting of the fluid against the cells. The epithelial cells remaining can be removed by incubation at 37°C in 0.05% trypsin and 0.01% EDTA for 15–30 min. Tumor cells recovered in this manner are replated on fresh feeder layers.

Of the six SCC lines cultivated by this method by Rheinwald and Beckett (1981), one, SCC-13, was reported to remain feeder layer-dependent whereas others grew both on the feeder cells and on conventional tissue culture flasks or dishes.

3. Extracellular Matrices

We performed a number of early experiments using fibronectin, laminin, or rat tail collagen matrices to coat flasks prior to attempting culture of head and neck tumors. In each case, we used uncoated culture flasks as well. Although the number of cases we tested were small and the matrix materials were crude, we had no examples in which the tumor grew in the matrix-coated flask but not in the untreated flask. Because of the lack of promise for increased success and the additional work and expense required to coat the flasks, we abandoned these attempts. In recent years, however, an explosion of knowledge about extracellular matrix components has occurred; now purified matrix proteins, even of human origin, are commercially available. In addition, flasks coated with complex matrices for tumor cultivation and preformulated complex matrices consisting of collagen blocks impregnated with laminin or with fibroblasts are being sold for cultivation of human skin in a more physiological setting than that afforded by conventional culture methods. These materials may allow increased success in the long-term cultivation of head and neck squamous carcinomas. However, to date I am unaware of any studies that have compared these new substrates with conventional culture methods for cultivation of head and neck cancer.

4. Growth Factors

Growth factors, notably epidermal growth factor (EGF) but also other hormones such as insulin, transferrin, hydrocortisone, and bovine pituitary extracts, have been shown to be important components of media for cultivating normal keratinocytes (Rheinwald and Green, 1977; Boyse and Ham, 1985). This dependence suggested the possibility that such media also might be beneficial for cultivation of squamous cell carcinomas. Controlled tests comparing this type of medium with conventional media such as MEM, DMEM, Ham's F12, and so on for the cultivation of primary head and neck tumors have not been carried out. We compared keratinocyte growth medium with conventional medium for primary tumor cultivation in a few cases. To date, we have not succeeded in cultivating tumor cells in specialized medium when conventional medium failed to support growth. We also have not determined whether EGF-containing medium is contraindicated for cultivation of squamous cancers. Initially, we were concerned that, for tumors with overexpression of the EGF receptor, the presence of EGF in the medium in high concentration actually might suppress growth of the population of interest. Growth suppression by EGF of the A431 vulvar squamous cell carcinoma line, which expresses more than 10^6 EGF receptors per cell, has been demonstrated (Barnes, 1982). We observed that EGF inhibited growth of previously established UM-SCC cell lines with both high and low EGF receptor expression (H. T. Hoffman, T. E. Carey, L. Kidd, and M. Subnani, unpublished data).

5. Success Rate

The reported success rate from Rheinwald and Beckett (1981) and from Easty *et al.* (1981) is between 27 and 30% of attempts. We had a similar rate of success in our early series and a consistent rate of about 35% for all attempts since then.

The reasons for culture failure are not always apparent, but we ascribe our losses of promising cultures to the following categories. Although head and neck tumors are of epithelial surface origin, a surprisingly small percentage (~2–5%) of primary cultures in our hands is lost because of bacterial or fungal contamination. This low rate of contamination can be attributed to careful selection by the surgeon of regions of healthy tumor tissue for cultivation and careful washing of each specimen with triple antibiotic solution (PSA). Retrieving sterile viable cultures from tumors that have harbored infections *in vivo* is very difficult. These cultures are almost always lost to overwhelming outgrowth of organisms after 1–2 days.

A few promising cultures (~5%) are lost to fibroblast outgrowth, but in most cases this loss can be controlled by selective trypsinization, as illustrated in Fig. 1. Fibroblasts are almost always less strongly attached to the substrate than squamous cells, so they are detached more rapidly in trypsin and EDTA.

More frustrating are the numerous tissue samples (20–25%) that grow out only differentiating keratinocytes. At the early stages of culture, these cells are not easily distinguished from tumor cells. Examples of three cultures that contained normal epithelial cells are illustrated in Fig. 2a,b,c. Normal epithelial cells initially grow rapidly with many mitotic figures and rapidly expanding cell islands (Fig. 2a). Then, after 2–6 wk, the evidence of terminal differentiation appears, manifest in some cases by the increasing appearance of large flattened cells with shrunken pycnotic nuclei and *in vitro* desquamation (Fig. 2b) or in other cases by accumulation of cytoplasmic fibrous bundles in cells with a morphology reminiscent of the epithelial "prickle cell" layer of stratified squamous epithelium (Fig. 2c). Some tumor lines have a morphological appearance very similar to the normal epithelial cells shown in Fig. 2a. UM-SCC-13, shown in Fig. 2d, is a slowly growing tumor cell line from a patient with laryngeal cancer that has an appearance similar to that of normal epithelial cells.

About 15% of the tumor samples that exhibit initial cell growth stop growing or die, perhaps because some essential nutrient or growth factor is missing because of some lytic process. UM-SCC-15 is a rare example of a culture that

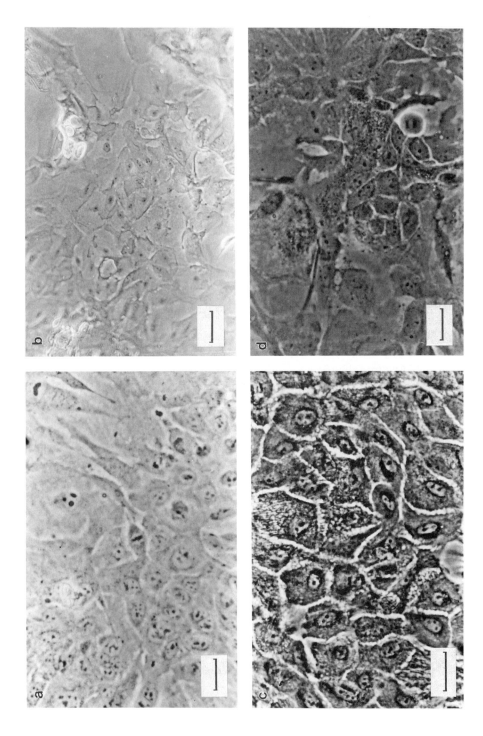

undergoes this kind of process. This culture was given a premature designation as a cell line when it was still in early passages. However, these cells exhibit vigorous *in vitro* growth until passage 6–8, when large vaculoated cells begin to appear as an increasing proportion of the population. The vacuolated cells get larger, accumulate more vacuoles, and eventually disintegrate; this process occurred each time we removed a low passage from the bank and tried to carry it past passage 8. We speculate that some such tumors may harbor a latent lytic virus that becomes activated *in vitro*, but to date we have no firm evidence to support this speculation. Examples such as this one support the traditionally held concept that a culture should not be considered a cell line until it has been passed 20 times. All our other cell lines have achieved this criterion.

Promising cultures also sometimes are lost as a result of what appear to be lytic events associated with white blood cells, either lymphocytes or other leukocytes that are present in some cultures in great abundance. These cases are relatively rare but are notable in the way the small lymphocytes cling to the surface of the attached epithelial and stromal cells.

The remaining 15% of specimens that do not give rise to cell cultures simply fail to grow any epithelial cells. These tissue samples usually appear, under the microscope, to consist mostly of scirrhous connective tissue. Such samples probably contain few or no tumor cells, but may be selected by the surgeon because the hard or indurated characteristics of such tissue in the operative field may mimic tumor tissue.

Fig. 2. Inverted phase-contrast photomicrographs of normal and malignant squamous cell cultures. (a) Early epithelial cell outgrowth from normal mucosa containing rapidly dividing and migrating normal epithelial cells. Note the high cytoplasmic to nuclear ratio and the normal mitotic figure in the center of the field. (b) Early culture from tumor tissue of a young patient with oral cancer after 6 wk in culture. This flask contained cells that looked like those in a and exhibited very rapid growth during the first several weeks in culture. Then the cells stopped dividing and became very large; the nuclei became pycnotic and the cytoplasm became clear, as shown. At this point cells began to detach, much like normal oral desquamating mucosal epithelial cells. (c) Epithelial cells from normal mucosa of the donor of the UM-SCC-18 cell line shown in Fig. 1. These cells grew rapidly during early culture but, after several weeks, began to slow in growth and show evidence of terminal differentiation similar to that seen in keratinizing epithelium with accumulation of cytoplasmic fibrous bundles and prominent surface microvilli connecting adjacent cells. (Fig. 2c is reproduced from Carey *et al.*, 1983, with permission of Mosby Yearbook, St. Louis, Missouri). (d) Passage 15 of UM-SCC-13 illustrates the similarity of some squamous carcinoma cell cultures to early epithelial cell outgrowth from normal mucosa. Bars, 10 μm.

C. Cell Line Nomenclature

As alluded to earlier, a culture can be considered established when it has passed 20 passage generations. This cutoff is arbitrary but effective. We believe that human tumor cell lines should be given a logical name that helps identify the origin of the line but protects the identity of the donor. Thus, cell lines established at Roswell Park Memorial Institute were given the RPMI prefix, whereas those from Sloan–Kettering Institute have the prefix SK (as in the SK-MEL series of melanoma lines established there). We give our University of Michigan cell lines the prefix UM. The middle part of the cell line designation refers to the tumor type, for example, MEL for melanoma and SCC for squamous cell carcinoma. Although individual cell line number designations are not easy to remember, we use a sequential numbering system assigned once a culture becomes an established cell line. Thus UM-SCC-18 is the eighteenth cell line we established and UM-SCC-82 is the most recent cell line in our SCC series. When more than one cell line is established from the same individual, but from a different tumor specimen such as a recurrent or metastatic tumor, the lines are given a letter designation. For example, UM-SCC-10A was from a primary carcinoma of the larynx; UM-SCC-10B was established from a recurrent tumor in the same patient 1 yr later (see Table I).

III. Morphology

A. Squamous Carcinomas: UM-SCC Series

Cultured SCCs primarily share an epithelioid morphology that has relatively few subtypes. Nevertheless, many SCC cell lines have very stable and distinctive microscopic features. At the individual cell level, we find two types of cellular appearance: cuboidal and spreading. Either type may grow as a monolayer or as stratifying colonies. Cuboidal cells are characterized by high nucleus : cytoplasm ratios. The large nuclei usually contain several prominent nucleoli and the scant cytoplasm is often highly granular. The spreading cell type has a lower nucleus : cytoplasm ratio, smaller less prominent nucleoli, relatively agranular cytoplasm, and an elongated spread-out appearance. We also distinguish among our head and neck SCC cell lines by two types of *in vitro* growth pattern that are morphologically distinct: (1) cell lines that grow as monolayers and (2) those that stratify. Each of these subgroups has two divisions. In the monolayer group, most SCC lines grow in islands but a few show loss of cell-to-cell adhesion and migrate as single cells. In the stratifying group, some cell lines exhibit multilayer stratification patterns involving most or all cell islands, and others have patches of dome formation.

1. Monolayer Nonstratifying Types

Most SCC lines are colony formers to a greater or lesser extent. Most have a cuboidal epithelioid morphology, are nonstratifying, and tend to grow in clusters in which the individual cells maintain close cell-to-cell contacts.

a. *Island-forming monolayer cultures.*   Examples of this predominant SCC cell type are shown in Figs. 3, 4, 5. UM-SCC-5 (Fig. 3a,b), from a supraglottic carcinoma in a 71-year-old male, consists mostly of spreading cells with few or no cells migrating independently from the large islands. This appearance is highly stable *in vitro* since it is present at passage 4 (Fig. 3a) and at passage 77 (Fig. 3b). As shown in Fig. 3c–f, UM-SCC-22A (Fig. 3c,d) and 22B (Fig. 3e,f), which were derived from the primary tumor and lymph node metastasis in a 69-year-old female patient with squamous cancer of the hypopharynx, have very distinctive characteristic features, both in the morphology of the individual cells, which are cuboidal, and in the morphology of the colonies the cells form. Note that, although most of the cells are in colonies, some are breaking off to form small new colonies.

Figure 4 shows additional examples of monolayer colonies. UM-SCC-25 (Fig. 4a,b), from a neck metastasis in a 52-year old male with recurrence of a laryngeal cancer after radiotherapy, consists mostly of large cuboidal cells. UM-SCC-38 (Fig. 4c,d), from a primary squamous cancer of the base of tongue in a 58-year-old male, is pleomorphic with cuboidal and spreading cell morphologies and a tendency for some cells to migrate independently of the colonies. UM-SCC-30 (Fig. 4e), from a persistent tumor in a 53-year-old female with squamous cancer of the larynx that initially was treated by chemotherapy, is predominantly of the spreading cell type, whereas UM-SCC-55 (Fig. 4f), established from a metastatic lesion in the neck of a 68-year-old male with recurrent cancer of the alveolar ridge, has both cuboidal and spreading morphologies.

b. *Migrating individual cells.* (This classification includes rare attachment-defective types). The tumor cells in some SCC lines have a tendency to migrate independently of one another in culture, indicating a decreased degree of cell-to-cell adhesion compared with other squamous cancers or normal keratinocytes grown under the same conditions. In few SCC cell lines the cells grow and move independent of colonies. Two examples of cell lines with this phenotype, UM-SCC-11B and UM-SCC-20, are shown in Fig.5 and 6. This phenotype may be associated with tumor progression, since we observe a greater degree of independent cell migration in some secondary tumor lines established from patients from whom an earlier tumor line had been obtained already. In Fig. 5 we compare UM-SCC-11A, taken prior to therapy from a primary tumor of the hypopharynx in a 76-year-old male, which grows primarily in islands (Fig. 5a,b), with the loose histological appearance of the persistent tumor in the same patient after failed induction chemotherapy (Fig.

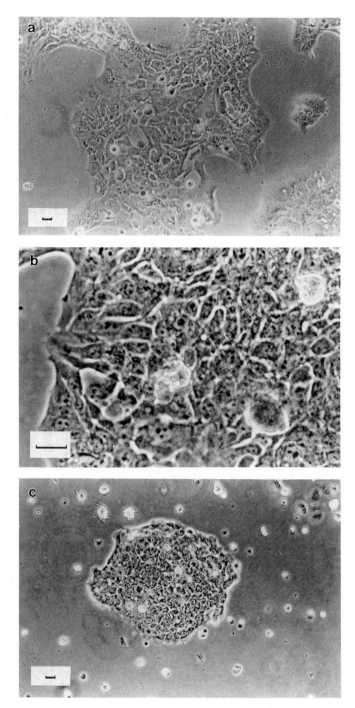

Fig. 3. Inverted phase-contrast photomicrographs of monolayer-forming squamous carcinoma cell lines at high and low magnification. (a) UM-SCC-5 at passage 4 (100×). (b) UM-SCC-5 at passage 77 (320×). Note the similarity to the early passage culture in cell morphology and growth pattern.

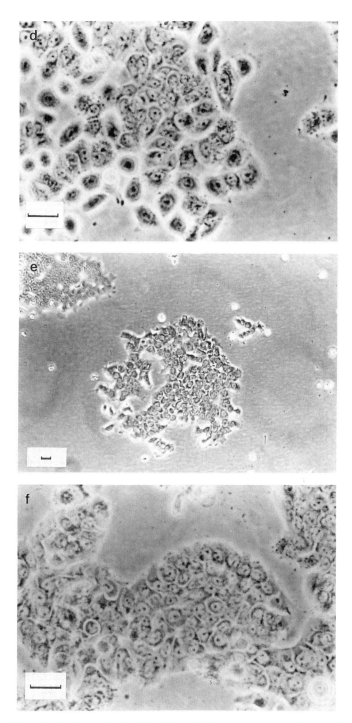

Fig. 3. *Continued* (c–f) Cuboidal epithelial appearance of two cell lines from a patient with squamous cancer of the hypopharynx. (c) UM-SCC-22A at passage 1 (100×). (d) UM-SCC-22A at passage 18 (320×). (e) UM-SCC-22B at passage 1 (100×). (f) UM-SCC-22B at passage 27 (320×). Bars, 10 μm.

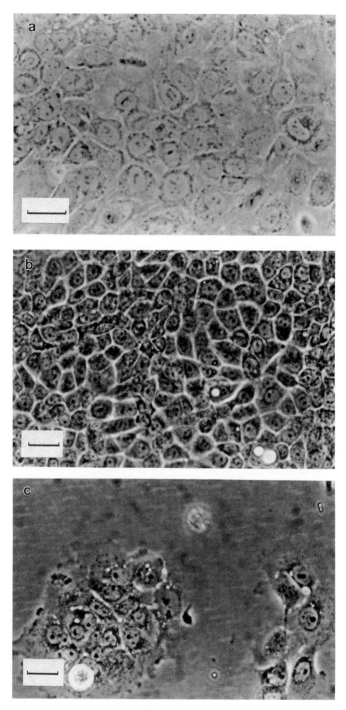

Fig. 4. Inverted phase-contrast photomicrographs of monolayer-forming squamous carcinoma cell lines. (a) UM-SCC-25 at passage 16 (400×). (b) UM-SCC-25 at passage 69 (320×). (c) UM-SCC-38 at passage 11, shown at low density to illustrate island formation (320×).

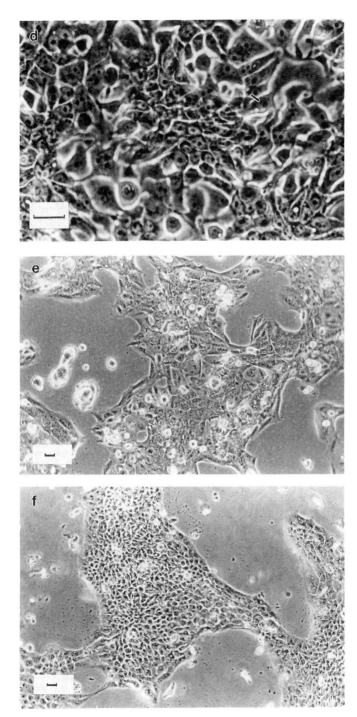

Fig. 4. *Continued* (d) UM-SCC-38 at passage 19, shown at higher density (320×). Note the pleomorphic cell types in this culture. (e) UM-SCC-30 at passage 58 (100×). (f) UM-SCC-55 at passage 26 (100×). Bars, 10 μm.

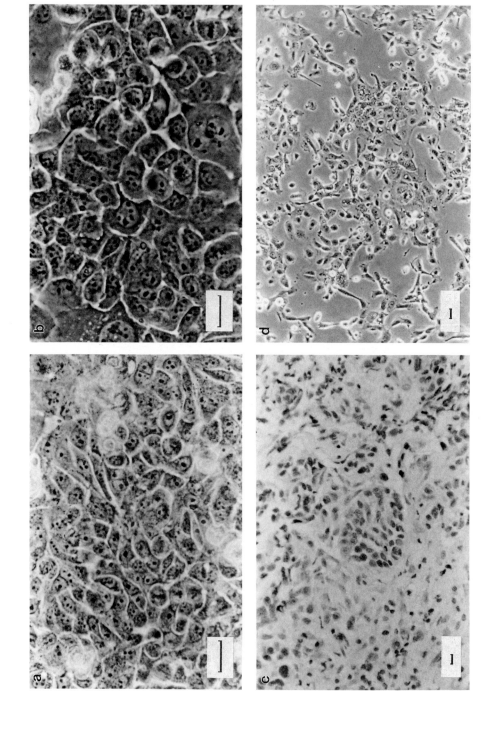

5c) and the independently migrating cells in cultures of UM-SCC-11B, derived from the persistent tumor (Fig. 5d).

UM-SCC-20, derived from a lymph node metastasis of laryngeal cancer in a 62-year-old male previously treated by radiation and subsequent laryngec-tomy, is the only case we have observed of a human SCC cell line that had a component that grew in suspension independent of adhesion to a substrate. Photographs of this tumor culture, from which we eventually established the UM-SCC-20 cell line, are shown in Fig. 6. The histology of the tumor specimen (Fig. 6a) shows that it was poorly differentiated with loose clusters of cells. *In vitro*, the tumor cells initially grew as a mixed culture of poorly adherent and floating cells (Fig. 6b, passage 2). The suspension cells were separated at passage 5 (Fig. 6c), but these did not grow. Eventually only attached cells could be propagated in culture (Fig. 6d, passage 55).

2. Stratifying Colonies

a. Domes or mounds and stellate domes.   Several of the cell lines in our bank are immediately recognizable under the microscope or even by visual inspection because these cells have a very strong propensity to stratify and form mounds or domes. UM-SCC-9, from a second recurrence in a 75-year-old female with a primary carcinoma of the tongue, has limited propensity for mound or dome formation. The patient's tumor (Fig. 7a) contained areas of no differentiation and other areas with dyskeratosis and involution of the nuclei (see arrows). In the cell line, the areas of dome formation (Fig. 7b, arrows) might represent areas in which similar aberrant differentiation is occurring. The most distinct example of dome formation among our SCC lines consists of the stellate clusters and domes formed by UM-SCC-16 (Fig.7c–f). In primary culture this cell line, which was derived from a primary carcinoma of the larynx in a 72-year-old female formed a single large stratifying mound with very little lateral spread (Fig. 7c, white arrowheads). With time in culture and repeated exposure to limited trypsinization, some cells were separated that reattached in the same flask and formed small islands (white arrows) that also stratified. This cell line consists of multipolar cells that form a monolayer in which small stellate clusters or rosettes form at regular intervals. Examples of these rosettes are apparent in Fig. 7d–f. From the stellate clusters, stratified domes arise.

Fig. 5. Photomicrographs of two cell lines and a tumor section from the same patient. (a) UM-SCC-11A cultured from the diagnostic biopsy of primary larynx cancer in this patient, shown in primary culture (320×). (b) UM-SCC-11A shown at passage 49 (320×). These large tumor cells in UM-SCC-11A were tetraploid by flow cytometry and karyotype. (c) Hematoxylin- and eosin-stained tissue section from the persistent cancer in the same patient, removed 1.5 months after the biopsy (100×). (d) UM-SCC-11B at passage 20, showing independently migrating tumor cells (100×). The UM-SCC-11B cells were pseudo diploid with a karyotype that reflected a 2N version of the UM-SCC-11A tumor line. Bars, 10 μm.

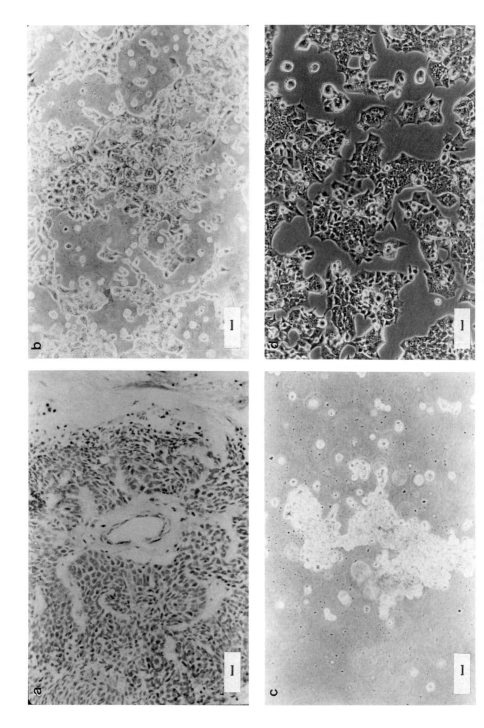

This phenotype has remained constant over more than 50 *in vitro* passages, making this cell line immediately recognizable under the microscope.

Other cell lines with a propensity for dome formation include UM-SCC-31 (Fig. 8a,b), from a recurrent oral carcinoma in a 58-year-old male; UM-SCC-32 (Fig. 8c,d), from a primary carcinoma of the oral pharynx in a 58-year-old male; and UM-SCC-37 (Fig. 8e,f), from a recurrent cancer of the hypopharynx in a 58-year-old male treated previously with chemotherapy and radiation.

b. Dense Islands. Some SCC cell lines grow in islands that reach a certain size and then do not expand in area although the cells within the island proliferate. These cell lines become densely packed within the islands without forming confluent monolayers and give the impression that the cells are not proliferating. Notable examples are the UM-SCC-17A and 17B cell lines (Fig. 9a,b), from a 47-year-old female with persistent cancer of the larynx (UM-SCC-17A) that invaded into the soft tissue of the neck (UM-SCC-17B) after radiation therapy to the primary tumor. As previously reported (Carey *et al.*, 1989), these cells have retained this phenotype over more than 100 *in vitro* passages. UM-SCC-39 (Fig. 9c,d), derived from a laryngeal pyriform sinus cancer in a 50-year-old male is another example of a cell line with this densely packed phenotype.

B. SCC Cell Lines from the American Type Culture Collection Repository

For comparative purposes, four head and neck SCC lines established in other laboratories are shown in Fig. 10, including RPMI 2650, a cell line reported to have a near normal karyotype, established from a pleural effusion of a patient with advanced squamous cancer of the nasal septum (Moore and Sandberg, 1964; Fig. 10a); FaDu, established from a biopsy of a hypopharyngeal carcinoma in a 56-year-old male (Rangan, 1972; Fig. 10b); Detroit 562, established from a pleural effusion in a female patient with an advanced

---

Fig. 6. Photomicrographs of the UM-SCC-20 cell line and the tumor specimen from which it was established. (a) Hematoxylin- and eosin-stained tissue section from the tumor used to establish UM-SCC-20 (100×). (b) Early culture of UM-SCC-20, showing both attached and floating cells (100×). (c) UM-SCC-20 at passage 5, showing cells in free suspension culture (100×). (d) UM-SCC-20 at passage 55, showing only attached cells (100×). Bars, 10 μm.

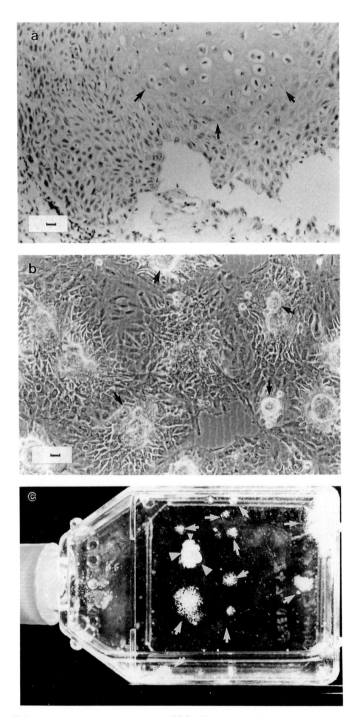

Fig. 7. Stratifying squamous carcinoma cultures (SCC). Photographs of tumor tissue, cell cultures, and a tissue culture flask containing stratifying SCC cells. (a) Hematoxylin- and eosin-stained section from the tumor used to establish UM-SCC-9 (100×). (Arrows show areas of dyskeratosis). (b) UM-SCC-9 at passage 52, showing areas of stratification (100×). (c) A T-25 tissue culture flask containing the primary culture of UM-SCC-16. The large white areas are stratifying tumor colonies. For the first 6 months in culture, only the large colony in the upper left (white arrowheads), which grew from a small tumor fragment, was present and nearly all the growth was vertical. After many attempts at partial trypsinization, a few cells that had the capacity to reattach and grow broke off this colony. These cells grew well over the next several weeks; repeated partial tryp-

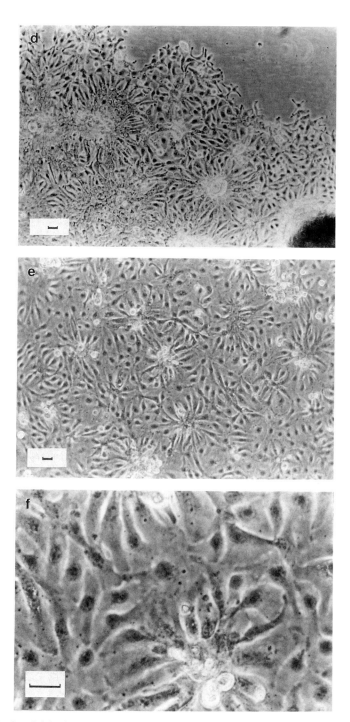

Fig. 7. *Continued* sinization created additional colonies (white arrows). Shortly after this photograph was taken, a successful passage into a new flask was also possible. (d) UM-SCC-16 at passage 1 (100×). These cells have the same appearance as the cells that made up the large and small colonies shown in c. Note the areas of stellate clusters and the stratifying areas around these clusters. (e) UM-SCC-16 at passage 59, showing multiple stellate clusters and illustrating the stability of this phenotype over long-term *in vitro* culture (100×). (f) Higher power magnification of the culture shown in e (320×). After an additional 2–3 days, stratifying domes were present over each stellate cluster. Bars, 10 µm.

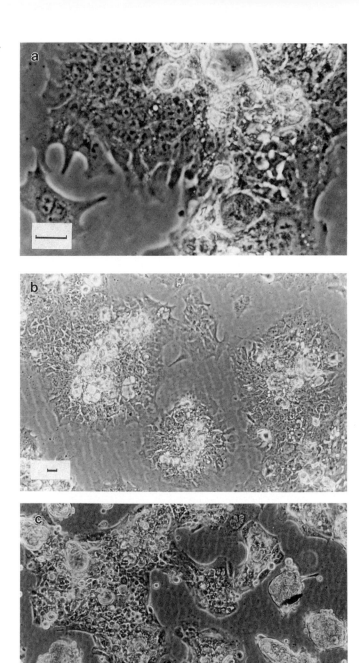

Fig. 8. Inverted phase-contrast photomicrographs of squamous carcinoma cultures with the stratifying phenotype. (a,b) UM-SCC-31 shown as passage 48, illustrating stratifying areas in this nonconfluent culture at two magnifications (a, 320×; b, 100×). (c,d) UM-SCC-32 shown at passage 13, illustrating dome formation (c, 100×; d, 320×).

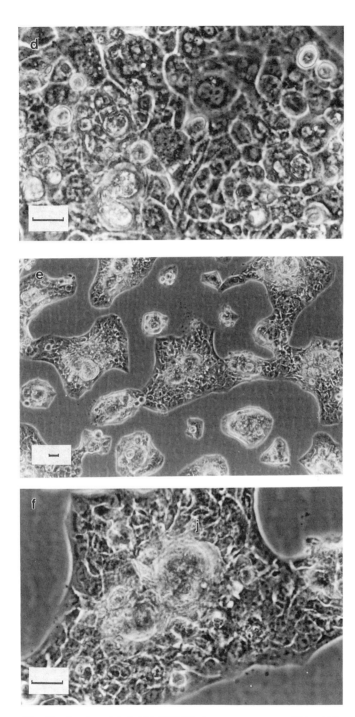

Fig. 8. *Continued* (e,f) UM-SCC-37 shown at passage 17 (e, 100×; f, 320×). Note that, for these cultures, stratification begins from relatively small islands. Bars, 10 μm.

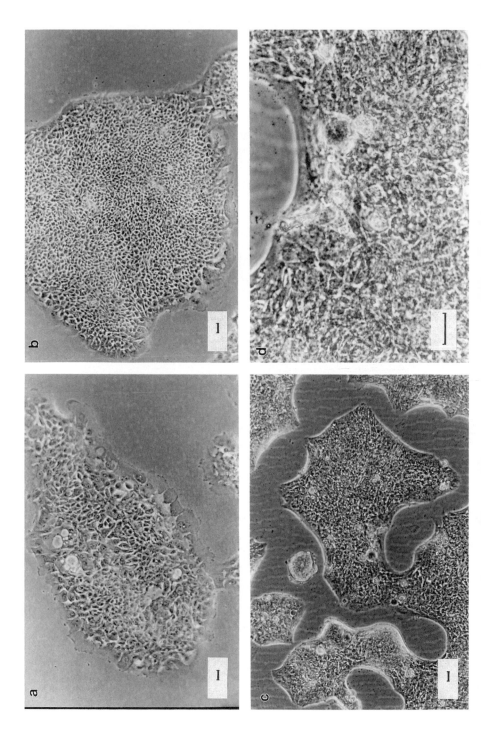

primary carcinoma of the hypopharynx (Peterson *et al.*, 1971; Fig. 10c); and SCC-15, established from a 55-year-old male patient with a carcinoma of the tongue (Rheinwald and Beckett, 1981; Fig. 10d). All these lines exhibit mono-layer colony-forming growth with little tendency to stratify.

C. Other Tumors of the Head and Neck Region

1. Thyroid Tumor Cell Lines

Tumors of the thyroid are represented by few cell lines in the American Type Culture Collection (ATCC), CRL 1803, a thyroid tumor cell line also known as TT, was established from a needle biopsy of a medullary carcinoma of the thyroid in a 77-year-old female by Leong (1981). As shown in Fig. 11a,b, this line is characterized by triangular spindle-shaped cells that grow in clusters and tend to stratify. These cells are reported to produce calcitonin and car-cinoembryonic antigen (CEA). The other cell line from the ATCC catalog that is categorized as a thyroid tumor cell line is SW579, established by Liebovitz in 1973 (see also Fogh *et al.*, 1977). However, since this line was from a tumor described as a squamous carcinoma in the thyroid gland, whether it really represents a primary tumor of the thyroid gland is questionable. The morpho-logical appearance of this cell line is shown in Fig. 11b,c. Note that these cells are more cuboidal and less elongated than the TT cells and grow as nonstra-tifying monolayers.

2. Salivary Gland Tumor Cell Lines

A-253 is one of a series of cell lines developed by Giard *et al.* (1973) from a variety of epithelial tumors. This line was established from an epidermoid carcinoma of the submaxillary gland of a 54-year-old male patient (Fig. 12a,b). A-253 cells have a cuboidal morphology with a relatively high nucleus:cy-toplasm ratio. UM-SCC-64 is a cell line we established from a female patient with a mucoepidermoid carcinoma of the parotid gland. This cell line has a multipolar morphology with a relatively low nucleus:cytoplasm ratio (Fig. 12c,d).

Fig. 9. Inverted phase-contrast photomicrographs of squamous carcinoma cultures that form islands and then grow to very high density within these islands without becoming confluent cultures. (a) UM-SCC-17A shown at passage 24 (×). (b) UM-SCC-17B shown at passage 21 (100×). (Fig. 8a,b reproduced from Carey *et al.*, 1989, with permission of the American Association for Cancer Research.) (c,d) UM-SCC-39 shown at passage 8 at two different magnifications, illustrating the dense packing of cells within cell islands (c, 100×; d, 320×). Bars, 10 μm.

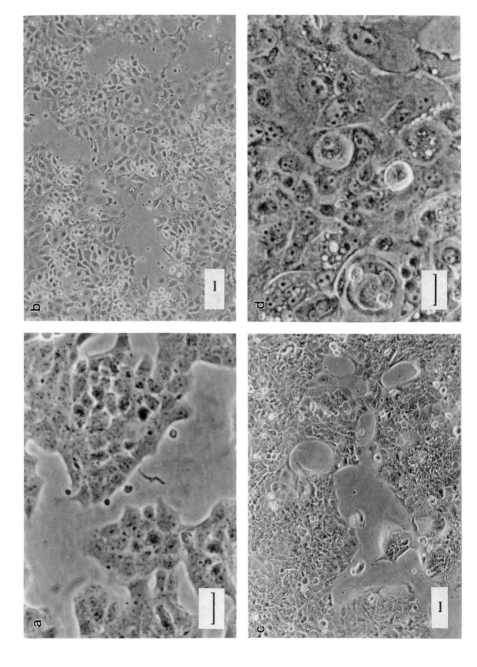

IV. Other Characteristics

A. High and Low Malignant Types

The overall success rate for SCC cell line cultivation from head and neck tumors is 30–35%. From our experience and from review of the literature, regardless of the tumor type, establishing cell lines from advanced tumors or primary tumors that have very aggressive *in vivo* behavior appears to be easiest. We have the impression that the ability of a tumor to grow *in vitro* is probably a poor prognostic sign. Proving this type of association is difficult because the failure of a tumor to grow *in vitro* can be ascribed to many factors other than the lack of biological aggressiveness. Generally the most advanced tumors, that is, tumors from patients with recurrent disease, metastatic disease, or locally advanced primary tumors, result in cell lines with a high frequency of success. Cell lines derived from donors that, soon after treatment, suffer from tumor progression often have *in vitro* characteristics that we have come to associate with highly malignant behavior *in vivo*. These tumors often show rapid proliferation early *in vitro* culture; the cells show a high capacity to migrate *in vivo,* usually have high nonpolar expression of the A9 antigen/$\alpha6\beta4$ integrin (Kimmel and Carey, 1986; Van Waes and Carey, 1992), and tend to be capable of growth in growth factor- and serum-free medium (Hoffman *et al.,* 1990). These tumors contain numerous complex chromosome rearrangements that include most of the consistent changes seen in SCCs as a group (Bradford *et al.,* 1991). The lines from less aggressive tumors are less likely to show evidence of rapid growth and migration in early culture. They tend to retain some restrictions exhibited by normal cells, such as polarization of the $\alpha6\beta4$ integrin and the inability to grow in serum- and growth factor-deficient medium, regardless of the length of time in culture or the number of *in vitro* passages. These lines also are more likely to lack some of the consistent chromosome changes found in the highly malignant group. Most important, the donors of these lines have prolonged disease-free survival.

B. Antigen Phenotype: Blood Group, Pemphigus, Pemphigoid, and the $\alpha6\beta4$ Integrin

SCCs express a membrane antigen phenotype that is useful for *in vitro* characterization and for confirming the identity of a cell line by comparison

---

Fig. 10. Inverted phase-contrast photomicrographs of squamous carcinoma cell lines from the American Type Culture Collection Repository. (a) RPMI 2650 shown at passage 25 (320×). (b) FaDu shown at passage 117 (100×). (c) Detroit 562 shown at passage 48 (100×). (d) SCC-15 shown at passage 23 (320×). Bars, 10 μm.

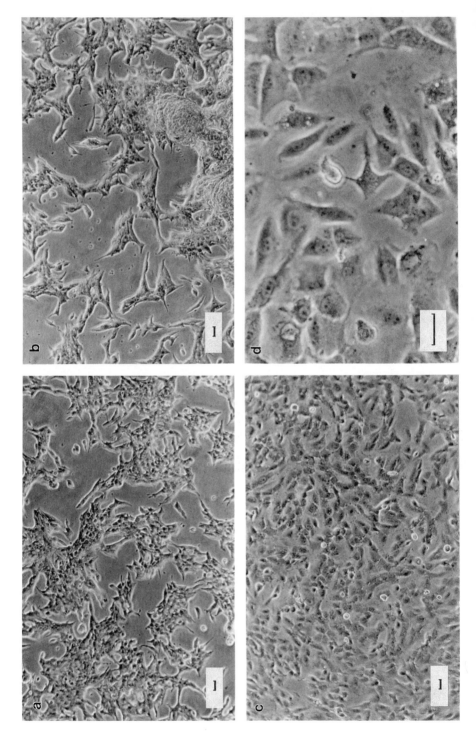

with the antigenic phenotype of the donor. Blood group antigens corresponding to the patient's blood type often are expressed on SCC cells *in vitro* and are
useful for characterizing and confirming cell line identity *in vitro* (Kimmel *et al.*,
1986). Antibodies from patients with autoimmune blistering skin diseases such
as pemphigus vulgaris (PV) and bullous phemphigoid (BP) are useful for
differentiating between squamous carcinomas and several other tumor types
(Yuspa *et al.*, 1980; Carey *et al.*, 1983). For example, we established two cell
lines from cervical lymph node biopsies from patients with no known primary
tumors. Each patient was presumed to have occult squamous cancer of the
head and neck region, but when we saw that the tumor cells in each case were
negative for reactivity with blood group, PV, and BP sera, we were suspicious
that this assumption was not correct. Subsequent clinical study of the patients
revealed that one patient had an adenocarcinoma of the lung and the other had
an adenocarcinoma of the pancreas, neither of which expressed the squamous
epithelial markers. The $\alpha6\beta4$ integrin (Van Waes and Carey, 1992) is another
antigen that usually is expressed strongly by most squamous cancers, although it cannot be used to distinguish squmous cancers from all other epithelial tumor types since some adenocarcinomas and transitional carcinomas
also express this structure (Kimmel and Carey, 1986).

V. Discussion

An ongoing debate exists about the validity of cell lines as experimental
models of human disease. Some investigators think that human cancer cell
lines are poor tools that are genetically unstable and susceptible to continuous
in vitro evolution. In contrast to this line of thought, we have been surprised
continually by the degree of *in vitro* stability exhibited by our head and neck
cancer lines. As shown in several figures in this chapter, the cell lines retain
characteristic morphological features over long periods of *in vitro* cultivation.
Similarly, the karyotypic stability of these lines is remarkable. We have studied
direct harvests, primary tumor cultures, and established cell lines from the
same tumors and found the same karyotype in all harvests (Worsham *et al.*,
1991,1993). Similarly, study of the same cell lines over numerous *in vitro*
passages found few or no karyotypic changes (Carey *et al.*, 1989; Bradford *et
al.*, 1991). In contrast, the *in vivo* evolution that occurs in tumors in patients is
striking. We have analyzed several cell lines established from the primary

Fig. 11. Cell lines from tumors of the thyroid gland (cells obtained from the ATCC Repository).
(a,b) TT (CRL medullary carcinoma of the thyroid, shown at passage 33 (100×). (c,d) SW579 tumor
of the thyroid at passage 23, shown at two magnifications (c, 100×; d, 320×). Bars, 10 μm.

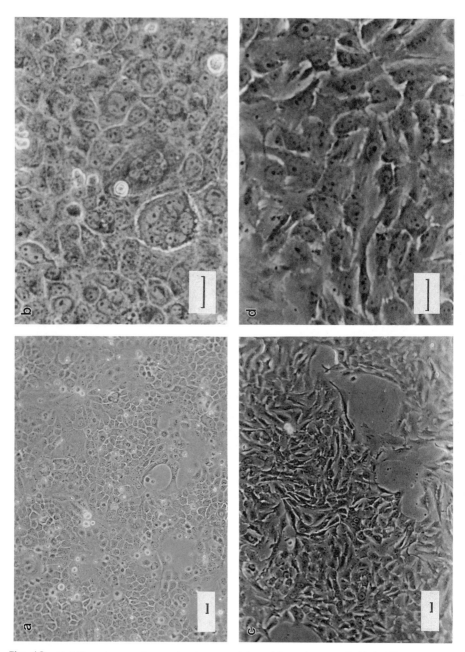

Fig. 12. Cell lines from salivary gland mucoepidermoid carcinomas. (a,b) A-253 epidermoid carcinoma of the submaxillary gland shown at passages 18 (a, 100×) and 19 (b, 320×). (Cells obtained from the ATCC repository.) (c,d) UM-SCC-64, mucoepidermoid carcinoma of the parotid gland at passage 4, shown at two magnifications (c, 100×; d, 320×). Bars, 10 μm.

tumor and a second cell line obtained from a recurrence or metastasis in the same patient and found that the secondary tumor cell line has gained additional changes not represented in the culture from the primary tumor. Similarly, we found that cell lines derived from patients with long-term survival often fail to proliferate in serum-free, growth factor-free media even after more than 100 *in vitro* passages. In contrast, cell lines from more aggressive tumors in patients who fail to survive often proliferate in serum-free medium, even at very low passage (Hoffman *et al.*, 1990). Sensitivity to radiation (Grenman *et al.*, 1991) and polar or nonpolar expression of the $\alpha6\beta4$ integrin (Kimmel and Carey, 1986) are other characteristics of human SCC cell lines that we have found to be stable over long-term *in vitro* propagation. In summary, we feel that human cancer cell lines are valuable *in vitro* tools that closely represent the *in vivo* condition at the time of culture, and that much has been and will continue to be learned about the phenomenon of cancer from their use.

VI. Future Prospects

Head and neck cancer cell lines are providing important information about consistent chromosome changes in squamous cell cancer that identify the loci of genes involved in the malignant process (Van Dyke *et al.*, 1993). Examples are the consistent gains of 11q13-q24 that appear in about one-third of squamous carcinomas. Cell lines such as UM-SCC-11, UM-SCC-14, UM-SCC-22, and UM-SCC-38 that have amplification of this region are being used to examine the genes contained within this amplicon, for example, *bcl-1, prad*1 (cyclin D), and the folate receptor. Likewise, consistent losses of chromosome 3p, 8p, 9p, 17p, and 18q are being scrutinized for genes important in tumor development and progression. These cell lines also provide important tools for investigating the basis for altered expression of the $\alpha6\beta4$ integrin in aggressive squamous cancers and for detecting the mechanisms by which this integrin passes signals across the membrane of migrating and stationary cells. *In vitro* cell lines are valuable resources for investigating the abnormalities that lead to malignant behavior and metastasis.

References

Baker, S. R. (1985). An in vivo model for squamous cell carcinoma of the head and neck. *Laryngoscope* **96,** 43–56.
Barnes, D. W. (1982). Epidermal growth factor inhibits growth of A431 human epidermoid carcinoma in serum-free cell culture. *J. Cell Biol.* **93,** 1–4.

Boyse, S. T., and Ham, R. G. (1985). Cultivation, frozen storage, and clonal growth of normal human epidermal keratinocytes in serum-free media. *J. Tissue Culture Meth.* **9**, 83–93.

Bradford, C. R., Kimmel, K. A., Van Dyke, D. L., Worsham, M. J., Tilley, B. J., Burk, D., del Rosario, F., Lutz, S., Tooley, R., Hayashida, D. J. S., and Carey, T. E. (1991). 11p deletions and breakpoints in squamous cell carcinoma: Association with altered reactivity with the UM-E7 antibody. *Genes Chromosomes Cancer* **3**, 272–282.

Carey, T. E. (1985). Establishment of epidermoid carcinoma cell lines. *In* "Head and Neck Cancer" (R. E. Wittes, ed.), pp. 287–314. John Wiley & Sons, New York.

Carey, T. E., Kimmel, K. A., Schwartz, D. R., Richter, D. E., Baker, S. R., and Krause, C. J. (1983). Antibodies to human squamous cell carcinoma. *Otolaryngol. Head Neck Surg* **91**, 482–491.

Carey, T. E., Van Dyke, D. L., Worsham, M. J., Bradford, C. R., Babu, V. R., Schwartz, D. R., Hsu, S., and Baker, S. R. (1989). Characterization of human laryngeal primary and metastatic squamous cell carcinoma cell lines UM-SCC-17A and UM-SCC-17B. *Cancer Res.* **49**, 6098–6107.

Eagle, H. (1955). Nutrition needs of mammalian cells in tissue culture. *Science* **122**, 501–504.

Easty, D. M., Easty, G. C., Carter, R. L., Monaghan, P., and Butler, L. J. (1981). Ten human carcinoma cell lines derived from squamous carcinomas of the head and neck. *Br. J. Cancer* **43**, 772–785.

Fogh, J., Fogh, J. M., and Orfeo, T. (1977). One hundred and twenty-seven cultured human tumor cell lines producing tumors in nude mice. *J. Natl. Cancer Inst.* **59**, 221–226.

Fogh, J., Tiso, J., Orfeo, T., Sharkey, F. E., Daniels, W. P., and Fogh, J. M. (1980). Thirty-four lines of six human tumor categories established in nude mice. *J. Natl. Cancer Inst.* **64**, 745–751.

Frenette, G. P., Carey, T. E., Varani, J., Schwartz, D. R., Fligiel, S. E. F. G., Ruddon, R. W., and Peters, B. P. (1988). Biosynthesis and secretion of laminin and laminin-associated glyco-proteins by non-malignant and malignant human keratinocytes: Comparison of cell lines from primary and secondary tumors in the same patient. *Cancer Res.* **48**, 5193–5202.

Giard, D. J., Aaronson, S. A., Todaro, G., Arnstein, P., Kersey, J. H., Dosik, H., and Parks, W. P. (1973). In vitro cultivation of human tumors: Establishment of cell lines derived from a series of solid tumors. *J. Natl. Cancer. Inst.* **51**, 1417–1423.

Grenman, R., Carey, T. E., McClatchey, K. D., Wagner, J. G., Pekkola-Heino, K., Schwartz, D. R., Wolf, G. T., Lacivita, L. P., Ho, L., Baker, S. R., Krause, C. J., and Lichter, A. S. (1991). In vitro radiation resistance among cell lines established from patients with squamous cell carcinoma of the head and neck. *Cancer* **67**, 2741–2747.

Grenman, R., Pekkola-Heino, K., Joensuu, H., Aitasolo, K., Klemi, P., and Lakkala, T. (1992). UT-MUC-1, a new mucoepidermoid carcinoma cell line, and its radiosensitivity. *Arch. Otolaryngol. Head Neck Surg.* **118**, 542–547.

Heo, D. S., Synderman, C., Gollin, S. M., Pan, S., Walker, E., Deka, R., Barnes, E. L., Johnson, J. T., Heberman, R. B., and Whiteside, T. L. (1989). Biology, cytogenetics, and sensitivity to immunologic effector cells of new head and neck squamous carcinoma cell lines. *Cancer Res.* **49**, 5167–5175.

Hoffman, H. T., Merkle, S., and Carey, T. E. (1990). Head and neck squamous carcinoma cell lines capable of serum-free growth in vitro. *In* "Head and Neck Cancer" (W. E. Fee, H. Goepfert, M. E. Johns, E. W. Strong, and P. H. Ward, eds.), Vol. 2, pp. 94–97. Decker, Toronto.

Kimmel, K. A., Carey, T. E. (1986). Altered expression in squamous carcinoma cells of an orientation restricted epithelial antigen detected by monoclonal antibody A9. *Cancer Res.* **46**, 3614–1623.

Kimmel, K. A., Carey, T. E., Judd, W. J., and McClatchey, K. D. (1986). Monoclonal antibody (GIO) to a common antigen of human squamous cell carcinoma: Binding of the antibody to the H type 2 blood group determinant. *J. Natl. Cancer Inst.* **76**, 9–19.

Komiyama, S., Matsui, K., Kudoh, S., Nogae, I., Kuratomi, Y., Saburi, Y., Asoh, K.-I., Kohno, K., Kuwano, M. (1989). Establishment of tumor cell lines from a patient with head and neck cancer and their different sensitivities to anti-cancer agents. *Cancer* **63**, 675–681.

Krause, C. J., Carey, T. E., Ott, R. W., Hurbis, C., McClatchey, K. D., and Regezi, J. A. (1981). Human squamous cell carcinoma: Establishment and characterization of new permanent cell lines. *Arch. Otolaryngol.* **107,** 703–710.

Leibovitz, A. (1973). Reported in "ATCC Catalogue of Cell Lines and Hybridomas," 7th Ed. (1992), pp. 253. American Type Culture Collection, Rockville, Maryland.

Leong, S. S. (1981). Advances in thyroid neoplasia. *Field Educational Italia,* 95–105.

Moore, A. E., Sabachewsky, L., and Toolan, H. W. (1955). Culture characteristics of four permanent lines of human cancer cells. *Cancer Res.* **13,** 598–605.

Moore, G. E., and Sandberg, A. A. (1964). Studies of a human tumor cell line with a diploid karyotype. *Cancer* **17,** 170–175.

Nakashima, T., Makishima, K., and Hiroto, I. (1980). Establishment of a new cell line from maxillary sinus carcinoma. *Ann. Otol.* **89,** 24–28.

Okabe, T., Sato, N., Kondo, Y., Asano, S., Ohsawa, N., Kosaka, K., and Ueyama, Y. (1978). Establishment and characterization of a human cancer cell line that produces human colony-stimulating factor. *Cancer Res.* **38,** 3910–3917.

Peterson, W. D., Jr., Stulberg, C. S., and Simpson, W. F. (1971). A permanent heteroploid cell line with type B glucose-6-phosphatase dehydrogenase. *Proc. Soc. Exp. Biol. Med.* **136,** 1187–1191.

Rangan, S. R. (1972). A new human cell line (FaDu) from a hypopharyngeal carcinoma. *Cancer* **29,** 117–121.

Rheinwald, J. G., and Beckett, M. A. (1981). Tumorigenic keratinocyte lines requiring anchorage and fibroblast support cultured from human squamous cell carcinomas. *Cancer Res.* **41,** 1657–1663.

Rheinwald, J. G., and Green, H. (1975). Serial cultivation of strains of human epidermal keratinocytes: The formation of keratinizing colonies from single cells. *Cell* **6,** 331–344.

Rheinwald, J. G., and Green, H. (1977). Epidermal growth factor and the multiplication of cultured human epidermal keratinocytes. *Nature (London)* **265,** 421–424.

Rupniak, H. T., Rowlatt, C., Lane, E. B., Steele, J. G., Trejdosiewicz, L. K., Laskiewicz, B., Povey, S., and Hill, B. T. (1985). Characteristics of four new human cell lines derived from squamous cell carcinomas of the head and neck. *J. Natl. Cancer Inst.* **75,** 621–635.

Sacks, P. G., Parnes, S. M., Gallick, G. E., Mansouri, Z., Lichtner, R., Satya-Prakash, K. L., Pathak, S., and Parsons, D. F. (1988). Establishment and characterization of two new squamous cell carcinoma cell lines derived from tumors of the head and neck. *Cancer Res.* **48,** 2858–2866.

Sacks, P. G., Oke, V., Amos, B., Vasey, T., and Lotan, R. (1989). Modulation of growth, differentiation, and glycoprotein synthesis by β-all-trans retinoic acid in a multicellular tumor spheroid model for squamous carcinoma of the head and neck. *Int. J. Cancer* **44,** 926–933.

Van Dyke, D. L., Worsham, M. J., Benninger, M. S., Krause, C. J., Baker, S. R., Wolf, G. I., Drumheller, T., Tilley, B. C., and Carey, T. E. (1993). Recurrent cytogenetic abnormalities in squamous cell carcinomas of the head and neck region. *Genes Chromosomes Cancer,* in press.

Van Waes, C., and Carey, T. E. (1992). Overexpression of the A9 antigen/α6β4 integrin in head and neck cancer. *Otolaryngol. Clin. North Am.* **25,** 1117–1139.

Van Waes, C., Kozarsky, K. F., Warren, A. B., Kidd, L., Paugh, D., Liebert, M., and Carey, T. E. (1991). The A9 antigen associated with aggressive human squamous carcinoma is structurally and functionally similar to the newly defined integrin α6β4. *Cancer Res.* **51,** 2395–2402.

Vlock, D. R., Scalise, D., Schwartz, D. R., Richter, D. E., Krause, C. J., Baker, S. R., and Carey, T. E. (1989). Incidence of serum antibody reactivity to autologous head and neck cancer cell lines and augmentation of antibody reactivity following acid dissociation and ultrafiltration. *Cancer Res.* **49,** 1361–1365.

Weichselbaum, R. R., Dahlberg, W., Beckett, M., Karrison, T., Miller, D., Carl, J., and Ervin, T. J. (1986). Radiation-resistant and repair-proficient human tumor cells may be associated with radiotherapy failure in head- and neck-patients. *Proc. Natl. Acad. Sci. U.S.A.* **83,** 2684–2688.

Worsham, M. J., Van Dyke, D. L., Grenman, S. E., Grenman, R., Hopkins, M. P., Roberts, J. A.,

Gasser, K. M., Schwartz, D. R., and Carey, T. E. (1991). Consistent chromosome abnormalities in squamous cell carcinoma of the vulva. *Genes Chrom. Cancer* **3,** 420–432.

Worsham, M. J., Carey, T. E., Benniger, M. S., Gasser, K. M., Kelker, W., Zarbo, R. J., and Van Dyke, D. L. (1993). Clonal cytogenetic evolution in a squamous cell carcinoma of the skin in a xeroderma pigmentosum patient, HFH-SCC-XP-1. *Genes Chrom. Cancer* (*in Press*).

Yanagawa, T., Hayashi, Y., Yoshida, H., Yuri, Y., Nagamine, S., Bando, T., and Sato, M. (1986). An adenoid squamous carcinoma-forming cell line established from an oral keratinizing squamous cell carcinoma expressing carcinoembryonic antigen. *Am. J. Pathol.* **124,** 496–509.

Yuspa, S. H., Hawley-Nelson, P., Koehler, B., and Stanley, J. R. (1980). A survey of transformation markers in differentiating epidermal cell lines in culture. *Cancer Res.* **40,** 4694–4703.

Zenner, H. P., Lehner, W., and Herrmann, I. F. (1979). Establishment of carcinoma cell lines from larynx and submandibular gland. *Arch. Otorhinolaryngol.* **225,** 269–277.

Zenner, H. P., Herrmann, I. F., Bremer, W., and Stahl-Mauge, C. (1983). Head and neck carcinoma models: In vivo reproduction in athymic mice and in vitro culture. *Acta Otolaryngol.* **96,** 371–381.

Cell Culture of Lung Cancers

5

Adi Gazdar

Simmons Cancer Center and Department of Pathology
University of Texas Southwestern Medical Center
Dallas, Texas 75235-8593

I. Cellular Origins of Lung Cancers 121

II. Major Forms of Lung Cancer 123

III. Defined Media for the Culture of Lung Tumors 124

IV. Establishment of Small-Cell Lung Carcinoma Cell Lines 125
  A. Classic Subtype 132
  B. Variant Form 133

V. Other Neuroendocrine Tumors 134

VI. Cell Lines Derived from Non-Small-Cell Lung Carcinoma 136
  A. Squamous Cell Carcinoma 137
  B. Adenocarcinomas 138
  C. Large-Cell Carcinomas 145

VII. Mesotheliomas 146

VIII. Conclusions 146
  References 146

Lung cancer is the most common form of cancer in the world, and is the most common cause of cancer deaths in the United States (Minna *et al.*, 1989). Most lung cancers arise from the epithelial lining of the airways. The several types of lung cancer vary in their properties, *in vitro* growth characteristics, and appearances. Thus, understanding their cellular origin is important.

I. Cellular Origins of Lung Cancers

The major function of the airways is gas exchange. The airways may be divided into the conducting part (the bronchi and bronchioles), the major function of which is gas transport, and the respiratory part (alveolar ducts and sacs), where gas exchange occurs. Inspired air enters the lungs via the right and left main bronchi, which divide several times. Bronchi have cartilaginous rings, a muscle coat, submucosal glands, and a pseudostratified epithelium. The epithelium consists of basal cells, mucous cells, and tall columnar ciliated cells. Scattered among the basal cells are neuroendocrine (NE) cells. After several divisions, the bronchi lose their cartilaginous rings and glands and give rise to smaller conducting airways known as bronchioles. The mucosa of the larger bronchioles closely resembles that of the bronchi. After several divisions, as the bronchioles near the respiratory part of the lungs, their lining changes to a simple single layer of cells consisting of short stubby ciliated cells and Clara cells. The latter are columnar cells containing large secretory granules in their apices, which discharge into the lumen. Although the function of their secretion is not fully understood, the fluid is thought to be related to surfactant and may have a detergent-like action. Clara cells contain a characteristic 10-kDa protein similar or identical to uteroglobin. Mucous cells are relatively rare or absent. The terminal bronchioles end in thin-walled respiratory ducts which communicate with several respiratory sacs. These structures are the sites of gas exchange and also consist of two cell types. Type I pneumocytes are large flattened cells, the major function of which is to permit gas exchange with the underlying respiratory capillaries. Type II cells are globular cells, the major function of which is secretion of surfactant. Surfactant has a detergent-like function, lowering the surface tension and permitting the thin-walled alveoli to maintain their normal spherical shape in the face of strong counterforces. Surfactant is a complex phospholipid containing at least three important surfactant-associated proteins known as SP-A, SP-B, and SP-C (Possmayer, 1988). The cytoplasmic storage form of this secretion is in multi-lamellar bodies.

NE cells are found scattered throughout the bronchi and larger bronchioles. They are much more numerous in fetal and early postnatal life, suggesting that their secretory products play a role in the development and maturation of the respiratory tree. NE cells are found in many organ systems, either as organoid masses (e.g., the adrenal medulla) or scattered (epithelial linings of the gastrointestinal and respiratory tracts). In the latter sites, their secretions are of the paracrine type (i.e., they nourish adjacent cells), in contrast to hemocrine secretions which are transported to distant sites via the bloodstream. The function of NE cells is the production, storage, and secretion of biogenic amines and peptide hormones. In addition to these specific products (which characterize the specific NE cell types), all NE cells express several general markers, including dense core granules (the cytoplasmic storage sites of the products), chromogranin A (structural component of the granules), L-DOPA decarboxylase (an enzyme essential for amine production), and many others (Gazdar et al., 1988; Jensen et al., 1990). The specific products of paracrine cells are secreted basolaterally, in contrast to the apical luminally directed secretions of mucous, Clara, and Type II cells. Specific products of the pulmonary endocrine cells include the amine serotonin and the peptides calcitonin and gastrin releasing peptide (GRP) (Becker and Gazdar, 1984).

The major functions of the bronchial epithelium include lubrication (mucin production) and removal of inhaled particles (ciliary movement). However, a large number of mechanical, infectious, and chemical events may result in replacement of this pseudostratified epithelium with a tough, protective, water-impermeable, stratified layer of nonkeratinizing squamous cells similar to those present in the esophagus.

II. Major Forms of Lung Cancer

The WHO classification of lung cancers (Anonymous, 1982) describes four major types and several minor ones. The major forms include squamous cell carcinoma, adenocarcinoma, small-cell carcinoma, and large-cell undifferentiated carcinoma. A decrease in the incidence of squamous cell carcinomas, and a relative and absolute increase in the incidence of adenocarcinomas, appears to be taking place in the United States (and, perhaps, in other parts of the world) (Gazdar and Linnoila, 1988).

Squamous cell carcinomas arise in the major bronchi, presumably from metaplastic squamous cells. They frequently are associated with foci of dysplasia and carcinoma *in situ*.

Small-cell lung carcinomas (SCLC) are NE tumors that express all the general NE cell properties. In addition, they frequently express calcitonin and GRP, as well as several other ectopic peptide products. SCLC is a rapidly growing tumor that metastasizes early. Common metastatic sites include bone marrow, lymph nodes, the pleura, liver, brain, and adrenal glands. In appearance, these tumors are relatively undifferentiated, containing cells with scant cytoplasm with few organelles and a few relatively small dense core granules. A rare form of NE lung tumor is the bronchial carcinoid. Unlike SCLC, carcinoids grow relatively slowly, metastasize late, and usually are cured by surgical treatment. These cells are much larger than those of SCLC, and the presence of abundant relatively large dense core granules confirms that these are relatively well differentiated NE tumors.

Adenocarcinomas arise from glandular epithelia. The typical gland-forming or acinar adenocarcinomas appear to be decreasing in number, but peripheral airway tumors (which include the papillary and bronchioloalveolar subtypes of adenocarcinomas) appear to be increasing (Gazdar and Linnoila, 1988). These tumors are believed to arise from the peripheral airways (bronchioles and alveoli), and frequently show ultrastructural and biochemical markers of Clara or Type II cells (Gazdar *et al.*, 1990a; Linnoila *et al.*, 1992). In contrast, acinar tumors probably arise from the surface epithelium of bronchi and larger bronchioles. Mucin secretion may be a prominent feature of all forms of adenocarcinoma. Since mucin-secreting cells are not normally present in the peripheral airways, mucus production by these cells may be regarded as an ectopic feature. Poorly differentiated adenocarcinomas consist of sheets of cells containing small amounts of intra- or extracellular mucin.

The fourth major form of lung cancer is large-cell undifferentiated carcinoma. These tumors consist of relatively large cells without obvious pathways of differentiation as detected by routine histochemical techniques. However, ultrastructural and other special studies demonstrate that most of them represent extremely undifferentiated versions of some other form of lung cancer, especially adenocarcinoma, although the possibility remains that some represent tumors derive from truly undifferentiated "stem cells."

Because of major differences in their natural history, clinical behavior, response to therapy, and biological properties, lung cancers frequently are divided into SCLC and non-SCLC (NSCLC) types. The latter consist of all the forms of lung cancer other than SCLC (approximately 75% of lung cancers in the United States) (Minna et al., 1989). Whereas SCLCs (and carcinoids) express NE markers, most NSCLC tumors fail to do so. However, about 15% of NSCLC tumors and cell lines express multiple NE markers (Gazdar and Linnoila, 1988; Gazdar et al., 1992). These NSCLC–NE tumors may be relatively chemosensitive and, thus, important to identify.

Although some controversy remains over the origin of SCLC, researchers generally believe that all lung cancers arise from the endodermally derived bronchial epithelium. At 16 weeks of gestation, the tracheobronchial tree is lined by a single cell type—columnar cells distended with glycogen. Thus, all the cell types of the bronchial tree arise from this single multipotent precursor cell, which may explain why many lung cancers demonstrate more than one form of differentiation.

Mesotheliomas are cancers arising from the simple squamous epithelia of the pleural or abdominal cavities. Unlike the respiratory epithelium, these cells are mesodermally derived. Thus, strictly speaking, they are not of bronchogenic origin and not lung cancers. However, these two different tumor types are sometimes discussed together. Unlike lung cancers, mesotheliomas are not linked to smoking. The single most important etiologic feature associated with mesotheliomas is asbestos exposure (also linked to some lung cancers).

III. Defined Media for the Culture of Lung Tumors

Many human tumors have been established in serum-supplemented media (SSM). However, the use of SSM has many disadvantages: (1) SSM supports the replication of stromal cells which may overgrow the tumor cells; (2) SSM may select the rapidly proliferating, relatively undifferentiated tumor subpopulation; (3) SSM may contain antibodies and other factors that inhibit tumor cell proliferation; and (4) SSM permits the outgrowth of B lymphoblastoid cells (see subsequent discussion). Sato and others have proposed that each differentiated cell type has its own unique growth requirements, and that differentiated tumor cells would have nutritional requirements similar to those of their normal counterparts (Cuttitta et al., 1990).

By trial and error, we have devised media for the selective growth of most forms of lung cancer (Simms *et al.*, 1980; Carney *et al.*, 1981; Brower *et al.*, 1986; Gazdar and Oie, 1986b). The major types of lung cancer differ in their growth factor requirements. We originally devised a medium, HITES, for the selective growth of SCLC. This medium has been used successfully by us and by many other investigators for the establishment of numerous SCLC lines (Carney *et al.*, 1981; Gazdar *et al.*, 1990a). Similarly, ACL-4 medium has been used for the culture of NSCLC, especially adenocarcinoma and large-cell carcinoma. For squamous cell carcinomas, we have modified a complex medium originally devised by Rheinwald and associates (Erim *et al.*, 1982). However, this medium is characterized only partially. Tumors, especially SCLC, frequently secrete their own growth factors (autocrine secretion), and may be maintained in very simple medium containing selenium, insulin, and transferrin. In fact, some tumors have been maintained for long periods of time in RPMI-1640 medium without any further supplementation (Cuttitta *et al.*, 1990). Presumably, these cultures represent outgrowths of subpopulations that make more factors than other cells in the tumor population and, therefore, have a growth advantage. To permit these cells to reach a critical mass, primary cultures must be cultured in fully defined medium, with or without supplementation with small amounts of serum (we use 2.5%). We have documented the advantage of using fully defined media for the establishment of SCLC (HITES) as well as for adenocarcinomas of the lung and other organ sites (ACL-4) (Carney *et al.*, 1981; Park *et al.*, 1987; Gazdar *et al.*, 1990). These findings have been confirmed independently (Masuda *et al.*, 1991). However, the medium for squamous cell carcinomas is of only modest advantage over SSM.

IV. Establishment of Small-Cell Lung Carcinoma Cell Lines

Because of the enormous interest in the unique biological and clinical features of SCLC, more than 200 cell lines have been established and characterized. In fact, most of our current knowledge about the biology and molecular genetics of SCLC stems from the study of these lines. The establishment of a floating cell line from an SCLC tumor in 1971 (Oboshi *et al.*, 1971) initiated these studies and proved to be a major surprise. Until then, most human epithelial tumor cultures grew as attached epithelial cells; the finding of a floating line was greeted with skepticism. A few scattered reports of single cell lines appeared during the next decade. However, nearly a decade later, groups of scientists at Dartmouth Medical School and the National Cancer Institute (NCI) reported the establishment of banks of cell lines (Gazdar *et al.*, 1980; Pettengill *et al.*, 1980b). Since that time, many groups have reported the establishment of individual and multiple cell lines (see Table 1).

Table I

Advances in Lung Cancer Cell Culture

Topic	Reference
Cell culture and characterization SCLC[a]	Oboshi et al. (1971); Pettengill et al. (1977,1980b,1984); Fisher and Paulson (1978); Gazdar et al. (1980); Simms et al. (1980); Bergh et al. (1982); Carney et al. (1982); Baillie-Johnson et al. (1985); Duchesne et al. (1987); Graziano et al. (1987); Berendsen et al. (1988); Postmus et al. (1988); Watanabe et al. (1988); Hay et al. (1991)
SCLC variant	Fisher and Paulson (1978); Carney et al. (1983,1985); Morstyn et al. (1984); Leij et al. (1985); Gazdar et al. (1985); Broers et al. (1986); Johnson et al. (1986); Seifter et al. (1986); Kiefer et al. (1987); Berendsen et al. (1988); Mabry et al. (1988); Watanabe et al. (1988); Bepler et al. (1989); Doyle et al. (1989,1990); Falco et al. (1990); Gamou et al. (1990)
Bronchial carcinoids	Funa et al. (1986); Gazdar et al. (1988); Lai et al. (1989); Moody et al. (1990)
NSCLC[b]	Luduena et al. (1974); Fogh (1975); Lieber et al. (1976); Takaki (1980); Anger et al. (1992); Okabe et al. (1984); Baillie-Johnson et al. (1985); Bergh et al. (1985); Brower et al. (1986); Cole et al. (1986); Gazdar and Oie (1986a); Duchesne et al. (1987); Bepler et al. (1988a); Stark et al. (1988); Gazdar et al. (1990); Masuda et al. (1991)
Mesothelioma	LaRocca and Rheinwald (1984); Bepler et al. (1988a); Demetri et al. (1989); Flejter et al. (1989); Ke et al. (1989); Klominek et al. (1989); Versnel et al. (1989,1991); Manning et al. (1991)
Bronchial epithelium	Amstad et al. (1988); Bonofil et al. (1989); Ura et al. (1989); Nervi et al. (1991); Pfeifer et al. (1991)
Defined media	Simms et al. (1980); Carney et al. (1981,1982,1985); Minna et al. (1982); Brower et al. (1986); Gazdar and Oie (1986a,b); Cuttitta et al. (1987,1990); Stevenson et al. (1989,1990); Gazdar et al. (1990a,b); Masuda et al. (1991)
Correlations between in vitro growth and clinical outcome	Stevenson et al. (1989,1990)

Autocrine and other growth factors (see also "Defined media")	
Gastrin releasing peptide	Sorenson et al. (1982); Spindel et al. (1984); Cuttitta et al. (1985,1988); Moody et al. (1985); Sausville et al. (1986); Heikkila et al. (1987); Minna et al. (1988); Polak et al. (1988); Sunday et al. (1988,1990); Trepel et al. (1988); Woll and Rozengurt (1988); Aguayo et al. (1989); Mahmoud et al. (1989); Wiedermann (1989); Avis et al. (1991); Battey and Wada (1991); Battey et al. (1991)
IGF-1[c]	Jaques et al. (1988,1989); Nakanishi et al. (1988); Havemann et al. (1990); Macaulay et al. (1990); Reeve et al. (1990,1992); Shigematsu et al. (1990); Guillemin et al. (1991); Macaulay (1992)
Other	Sorenson et al. (1981b); Becker et al. (1983,1984); Bepler et al. (1988b); Browder et al. (1989); Chang et al. (1989); Deftos et al. (1989); Heymanns et al. (1989); Stewart et al. (1989); Bliss et al. (1990); Mendelsohn (1990); Moody et al. (1990); Brandt et al. (1991); Cardona et al. (1991); Hibi et al. (1991); Quinn et al. (1991); Schuller (1991); Sekido et al. (1991); Giaccone et al. (1992)
Neuroendocrine properties of SCLC and other lung cancers	Baylin et al. (1980); Gazdar et al. (1980,1992); McCann et al. (1981); Sorenson et al. (1981a); Pettengill et al. (1982); Gazdar and Carney (1984); McMahon et al. (1984); Becker and Gazdar (1985); Bunn et al. (1985); Funa et al. (1986); Bergh et al. (1989); Johansson et al. (1989); Doyle et al. (1990); Jensen et al. (1990); Moody et al. (1990); Carbone et al. (1991); Schuller (1991); Moolenaar et al. (1992)
Cytogenetics	Whang-Peng et al. (1982,1983,1991); Graziano et al. (1987); Morstyn et al. (1987); Yokota et al. (1987,1988,1989a); Harbour et al. (1988); Waters et al. (1988); Becker and Sahin (1989); De Fusco et al. (1989); Flejter et al. (1989); Shiraishi and Sekiya (1989); van der Hout et al. (1989); Weston et al. (1989); Lukeis et al. (1990); Miura et al. (1990); Mori et al. (1990); Kuo et al. (1991); Sozzi et al. (1991); Xu et al. (1991)

(continues)

Table I
Continued

Topic	Reference
Activation of dominant oncogenes	
myc family	Little et al. (1983); Gazdar et al. (1985); Nau et al. (1985,1986); Johnson et al. (1986,1988); Seifter et al. (1986); Wong et al. (1986); Bepler et al. (1987); Brooks et al. (1987); Keifer et al. (1987); Gemma et al. (1988); Ibson et al. (1988); Krystal et al. (1988); Saksela et al. (1988); Sausville et al. (1988); Chauvin et al. (1989); Doyle et al. (1989,1991); Takahashi et al. (1989b); Bergh (1990); Gazzeri et al. (1990); Saksela (1990); Tefre et al. (1990); Brennan et al. (1991); Pfeifer et al. (1991); Sekido et al. (1992)
ras family	Mabry et al. (1989); Rodenhuis and Slebos (1990); Mitsudomi et al. (1991a,b,1992); Sugio et al. (1992)
EGF[d] receptor	Dazzi et al. (1989); Levitt et al. (1990); Mendelsohn (1990); Tateishi et al. (1990)
Other oncogenes	Mellstrom et al. (1987); Bergh et al. (1988); Soderdahl et al. (1988); Chauvin et al. (1989); Yoshida et al. (1989); Bergh (1990); Kern and Weisenthal (1990); Weiner et al. (1990); Tyson et al. (1991); Versnel et al. (1991); Kern et al. (1992)
Loss of DNA and recessive oncogenes	
Chromosome 3p	Whang-Peng et al. (1982); Brauch et al. (1987,1990); Buys et al. (1987); Graziano et al. (1987); Kok et al. (1987); Mooibroek et al. (1987); Naylor et al. (1987); Yokota et al. (1987,1989b); Drabkin et al. (1988); Waters et al. (1988); Becker and Sahin (1989); Flejter et al. (1989); Leduc et al. (1989); Zbar (1989); Cole et al. (1992)
Chromosome 13 and the *rb* gene	Harbour et al. (1988); Yokota et al. (1988,1989a); Hensel et al. (1990); Kaye et al. (1990); Mori et al. (1990); Hay et al. (1991); Murakami et al. (1991); Xu et al. (1991)
Chromosome 17 and the *p53* gene	Lavigueur et al. (1989); Takahashi et al. (1989a,1990,1991,1992); Sameshima et al. (1990,1992); Hollstein et al. (1991); Bodner et al. (1992); D'Amico et al. (1992b); Hiyoshi et al. (1992); Miller et al. (1992); Mitsudomi et al. (1992)
Other loci	Shiraishi et al. (1987); Weston et al. (1987,1989); Skinner et al. (1990); D'Amico et al. (1992a)
Xenografts	Pettengill et al. (1980); Chambers et al. (1981); Gazdar et al. (1981,1987); Sorensen et al. (1981); Carney et al. (1982); Duchesne et al. (1987); Gemma et al. (1988); McLemore et al. (1988)

[a] SCLC, small-cell lung carcinoma.
[b] NSCLC, non-small-cell lung carcinoma.
[c] IFG-1, insulin-like growth factor 1.
[d] EGF, epidermal growth factor.

Because SCLC tumors seldom are resected, most lines have been established from metastatic lesions (Gazdar *et al.*, 1980; Carney *et al.*, 1985). Frequent sources include bone marrow aspirates, malignant pleural effusions, and lymph node metastases, but a variety of other sites have been cultured successfully, including needle aspirations of liver. SCLC frequently involves the marrow in a pattern similar to leukemic spread; bone marrow aspirates are easier to culture than osseous metastases. Even samples containing minute numbers of tumor cells may be cultured readily. Similarly, malignant SCLC effusions usually contain relatively small numbers of cells, but these are in small aggregates suspended in fluid (i.e., they are a form of *in vivo* culture). In our experience, marrow aspirates and effusions containing small numbers of tumor cells are much easier to culture than solid lesions containing much larger cell numbers. Whether differences exist in the culture success rates of primary versus metastatic lesions is not known. Most SCLC lines have been established from recurrent tumors from previously treated patients (Gazdar *et al.*, 1980; Carney *et al.*, 1985). These samples appear to have a growth advantage over tumors from untreated patients. However, at least one relatively large series of cell lines has been established from previously untreated patients (Gazdar *et al.*, 1990). In contrast to NSCLC (see subsequent discussion), the ability to culture an SCLC sample successfully is not a negative prognostic sign (Stevenson *et al.*, 1989). Because rigorous disaggregation of SCLC results in the loss of large numbers of cells, we use simple mechanical disaggregation followed (if necessary) by Ficoll–Hypaque centrifugation. Characteristic SCLC clumps can be recognized by examination with a phase contrast inverted microscope.

Most SCLC lines grow as floating cell aggregates (Fig. 1), lacking substrate attachment (Gazdar *et al.*, 1980; Baillie-Johnson *et al.*, 1985; Bergh *et al.*, 1985; Carney *et al.*, 1985; de Leij *et al.*, 1985). In contrast, most lines established by the Dartmouth group demonstrate substrate attachment (Fig. 2) (Pettengill *et al.*, 1980b, 1984). The reasons for the latter result are not apparent, but may be related to the use of Waymouth's MB 752/1 medium supplemented with 20% fetal bovine serum (FBS) by the Dartmouth group, and the high calcium content of this medium. Most other investigators have used either R10 or HITES medium (the latter, of course, lacks the high molecular weight attachment factors present in serum). The floating lines grow as true spheroids (the larger ones having central necrosis) or as irregular cellular aggregates. Some cell lines demonstrate partial loose substrate attachment and can be detached by vigorous agitation of the culture dishes. However, predominantly attached subpopulations may be selected by removing the floating cells and refeeding the attached cells. Lack of substrate attachment and cell-to-cell aggregation by SCLC and many NE cultures appears to be related to expression of neural cell adhesion molecule (NCAM) by these cells (Doyle *et al.*, 1990; Carbone *et al.*, 1991). Cell aggregation is a characteristic feature of SCLC, both *in vivo* and *in vitro*. Although numerous single cells are present in culture, especially those that have not been fed for a few days, trypan blue staining will reveal that most

TYPE I   TYPE II

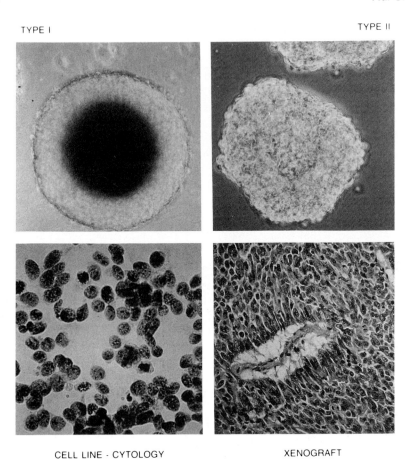

CELL LINE - CYTOLOGY   XENOGRAFT

Fig. 1. Small-cell lung cancer, classic type (SCLC-c); NCI series of cell lines. The tumor cells usually lack substrate adherence, and grow either as true spheroids (Type I pattern) or as irregular floating clumps (Type II pattern). Cytological examination of cultured cells or xenograft of the cells in athymic nude mice demonstrates the typical morphology of SCLC cells—small cells with scant cytoplasm, inconspicuous nucleoli, and chromatin dispersed as finely granular clumps.

of these cells are dead. Preparation of a strict single-cell suspension will result in death of many or most cells. The best method to passage floating cultures is to tilt the culture dish so the aggregates settle to the bottom, decant the supernatant containing (mainly dead) single cells, add a small quantity of fresh medium, and gently pipette until small clusters are formed; then pass these clusters into new medium.

Typical (or classic) SCLC lines have relatively long population doubling times, which may seem surprising considering the explosive *in vivo* growth and spread of this tumor. In contrast, individual cell doubling times are rela-

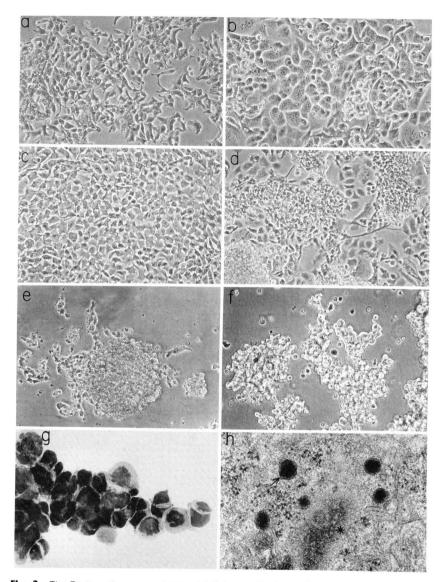

Fig. 2. The Dartmouth group, which establishes and cultures SCLC cells in Waymouth's medium, has established cell lines that show greater heterogeneity of growth appearances than the NCI series of cell lines. (a–f) A variety of morphologies ranges from epithelioid attached cells to floating clusters, as well as some with intermediate patterns. (g) Morphology of one of the cultures. (h) Ultrastructural examination reveals small membrane-bound dense core granules in the cytoplasm, a characteristic feature of neuroendocrine cells. (Figure provided courtesy of Dr. Olive Pettengill, Dartmouth Medical School, Hanover, New Hampshire.)

tively short, as demonstrated by the numerous mitotic figures present in cultures. However, the high spontaneous death rate (apoptosis?) results in long population doubling times.

A. Classic Subtype

Classic cultures express all the features characteristic of NE cells, including dense core granules (Fig. 2) (Gazdar et al., 1980; Carney et al., 1985; Jensen et al., 1990; Carbone et al., 1991). In addition, cultures express several peptide products characteristic of SCLC; individual lines may express as many as 10 peptides (Moody et al., 1981; Sorenson et al., 1981b,1982; Giaccone et al., 1992). In fact, the cultures frequently express these markers at higher concentrations and at high frequencies than the tumors from which they were derived. These findings have several causes: (1) cell lines are pure tumor cell populations and lack stromal cells; (2) tumors often have extensive areas of necrosis; and (3) cell lines contain higher percentages of cycling cells than do tumors.

Cytologic morphologies and the histological appearances of xenografts of classic lines are similar to those of the so-called intermediate subtype of SCLC, irrespective of the subtype of the original tumor. These findings suggest that the intermediate subtype is the true morphological appearance of SCLC tumors, and that the so-called oat-cell subtype is an ischemic artifact. Many, but not all, cell lines are tumorigenic in athymic nude mice, although the latent periods may be long (Pettengill et al., 1980a; Gazdar et al., 1981). Intracranial inoculation into nude mice is a more sensitive method to demonstrate tumorigenicity, and results in a fatal form of meningo-encephalic spread (Chambers et al., 1981; Gazdar et al., 1981) that is not dissimilar to the brain metastases present at autopsy in a high percentage of patients.

Because of the floating growth pattern of many SCLC cultures, their in vitro appearances resemble those of B-lymphoblastoid (BL) cultures. BL lines are Epstein–Barr virus (EBV)-transformed B cells, and may be generated electively or may appear spontaneously, especially from samples containing large numbers of lymphocytes (lymph node biopsies, effusions, and marrow aspirates). The aggregates of BL cells have a characteristic appearance; careful examination often will distinguish them. BL cells are loosely aggregated, never form true spheroids, and have cytoplasmic protusions ("uropods") radially oriented from the aggregates. If identification is still uncertain, Giemsa-stained smears will demonstrate abundant cytoplasm and large nucleoli, unlike SCLC. In addition, BL cells express surface polyclonal immunoglobulins and EBV nuclear antigen (EBNA). Paired tumor and BL lines are very useful reagents for a number of molecular genetic studies, since the BL cells can be used as a source of constitutive DNA and the derivation of both lines from the same individual can be proven using polymorphic DNA probes. Spontaneous outgrowth of BL cells occurs in 30–40% of lymphoid cell-containing samples from EBV-positive patients cultured in serum-containing medium (A. F. Gazdar, unpublished data). Growth of BL cells is not immediate, but usually is notice-able after a few weeks of culture. Because of the rapid growth of BL cells, they often outgrow the relatively slowly dividing SCLC cells. Switching the culture to serum-free HITES medium often suppresses growth. However, if establishing both SCLC and BL cell populations is desirable, the sample should be sub-cultured simultaneously in both media.

B. Variant Form

During our initial experience with culture of SCLC, we noted that some cultures grew as very loose aggregates, sometimes serpentine, which we termed as the variant form of SCLC (SCLC-v), in contrast to the more typical or classic form (SCLC-c). In contrast to SCLC-c, in which individual cell details cannot be observed by phase contrast microscopy, SCLC-v cells have larger cytoplasmic and nuclear areas and prominent nucleoli. These features may be recognized by phase contrast microscopy (Fig. 3; Carney et al., 1985). SCLC-v

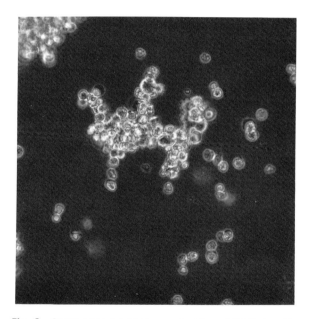

Fig. 3. Small-cell lung cancer, variant type (SCLC-v), cell line NCI-H82. Although the cells lack substrate attachment, in contrast to SCLC-c lines the cells grow as loose aggre-gates or in cords. Individual cell outlines can be discerned. These cells express only part of the neuroendocrine marker profile. However, perhaps because of the expression of neu-ral cell adhesion molecule (NCAM), the cells lack substrate attachment.

cells have a relatively high cloning efficiency, shorter doubling times, partial loss of NE markers, and rapid growth in athymic mice, and are relatively radioresistant. The finding that many but not all SCLC-v lines have amplified and overexpressed copies of the c-*myc* gene helped explain some of the unusual features (Little *et al.*, 1983; Gazdar *et al.*, 1985). Most lines have been established from patients who had received prior cytotoxic therapy. Overexpression of other members of the *myc* gene family is not associated consistently with the SCLC-v phenotype. Originally, about 25% of our SCLC cultures had variant features, but this proportion has dropped to less than 5%. This change may be the result of our improved ability to establish lines from untreated patients, as well as of the decreased incidence of c-*myc* amplification noted after switching patients to etoposide–cisplatin-based regimens for initial treatment (Johnson *et al.*, 1988,1992). Other investigators also have established SCLC-v lines, although their findings regarding the phenotypes are more variable than ours (de Leij *et al.*, 1985; Bepler *et al.*, 1989).

V. Other Neuroendocrine Tumors

In contrast to SCLC, bronchial carcinoids are well-differentiated NE tumors. These tumors are slow growing and relatively benign, seldom resulting in distant metastases or death. Very few cell lines have been established from these tumors. One of the lines we established, NCI-H727, replicates relatively slowly and grows attached to the substrate (Fig. 4; Gazdar *et al.*, 1988; Deftos *et al.*, 1990; Moody *et al.*, 1990). Others lack substrate attachment and grow in a manner that resembles SCLC. These lines probably arise from atypical carcinoids, tumors with appearances and clincial properties intermediate between SCLC and typical carcinoids (Arrigoni *et al.*, 1972).

About 15% of otherwise typical NSCLC tumors express many or all of the NE cell markers and are referred to as NSCLC–NE tumors (Fig. 5; Gazdar *et al.*, 1988,1992). These tumors are of potential clinical interest since they are more chemosensitive than other NSCLC tumors, both *in vitro* and *in vivo* (Graziano *et al.*, 1989; Gazdar *et al.*, 1992). Paradoxically, these tumors express high levels of the multidrug resistance-associated *mdr*1 gene product (Lai *et al.*, 1989). Cell lines established from NSCLC–NE tumors frequently lack substrate attachment (presumably because they express NCAM) (Carbone *et al.*, 1991), and appear to represent a distinct subtype of NSCLC tumors.

Although not a form of lung cancer, tumors morphologically identical to SCLC may arise in a number of organs, endocrine or otherwise (Remick *et al.*, 1987), and are referred to as extrapulmonary small-cell carcinomas (ExPuSc). Cultures established from ExPuSc tumors have the growth properties of SCLC lines and express NE properties. However, molecular genetic studies indicate that these tumors are entities distinct from SCLCs (Johnson *et al.*, 1989).

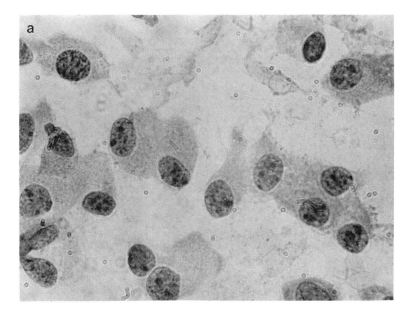

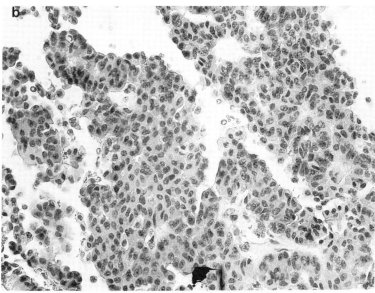

Fig. 4. Bronchial carcinoid cell line NCI-H727. This cell line is the only well-characterized one established from a tumor that approaches a "typical" carcinoid. Several others have been established from more atypical carcinoid tumors. The latter usually express NCAM and grow as floating aggregates (Type II morphology). However, NCI-H727 cells lack NCAM expression and grow as attached cells. (a) Stained morphology of the cells demonstrates bland uniform cells with small nucleoli and modest amounts of cytoplasm. (b) A xenograft of the cells confirms the resemblance to a carcinoid tumor. However, in other areas the xenografted cells demonstrate focal mucin production. Although mucin production is not a characteristic of carcinoid cells per se, many central carcinoids (i.e., carcinoids arising from larger bronchi) have focal areas of mucin production. Carcinoids are neuroendocrine tumors that demonstrate greater degrees of differentiation than SCLC tumors.

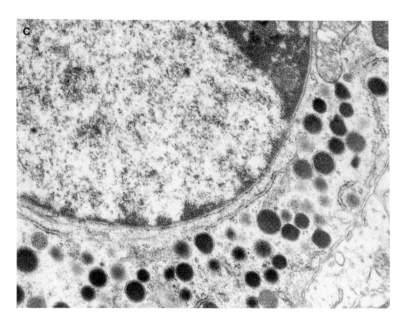

Fig. 4. *Continued* (c) Electron microscopic demonstration of numerous large, dense core cytoplasmic granules.

VI. Cell Lines Derived from Non-Small-Cell Lung Carcinoma

Although more than 100 cell lines have been established from NSCLC tumors (see Table I), most of them either have not been characterized properly or lack evidence of squamous or glandular differentiation. Such tumors would be regarded as large-cell carcinomas, although many were derived from morphologically differentiated tumors (Bepler *et al.*, 1988a). One interpretation of these findings is that selection of rapidly growing undifferentiated cells occurred during culture. Most of these lines have been established in SSM and grow attached to the substrate (Bepler *et al.*, 1988a). Establishing cell lines from metastatic lesions (about 35%) is considerably easier than establishing them from primary tumors (about 10%). The ability to establish tumor cell lines is an independent negative prognostic factor for survival in NSCLC (Stevenson *et al.*, 1990).

We have used SSM and defined media to establish NSCLC lines that express differentiated properties. Lines have been established from all the major histological varieties (squamous, large-cell, and adenocarcinoma), as well as from many of the rarer forms such as NSCLC–NE (see preceding discussion) and mucoepidermoid carcinoma (Yoakum *et al.*, 1983). The *in vitro* properties of

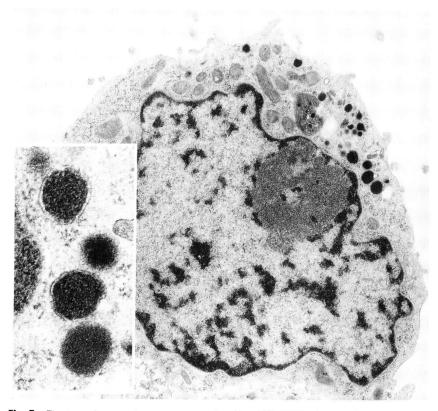

Fig. 5.   Electron microscopic appearances of cell line NCI-H810, a large-cell carcinoma that expresses neuroendocrine (NE) markers (NSCLC–NE tumor). These tumors resemble other NSCLCs in morphological appearances, but express part or all of the NE cell program, including NCAM. Hence, most lines lack substrate attachment. The cell line contains many large, dense core granules (inset), as would a carcinoid cell line.

these lines demonstrate much greater heterogeneity than SCLC, reflecting the diverse nature of the collection of tumor cell types referred to as NSCLC. Although many lines demonstrate substrate attachment, others lack this feature and grow as floating aggregates. Some of the better defined types of lines are described in the following sections.

A. Squamous Cell Carcinoma

Although we have had relatively little success in culturing well-differentiated squamous cell carcinomas, we have established several lines from adneo-squamous carcinomas that continue to express all the properties characteristic of squamous cells (Levitt *et al.*, 1990). Keratinizing (squamous) epithelia are characterized by many features, including production of high molecular weight keratins, prominent desmosomes, transglutaminase enzyme activity, involucrin production, and formation of tough water-impermeable cornified envelopes (also known as cross-linked envelopes) in the presence of high calcium concentrations. Synthesis of the cornified envelope is the ultimate step in squamous differentiation. This highly insoluble structure, produced on the inside of the plasma membrane, protects against loss of body fluids, entrance of toxic agents and microorganisms, and damage by physical forces. The extraordinary insolubility of the envelope is generated by the action of the enzyme transglutaminase, which forms cross-links between precursor proteins including involucrin and loricrin. Several of the lines express these characteristic features (Levitt et al., 1990). Squamous epithelia are stratified; the basal cells are involved in replication, and have the appearance of progressive differentiation toward the more superficial cells. Thus, full differentiation *in vitro* is seen only in multilayered cultures and not in rapidly dividing cells (Fig. 6). Also, because the final stage of differentiation (e.g., cornified envelope formation) is a terminal event, fully differentiated squamous cells are essentially dead cells designed to form a water-insoluble protective barrier layer. These considerations are important in determining the degree of differentiation in cultured squamous cells. In addition, cell culture methods must tread a fine line between replicating undifferentiated cells and nonreplicating fully differentiated cells.

B. Adenocarcinomas

As previously discussed, adenocarcinomas (ADC) of the lung consist of several varieties, partially reflecting their origin from different parts of the bronchial tree. More central tumors frequently form glandular structures and elaborate mucin. These gland-forming tumors, although considered the prototype of lung ADC, are decreasing in incidence; only a few cell lines exist. We have established a few mucin-producing tumor lines, most of which demonstrate substrate attachment. Mucin production is variable, ranging from a few vacuoles to goblet cell production. Considerably more cell lines have been established from peripheral airway cells (PAC) (Gazdar et al., 1990a). These cell lines are also heterogeneous and demonstrate differentiation toward Clara cells, Type II pneumocytes, mucin production, or some combination of these features. We investigated the expression of PAC markers in a bank of 41 lung cancer cell lines. Ultrastructural studies demonstrated the presence of cytoplasmic structures characteristic of Clara cells or of Type II pneumocytes in 9 of 34 (26%) NSCLC cell lines, including 7 of 17 (41%) adenocarcinomas, one squamous cell carcinoma, and one large-cell carcinoma. Of interest was the finding that the cytoplasmic structures were present in 5 of 6 (83%) cell lines initiated from papillolepidic adenocarcinomas. In addition, we examined the lines for expression of the surfactant-associated proteins SP-A, SP-B, and

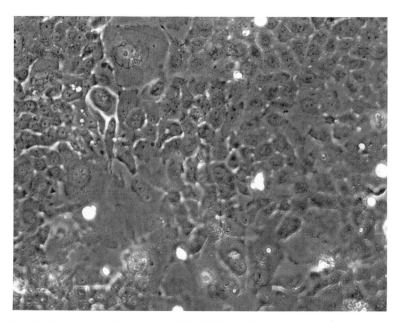

Fig. 6. Squamous cell carcinoma SCC-15. The cells grow as an adherent sheet that becomes multilayered after confluence. Because of the multilayered nature of the culture, some cells appear to be out of focus in this micrograph. The more superficial cells undergo terminal differentiation, although cornified envelope formation is present only after exposure to high calcium concentrations, preferably in the presence of an ionophore.

SP-C. Of the 9 cell lines containing cytoplasmic inclusions characteristic of PAC cells, 8 also expressed protein or RNA of SP-A, the major surfactant-associated protein. Of these lines, 5 expressed SP-B RNA (either constitutively or after dexamethasone induction), whereas a single line expressed SP-C only after dexamethasone induction. Of 6 SCLC cell lines examined, none expressed any of the PAC markers. Thus, PAC markers are expressed frequently (but not exclusively) in pulmonary adenocarcinoma tumors and cell lines, especially in those initiated from tumors having papillolepidic growth patterns (Gazdar et al., 1990a; Linnoila et al., 1992).

PAC cell lines are derived from tumors with the characteristic feature of papillary growth (Figs. 7–9). These characteristics are maintained in vitro. Many PAC lines are poorly adherent or grow as floating masses (Fig. 8). However, their growth patterns are considerably different from those of SCLC. Papillary in vitro growth is characterized by rounded cellular mases with their apical surfaces oriented outward (in contrast to glandular structures, in which the apices are oriented toward centrally located lumina). PAC cultures frequently elaborate and secrete markers characteristic of PAC cells, including one or more surfactant-associated proteins, Clara cell 10-kDa protein, and

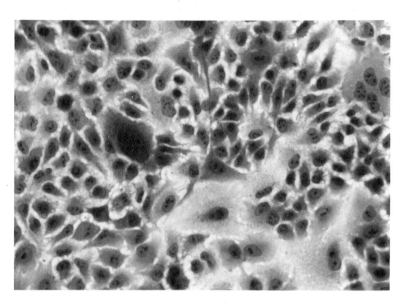

Fig. 7. Adenocarcinoma cell line A549. This widely used cell line was established from a peripheral adenocarcinoma of the bronchioloalveolar subtype. Although it is claimed that line expresses markers characteristic of Type II cell differentiation, the author and others have not been able to confirm these findings.

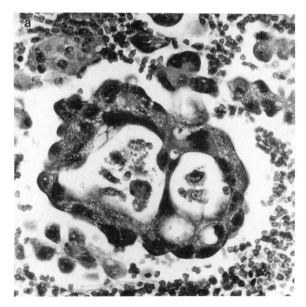

Fig. 8. Culture characteristics of peripheral adenocarcinomas (PAC). A characteristic growth feature of these tumors is the formation of papillary structures both *in vivo* and *in vitro*. Cell lines established from these tumors may or may not

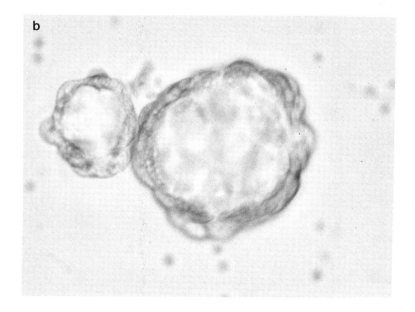

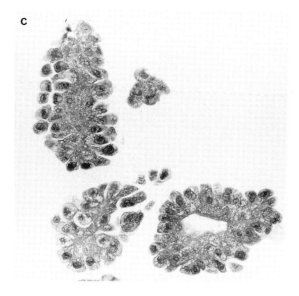

demonstrate substrate attachment. (a) Papillary tumor cells in a malignant effusion caused by tumor invasion of the pleura. (b,c) Similar appearances of floating
cell line NCI-H1404 in culture (phase-contrast photomicrograph and stained
cells, respectively). The papillary-like structures have been referred to as "reverse glands" because the apical surfaces of the cells are oriented outward. The
centers of the structures often are distended with vectorially transported fluid.

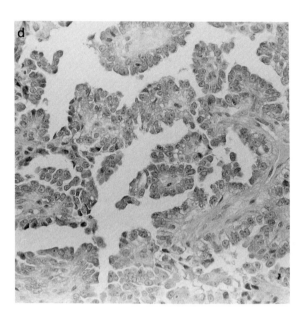

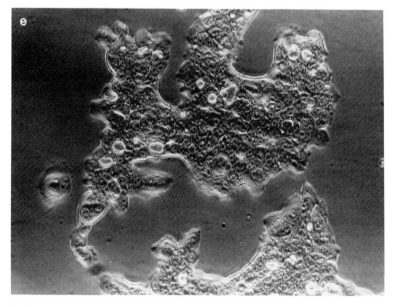

Fig. 8. *Continued* (d) A xenograft of the cultured cells demonstrates characteristic papillary structures. (e) An attached cell line, NCI-H441. This cell line grows as irregular islets of tight epitheloid cells.

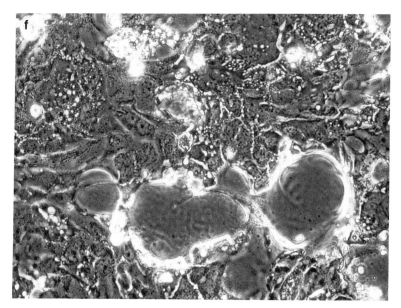

Fig. 8. *Continued* (f) Another attached PAC line demonstrates dome formation, which is the morphological appearance of vectorially transported, fluid-filled structures in attached cells.

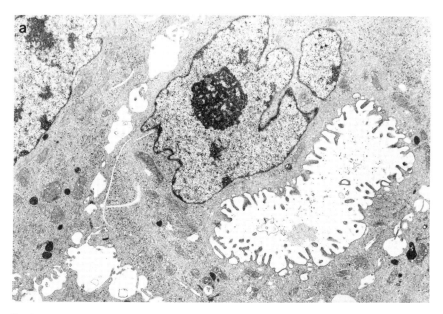

Fig. 9. Ultrastructural appearances of PAC tumor cell lines. (a) NCI-H441 demonstrates the presence of an intracellular glandlike structure. Note the presence of microvilli.

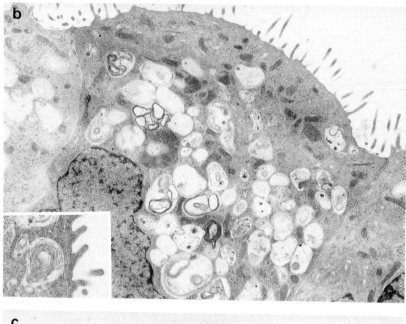

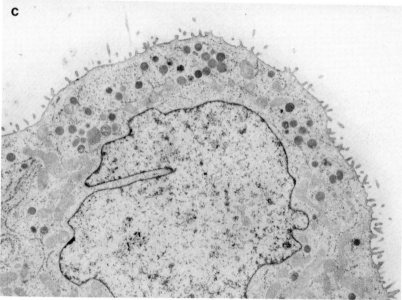

Fig. 9. *Continued* (b) Cell line NCI-H920 demonstrates abundant multilamellar bodies, the intracytoplasmic storage sites of surfactant. Because the lipid contents of the multilamellar bodies do not always survive fixation, many of the bodies have an empty appearance. However, the inset shows the characteristic appearance of a body about to be extruded from the cell. (c) NCI-1404 shows large granules of varying density. Although they are membrane bound, these granules do not have the "bull's eye" appearance of neuroendocrine granules. They represent the cytoplasmic storage sites of Clara cell secretions and are concentrated on the apical side.

intracytoplasmic structures characteristic of these cells (Fig. 9; Gazdar *et al.*, 1990a). Because the characteristic cellular products are secreted apically, the center of glandular structures may contain mucin. In contrast, the products of PAC cells are secreted into the supernatant fluids. However, these cells are involved in vectorial transport of fluids, so the centers of floating papillary structures often are filled with supernatant fluids that have been absorbed into the cell from the apical surface and transported out of the cell at the basolateral surface. In adherent PAC cultures, vectorial transport is manifested by dome formation, which consists of fluid-filled dome-shaped structures that lift the cells off the substrate. Eventually, these domes rupture, leaving roughly circular gaps in the cell sheath.

C. Large-Cell Carcinomas

By definition, large-cell carcinomas have no distinguishing features, nor do cell lines derived from them (Fig. 10). Most lines grow as attached epitheloid cells, without evidence of squamous, glandular, or papillary differentiation. Many (perhaps most) cell lines derived from other types of NSCLC also lack differentiation, and are indistinguishable from large-cell carcinoma cell lines, indicating selective growth in culture of the least differentiated tumor.

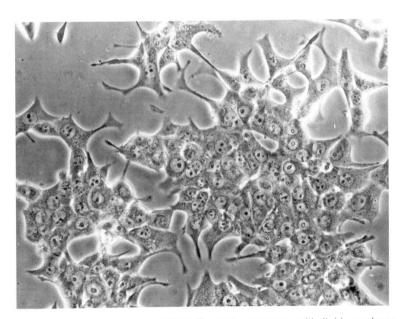

Fig. 10. Large-cell carcinoma SCC-5. The cells grow as an epithelioid monolayer and do not express any evidence of differentiated features.

VII. Mesotheliomas

Malignant mesothelioma (MM) is an aggressive tumor of the mesothelial cells lining serosal cavities and frequently is associated with exposure to asbestos. Mesothelial cells are characteristically biphasic, assuming both fibroblastic and epithelioid morphologies. Until recently, studies of this tumor have been limited by an absence of well-characterized human MM cell lines. However, several series have been described (Ke *et al.*, 1989; Versnel *et al.*, 1989,1991; Manning *et al.*, 1991). Growth characteristics and cytological examination demonstrate characteristic mesothelial cell morphology. Epithelial membrane antigen (EMA) and cytokeratin were demonstrated in cells from all 5 lines. Cultured cells are biphasic, appearing more fibroblastic during growth and more epithelioid as confluence is approached. Adjacent cells frequently have small spaces between them, in contrast to true epithelioid cells, and are appreciated best in sparse cultures. These cells lack CEA and epithelial mucin. The presence of cell junctions, glycogen, and numerous long, thin, branching microvilli was readily demonstrable by electron microscopy. All lines had abnormal karyotypes; the modal chromosome number varied from 40 to 80. Variable chromosome numbers, numerous structural rearrangements, and unrecognizable marker chromosomes were observed readily. Characteristic chromosomal changes include del 6q21 (Manning *et al.*, 1991; Meloni *et al.*, 1992), but several others have been reported (Tiainen *et al.*, 1992). Cultures may grow in defined media devised for adenocarcinoma and may secrete a number of growth factors, including autocrine ones (Lauber *et al.*, 1992; Schmitter *et al.*, 1992).

VIII. Conclusions

The large number of characterized cell lines established from patients with a wide range of lung cancers reflects the intense interest in studying the tumor that is the primary cause of cancer deaths in the United States. Culture of these tumors has been complicated by the multiple different phenotypes, by the initial growth of mainly undifferentiated lines from NSCLC, and by the initial inability to culture SCLC tumors routinely. As discussed in this chapter, these problems have, for the most part, been corrected. The explosive growth in our knowledge about the biology and molecular genetics of lung cancer is largely the result of the intense study of appropriate cell lines.

References

Aguayo, S. M., Kane, M. A., King, T. E., Schwartz, M. I., Grauer, L., and Miller, Y.E. (1989). Increased levels of bombesin-like peptides in the lower respiratory tract of asymptomatic cigarette smokers. *J. Clin. Invest.* **84,** 1105–1113.

Amstad, P., Reddel, R. R., Pfeifer, A., Malan, S. L., Mark, G., and Harris, C. C. (1988). Neoplastic transformation of a human bronchial epithelial cell line by a recombinant retrovirus encoding viral Harvey ras. *Mol. Carcinogen.* **1,** 151–160.

Anger, B., Bockman, R., Andreeff, M., Erlandson, R., Jhanwar, S., Kameya, T., Saigo, P., Wright, W., Beattie, E. J., Oettgen, H. F., and Old, L. J. (1982). Characterization of two newly established human cell lines from patients with large-cell anaplastic lung carcinoma. *Cancer* **50,** 1518–1529.

Anonymous (1982). The World Health Organization histological typing of lung tumours. *Am. J. Clin. Pathol.* **77,** 123–136.

Arrigoni, M. G., Woolner, L. B., and Bernatz, P. E. (1972). Atypical carcinoid tumors of the lung. *J. Thorac. Cardiovasc. Surg.* **64,** 413–421.

Avis, I. L., Kovacs, T. O., Kasprzyk, P. G., Treston, A. M., Bartholomew, R., Walsh, J. H., Cuttitta, F., and Mulshine, J. L. (1991). Preclinical evaluation of an anti-autocrine growth factor monoclonal antibody for treatment of patients with small-cell lung cancer. *J. Natl. Cancer Inst.* **83,** 1470–1476.

Baillie-Johnson, H., Twentyman, P. R., Fox, N. E., Walls, G. A., Workman, P., Watson, J. V., Johnson, N., Reeve, J. G., and Bleehen, N. M. (1985). Establishment and characterisation of cell lines from patients with lung cancer (predominantly small cell carcinoma). *Br. J. Cancer* **52,** 495–504.

Battey, J., and Wada, E. (1991). Two distinct receptor subtypes for mammalian bombesin-like peptides. *Trends Neurosci.* **14,** 524–528.

Battey, J. F., Way, J. M., Corjay, M. H., Shapira, H., Kusano, K., Harkins, R., Wu, J. M., Slattery, T., Mann, E., and Feldman, R. I. (1991). Molecular cloning of the bombesin/gastrin-releasing peptide receptor from Swiss 3T3 cells. *Proc. Natl. Acad. Sci. U.S.A.* **88,** 395–399.

Baylin, S. B., Abeloff, M. D., Goodwin, G., Carney, D. N., and Gazdar, A. F. (1980). Activity of L-dopa decarboxylase and diamine oxidase (histaminase) in human lung cancer: The decarboxylase as a marker for small (oat) cell cancer in tissue culture. *Cancer Res.* **40,** 190–194.

Becker, D., and Sahin, A. A. (1989). Loss of heterozygosity at chromosomal regions 3p and 13q in non-small-cell carcinoma of the lung represents low-frequency events. *Genomics* **4,** 97–100.

Becker, K. L., and Gazdar, A. F. (1984). "The Endocrine Lung in Health and Disease." Saunders, Philadelphia.

Becker, K. L., and Gazdar, A. F. (1985). What can biology of lung cancer teach us about the endocrine lung? *Biochem. Pharmacol.* **34,** 155–159.

Becker, K. L., Gazdar, A. F., Carney, D. N., Snider, R. H., Moore, C. F., and Silva, O. L. (1983). Calcitonin secretion by continuous cultures of small cell carcinoma of the lung: Incidence and heterogeneity studies. *Cancer Lett.* **18,** 179–185.

Becker, K. L., Silva, O. L., Gazdar, A. F., Snider, R. H., and Moore, C. F. (1984). Calcitonin and small cell cancer of the lung. *In* "The Endocrine Lung in Health and Disease" (K. L. Becker and A. F. Gazdar, eds.), pp. 528–548. Saunders, Philadelphia.

Bepler, G., Jaques, G., Havemann, K., Koehler, A., Johnson, B. E., and Gazdar, A. F. (1987). Characterization of two cell lines with distinct phenotypes established from a patient with small cell lung cancer. *Cancer Res.* **47,** 1883–1891.

Bepler, G., Koehler, A., Kiefer, P., Havemann, K., Beisenherz, K., Jaques, G., Gropp, C., and Haeder, M. (1988a). Characterization of the state of differentiation of six newly established human non-small-cell lung cancer cell lines. *Differentiation* **37,** 158–171.

Bepler, G., Zeymer, U., Mahmoud, S., Fiskum, G., Palaszynski, E., Rotsch, M., Willey, J., Koros, A., Cuttitta, F., and Moody, T. W. (1988b). Substance P analogues function as bombesin receptor antagonists and inhibit small cell lung cancer clonal growth. *Peptides* **9,** 1367–1372.

Bepler, G., Bading, H., Heimann, B., Kiefer, P., Havemann, K., and Moelling, K. (1989). Expression of p64c-*myc* and neuroendocrine properties define three subclasses of small cell lung cancer. *Oncogene* **4,** 45–50.

Berendsen, H. H., De Leij, L., de Vries, E. G., Mesander, G., Mulder, N. H., de Jong, B., Buys, C. H., Postmus, P. E., Poppema, S., and Sluiter, H. J. (1988). Characterization of three small cell lung cancer cell lines established from one patient during longitudinal follow-up. *Cancer Res.* **48,** 6891–6899.

Bergh, J. C. (1990). Gene amplification in human lung cancer. The *myc* family genes and other proto-oncogenes and growth factor genes. *Am. Rev. Respir. Dis.*

Bergh, J., Larsson, E., Zech, L., and Nilsson, K. (1982). Establishment and characterization of two neoplastic cell lines (U-1285 and U-1568) derived from small cell carcinoma of the lung. *Acta Pathol. Microbiol. Immunol. Scand. A* **90,** 149–158.

Bergh, J., Nilsson, K., Ekman, R., and Giovanella, B. (1985). Establishment and characterization of cell lines from human small cell and large cell carcinomas of the lung. *Acta Pathol. Microbiol. Immunol. Scand. A* **93,** 133–147.

Bergh, J., Björk, P., Westlin, J. E., and Nilsson, S. (1988). Expression of an estramustine-binding associated protein in human lung cancer cell lines. *Cancer Res.* **48,** 4615–4619.

Bergh, J., Arnberg, H., Eriksson, B., and Lundqvist, G. (1989). The release of chromogranin A and B like activity from human lung cancer cell lines. A potential marker for a subset of small cell lung cancer. *Acta Oncol.* **28,** 651–654.

Bliss, D. P., Battey, J. F., Linnoila, R. I., Birrer, M. J., Gazdar, A. F., and Johnson, B. E. (1990). Expression of the atrial natriuretic factor gene in small cell lung cancer tumors and tumor cell lines. *J. Natl. Cancer Inst.* **82,** 305–310.

Bodner, S. M., Minna, J. D., Jensen, S. M., D'Amico, D., Carbone, D., Mitsudomi, T., Fedorko, J., Buchhagen, D. L., Nau, M. M., Gazdar, A. F., and Linnoila, R. I. (1992). Expression of mutant p53 proteins in lung cancer correlates with the class of p53 gene mutation. *Oncogene* **7,** 743–749.

Bonofil, R. D., Reddel, R. R., Ura, H., Reich, R., Fridman, R., Harris, C. C., and Klein-Szanto, J. P. (1989). Invasive and metastatic potential of v-Ha-*ras* transformed human bronchial epithelial cell line. *J. Natl. Cancer Inst.* **81,** 587–594.

Brandt, D. W., Burton, D. W., Gazdar, A. F., Oie, H. E., and Deftos, L. J. (1991). All major lung cancer cell types produce parathyroid hormone-like protein: Heterogeneity assessed by high performance liquid chromatography. *Endocrinology* **129,** 2466–2470.

Brauch, H., Johnson, B., Hovis, J., Yano, T., Gazdar, A., Pettengill, O. S., Graziano, S., Sorenson, G. D., Poiesz, B. J., Minna, J. D., Linehan, M., and Zbar, B. (1987). Molecular analysis of the short arm of chromosome 3 in small-cell and non-small cell carcinoma of the lung. *N. Engl. J. Med.* **317,** 1109–1113.

Brauch, H., Tory, K., Kotler, F., Gazdar, A. F., Pettengill, O. S., Johnson, B., Graziano, S., Winton, T., Buys, C. H., Sorenson, G. D., Minna, J., and Zbar, B. (1990). Molecular mapping of deletion sites in the short arm of chromosome 3 in human lung cancer. *Genes Chrom. Cancer* **1,** 240–246.

Brennan, J., O'Connor, T., Makuch, R. W., Simmons, A. M., Russell, E., Linnoila, R. I., Phelps, R. M., Gazdar, A. F., Ihde, D. C., and Johnson, B. E. (1991). *myc* family DNA amplification in 107 tumors and tumor cell lines from patients with small cell lung cancer treated with different combination chemotherapy regimens. *Cancer Res.* **51,** 1708–1712.

Broers, J. L., Carney, D. N., Klein Rot, M., Schaart, G., Lane, E. B., Vooijs, G. P., and Ramaekers, F. C. (1986). Intermediate filament proteins in classic and variant types of small cell lung carcinoma cell lines: A biochemical and immunochemical analysis using a panel of monoclonal and polyclonal antibodies. *J. Cell Sci.* **83,** 37–60.

Brooks, B. J., Battey, J., Nau, M. M., Gazdar, A. F., and Minna, J. D. (1987). Amplification and expression of the *myc* gene in small cell lung cancer. In " Advances in Viral Oncology" (G. Klein, ed.), pp. 155–172. Raven Press, New York.

Browder, T. M., Dunbar, C. E., and Nienhuis, A. W. (1989). Private and public autocrine loops in neoplastic cells. *Cancer Cells* **1,** 9–17.

Brower, M., Carney, D. N., Oie, H. K., Gazdar, A. F., and Minna, J. D. (1986). Growth of cell lines and clinical specimens of human non-small cell lung cancer in a serum-free defined medium. *Cancer Res.* **46,** 798–806.

Bunn, P. A., Linnoila, I., Minna, J. D., Carney, D., and Gazdar, A. F. (1985). Small cell lung cancer, endocrine cells of the fetal bronchus, and other neuroendocrine cells express the Leu-7 antigenic determinant present on natural killer cells. *Blood* **65**, 764–768.

Buys, C. H., Osinga, J., van der Veen, A. Y., Mooibroek, H., van der Hout, A. H., De, L. L., Postmus, P. E., and Carritt, B. (1987). Genome analysis of small cell lung cancer (SCLC) and clinical significance. *Eur. J. Respir. Dis. (Suppl.)* **149**, 29–36.

Carbone, D. P., Koros, A. M., Linnoila, R. I., Jewett, P., and Gazdar, A. F. (1991). Neural cell adhesion molecule expression and messenger RNA splicing patterns in lung cancer cell lines are correlated with neuroendocrine phenotype and growth morphology. *Cancer Res.* **51**, 6142–6149.

Cardona, C., Rabbitts, P. H., Spindel, E. R., Ghatei, M. A., Bleehen, N. M., Bloom, S. R., and Reeve, J. G. (1991). Production of neuromedin B and neuromedin B gene expression in human lung tumor cell lines. *Cancer Res.* **51**, 5205–5211.

Carney, D. N., Bunn, P. A., Gazdar, A. F., Pagan, J. F., and Minna, J. D. (1981). Selective growth in serum-free hormone supplemented medium of tumor cells obtained by biopsy from patients with small cell carcinoma of the lung. *Proc. Natl. Acad. Sci. U.S.A.* **78**, 3185–3189.

Carney, D. N., Gazdar, A. F., Bunn, P. A., and Guccion, J. G. (1982). Demonstration of the stem cell nature of clonogenic tumor cells from lung cancer patients. *Stem Cells* **1**, 149–164.

Carney, D. N., Mitchell, J. B., and Kinsella, T. J. (1983). In vitro radiation and chemotherapy sensitivity of established cell lines of human small cell lung cancer and its large cell morphological variants. *Cancer Res.* **43**, 2806–2811.

Carney, D. N., Gazdar, A. F., Bepler, G., Guccion, J., Marangos, P. J., Moody, T. W., Zweig, M. H., and Minna. J. D. (1985). Establishment and identification of small cell lung cancer cell lines having classic and variant features. *Cancer Res.* **45**, 2913–2923.

Chambers, W. F., Pettengill, O. S., and Sorenson, G. D. (1981). Intracranial growth of pulmonary small cell carcinoma cells in nude athymic mice. *Exp. Cell Res.* **49**, 90–97.

Chang, A. C., Israel, A., Gazdar, A., and Cohen, S. N. (1989). Initiation of pro-opiomelanocortin mRNA from a normally quiescent promoter in a human small cell lung cancer cell line. *Gene* **84**, 115–126.

Chauvin, C., Jacrot, M., Riondel, J., Brambilla, E., Foote, A. M., Brambilla, C., and Benabid, A. L. (1989). Amplification of oncogenes in lung carcinoma grafted in nude mice. *Anticancer Res.* **9**, 449–452.

Cole, S. P., Campling, B. G., Dexter, D. F., Holden, J. J., and Roder, J. C. (1986). Establishment of a human large cell lung tumor line (QU-DB) with metastatic properties in athymic mice. *Cancer* **58**, 917–923.

Cole, T. R., Hughes, H. E., Jeffreys, M. J., Williams, G. T., and Arnold, M. M. (1992). Small cell lung carcinoma in a patient with Sotos syndrome: Are genes at 3p21 involved in both conditions? *J. Med. Genet.* **29**, 338–341.

Cuttitta, F., Carney, D. N., Mulshine, J., Moody, T. W., Fedorko, J., Fischler, A., and Minna, J. D. (1985). Bombesin-like peptides can function as autocrine growth factors in human small cell lung cancer. *Nature (London)* **316**, 823–826.

Cuttitta, F., Levitt, M. L., Kasprzyk, P. G., Nakanishi, Y., Reeve, J., and Walsh, J. (1987). Growth of human cancer cell lines in unsupplemented basal media as a means of identifying autocrine growth factors. *Proc. Am. Assoc. Cancer Res.* **28**, 27.

Cuttitta, F., Fedorko, J., Gu, J., Lebacq-Verheyden, A. M., Linnoila, R. I., and Battey, J. F. (1988). Gastrin-releasing peptide gene-associated peptides are expressed in normal fetal lung and small cell lung cancer: A novel peptide family found in man. *J. Clin. Endocrinol. Metab.* **67**, 576–583.

Cuttitta, F., Kasprzyk, P. G., Treston, A. M., Avis, I., Jensen, S., Levitt, M., Siegfried, J., Mobley, C., and Mulshine, J. L. (1990). Autocrine growth factors that regulate the proliferation of pulmonary malignancies in man. In "Respiratory Epithelium" (D. G. Thomassen and P. Nettesheim eds.), pp. 228–270. Hemisphere, New York.

D'Amico, D., Carbone, D., Johnson, B. E., Meltzer, S. J., and Minna, J. D. (1992a). Polymorphic sites within the MCC and APC loci reveal very frequent loss of heterozygosity in human small cell lung cancer. *Cancer Res.* **52,** 1996–1999.

D'Amico, D., Carbone, D., Mitsudomi, T., Nau, M., Fedorko, J., Russell, E., Johnson, B., Buchhagen, D., Bodner, S., Phelps, R., Gazdar, A., and Minna, J. D. (1992b). High frequency of somatically acquired p53 mutations in small cell lung cancer cell lines and tumors. *Oncogene* **7,** 339–346.

Dazzi, H., Hasleton, P. S., Thatcher, N., Barnes, D. M., Wilkes, S., Swindell, R., and Lawson, R. A. (1989). Expression of epidermal growth factor receptor (EGF-R) in non-small cell lung cancer. Use of archival tissue and correlation of EGF-R with histology, tumour size, node status and survival. *Br. J. Cancer* **59,** 746–749.

Deftos, L., Gazdar, A. F., Ikeda, K., and Broadus, A. E. (1989). The parathyroid hormone-related protein associated with malignancy is secreted by neuroendocrine tumors. *Mol. Endocrinol.* **3,** 503–508.

Deftos, L. J., Gazdar, A. F., Hogue, A. R., Mullen, P. S., and Burton, D. W. (1990). Distinct patterns of chromogranin A-related species can be demonstrated in endocrine cells. *Bone Min.* **9,** 169–178.

De Fusco, P. A., Frytak, S., Dahl, R. J., Weiland, L. H., Unni, K. K., and Dewald, G. W. (1989). Cytogenetic studies in 11 patients with small cell carcinoma of the lung. *Mayo Clin. Proc.* **64,** 168–176.

de Leij, L., Postmus, P. E., Buys, C. H., Elema, J. D., Ramaekers, F., Poppema, S., Brouwer, M., van der Veen, A. Y., Mesander, G., and The, T. H. (1985). Characterization of three new variant type cell lines derived from small cell carcinoma of the lung. *Cancer Res.* **45,** 6024–6033.

Demetri, G. D., Zenzie, B. W., Rheinwald, J. G., and Griffin, J. D. (1989). Expression of colony-stimulating factor genes by normal human mesothelial cells and human malignant mesothelioma cells lines in vitro. *Blood* **74,** 940–960.

Doyle, L. A., Giangiulo, D., Hussain, A., Park, H. J., Yen, R. W., and Borges, M. (1989). Differentiation of human variant small cell lung cancer cell lines to a classic morphology by retinoic acid. *Cancer Res.* **49,** 6745–6751.

Doyle, L. A., Borges, M., Hussain, A., Elias, A., and Tomiyasu, T. (1990). An adherent subline of a unique small-cell lung cancer cell line downregulates antigens of the neural cell adhesion molecule. *J. Clin. Invest.* **86,** 1848–1854.

Doyle, L. A., Mabry, M., Stahel, R. A., Waibel, R., and Goldstein, L. H. (1991). Modulation of neuroendocrine surface antigens in oncogene-activated small cell lung cancer lines. *Br. J. Cancer Suppl.* **14,** 39–42.

Drabkin, H., Kao, F. T., Hartz, J., Hart, I., Gazdar, A., Weinberger, C., Evans, R., and Gerber, M. (1988). Localization of human ERBA2 to the 3p22—3p24.1 region of chromosome 3 and variable deletion in small cell lung cancer. *Proc. Natl. Acad. Sci. U.S.A.* **85,** 9258–9262.

Duchesne, G. M., Eady, J. J., Peacock, J. H., and Pera, M. F. (1987). A panel of human lung carcinoma lines: Establishment, properties and common characteristics. *Br. J. Cancer* **56,** 287–293.

Erim, T. J., Beckett, M. A., Miller, D., and Rheinwald, J. G. (1982). Clonal growth and serial passage of head and neck squamous cell carcinoma. *Proc. Am. Soc. Clin. Oncol.* **195.**

Falco, J. P., Baylin, S. B., Lupu, R., Borges, M., Nelkin, B. D., Jasti, R. K., Davidson, N. E., and Mabry, M. (1990). v-*ras*H induces non-small cell phenotype, with associated growth factors and receptors, in a small cell lung cancer cell line. *J. Clin. Invest.* **85,** 1740–1745.

Fisher, E. R., and Paulson, J. D. (1978). A new in vitro cell line established from human large cell variant of oat cell lung cancer. *Cancer Res.* **38,** 3830–3835.

Flejter, W. L., Li, F. P., Antman, K. H., and Testa, J. R. (1989). Recurring loss involving chromosomes 1, 3, and 22 in malignant mesothelioma: Possible sites of tumor suppressor genes. *Genes Chrom. Cancer* **1,** 148–154.

Fogh, J. (1975). "Human Tumor Cells *In Vitro.*" Plenum Press, New York.

Funa, K., Gazdar, A. F., Minna, J, D., and Linnoila, R. I. (1986). Paucity of β2-microglobulin expression on small cell lung cancer, bronchial carcinoids, and certain other neuroendocrine tumors. *Lab. Invest.* **55,** 186–190.

Gamou, S., Shimosato, Y., and Shimizu, N. (1990). Regulation of the epidermal growth factor receptor gene expression in a morphological variant isolated from an epidermal growth factor receptor-deficient small cell lung carcinoma cell line. *Cell Growth Diff.* **1**, 351–359.

Gazdar, A. F., Carney, D. N., Russell, E. K., Sims, H. L., Baylin, S. B., Bunn, P. J., Guccion, J. G., and Minna, J. D. (1980). Establishment of continuous, clonable cultures of small-cell carcinoma of lung which have amine precursor uptake and decarboxylation cell properties. *Cancer Res* **40**, 3502–3507.

Gazdar, A. F., and Carney, D. N. (1984). Endocrine properties of small cell lung carcinoma. *In* "The Endocrine Lung in Health and Disease" (K. L. Becker and A. F. Gazdar eds.), pp. 501–508. Saunders, Philadelphia.

Gazdar, A. F., and Linnoila, R. I. (1988). The pathology of lung cancer—Changing concepts and newer diagnostic techniques. *Sem. Oncol.* **15**, 215–225.

Gazdar, A. F., and Oie, H. K. (1986a). Cell culture methods for human lung cancer. *Cancer Genet. Cytogenet.* **19**, 5–10.

Gazdar, A. F., and Oie, H. K. (1986b). Growth of cell lines and clinical specimens of human non-small cell lung cancer in a serum-free defined medium. *Cancer Res.* **46**, 6011.

Gazdar, A. F., Carney, D. N., Russell, E. K., Sims, H. L., Baylin, S. B., Bunn, P. J., Guccion, J. G., and Minna, J. D. (1980). Establishment of continuous, clonable cultures of small-cell carcinoma of lung which have amine precursor uptake and decarboxylation cell properties. *Cancer Res.* **40**, 3502–3507.

Gazdar, A. F., Carney, D. N., Sims, H. L., and Simmons, A. (1981). Heterotransplantation of small cell carcinoma of the lung into nude mice: Comparison of intracranial and subcutaneous routes. *Int. J. Cancer* **28**, 777–783.

Gazdar, A. F., Carney, D. N., Nau, M. M., and Minna, J. D. (1985). Characterization of variant subclasses of cell lines derived from small cell lung cancer having distinctive biochemical, morphological and growth properties. *Cancer Res.* **45**, 2924–2930.

Gazdar, A. F., Shoemaker, R., Mayo, J., Oie, H. K., Donovan, P., and Fine, D. (1987). Human lung cancer xenografts and metastases in athymic (nude) mice. *In* "Immune-Deficient Animals in Biomedical Research, 5th International Workshop" (N. Rygaard, N. Brunner, N. Graem, and M. Spang-Thomsen, eds.), pp. 277–280. Karger, Basel.

Gazdar, A. F., Helman, L. J., Israel, M. A., Russell, E. K., Linnoila, R. I., Mulshine, J. L. Schuller, H. M., and Park, J. G. (1988). Expression of neuroendocrine cell markers L-DOPA decarboxylase, chromogranin A, and dense core granules in human tumors of endocrine and nonendocrine origin. *Cancer Res.* **48**, 4078–4082.

Gazdar, A. F., Linnoila, R. I., Kurita, Y., Oie, H. K., Mulshine, J. L., Clark, J. C., and Whitsett, J. A. (1990a). Peripheral airway cell differentiation in human lung cancer cell lines. *Cancer Res.* **50**, 5481–5487.

Gazdar, A. F., Park, J. G., and Oie, H. K. (1990b). Characteristics of human colorectal cell lines established in defined and serum-supplemented media. *In* "Colon Cancer Cells" (M. P. Moyer and G. Poste, eds.), pp. 227–251. Academic Press, New York.

Gazdar, A. F., Steinberg, S. M., Russell, E. K., Linnoila, R. I., Oie, H. K., Ghosh, B. C., Cotelingam, J. D., Johnson, B. E., Minna, J. D., and Ihde, D. C. (1990c). Correlation of in vitro drug-sensitivity testing results with response to chemotherapy and survival in extensive-stage small cell lung cancer: A prospective clinical trial. *J. Natl. Cancer Inst.* **82**, 117–124.

Gazdar, A. F., Kadoyama, C., Venzon, D., Park, J.-G., Tsai, C.-M., Linnoila, R. I., Mulshine, J. L., Ihde, D. C., and Giaccone, G. (1992). The association between histological type and neuroendocrine differentiation on drug sensitivity of lung cancer cell lines. *J. Natl. Cancer Inst. Monogr.* **13**, 23–29.

Gazzeri, S., Brambilla, E., Chauvin, C., Jacrot, M., Benabid, A. L., and Brambilla, C. (1990). Analysis of the activation of the *myc* family oncogene and of its stability over time in xenografted human lung carcinomas. *Cancer Res.* **50**, 1566–1570.

Gemma, A., Nakajima, T., Shiraishi, M., Noguchi, M., Gotoh, M., Sekiya, T., Niitani, H., and Shimosato, Y. (1988). myc family gene abnormality in lung cancers and its relation to xenotransplantability. *Cancer Res.* **48**, 6025—6028.

Giaccone, G., Battey, J., Gazdar, A. F., Oie, H., Draoui, M., and Moody, T. W. (1992). Neuromedin B is present in lung cancer cell lines. *Cancer Res. (Suppl.)* **52,** 2732s-2736s.

Graziano, S. L., Cowan, B. Y., Carney, D. N., Bryke, C. R., Mitter, N. S., Johnson, B. E., Mark, G. E., Planas, A. T., Catino, J. J., and Comis, R. L. (1987). Small cell lung cancer cell line derived from a primary tumor with a characteristic deletion of 3p. *Cancer Res.* **47,** 2148-2155.

Graziano, S. L., Mazid, R., Newman, N., Tatum, A., Oler, A., Mortimer, J. A., Gullo, J. L., DiFino, S. M., and Scalzo, A. J. (1989). The use of neuroendocrine immunoperoxidase markers to predict chemotherapy response in patients with non-small cell lung cancer. *J. Clin. Oncol.* **7,** 1398-1406.

Guillemin, B., Zhang, Y., Lee, T. C., and Rom, W. N. (1991). Role of peptide growth factors in asbestos-related human lung cancer. *Ann. N. Y. Acad. Sci.* **643,** 245-257.

Harbour, J. W., Sali, S. L., Whang-Peng, J., Gazdar, A. F., Minna, J. D., and Kaye, F. J. (1988). Abnormalities in structure and expression of the human retinoblastoma gene in SCLC. *Science* **241,** 353-357.

Havemann, K., Rotsch, M., Schoneberger, H. J., Erbil, C., Hennig, C., and Jaques, G. (1990). Growth regulation by insulin-like growth factors in lung cancer. *J. Steroid Biochem. Mol. Biol.* **37,** 877-882.

Hay, F. G., Duncan, L. W., and Leonard, R. C. (1991). Establishment and characterisation of two new small cell lung cancer cell lines—One from a patient with previous familial retinoblastoma. *Br. J. Cancer Suppl.* **14,** 43-45.

Heikkila, R., Trepel, J. B., Cuttitta, F., Neckers, L. M., and Sausville, E. A. (1987). Bombesin-related peptides induce calcium mobilization in a subset of human small cell lung cancer cell lines. *J. Biol. Chem.* **262,** 16456-16460.

Hensel, C. H., Hsieh, C. L., Gazdar, A. F., Johnson, B. E., Sakaguchi, A. Y., Naylor, S. L., Lee, W. H., and Lee, E. Y. (1990). Altered structure and expression of the human retinoblastoma susceptibility gene in small cell lung cancer. *Cancer Res.* **50,** 3067-3072.

Heymanns, J., Neumann, K., and Havemann, K. (1989). Tetanus toxin as a marker for small-cell lung cancer cell lines. *J. Cancer Res. Clin. Oncol.* **115,** 537-542.

Hibi, K., Takahashi, T., Sekido, Y., Ueda, R., Hida, T., Ariyoshi, Y., Takagi, H., and Takahashi, T. (1991). Coexpression of the stem cell factor and the c-*kit* genes in small-cell lung cancer. *Oncogene* **6,** 2291-2296.

Hiyoshi, H., Matsuno, Y., Kato, H., Shimosato, Y., and Hirohashi, S. (1992). Clinicopathological significance of nuclear accumulation of tumor suppressor gene p53 product in primary lung cancer. *Jpn. J. Cancer Res.* **83,** 101-106.

Hollstein, M., Sidransky, D., Vogelstein, B., and Harris, C. C. (1991). p53 mutations in human cancers. *Science* **253,** 49-53.

Ibson, J. M., Waters, J. J., Twentyman, P. R., Bleehan, N. M., and Rabbits, P. H. (1988). Oncogene amplification and chromosomal abnormalities in small cell lung cancer. *J. Cell Biochem.* **33,** 267-288.

Jaques, G., Rotsch, M., Wegmann, C., Worsch, U., Maasberg, M., and Havemann, K. (1988). Production of immunoreactive insulin-like growth factor I and response to exogenous IGF-I in small cell lung cancer cell lines. *Exp. Cell Res.* **176,** 336-343.

Jaques, G., Kiefer, P., Rotsch, M., Hennig, C., Göke, R., Richter, G., and Havemann, K. (1989). Production of insulin-like growth factor binding proteins by small-cell lung cancer cell lines. *Exp. Cell Res.* **184,** 396-406.

Jensen, S. M., Gazdar, A. F., Cuttitta, F., Russell, E. K., and Linnoila, R. I. (1990). A comparison of synaptophysin, chromogranin, and L-DOPA decarboxylase as markers for neuroendocrine differentiation in lung cancer cell lines. *Cancer Res.* **50,** 6068-6074.

Johansson, S., Rydqvist, B., Swerup, C., Heilbronn, E., and Arhem, P. (1989). Action potentials of cultured human oat cells: Whole-cell measurements with the patch-clamp technique. *Acta Physiol. Scand.* **135,** 573-578.

Johnson, B. E., Battey, J., Linnoila, I., Becker, K. L., Makuch, R., Snider, R., Carney, D. N., and

Minna, J. D. (1986). Changes in the phenotype of human small cell lung cancer cell lines following transfection and expression of the c-myc proto-oncogene. *J. Clin. Invest.* **78**, 525–532.

Johnson, B. E., Makuch, R. W., Simmons, A. D., Gazdar, A. F., Burch, D., and Cashell, A. W. (1988). *myc* family DNA amplification in small cell lung cancer patients' tumors and corresponding cell lines. *Cancer Res.* **48**, 5163–5166.

Johnson, B. E., Whang-Peng, J., Naylor, S. L., Zbar, B., Brauch, H., Lee, E., Simmons, A., Russell, E., Nam, M. H., and Gazdar, A. F. (1989). Retention of chromosome 3 in extrapulmonary small cell cancer shown by molecular and cytogenetic studies. *J. Natl. Cancer Inst.* **81**, 1223–1228.

Johnson, B. E., Brennan, J. F., Ihde, D. C., and Gazdar, A. F. (1992). *myc* family DNA amplification in tumors and tumor cell lines from patients with small cell lung cancer. *J. Natl. Cancer Inst. Monogr.* **13**, 39–43.

Kawashima, K., Shikama, H., Imoto, K., Izawa, M., Naruke, T., Okabayashi, K., and Nishimura, S. (1988). Close correlation between restriction fragment length polymorphism of the L-*myc* gene and metastasis of human lung cancer to the lymph nodes and other organs. *Proc. Natl. Acad. Sci. U.S.A.* **85**, 2353–2356.

Kaye, F. J., Kratzke, R. A., Gerster, J. L., and Horowitz, J. M. (1990). A single amino acid substitution results in a retinoblastoma protein defective in phosphorylation and oncoprotein binding. *Proc. Natl. Acad. Sci. U.S.A.* **87**, 6922–6926.

Ke, Y., Reddel, R. R., Gerwin, B. I., Reddel, H. K., Somers, A. N., McMenamin, M. G., LaVeck, M. A., Stahel, R. A., Lechner, J. F., and Harris, C. C. (1989). Establishment of a human in vitro mesothelial cell model system for investigating mechanisms of asbestos-induced mesothelioma. *Am. J. Pathol.* **134**, 979–991.

Kern, D. H., and Weisenthal, L. M. (1990). Highly specific prediction of antineoplastic drug resistance with an in vitro assay using suprapharmacologic drug exposures. *J. Natl. Cancer Inst.* **82**, 582–588.

Kern, J. A., Robinson, R. A., Gazdar, A., Torney, L., and Weiner, D. B. (1992). Mechanisms of p185HER2 expression in human non-small cell lung cancer cell lines. *Am. J. Respir. Cell. Mol. Biol.* **6**, 359–363.

Kiefer, P. E., Bepler, G., Kubasch, M., and Havemann, K. (1987). Amplification and expression of protooncogenes in human small cell lung cancer cell lines. *Cancer Res.* **47**, 6236–6242.

Klominek, J., Robert, K. H., Hjerpe, A., Wickström, B., and Gahrton, G. (1989). Serum-dependent growth patterns of two, newly established human mesothelioma cell lines. *Cancer Res.* **49**, 6118–6122.

Kok, K., Osinga, J., Carritt, B., Davis, M. B., van der Hout, A. H., van der Veen, A. Y., Landsvater, R. M., de Leij, L. F., Berendsen, H. H., Postmus, P. E., Poppema, S., and Buys, C. H. (1987). Deletion of a DNA sequence at the chromosomal region 3p21 in all major types of lung cancer. *Nature (London)* **330**, 578–581.

Krystal, G., Birrer, M., Way, J., Nau, M., Sausville, E., Thompson, C., Minna, J., and Battey, J. (1988). Multiple mechanisms for transcriptional regulation of the myc gene family in small-cell lung cancer. *Mol. Cell. Biol.* **8**, 3373–3381.

Kuo, W. L., Tenjin, H., Segraves, R., Pinkel, D., Golbus, M. S., and Gray, J. (1991). Detection of aneuploidy involving chromosomes 13,18, or 21 by fluorescence in situ hybridization (FISH) to interphase and metaphase amniocytes. *Am. J. Hum. Genet.* **49**, 112–119.

Lai, S. L., Goldstein, L. J., Gottesman, M. M., Pastan, I., Tsai, C. M., Johnson, B. E., Mulshine, J. L., Ihde, D. C., Kayser, K., and Gazdar, A. F. (1989). MDR1 gene expression in lung cancer. *J. Natl. Cancer Inst.* **81**, 1144–1150.

LaRocca, P. J., and Rheinwald, J. B. (1984). Coexpression of simple epithelial keratins and vimentin by human mesothelium and mesothelioma in vivo and in culture. *Cancer Res.* **44**, 2991–2999.

Lauber, B., Leuthold, M., Schmitter, D., Cano-Santos, J., Waibel, R., and Stahel, R. A. (1992). An autocrine mitogenic activity produced by a pleural human mesothelioma cell line. *Int. J. Cancer* **50**, 943–950.

Lavigueur, A., Maltby, V., Mock, D., Rossant, J., Pawson, T., and Bernstein, A. (1989). High incidence of lung, bone, and lymphoid tumors in transgenic mice overexpressing mutant alleles of the p53 oncogene. *Mol. Cell. Biol.* **9**, 3982–3991.

Leduc, F., Brauch, H., Hajj, C., Dobrovic, A., Kaye, F., Gazdar, A., Harbour, J. W., Pettengill, O. S., Sorenson, G. D., and van den Berg, A. (1989). Loss of heterozygosity in a gene coding for a thyroid hormone receptor in lung cancers. *Am. J. Hum. Genet.* **44**, 282–287.

Levitt, M. L., Gazdar, A. F., Oie, H. K., Schuller, H., and Thatcher, S. M. (1990). Cross-linked envelope-related markers for squamous differentiation in human lung cancer cell lines. *Cancer Res.* **50**, 120–128.

Lieber, M., Smith, B., Szakal, A., Nelson, R. W., and Todaro, G. (1976). A continuous tumor-cell line from a human lung carcinoma with properties of type II alveolar epithelial cells. *Int. J. Cancer* **17**, 62–70.

Linnoila, R. I., Jensen, S. M., Steinberg, S. M., Mulshine, J. L., Eggleston, J. C., and Gazdar, A. F. (1992). Peripheral airway cell marker expression in non-small cell lung carcinoma. Association with distinct clinicopathologic features. *Am. J. Clin. Pathol.* **97**, 233–43.

Little, C. D., Nau, M. M., Carney, D. N., Gazdar, A. F., and Minna, J. D. (1983). Amplification and expression of the c-*myc* oncogene in human lung cancer cell lines. *Nature (London)* **306**, 194–196.

Luduena, M. A., Sussman, H. H., and Rabson, A. S. (1974). Synthesis of human placental alkaline phosphatase in vitro by the ChaGo cell line. *J. Natl. Cancer Inst.* **52**, 1705–1709.

Lukeis, R., Irving, L., Garson, M., and Hasthorpe, S. (1990). Cytogenetics of non-small cell lung cancer: Analysis of consistent non-random abnormalities. *Genes Chrom. Cancer* **2**, 116–124.

Mabry, M., Nakagawa, T., Nelkin, B. D., McDowell, E., Gesell, M., Eggleston, J. C., Casero, R. A., and Baylin, S. B. (1988). v-Ha-ras oncogene insertion: A model for tumor progression of human small cell lung cancer. *Proc. Natl. Acad. Sci. U.S.A.* **85**, 6523–6527.

Mabry, M., Nakagawa, T., Baylin, S., Pettengill, O., Sorenson, G., and Nelkin, B. (1989). Insertion of the v-Ha-*ras* oncogene induces differentiation of calcitonin-producing human small cell lung cancer. *J. Clin Invest* **84**, 194–199.

Macaulay, V. M. (1992). Insulin-like growth factors and cancer. *Br. J. Cancer* **65**, 311–20.

Macaulay, V. M., Everard, M. J., Teale, J. D., Trott, P. A., Van Wyk, J. J., Smith, I. E., and Millar, J. L. (1990). Autocrine function for insulin-like growth factor I in human small cell lung cancer cell lines and fresh tumor cells. *Cancer Res.* **50**, 2511–2517.

McCann, F. V., Pettengill, O. S., Cole, J. J., Russell, J. A., and Sorenson, G. D. (1981). Calcium spike electrogenesis and other electrical activity in continuously cultured small cell carcinoma of the lung. *Science* **212**, 1155–1157.

McLemore, T. L., Eggleston, J. C., Shoemaker, R. H., Abbott, B. J., Bohlman, M. E., Liu, M. C., Fine, D. C., Mayo, J. G., and Boyd, M. R. (1988). Comparison of intrapulmonary, percutaneous intrathoracic, and subcutaneous models for the propagation of human pulmonary and non-pulmonary cancer cell lines in athymic nude mice. *Cancer Res.* **48**, 2880–2886.

McMahon, J. B., Schuller, H. M., Gazdar, A. F., and Becker, K. L. (1984). Influence of priming with 5-hydroxytryptophan and APUD characteristics in human small cell lung cancer cell lines. *Lung* **162**, 261–269.

Mahmoud, S., Palaszynski, E., Fiskum, G., Coy, D. H., and Moody, T. W. (1989). Small cell lung cancer bombesin receptors are antagonized by reduced peptide bond analogues. *Life Sci.* **44**, 367–373.

Manning, L. S., Whitaker, D., Murch, A. R., Garlepp, M. J., Davis, M. R., Musk, A. W., and Robinson, B. W. (1991). Establishment and characterization of five human malignant mesothelioma cell lines derived from pleural effusions. *Int. J. Cancer* **47**, 285–290.

Masuda, N., Fukuoka, M., Takada, M., Kudoh, S., and Kusunoki, Y. (1991). Establishment and characterization of 20 human non-small cell lung cancer lines in a serum-free defined medium (ACL-4). *Chest* **100**, 429–438.

Mellstrom, K., Bjelfman, C., Hammerling, U, and Pahlman, S. (1987). Expression of c-*src* in cultured human neuroblastoma and small cell lung carcinoma cell lines correlates with neuro-crine differentiation. *Mol. Cell. Biol.* **7,** 4178–4184.

Meloni, A. M., Stephenson, C. F., Li, F. P., and Sandberg, A. A. (1992). del(6q) as a possible primary change in malignant mesothelioma. *Cancer Genet. Cytogenet.* **59,** 57–61.

Mendelsohn, J. (1990). Anti-epidermal growth factor receptor monoclonal antibodies as potential anti-cancer agents. *J. Steroid Biochem. Mol. Biol.* **37,** 889–892.

Miller, C. W., Simon, K., Aslo, A., Kok, K., Yokota, J., Buys, C. H., Terada, M., and Koeffler, H. P. (1992). p53 mutations in human lung tumors. *Cancer Res.* **52,** 1695–1698.

Minna, J. D., Carney, D. N., Oie, H. K., Bunn, P. A., and Gazdar, A. F. (1982). Growth of human small cell lung cancer in defined medium. *In* "Growth of Cells in Hormonally Defined Media" (G. Sato and A. Pardee, eds.), pp. 627–639. Cold Spring Harbor Press, Cold Spring Harbor, New York.

Minna, J. D., Cuttitta, F., Battey, J. F., Mulshine, J. L., Linnoila, I., Gazdar, A. F., Trepel, J., and Sausville, E. A. (1988). Gastrin-releasing peptide and other autocrine growth factors in lung cancer: Pathogenetic and treatment implications. *In* "Important Advances in Oncology 1988" (V. T. DeVita, S. Hellman, and S. A. Rosenberg, eds.), pp. 55–64. Lippincott, Philadelphia.

Minna, J. D., Glatstein, E. J., Pass, H. I., and Ihde, D. C. (1989). Cancer of the lung. *In* "Cancer: Principles and Practice of Oncology" (V. T. DeVita, Jr., S. Hellman, and S. A. Rosenberg, eds.), pp. 591–705. Lippincott, Philadelphia.

Mitsudomi, T., Steinberg, S. M., Oie, H. K., Mulshine, J. L., Phelps, R., Viallet, J., Pass, H., Minna, J. D., and Gazdar, A. F. (1991a). *ras* gene mutations in non-small cell lung cancers are associated with shortened survival irrespective of treatment intent. *Cancer Res.* **51,** 4999–5002.

Mitsudomi, T., Viallet, J., Mulshine, J. L., Linnoila, R. I., Minna, J. D., and Gazdar, A. F. (1991b). Mutations of *ras* genes distinguish a subset of non-small-cell lung cancer cell lines from small-cell lung cancer cell lines. *Oncogene* **6,** 1353–1362.

Mitsudomi, T., Steinberg, S. M., Nau, M. M., Carbone, D., D'Amico, D., Bodner, S., Oie, H. K., Linnoila, R. I., Mulshine, J. L., Minna, J. D., and Gazdar, A. F. (1992). p53 gene mutations in non-small-cell lung cancer cell lines and their correlation with the presence of *ras* mutations and clinical features. *Oncogene* **7,** 171–80.

Miura, I., Siegfried, J. M., Resau, J., Keller, S. M., Zhou, J. Y., and Testa, J. R. (1990). Chromosome alterations in 21 non-small cell lung carcinomas. *Genes Chrom. Cancer* **2,** 328–338.

Moody, T. W., Pert, C. B., Gazdar, A. F., Carney, D. N., and Minna, J. D. (1981). High levels of intracellular bombesin characterize human small cell lung carcinoma. *Science* **214,** 1246–1248.

Moody, T. W., Carney, D. N., Cuttitta, F., Quattrocchi, K., and Minna, J. D. (1985). High affinity receptors for bombesin/GRP-like peptides on human small cell lung cancer. *Life Sci.* **37,** 105–113.

Moody, T. W., Lee, M., Kris, R. M., Bellot, F., Bepler, G., Oie, H., and Gazdar, A. F. (1990). Lung carcinoid cell lines have bombesin-like peptides and EGF receptors. *J. Cell Biochem.* **43,** 139–147.

Mooibroek, H., Osinga, J., Postmus, P. E., Carritt, B., and Buys, C. H. (1987). Loss of heterozygosity for a chromosome 3 sequence presumably at 3p21 in small cell lung cancer. *Cancer Genet. Cytogenet.* **27,** 361–365.

Moolenaar, C. E., Pieneman, C., Walsh, F. S., Mooi, W. J., and Michalides, R. J. (1992). Alternative splicing of neural-cell-adhesion molecule mRNA in human small-cell lung-cancer cell line H69. *Int. J. Cancer* **51,** 238–243.

Mori, N., Yokota, J., Akiyama, T., Sameshima, Y., Okamoto, A., Mizoguchi, H., Toyoshima, K., Sugimura, T., and Terada, M. (1990). Variable mutations of the RB gene in small cell lung cancer. *Oncogene* **5,** 1713–1717.

Morstyn, G., Russo, A., Carney, D. N., Karawya, E., Wilson, S. H., and Mitchell, J. B. (1984). Heterogeneity in the radiation survivial curves and biochemical properties of human lung cancer cell lines. *J. Natl. Cancer Inst.* **73,** 801–807.

Morstyn, G., Brown, J., Novak, U., Gardner, J., Bishop, J., and Garson, M. (1987). Heterogeneous cytogenetic abnormalities in small cell lung cancer cell lines. *Cancer Res.* **47,** 3322–3327.

Murakami, Y., Katahira, M., Makino, R., Hayashi, K., Hirohashi, S., and Sekiya, T. (1991). Inactivation of the retinoblastoma gene in a human lung carcinoma cell line detected by single-strand conformation polymorphism analysis of the polymerase chain reaction product of cDNA. *Oncogene* **6,** 37–42.

Nakanishi, Y., Mulshine, J. L., Kasprzyk, P. G., Natale, R. B., Maneckjee, R., Avis, I., Treston, A. M., Gazdar, A. F., Minna, J. D., and Cuttitta, F. (1988). Insulin-like growth factor-1 can mediate autocrine proliferation of human small cell lung cancer cell lines. *J. Clin. Invest.* **82,** 354–359.

Nau, M. M., Brooks, B. J., Battey, J., Sausville, E. A., Gazdar, A. F., and Minna, J. D. (1985). L-*myc:* A new *myc*-related gene amplified and expressed in human small cell lung cancer. *Nature (London)* **318,** 69–73.

Nau, M. M., Brooks, B. J., Carney, D. N., Gazdar, A. F., Battey, J. F., Sausville, E. A., and Minna, J. D. (1986). Human small cell lung cancers show amplification and expression of the N-*myc* gene. *Proc. Natl. Acad. Sci. U.S.A.* **83,** 1092–1096.

Naylor, S. L., Johnson, B. E., Minna, J. D., and Sakaguchi, A. Y. (1987). Loss of heterozygosity of chromosome 3p markers in small cell lung cancer. *Nature (London)* **329,** 451–454.

Nervi, C., Vollberg, T. M., George, M. D., Zelent, A., Chambon, P., and Jetten, A. M. (1991). Expression of nuclear retinoic acid receptors in normal tracheobronchial cells and in lung carcinoma cells. *Exp. Cell Res.* **195,** 163–170.

Oboshi, S., Tsugawa, S., Seido, T., Shimosato, Y., and Koide, T. (1971). A new floating cell line derived from human pulmonary carcinoma of oat cell type. *Gann* **62,** 505–514.

Okabe, T., Fujisawa, M., Kudo, H., Honma, H., Ohsawa, N., and Takaku, F. (1984). Establishment of a human colony-stimulating-factor-producing cell line from an undifferentiated large cell carcinoma of the lung. *Cancer* **54,** 1024–1029.

Park, J. G., Oie, H. K., Sugarbaker, P. H., Henslee, J. G., Chen, T. R., Johnson, B. E., and Gazdar, A. F. (1987). Characteristics of cell lines established from human colorectal carcinomas. *Cancer Res.* **47,** 6710–6718.

Pettengill, O. S., Faulkner, C. S., Wurster, H. D., Maurer, L. H., Sorenson, G. D., Robinson, A. G., and Zimmerman, E. A. (1977). Isolation and characterization of a hormone-producing cell line from human small cell anaplastic carcinoma of the lung. *J. Natl. Cancer Inst.* **58,** 511–518.

Pettengill, O. S., Curphey, T. J., Cate, C. C., Flint, C. F., Maurer, L. H., and Sorenson, G. D. (1980a). Animal model for small cell carcinoma of the lung. Effect of immunosuppression and sex of mouse on tumor growth in nude athymic mice. *Exp. Cell Biol.* **48,** 279–297.

Pettengill, O. S., Sorenson, G. D., Wurster-Hill, D. H., Curphey, T. J., Noll, W. W., Cate, C. C., and Maurer, L. H. (1980b). Isolation and growth characteristics of continuous cell lines from small-cell carcinoma of the lung. *Cancer* **45,** 906–918.

Pettengill, O. S., Bacopoulos, N. G., and Sorenson, G. D. (1982). Biogenic amine metabolites in human lung tumor cells: Histochemical and mass-spectrographic demonstration. *Life Sci.* **30,** 1355–1360.

Pettengill, O. S., Carney, D. N., Sorenson, G. D., and Gazdar, A. F. (1984). Establishment and characterization of cell lines from human small cell carcinoma of the lung. *In* "The Endocrine Lung in Health and Disease" (K. L. Becker and A. F. Gazdar, eds.), pp. 460–468. Saunders, Philadelphia.

Pfeifer, A. M., Jones, R. T., Bowden, P. E., Mann, D., Spillare, E., Klein-Szanto, A. J., Trump, B. F., and Harris, C. C. (1991). Human bronchial epithelial cells transformed by the c-*raf*-1 and c-*myc* protooncogenes induce multidifferentiated carcinomas in nude mice: A model for lung carcinogenesis. *Cancer Res.* **51,** 3793–3801.

Polak, J. M., Hamid, Q., Springall, D. R., Cuttitta, F., Spindel, E., Ghatei, M. A., and Bloom, S. R. (1988). Localization of bombesin-like peptides in tumors. *Ann. N.Y. Acad. Sci.* **547,** 322–335.

Possmayer, F. (1988). A proposed nomenclature for pulmonary surfactant-associated proteins. *Am. Rev. Respir. Dis.* **138,** 990–998.

Postmus, P. E., de Ley, L., Van der Veen, A. Y., Mesander, G., Buys, C. H., and Elema, J. D. (1988). Two small cell lung cancer cell lines established from rigid bronchoscope biopsies. *Eur. J. Cancer Clin. Oncol.* **24,** 753–763.

Quinn, K., Treston, A. M., Scott, F. M., Kasprzyk, P. G., Avis, I., Siegfried, J. M., Mulshine, J. L., and Cuttitta, F. (1991). α-Amidation of peptide hormones in lung cancer. *Cancer Cells* **3,** 504–510.

Reeve, J. G., Payne, J. A., and Bleehen, N. M. (1990). Production of immunoreactive insulin-like growth factor-I (IGF-I) and IGF-I binding proteins by human lung tumours. *Br. J. Cancer* **61,** 727–731.

Reeve, J. G., Brinkman, A., Hughes, S., Mitchell, J., Schwander, J., and Bleehen, N. M. (1992). Expression of insulinlike growth factor (IGF) and IGF-binding protein genes in human lung tumor cell lines. *J. Natl. Cancer Inst.* **84,** 628–634.

Remick, S. C., Hafez, G. R., and Carbone, P. P. (1987). Extrapulmonary small cell carcinoma: A review of the literature with emphasis on therapy and outcome. *Medicine* **66,** 457–471.

Rodenhuis, S., and Slebos, R. J. (1990). The ras oncogenes in human lung cancer. *Am. Rev. Respir. Dis.* **142,** S27–S30.

Saksela, K. (1990). *myc* genes and their deregulation in lung cancer. *J. Cell Biochem.* **42,** 153–180.

Sameshima, Y., Akiyama, T., Mori, N., Mizoguchi, H., Toyoshima, K., Sugimura, T., Terada, M., and Yokota, J. (1990). Point mutation of the p53 gene resulting in splicing inhibition in small cell lung carcinoma. *Biochem. Biophys. Res. Commun.* **173,** 697–703.

Sameshima, Y., Matsuno, Y., Hirohashi, S., Shimosato, Y., Mizoguchi, H., Sugimura, T., Terada, M., and Yokota, J. (1992). Alterations of the p53 gene are common and critical events for the maintenance of malignant phenotypes in small-cell lung carcinoma. *Oncogene* **7,** 451–457.

Sausville, E. A., Lebacq, V. A., Spindel, E. R., Cuttitta, F., Gazdar, A. F., and Battey, J. F. (1986). Expression of the gastrin-releasing peptide gene in human small cell lung cancer: Evidence for alternative processing resulting in three distinct mRNAs. *J. Biol. Chem.* **261,** 2451–2457.

Sausville, E. A., Moyer, J. D., Heikkila, R., Neckers, L. M., and Trepel, J. B. (1988). A correlation of bombesin-responsiveness with *myc*-family gene expression in small cell lung carcinoma cell lines. *Ann. N.Y. Acad. Sci.* **547,** 310–321.

Schmitter, D., Lauber, B., Fagg, B., and Stahel, R. A. (1992). Hematopoietic growth factors secreted by seven human pleural mesothelioma cell lines: Interleukin-6 production as a common feature. *Int. J. Cancer* **51,** 296–301.

Schuller, H. M. (1991). Neuroendocrine lung cancer: A receptor-mediated disease? *Exp. Lung Res.* **17,** 837–852.

Seifter, E., Sausville, E. A., and Battey, J. (1986). Comparison of amplified and unamplified c-*myc* gene structure and expression in human small cell lung carcinoma cell lines. *Cancer Res.* **46,** 2050–2055.

Sekido, Y., Obata, Y., Ueda, R., Hida, T., Suyama, M., Shimokata, K., Ariyoshi, Y., and Takahashi, T. (1991). Preferential expression of c-*kit* protooncogene transcripts in small cell lung cancer. *Cancer Res.* **51,** 2416–2419.

Sekido, Y., Takahashi, T., Makela, T. P., Obata, Y., Ueda, R., Hida, T., Hibi, K., Shimokata, K., Alitalo, K., and Takahashi, T. (1992). Complex intrachromosomal rearrangement in the process of amplification of the L-myc gene in small-cell lung cancer. *Mol. Cell. Biol.* **12,** 1747–1754.

Shigematsu, K., Kataoka, Y., Kamio, T., Kurihara, M., Niwa, M., and Tsuchiyama, H. (1990). Partial characterization of insulin-like growth factor I in primary human lung cancers using immunohistochemical and receptor autoradiographic techniques. *Cancer Res.* **50,** 2481–2484.

Shiraishi, M., and Sekiya, T. (1989). Change in methylation status of DNA of remaining allele in human lung cancers with loss of heterozygosity on chromosomes 3p and 13q. *Jpn. J. Cancer Res.* **80,** 924–927.

Shiraishi, M., Morinaga, S., Noguchi, M., Shimosato, Y., and Sekiya, T. (1987). Loss of genes on the short arm of chromosome 11 in human lung carcinomas. *Jpn. J. Cancer Res.* **78,** 1302–1308.

Simms, E., Gazdar, A. F., Abrams, P. G., and Minna, J. D. (1980). Growth of human small cell (oat cell) carcinoma of the lung in serum-free growth factor supplemented medium. *Cancer Res.* **40,** 4356–4363.

Skinner, M. A., Vollmer, R., Huper, G., Abbott, P., and Iglehart, J. D. (1990). Loss of heterozygosity for genes on 11p and the clinical course of patients with lung carcinoma. *Cancer Res.* **50,** 2303–2306.

Soderdahl, G., Betsholtz, C., Johansson, A., Nilsson, K., and Bergh, J. (1988). Differential expression of platelet-derived growth factor and transforming growth factor genes in small- and non-small-cell human lung carcinoma lines. *Int. J. Cancer* **41,** 636–641.

Sorensen, G. D., Pettengill, O. S., and Cate, C. C. (1981a). Studies on xenografts of small cell carcinoma of the lung. *In* "Small Cell Lung Cancer" (F. A. Greco, R. K. Oldham, and P. A. Bunn, eds.) pp. 95–121. Grune and Stratton, New York.

Sorenson, G. D., Pettengill, O. S., Brinck-Johnsen, T., Cate, C. C., and Mauer, L. H. (1981b). Hormone production by cultures of small cell carcinoma of the lung. *Cancer* **47,** 1289–1296.

Sorenson, G. D., Bloom, S. R., Ghatei, M. A., Del, P. S., Cate, C. C., and Pettengill, O. S. (1982). Bombesin production by human small cell carcinoma of the lung. *Reg. Peptides* **4,** 59–66.

Sozzi, G., Miozzo, M., Tagliabue, E., Calderone, C., Lombardi, L., Pilotti, S., Pastorino, U., Pierotti, M. A., and Della-Porta, G. (1991). Cytogenetic abnormalities and overexpression of receptors for growth factors in normal bronchial epithelium and tumor samples of lung cancer patients. *Cancer Res.* **51,** 400–404.

Spindel, E. R., Chin, W. W., Price, J., Rees, L. H., Bessmer, G. M., and Habener, J. F. (1984). Cloning and characterization of cDNAs encoding human gartin releasing peptide. *Proc. Natl. Acad. Sci. U.S.A.* **81,** 5699–5703.

Stark, M., Hohr, D., and Usmiani, J. (1988). Characterization of a new permanent squamous carcinoma line from human lung B 109. *Anticancer Res.* **8,** 263–268.

Stevenson, H. C., Gazdar, A. F., Linnoila, R. I., Russell, E. K., Oie, H. K., Steinberg, S. M., and Ihde, D. C. (1989). Lack of relationship between in vitro tumor cell growth and prognosis in extensive stage small cell lung cancer. *J. Clin. Oncol.* **7,** 923–931.

Stevenson, H., Gazdar, A. F., Phelps, R., Linnoila, R. I., Ihde, D. C., Ghosh, B., Walsh, T., Woods, E. L., Oie, H., O'Connor, T., and Mulshine, J. L. (1990). Tumor cell lines established *in vitro:* An independent prognostic factor for survival in non-small-cell lung cancer. *Ann. Intern. Med.* **113,** 764–770.

Stewart, M. F., Crosby, S., Gibson, S., Twentyman, P. R., and White, A. (1989). Small cell lung cancer cell lines secrete predominantly ACTH precursor peptides, not ACTH. *Br. J. Cancer* **60,** 20–24.

Sugio, K., Ishida, T., Yokoyama, H., Inoue, T., Sugimachi, K., and Sasazuki, T. (1992). *ras* gene mutations as a prognostic marker in adenocarcinoma of the human lung without lymph node metastasis. *Cancer Res.* **52,** 2903–2906.

Sunday, M. E., Kaplan, L. M., Motoyama, E., Chin, W. W., and Spindel, E. R. (1988). Biology of disease: Gastrin-releasing peptide (mammalian bombesin) gene expression in health and disease. *Lab. Invest.* **59,** 5–24.

Sunday, M. E., Hua, J., Dai, H. B., Nusrat, A., and Torday, J. S. (1990). Bombesin increases fetal lung growth and maturation in utero and in organ culture. *Am. J. Respir. Cell. Mol. Biol.* **3,** 199–205.

Takahashi, T., Nau, M. M., Chiba, I., Birrer, M. J., Rosenberg, R. K., Vinocour, M., Levitt, M., Pass, H., Gazdar, A. F., and Minna, J. D. (1989a). p53: A frequent target for genetic abnormalities in lung cancer. *Science* **246,** 491–494.

Takahashi, T., Obata, Y., Sekido, Y., Hida, T., Ueda, R., Watanabe, H., Ariyoshi, Y., and Sugiura, T. (1989b). Expression and amplification of *myc* gene family in small cell lung cancer and its relation to biological characteristics. *Cancer Res.* **49,** 2683–2688.

Takahashi, T., D'Amico, D., Chiba, I., Buchhagen, D. L., and Minna, J. D. (1990). Identification of intronic point mutations as an alternative mechanism for p53 inactivation in lung cancer. *J. Clin. Invest.* **86,** 363–369.

Takahashi, T., Takahashi, T., Suzuki, H., Hida, T., Sekido, Y., Ariyoshi, Y., and Ueda, R. (1991). The p53 gene is very frequently mutated in small-cell lung cancer with a distinct nucleotide substitution pattern. *Oncogene* **6**, 1775–1778.

Takahashi, T., Carbone, D., Takahashi, T., Nau, M. M., Hida, T., Linnoila, I., Ueda, R., and Minna, J. D. (1992). Wild-type but not mutant p53 suppresses the growth of human lung cancer cells bearing multiple genetic lesions. *Cancer Res.* **52**, 2340–2343.

Takaki, T. (1980). An epithelial cell line (KNS-62) derived from a brain metastasis of bronchial squamous cell carcinoma. *J. Cancer Res. Clin. Oncol.* **96**, 27–33.

Tateishi, M., Ishida, T., Mitsudomi, T., Kaneko, S., and Sugimachi, K. (1990). Immunohistochemical evidence of autocrine growth factors in adenocarcinoma of the human lung. *Cancer Res.* **50**, 7077–7080.

Tefre, T., Brresen, A. L., Aamdal, S., and Brgger, A. (1990). Studies of the L-*myc* DNA polymorphism and relation to metastasis in Norwegian lung cancer patients. *Br. J. Cancer* **61**, 809–812.

Tiainen, M., Kere, L., Tammilehto, L., Mattson, K., and Knuutila, S. (1992). Abnormalities of chromosomes 7 and 22 in human malignant pleural mesothelioma: Correlation between Southern blot and cytogenetic analyses. *Genes Chrom. Cancer* **4**, 176–182.

Trepel, J. B., Moyer, J. D., Cuttitta, F., Frucht, H., Coy, D. H., Natale, R. B., Mulshine, J. L., Jensen, R. T., and Sausville, E. A. (1988). A novel bombesin receptor antagonist inhibits autocrine signals in a small cell lung carcinoma cell line. *Biochem. Biophys. Res. Commun.* **156**, 1383–1389.

Tyson, F. L., Boyer, C. M., Kaufman, R., O'Briant, K., Cram, G., Crews, J. R., Soper, J. T., Daly, L., Fowler, W. J., Haskill, J. S., *et al.* (1991). Expression and amplification of the HER-2/neu (c-erbB-2) protooncogene in epithelial ovarian tumors and cell lines. *Am. J. Obstet. Gynecol.* **165**, 640–646.

Ura, H., Bonfil, R. D., Reich, R., Reddel, R., Pfeifer, A., Harris, C. C., and Klein-Szanto, A. J. (1989). Expression of type IV collagenase and procollagen genes and its correlation with the tumorigenic, invasive, and metastatic abilities of oncogene-transformed human bronchial epithelial cells. *Cancer Res.* **49**, 4615–4621.

van der Hout, A. H., Kok, K., van der Veen, A. Y., Osinga, J., de Leij, L. F., and Buys, C. H. (1989). Localization of amplified c-*myc* and n-*myc* in small cell lung cancer cell lines. *Cancer Genet. Cytogenet.* **38**, 1–8.

Versnel, M. A., Bouts, M. J., Hoogsteden, H. C., van der Kwast, T. H., Delahaye, M., and Hagemeijer, A. (1989a). Establishment of human malignant mesothelioma cell lines. *Int. J. Cancer* **44**, 256–260.

Versnel, M. A., Hoogsteden, H. C., Hagemeijer, A., Bouts, M. J., van der Kwast, T. H., Delahaye, M., Schaart, G., and Ramaekers, F. C. (1989b). Characterization of three human malignant mesothelioma cell lines. *Cancer Genet. Cytogenet.* **42**, 115–128.

Versnel, M. A., Claesson-Welsh, L., Hammacher, A., Bouts, M. J., van der Kwast, T. H., Eriksson, A., Willemsen, R., Weima, S. M., Hoogsteden, H. C., and Hagemeijer, A. (1991). Human malignant mesothelioma cell lines express PDGF beta-receptors whereas cultured normal mesothelial cells express predominantly PDGF alpha-receptors. *Oncogene* **6**, 2005–2011.

Watanabe, H., Takahashi, T., Ueda, R., Utsumi, K. R., Sato, T., Ariyoshi, Y., Ota, K., Obata, Y., and Takahashi, T. (1988). Antigenic phenotype and biological characteristics of two distinct sublines derived from a small cell lung carcinoma cell line. *Cancer Res.* **48**, 2544–2549.

Waters, J. J., Ibson, J. M., Twentyman, P. R., Bleehen, N. M., and Rabbitts, P. H. (1988). Cytogenetic abnormalities in human small cell lung carcinoma: Cell lines characterized for *myc* gene amplification. *Cancer Genet. Cytogenet.* **30**, 213–223.

Weiner, D. B., Nordberg, J., Robinson, R., Nowell, P. C., Gazdar, A., Greene, M. I., Williams, W. V., Cohen, J. A., and Kern, J. A. (1990). Expression of the *neu* gene-encoded protein (P185neu) in human non-small cell carcinomas of the lung. *Cancer Res.* **50**, 421–425.

Weston, A., Willey, J. C., Haugen, A., Krontiris, T., McDowell, E., Resau, J., Trump, B. F., and Harris, C. C. (1987). DNA-restriction fragment length polymorphisms in human bronchogenic carcinoma. *Proc. Am. Assoc. Cancer Res.* **28**, 251.

Weston, A., Willey, J. C., Modali, R., Sugimura, H., McDowell, E. M., Resau, J., Light, B., Haugen, A., Mann, D. L., Trump, B. F. (1989). Differential DNA sequence deletions from chromosomes 3, 11, 13, and 17 in squamous-cell carcinoma, large-cell carcinoma, and adenocarcinoma of the human lung MD 20892. *Proc. Natl. Acad. Sci. U.S.A.* **86**, 5099–5103.

Whang-Peng, J., Bunn, P. A., Kao, S. C., Lee, E. C., Carney, D. N., Gazdar, A. F., and Minna, J. D. (1982). A non-random chromosomal abnormality, del 3p(14-23) in human small cell lung cancer. *Cancer Genet. Cytogenet.* **6**, 119–134.

Whang-Peng, J., Carney, D. N., Lee, E. C., Kao, S. C., Bunn, P. A., Gazdar, A. F., and Minna, J. D. (1983). A non-random chromosomal abnormality, del 3(p14-23) in small cell lung cancer. *In* "Cancer: Etiology and Prevention" (R. G. Crispen, ed.), pp. 47–60. Elsevier Science, Amsterdam.

Whang-Peng, J., Knutsen, T., Gazdar, A., Steinberg, S. M., Oie, H., Linnoila, I., Mulshine, J., Nau, M., and Minna, J. D. (1991). Nonrandom structural and numerical chromosome changes in non-small-cell lung cancer. *Genes Chrom. Cancer* **3**, 168–188.

Wiedermann, C. J. (1989). Bombesin-like peptides as growth factors. *Wien. Klin. Wochenschr.* **101**, 435–440.

Woll, P. J., and Rozengurt, E. (1988). Bombesin and bombesin antagonists: Studies in Swiss 3T3 cells and human small cell lung cancer. *Br. J. Cancer* **57**, 579–586.

Wong, A. J., Ruppert, J. M., Eggleston, J., Hamilton, S. R., Baylin, S. B., and Vogelstein, B. (1986). Gene amplification of c-*myc* and N-*myc* in small cell carcinoma of the lung. *Science* **233**, 461–464.

Xu, H. J., Hu, S. X., Cagle, P. T., Moore, G. E., and Benedict, W. F. (1991). Absence of retinoblastoma protein expression in primary non-small cell lung carcinomas. *Cancer Res.* **51**, 2735–2739.

Yoakum, G. H., Korba, B. E., Lechner, J. F., Tokiwa, T., Gazdar, A. F., Leeman, M. S., Anutrup, H., and Harris, C. C. (1983). High frequency transfection of cytopathology of the hepatitis B virus core antigen in human cells. *Science* **222**, 385–389.

Yokota, J., Wada, M., Shimosato, Y., Terada, M., and Sugimura, T. (1987). Loss of heterozygosity on chromosomes 3, 13, and 17 in small-cell carcinoma and on chromosome 3 in adenocarcinoma of the lung. *Proc. Natl. Acad. Sci. U.S.A.* **84**, 9252–9256.

Yokota, J., Akiyama, T., Fung, Y. K., Benedict, W. F., Namba, Y., Hanaoka, M., Wada, M., Terasaki, T., Shimosato, Y., Sugimura, T., and Terada, M. (1988). Altered expression of the retinoblastoma (RB) gene in small cell carcinoma of the lung. *Oncogene* **3**, 471–475.

Yokota, J., Mori, N., Akiyama, T., Shimosato, Y., Sugimura, T., and Terada, M. (1989a). Multiple genetic alterations in small-cell lung carcinoma. *Int. Symp. Princess Takamatsu Cancer Res. Fund* **20**, 43–48.

Yokota, J., Tsukada, Y., Nakajima, T., Gotoh, M., Shimosato, Y., Mori, N., Tsunokawa, Y., Sugimura, H., and Terada, M. (1989b). Loss of heterozygosity of the short arm of chromosome 3 in carcinoma of the uterine cervix. *Cancer Res.* **49**, 3598–3601.

Yoshida, K., Tsuda, T., Matsumura, T., Tsujino, T., Hattori, T., Ito, H., and Tahara, E. (1989). Amplification of epidermal growth factor receptor (EGFR) gene and oncogenes in human gastric carcinomas. *Virchows Arch. B Cell Pathol.* **57**, 285–290.

Zbar, B. (1989). Chromosomal deletions in lung cancer and renal cancer. *In* "Important Advances in Oncology" (V. DeVita, S. Hellman, and S. A. Rosenberg, eds.) pp. 41–60. J. B. Lippincott, Philadelphia.

Cell Lines from Human Breast

Albert Leibovitz

Arizona Cancer Center, College of Medicine
University of Arizona, Tucson, Arizona 85724

I. Introduction 161

II. Methods of Establishment and Maintenance 162

III. Morphology 162
A. Cell Lines Established from Breast Tissue 162
B. Metastasis to Solid Tissues 173
C. Metastasis to Pleural Fluid 175
D. Metastasis to Ascitic Fluid 176

IV. Discussion 180
A. Establishment of Cell Lines from Primary Breast Carcinomas 180
B. Establishment of Cell Lines from Metastatic Sites 180

V. Future Prospects 181
A. Markers to Differentiate Tumor Cells from Normal Cells 181
B. New Media to Enhance Cell Growth of Normal Epithelial Cells and Carcinoma Cells 181

References 181

I. Introduction

This atlas of cell lines derived from human mammary tissues or from metastatic sites is obtained primarily from 22 cell lines furnished by the American Type Culture Collection (ATCC) and 11 cell lines established in my laboratory. Review of the literature does not reveal any significantly different morphological types. About 85% of the cell lines are derived from intraductal carcinomas; the remaining few include lines derived from ductal carcinomas with papillary, medullary, or lobular involvement; one characterized cell line from a probable cystosarcoma phyllodes; four cell lines from "normal" tissue or breast milk; and one xenograft. Readers desiring to compare the *in vivo* morphology to the *in vitro* morphology of breast carcinomas should consult Carter (1990).

Atlas of Human Tumor Cell Lines
161

II. Methods of Establishment and Maintenance

A significant review of cell lines from breast carcinomas was written in 1978 by Engel and Young. This presentation is also a good review of the problems associated with attempting to establish cell lines from primary breast tissue, the diversity of media utilized, and cross-contamination, especially by the HeLa cell line. This chapter updates the list of established cell lines that have been characterized.

Articles on technique in establishing cell lines from mammary tissues or metastases to solid tissues essentially follow the basic methods of Lasfargues (1975; Lasfargues *et al.*, 1958,1978). For establishment of cell lines metastatic to pleural fluids, the basic methods of Cailleau (1975; Cailleau *et al.*, 1974,1978; Young *et al.*, 1974; Pathak *et al.*, 1979; Brinkley *et al.*, 1980) yielded 16 cell lines over a 6-yr span. For a general review of the wide variety of methods in use to establish cell lines from solid tumors, see Leibovitz (1985,1986).

To date, no reviews are available of the wide variety of media in use today to establish cell lines from primary breast tissue. Cell lines have been established using relatively simple media (Yamane *et al.*, 1984; Minafra *et al.*, 1989), although most investigators use quite complex media (Meltzer *et al.*, 1991). The trend is toward developing chemically defined serum-free media. Other publications (Band and Sager, 1989; Band *et al.*, 1990; Petersen *et al.*, 1990) indicate that such media will be available soon.

Maintenance of established cell lines is relatively simple compared with establishing cell lines. In early passages, some lines are dependent on certain growth factors (e.g., MCF-7 required estradiol; Soule *et al.*, 1973), but once a cell line becomes well established it usually proliferates in a wide variety of media, unless it is completely dependent on an exogenous source of a specific growth factor (Meltzer *et al.*, 1991).

III. Morphology

A. Cell Lines Established from Breast Tissue

1. Intraductal Carcinoma Cell Lines

See Table I for a list of cell lines derived from breast tissue carcinomas.

These lines may be very tight colonies of epithelial cells in which visualizing the internal structure of the cells is difficult (Fig. 1). Most lines are spreading islands in which individual cells are discerned more readily (Figs. 2,3). These cells have a high nucleus: cytoplasm ratio (Fig. 3) and may have several nucleoli (Fig. 3). The cells usually do not form a complete monolayer, even when the flask is filled with islands (Fig. 4).

Table I

Cell Lines Established from Carcinomas in Breast Tissue

Cell line	Morphology	Pathology[a]	Reference
BT-20	Fig. 3	IDC	Lasfargues and Ozzello (1958)
CaMa	Fig. 3	IDC	Dobrynin (1963)
BTM-1	Fig. 7	IDC	Martorelli *et al.* (1969)
BOT-2	Fig. 8	IDC	Nordquist *et al.* (1975)
Hs578T	Fig. 5	CAR	Hackett *et al.* (1977)
BT-474	Figs. 2,3	IDC	Lasfargues *et al.* (1978)
BT-483	Figs. 2,3	IDC	Lasfargues *et al.* (1978)
BT-549	Fig. 7	PAP	Couthino and Lasfargues (1978)
YMB-1	Fig. 3	IDC	Yamane *et al.* (1984)
CAL-18A	Fig. 3	IDC	Gioanne *et al.* (1985)
CAL-18B	Fig. 8	IDC	Gioanne *et al.* (1985)
Ca2-83	Fig. 3	IDC	Rudland *et al.* (1985)
VHB-1	Fig. 3	IDC	Vandewalle *et al.* (1987)
RW-972	Fig. 6	SAR	Tibbetts *et al.* (1988)
UACC-812	Fig. 1	IDC	Leibovitz *et al.* (1988)
UACC-893	Fig. 3	IDC	Leibovitz *et al.* (1988)
8701-BC	Fig. 7	IDC	Minafra *et al.* (1989)
21 NT	Fig. 3	IDC	Band *et al.* (1989)
21 PT	Fig. 3	IDC	Band *et al.* (1989)
HMT-3909 S1	Fig. 7	MED	Petersen *et al.* (1990)
HMT-3909 S8	Fig. 3	IDC[b]	Petersen *et al.* (1990)
UACC-2116	Fig. 2	XEN	Leibovitz and Massey (1992)

[a] IDC, intraductal carcinoma; CAR, carcinosarcoma; PAP, papillary carcinoma; SAR, sarcoma; MED, medullary carcinoma; XEN, xenograft.
[b] The HMT-3909 cell lines are derived from breast tissue with both intraductal and medullary carcinomas.

2. Ductal Carcinomas with Papillary Involvement

Only one such cell line has been established from primary breast tissue (Couthino and Lasfargues, 1978), but the result was not published. The line was donated to the ATCC and included in their catalog (BT-549 under the ATCC number HTB 122). The repository describes this line as a papillary intraductal carcinoma that is polymorphic, with epithelial components and giant cells, similar to HMT-3909 shown in Fig. 7.

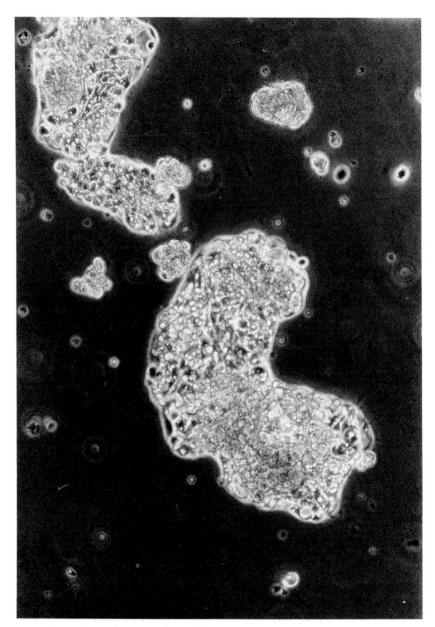

Fig. 1. UACC 812 (Leibovitz *et al.*, 1988; Meltzer *et al.*, 1991), established from a carcinoma in breast tissue, shows three-dimensional slow-growing islands of tightly packed epithelial cells. Observing morphology of individual cells is difficult. Phase-contrast microscopy, 160×.

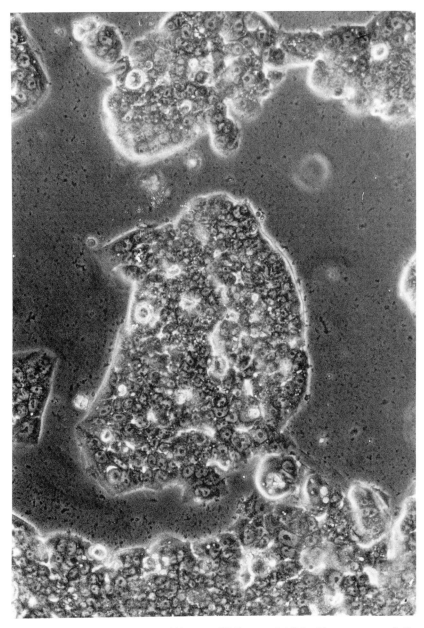

Fig. 2. UACC 2116 (Leibovitz and Massey, 1992) was established from a xenograft. The original specimen was a primary human breast carcinoma (B. Giovanella, personal communication). These cells are also slow spreading, forming moderately tight islands of epithelial cells, but the internal morphology of some of the cells can be seen. Phase-contrast microscopy, 160×.

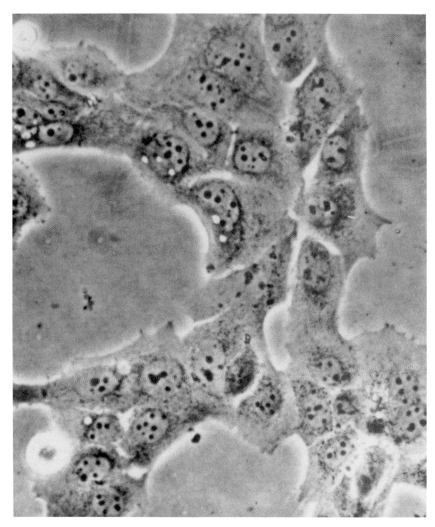

Fig. 3. HMT-3909S8 (Petersen *et al.*, 1990) are epithelial cells growing in islands, but the three-dimensional growth is minimal, allowing the internal morphology of most cells to be observed. The high nucleus : cytoplasm ratio is observed easily by phase-contrast microscopy; most cells have two or more nucleoli. This type of island growth is observed most commonly in cell cultures from primary breast carcinomas and isolates from metastatic carcinomas. Phase-contrast microscopy, 400×. (Reproduced from Petersen *et al.*, 1990, with permission of the American Association for Cancer Research and the Waverly Press.)

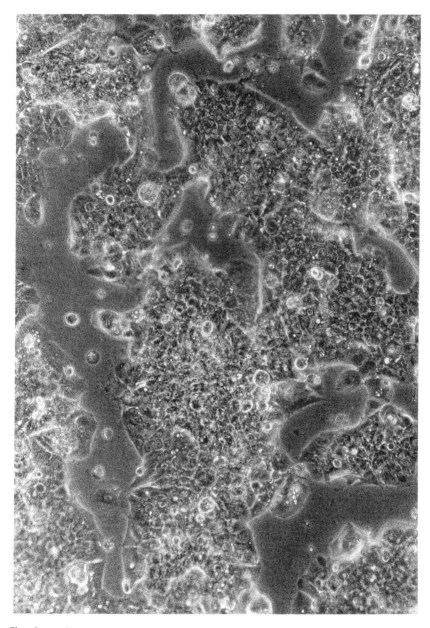

Fig. 4. UACC 893 (Leibovitz *et al.*, 1988; Meltzer *et al.*, 1991), isolated from a primary breast carcinoma. Almost all epithelial carcinoma isolates that have an island-type outgrowth grow three-dimensionally and persist as islands. Phase-contrast microscopy, 160×.

3. Ductal Carcinoma with Medullary Involvement

The HMT-3909 cell lines (Petersen *et al.*, 1990) were established from a primary breast carcinoma with both intraductal and medullary involvement. The HMT-3909S1 line was reticulate in morphology (see Fig. 7) and did not produce tumors in nude mice. The HMT-3909S8 cell line grew as islands (Fig. 3) and did produce tumors is nude mice.

4. Carcinosarcoma Cell Line

Hackett *et al.* (1977) established the Hs578T cell line from carcinosarcoma of the breast. The polymorphic epithelial cells grow as islands and form a monolayer (Fig. 5).

5. Cystosarcoma Phyllodes

Cell line RW-972 was established by Tibbetts *et al.* (1988), and may be the first cell line of its type to be published. The outgrowth is composed of a mixture of epithelial-like and stromal cells (Fig. 6). Of interest, some of the polymorphic cell lines established from metastatic epithelial carcinomas have a similar morphology (Fig. 7).

6. "Normal" Breast Tissues and Breast Milk

Hs57Bst was established by Hackett *et al.* (1977) from normal breast tissue of the patient with a carcinosarcoma noted earlier. This spindle-shaped cell line has characteristics of fibroblast cells and has a low nucleus:cytoplasm ratio, but was shown to be myoepithelial by electron microscopy. This line has a diploid karyogram. The morphology is similar to that of HMT-3909, shown in Fig. 7.

HBL-100 was established by Gaffney (1982) from the breast milk of a woman with no known pathology. The cell line is aneuploid, grows as spreading islands of epithelial cells (Fig. 3), forms colonies in soft agar, and is immortal.

HMT-3522 was established by Briand *et al.* (1987) in serum-free media from breast tissue of a woman with fibrocystic disease. This line is diploid, but has a few marker chromosomes and a loss of chromosomes in the D group. The line is epithelial, as determined by electron microscopy, and the cells grow as spreading islands (Fig. 3).

MCF-10 was established by Soule *et al.* (1990), also from a woman with fibrocystic disease. This isolate gave rise to two sublines: MCF-10A, which adheres to plastic and grows as a monolayer (Fig. 8), and MCF-10B, which grows as a suspension culture but attaches in low-calcium media and also grows as a monolayer. Epithelial luminal ductal characteristics were confirmed by transmission electron microscopy, scanning electron microscopy, and keratin staining.

Table II summarizes cell lines derived from normal breast tissue.

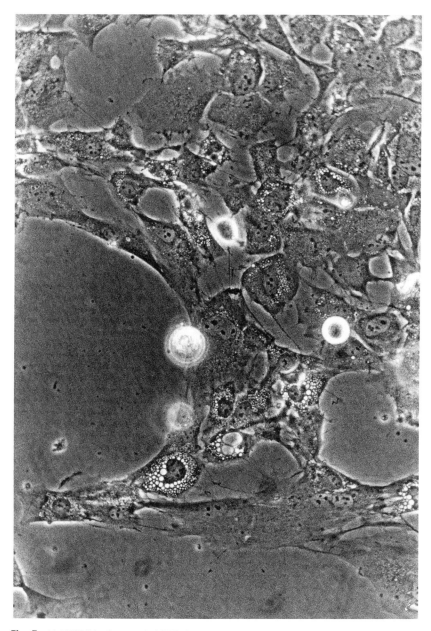

Fig. 5. Hs578T (Hackett *et al.*, 1977), isolated from a primary breast carcinosarcoma. The polymorphic epithelial cells grow as islands, but do form a monolayer. Phase-contrast microscopy, 160×.

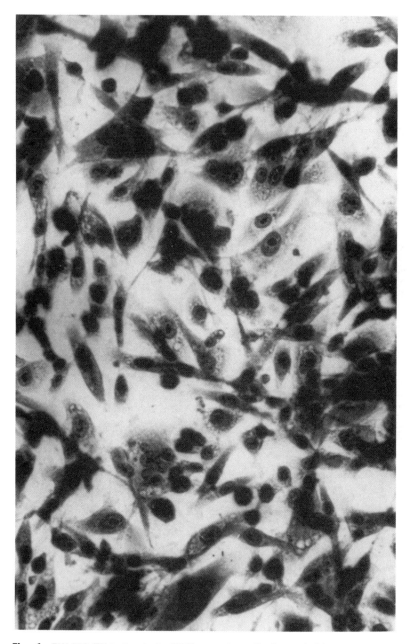

Fig. 6. RW-972 (Tibbetts *et al.*, 1988), an outgrowth of a probable cystosarcoma phyllodes, is composed of a mixture of epithelial-like cells and polymorphic stromal cells. Papanicolau stain, 240×. (Reproduced from Tibbetts *et al.*, 1988, with permission of J. B. Lippincott Company.)

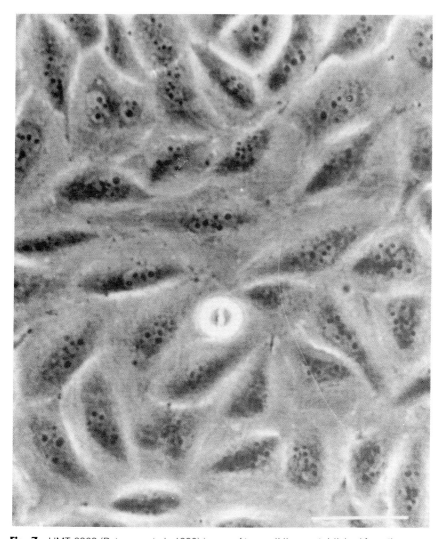

Fig. 7. HMT-3909 (Petersen *et al.*, 1990) is one of two cell lines established from the same patient with pathological evidence of both intraductal and medullary breast carcinoma. This cell line had a polymorphic outgrowth and formed monolayers. Cells with similar morphology have been established from metastatic sites of mammary carcinomas (see Table III–V). Phase-contrast microscopy, 400×. (Reproduced from Petersen *et al.*, 1990, with permission of the American Association for Cancer Research and the Waverly Press.)

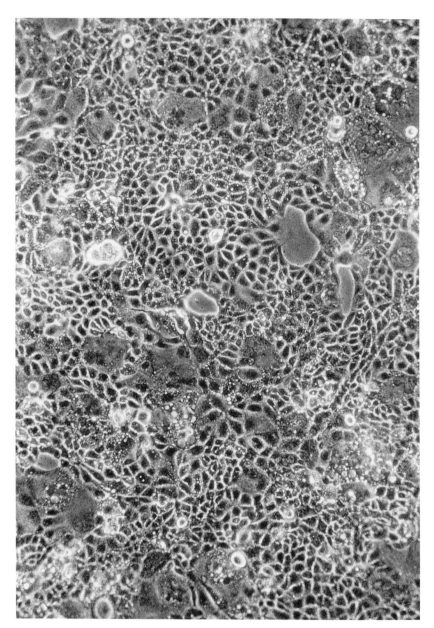

Fig. 8. UACC 265, established from a mammary intraductal carcinoma that has metastasized to the pleural field (Leibovitz and Massey, 1992). These cells did not form islands and grew out as a monolayer. Other investigators had a similar experience (see Table IV). Phase-contrast microscopy, 160×.

Table II
Cell Lines from "Normal" Breast Tissue or Breast Milk

Cell line	Morphology	Source	Reference
Hs57Bst	Fig. 5	Established from patient with a carcinosarcoma	Hackett *et al.* (1977)
HBL-100	Fig. 3	Established from breast milk of patient with no known pathology	Gaffney (1982)
HMT-3522	Fig. 3	Patient had fibrocystic disease	Briand *et al.* (1987)
MCF-10	Fig. 8	Patient had fibrocystic disease	Soule *et al.* (1990); Tait *et al.* (1990)

7. Xenografts

Xenografts also can be a source of tissue for establishing human breast carcinoma cells *in vitro*. UACC 2116 (Leibovitz and Massey, 1992) was established from a xenograft submitted by Giovanella (Stehlin Foundation for Cancer Research, Houston, Texas). The outgrowth appeared as moderately tight islands of epithelial cells (Fig. 2).

B. *Metastasis to Solid Tissues*

Four cell lines characterized in the literature each differ in their *in vitro* morphology (see Table III). The brain cell line MDA-MB-361 (Fig. 3) develops as slowly expanding islands of polygonal epithelial cells (Cailleau *et al.*, 1978). The lung cell line AlAb (Fig. 8) proliferates as spreading loose colonies of epithelial cells that form a monolayer (Reed and Gey, 1963). The skin nodule cell line DU 4475 (Fig. 9) appears as spherical cells that grow in suspension culture as grape-like clusters (Langlois *et al.*, 1979). The pubis cell line UACC 2436 (Fig. 7) consists of polymorphic cells. The line required an autologous

Table III
Cell Lines from Metastatic Carcinomas to Solid Tissues

Cell line	Tissue	Morphology	Reference
AlAb	Lung	Fig. 8	Reed and Gey (1963)
MDA-MB-361	Brain	Fig. 3	Cailleau *et al.* (1978)
DU 4475	Skin	Fig. 9	Langlois *et al.* (1979)
UACC-2436	Pubis	Fig. 7	Leibovitz and Massey (1992)

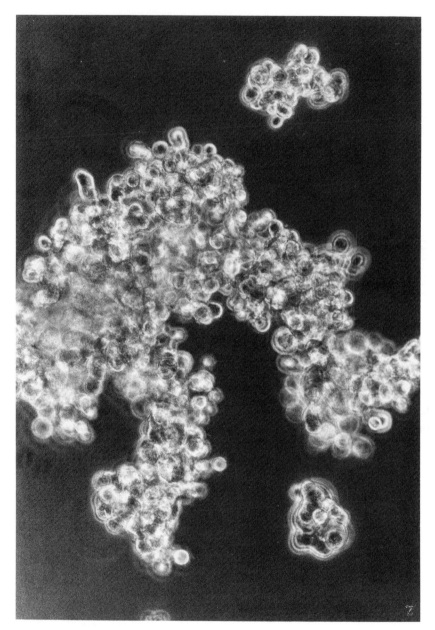

Fig. 9. DU4475 (Langlois *et al.*, 1979) was established from a mammary carcinoma that had metastasized to a skin nodule. The outgrowth is composed of spherical cells that only grow in suspension as grape-like clusters. Phase-contrast microscopy, 160×.

stromal-cell feeder layer for 3 months before the cells could grow independently *in vitro* (Leibovitz and Massey, 1992). Minafra *et al.* (1989) had a similar experience in establishing the 8701-BC cell line.

C. Metastasis to Pleural Fluid

As for metastases to solid tissues, cell lines derived from carcinomas that metastasize to the pleural fluid can be quite heterogeneous in morphology (see Table IV).

Table IV
Metastatic Carcinomas Established from Pleural Fluids

Cell line	Morphology	Pathology[a]	Reference
MCF-7	Fig. 3	IDC	Soule *et al.* (1973)
MDA-MB-134VI	Figs. 11,12	IDC	Cailleau *et al.* (1974)
MDA-MB-175VII	Fig. 3	PAP	Cailleau *et al.* (1974)
MDA-MB-231	Fig. 7	AC	Cailleau *et al.* (1974)
MDA-MB-157	Fig. 7	MED	Young *et al.* (1974)
SK-BR-3	Fig. 3	AC	Fogh and Trempe (1975)
MDA-MB-330	Fig. 3	LOB	Cailleau *et al.* (1978)
MDA-MB-415	Fig. 3	IDC	Cailleau *et al.* (1978)
MDA-MB-435S	Fig. 7	IDC	Cailleau *et al.* (1978)
MDA-MB-436	Fig. 7	AC	Cailleau *et al.* (1978)
MDA-MB-453	Figs. 11,12	AC	Cailleau *et al.* (1978)
T-47D	Fig. 3	IDC	Keydar *et al.* (1979)
ZR-75-27	Fig. 3	IDC	Engel *et al.* (1978)
MDA-MB-468	Fig. 3	IDC	Pathak *et al.* (1979)
PMC-42	Fig. 7	PAP	Whitehead *et al.* (1983)
EFM-19	Fig. 3	IDC	Simon *et al.* (1984)
EP	Fig. 3	IDC	Chu *et al.* (1985)
MW	Fig. 3	IDC	Chu *et al.* (1985)
600PE	Fig. 8	IDC	Smith *et al.* (1987)
21-MT-1	Fig. 3	IDC	Band *et al.* (1990)
21-MT-2	Fig. 8	IDC	Band *et al.* (1990)
MFM-223	Fig. 7	IDC	Hackenberg *et al.* (1991)
UACC-245	Fig. 10	IDC	Leibovitz and Massey (1992)
UACC-265	Fig. 8	IDC	Leibovitz and Massey (1992)
UACC-1179	Fig. 7	IDC	Leibovitz and Massey (1992)

[a] IDC, intraductal carcinoma; PAP, papillary carcinoma; AC, adenocarcinoma; MED, medullary carcinoma; LOB, lobular carcinoma.

1. Intraductal Carcinomas

a. Islands. Most cell lines derived from pleural fluid grow as slowly spreading islands of polygonal epithelial cells (Fig. 3).

b. Spheroids. UACC 245 (Leibovitz and Massey, 1992) grows only in suspension. The spheroid-like colonies bud off cells that form new spheroids (Fig. 10). Compare these cells to Fig. 9 (from a skin nodule), in which the suspension culture is composed of grape-like colonies.

c. Spheres. MDA-MB-453 (Cailleau *et al.*, 1978) contains small spherical cells that appear to reproduce by budding (Fig. 11). These cells do not form complete monolayers, even when the flask is quite full (Fig. 12).

2. Papillary Carcinomas

a. Spheres. MDA-MB-134 (Cailleau *et al.*, 1974) is a ductal carcinoma with papillary involvement that initially grew in suspension, but finally attached (Figs. 11,12).

b. Islands. MDA-MB-175 (Cailleau *et al.*, 1974) appears as large polygonal epithelial cells in slowly spreading islands (Fig. 3).

c. Polymorphic. MDA-MB-231 (Young *et al.*, 1974) consists of spindle-like cells that form a complete monolayer (Fig. 7).

3. Lobular Carcinomas

a. Islands. MDA-MB-330 (Cailleau *et al.*, 1978) was established as an intralobular carcinoma cell line that grew out as polygonal cells in islands (Fig. 3).

4. Medullary Carcinomas

a. Polymorphic. MDA-MB-175 VIII was the eighth cell line established by Cailleau *et al.* (1978) from the same patient with medullary carcinoma in sequential pleural effusion specimens. The growth was polymorphic and cells formed monolayers (Fig. 7).

D. *Metastasis to Ascitic Fluid*

Engel *et al.* (1978) established two cell lines from carcinomas metastasized to ascitic fluids (see Table V). ZR-75-1 grew as spreading islands of polygonal epithelial cells (Fig. 3), whereas ZR-75-30 grew as spherical cells (Fig. 11). Lippman *et al.* (1976) also established two cell lines from ascitic fluid (Evsa-E and Evsa-T) but did not characterize the morphology of these cells in the publication.

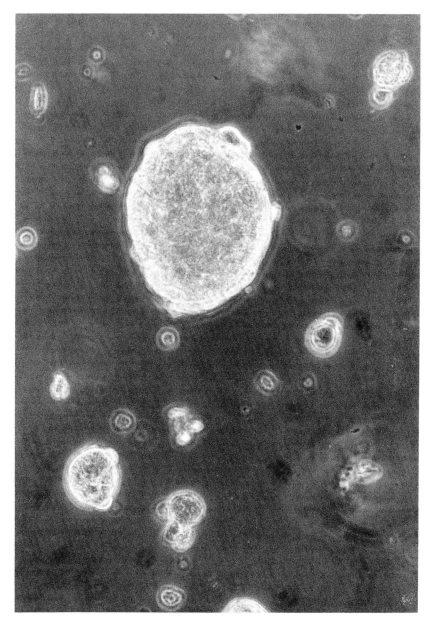

Fig. 10. UACC 245 (Leibovitz and Massey, 1992), established from an intraductal carcinoma that had metastasized to the pleural fluid, grew out as spheroid-like colonies in suspension. Cells bud off the spheroid-like colonies and form new spheroids. Phase-contrast microscopy, 160×.

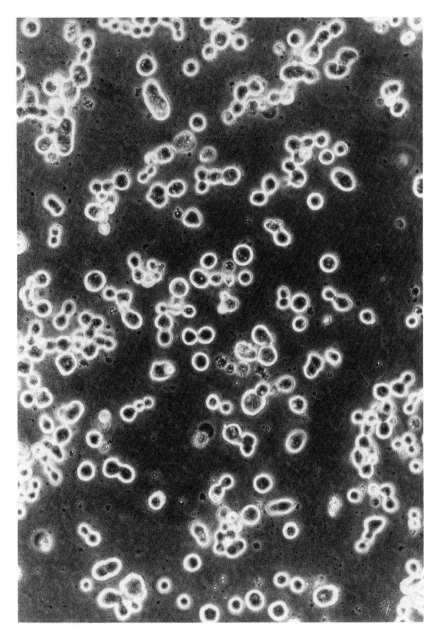

Fig. 11. MDA-MB-453 (Cailleau *et al.*, 1978), established from an intraductal mammary carcinoma that had metastasized to the pleural fluid, initially grew out as spherical cells in suspension, similar to those Fig. 9. However, these cells eventually attached to the substrate. The cells appear to reproduce by budding. Phase-contrast microscopy, 160×.

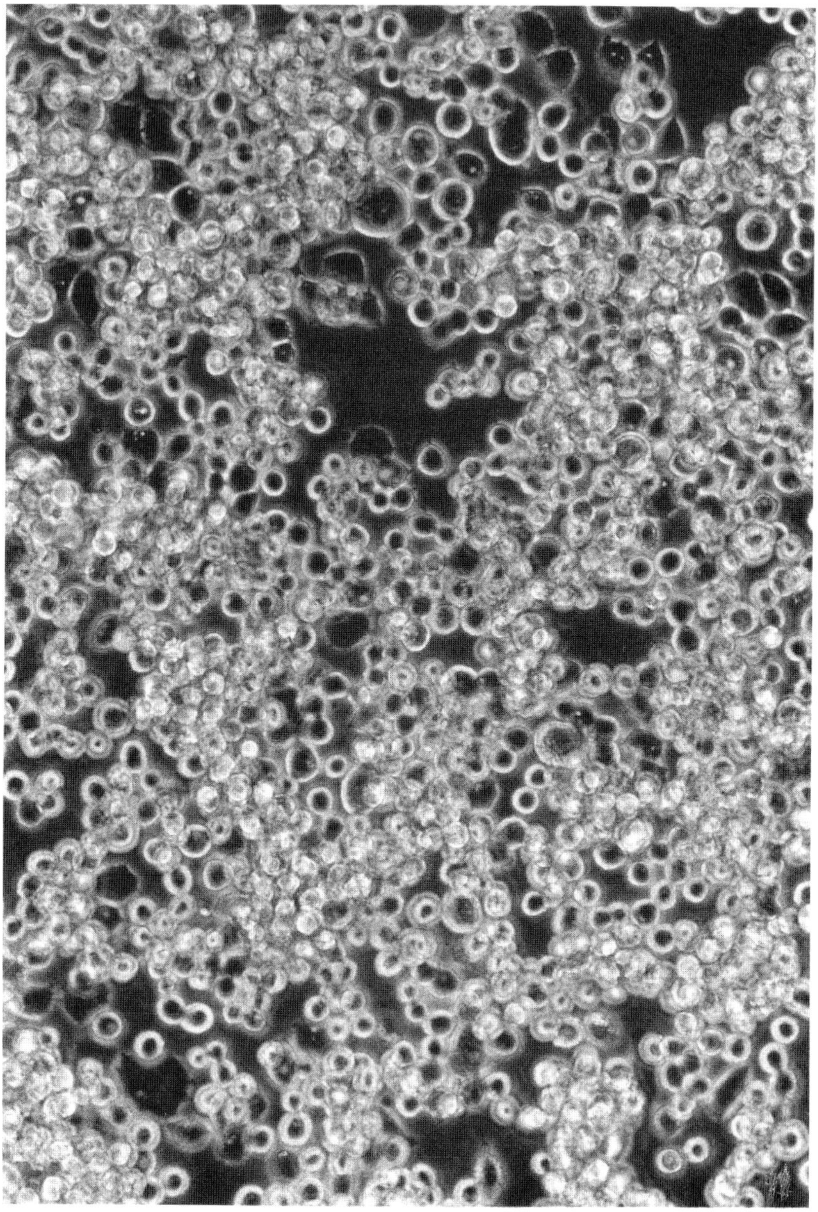

Fig. 12. MDA-MB-134 (Cailleau *et al.*, 1974) also yielded spherical cells from a ductal papillary mammary carcinoma that had metastasized to the pleural fluid. These cells fill the flask, forming attached grape-like clusters. Phase-contrast microscopy, 160×.

Table V

Metastatic Carcinoma Cell Lines Established from Ascitic Fluids

Cell line	Morphology	Pathology[a]	Reference
ZR-75-1	Fig. 3	IDC	Engel *et al.* (1978)
ZR-75-30	Fig. 3	IDC	Engel *et al.* (1978)

[a] IDC, intraductal carcinoma.

IV. Discussion

A. Establishment of Cell Lines from Primary Breast Carcinomas

Although short-term cultures of human breast carcinomas from primary tissues are relatively easy to obtain using a wide variety of media (Feller *et al.*, 1972; Hiratsuka *et al.*, 1982; Smith *et al.*, 1984; Petersen and van Deurs, 1986; Band and Sager, 1989), only rarely do they survive *in vitro* to become immortalized cell lines (Smith *et al.*, 1984). The list in Table I constitutes the composite effort of many investigators. Cailleau *et al.* (1974) reported no success after 200 attempts; Hackett *et al.* (1977) were unsuccessful after 100 tries. Thus, one might argue that existing cell lines may be fortuitous since so few have been established. Some lines were established using rather simple media (Vandewalle *et al.*, 1987) whereas other used quite complex media (Meltzer *et al.*, 1991). Tumor progression studies (Smith *et al.*, 1984,1987) caused Smith to hypothesize that "immortalization in culture occurs at a late stage in the progression of breast cancer." To date, no cell lines have been established from human breast carcinomas in which metastasis has not already occurred, at least to the axillary lymph nodes (Meltzer *et al.*, 1991).

B. Establishment of Cell Lines from Metastatic Sites

Cell lines are relatively easier to establish from metastatic sites, especially from the pleural fluid, than from primary tumors. Cailleau *et al.* (1978) had a success rate of ~ 10%. In general, cell lines established from metastatic sites are more heterogeneous than those from primary tissues. Although many lines produce slowly expanding islands of epithelial cells similar to the isolates from primary breast tissue (Figs. 2,3), others resemble spheroids (Fig. 10) or grape-like clusters growing in suspension (Fig. 9), are individual spherical cells that orginially grow in suspension but finally adhere (Figs. 11,12) or are polymorphic in morphology (Fig. 7).

V. Future Prospects

A. Markers to Differentiate Tumor Cells from Normal Cells

Numerous efforts have been made by many investigators to develop reagents to distinguish tumor cells from normal cells (reviewed by Smith *et al.*, 1984). Petersen and colleagues (Petersen and van Deurs, 1986,1987; Petersen *et al.*, 1990) have found that a significant percentage of tumor cells is NADPH–neotetrazolium reductase positive, whereas most normal cells are negative. Band and Sager (1989) used two markers to distinguish tumor cells from normal cells; expression of the human milk fat globule (HMFG-2) antigenic determinant on tumor cells and not on normal cells and preferential retention of the dye rhodamine 123 (R-123) in tumor cells.

B. New Media to Enhance Cell Growth of Normal Epithelial Cells and Carcinoma Cells

Petersen and colleagues (Petersen and van Deurs, 1987; Petersen *et al.*, 1990) and Briand *et al.* (1987) have developed a serum-free medium, CDM-3, for the establishment of cell lines from breast carcinomas and for long-term growth of normal cells. Similarly, Band and Sager (1989; Band *et al.*, 1990) have formulated a medium with 1% fetal bovine serum, DFCI-1, that enhances growth of both normal and tumor cells. This trend to develop chemically defined media for establishing cell lines and for growth of normal epithelial cells may enhance the testing of specific chemotherapeutic agents, biological response modifiers, and monoclonal antibodies and play a crucial role in the development of effective cancer therapy.

Acknowledgment

Research supported in part by National Cancer Institute, National Institutes of Health Grants CA 41183 and CA 23074.

References

Band, V., and Sager, R. (1989). Distinctive traits of normal and tumor-derived human mammary epithelial cells expressed in a medium that supports long-term growth of both cell types. *Proc. Natl. Acad. Sci. U.S.A.* **86**, 1249–1253.

Band, V., Zajchowski, D. A., Stenman, G., Morton, C. C., Kulesa, V., Connolly, J., and Sager, R. (1989). A newly established metastatic breast tumor cell line with integrated amplified copies of erbB2 and double minute chromosomes. *Genes Chromosomes Cancer* **1**, 48–58.

Band, V., Zajchowksi, D., Swisshelm, K., Trask, D., Kulesa, V., Cohen, C., Connolly, J., and Sager, R. (1990). Tumor progression in four mammary epithelial cell lines derived from the same patient. *Cancer Res.* **50**, 7351–7357,

Briand, P, Petersen, O. W., and van Deurs, B. (1987). A new diploid human breast epithelial cell line isolated and propagated in chemically defined medium. *In Vitro Cell. Dev. Biol.* **23,** 181–188.

Brinkley, B. R., Beall, P. T., Wible, J., Mace, M. L., Turner, D. S., and Cailleau, R. M. (1980). Variations in cell form and cytoskeleton human breast carcinoma cells *in vitro. Cancer Res.* **40,** 3118–3129.

Cailleau, R. M. (1975). Old and new problems in human tumor cell cultivation. *In* "Human Tumor Cells *in Vitro*" (J. Fogh, ed.), pp. 79–114. Plenum Press, New York.

Cailleau, R., Olivé, M., and Reeves, W. J., Jr. (1974). Breast tumor cell lines from pleural effusions. *J. Natl. Cancer Inst.* **53,** 661–674.

Cailleau, R., Olivé, M., and Cruciger, V. J. (1978). Breast carcinoma cell lines of metastatic origin: Preliminary characterization. *In Vitro* **14,** 911–915.

Carter, D. (1990). "Interpretation of Breast Biopsies," 2d Ed., Raven Press, New York.

Couthino, W. G., and Lasfargues, E. Y. (1978). Primary breast carcinoma cell lines BT-549. *In* "ATCC Catalog of Cell Lines and Hybridomas," 7th Ed. (1992), p. 259. American Type Culture Collection, Rockville, Maryland.

Chu, M. Y., Hagerty, M. G., Wiemann, M. C., Tibbetts, L. M., Sato, S., Cummings, F. J., Bogaars, H. A., Leduc, E. H., and Calabresi, P. (1985). Differential characteristics of two newly established human breast carcinoma cell lines. *Cancer Res.* **45,** 1357–1366.

Dobrynin, Y. V. (1963). Establishment and characteristics of cell strains from some epithelial tumors of human origin. *J. Natl. Cancer Inst.* **31,** 1173–1196.

Engel, L. W., and Young, N. A. (1978). Human breast carcinoma cells in continuous culture. A review. *Cancer Res.* **38,** 4327–4339.

Engel, L. W., Young, N. A., Tralka, T. S., Lippman, M. E., O'Brien, S. J., and Joyce, M. J. (1978). Establishment and characterization of three continuous cell lines derived from human breast carcinomas. *Cancer Res.* **38,** 3352–3364.

Feller, W. F., Stewart, S. E., and Kantor, J. (1972). Primary tissue explants of human breast cancer. *J. Natl. Cancer Inst.* **48,** 1117–1120.

Fogh, J., and Trempe, G. (1975). New human tumor cell lines. *In* "Human Tumor Cells *in Vitro*" (J. Fogh, ed.) pp. 115–141. Plenum, New York.

Gaffney, E. V. (1982). A cell line (HBL-100) established from breast milk. *Cell Tissue Res.* **227,** 563–568.

Gioanne, J., Courdi, A., Lelalanne, C. M., Fischel, J.-L., Zanghellini, E., Lambert, J.-C., Ettore, F., and Namer, M. (1985). Establishment, characterization, chemosensitivity, and radiosensitivity of two different cell lines derived from a human breast biopsy. *Cancer Res.* **45,** 1246–1258.

Hackenberg, R., Luttchens, S., Hoffman, J., Kunzman, R., Holzel, F., and Schulz, K.-D. (1991). Androgen sensitivity of the new breast cancer cell line MFM-223. *Cancer Res.* **51,** 5722–5727.

Hackett, A. J., Smith, H. S., Springer, E. L., Owens, R. B., Nelson-Rees, W. A., Riggs, J. L., and Gardner, M. B. (1977). Two syngeneic cell lines from human breast tissue: The aneuploid mammary epithelial (Hs578T) and the diploid myoepithelial (Hs578Bst) cell lines. *J. Natl. Cancer Inst.* **58,** 1795–1806.

Hiratsuka, M., Senoo, T., Kimoto, T., and Namba, M. (1982). An improved short-term culture method for human mammary epithelial cells. *Gann* **73,** 124–128.

Keydar, I., Chen, L., Karby, S., Weiss, F. R., Delarea, J., Radu, M., Chaitcik, S., and Brenner, H. J. (1979). Establishment and characterization of a cell line of human breast carcinoma origin. *Eur. J. Cancer* **15,** 659–670.

Langlois, A. J., Holder, W. D., Inglehart, J. D., Nelson-Rees, W. A., Wells, W. J., S. A., Jr., and Bolognesi, D. P. (1979). Morphological and biochemical properties of a new human breast cancer cell line. *Cancer Res.* **39,** 2604–2613.

Lasfargues, E. Y. (1975). New approaches to the cultivation of human breast carcinomas. *In* "Human Tumor Cells *in vitro*" (J. Fogh, ed.), pp. 51–77. Plenum Press, New York.

Lasfargues, E. Y., and Ozzello, L. (1958). Cultivation of human breast carcinomas. *J. Natl. Cancer Inst.* **21,** 1131–1147.

Lasfargues, E. Y., Couthino, W. G., and Redfield, E. S. (1978). Isolation of two human epithelial cell lines from solid breast carcinomas. *J. Natl. Cancer Inst.* **61,** 967–973.

Leibovitz, A. (1985). The establishment of cell lines from human solid tumors. *In* "Advances in Cell Culture" (K. Maramorosch, ed.), Vol. 4, pp. 249–260. Academic Press, New York.

Leibovitz, A. (1986). Development of tumor cell lines. *Cancer Genet. Cytogenet.* **19,** 11–19.

Leibovitz, A., and Massey, K. (1992). Previously unpublished.

Leibovitz, A., Dalton, W., Massey, K., Villar, H., and Trent, J. (1988). Establishment and character-ization of two new breast carcinoma cell lines from primary tumors. *Proc. Am. Assoc. Cancer Res.* **29,** 24.

Lippman, M. E., Bolan, G., and Huff, K. (1976). The effects of estrogen and antiestrogens on hormone responsive human breast cancer in long-term tissue culture. *Cancer Res.* **36,** 4595–4601.

Martorelli, B., Jr., Parshley, M. S., and Moore, J. G. (1969). Effects of chemotherapeutic agents on two cell lines of human breast carcinomas in tissue culture. *Surg. Gyn. Obs.* **128,** 1001–1006.

Meltzer, P., Leibovitz, A., Dalton, W., Villar, H., Kute, T., Davis, J., Nagle, R., and Trent, J. (1991). Establishment of two new cell lines derived from human breast carcinomas with HER-2/neu amplification. *Br. J. Cancer* **63,** 727–735.

Minafra, S., Morello, V., Glorioso, F., LaFiura, A. M., Tomasino, R. M., Feo, S., McIntosh, D., and Wooley, D. E. (1989). A new cell line (8701-BC) from primary ductal infiltrating carcinoma of human breast. *Br. J. Cancer* **60,** 185–192.

Nordquist, R. E., Ishmael, D. R., Loving, C. A., Hyder, D. M., and Hoge, A. (1975). The tissue culture and morphology of the human breast tumor cell line BOT-2. *Cancer Res.* **35,** 3100–3105.

Pathak, S., Siciliano, M. J., Cailleau, R., Wiseman, C. L., and Hsu, T. C. (1979). A human breast adenocarcinoma with chromosome and isoenzyme markers similar to those of the HeLa line. *J. Natl. Cancer Inst.* **62,** 263–271.

Petersen, O. W., and van Deurs, B. (1986). Demonstration of human breast carcinoma cells in cryosections and primary monolayer cultures of surgical biopsies by neotetrazolium reductase cytochemistry. *Cancer Res.* **46,** 2013–2020.

Petersen, O. W., and van Deurs, B. (1987). Preservation of defined phenotypic traits in short-termed cultured breast carcinoma derived epithelial cells. *Cancer Res.* **47,** 856–866.

Petersen, O. W., van Deurs, B., Vang Nielsen, K., Winkel Madsen, M., Laursen, I., Balslev, I., and Briand, P. (1990). Differential tumorigenicity of two autologous human breast carcinoma cell lines HMT-3909S1 and HMT-3909S8, established in serum-free medium. *Cancer Res.* **50,** 1257–1270.

Reed, M. V., and Gey, G. O. (1962). Cultivation of normal and malignant lung tissue. I. The establishment of three adenocarcinoma cell strains. *Lab. Invest.* **11,** 638–653.

Rudland, P. S., Hallowes, R. C., Cox, S. A., Omerod, E. J., and Warburton, M. J. (1985). Loss of production of myoepithelial cells and basement membrane proteins but retention of response to certain growth factors and hormones. *Cancer Res.* **45,** 3864–3877.

Simon, W. E., Hansel, M., Dietel, M., Matthiesen, L., Albrecht, M., and Holzel, F. (1984). Alteration of steroid hormone sensitivity during the cultivation of human mammary carcinoma cells. *In Vitro* **20,** 157–166.

Smith, H. S., Wolman, S. R., and Hackett, A. J. (1984). The biology of breast cancer at the cellular level. *Biochim. Biophys. Acta* **738,** 103–123.

Smith, H. S., Wolman, S. R., Dairkee, S. H., Hancock, M. C., Lippman, M., Leff, A., and Hackett, A. J. (1987). Immortalization in culture: Occurrence at a late stage in the progression of breast cancer. *J. Natl. Cancer Inst.* **78,** 611–615.

Soule, H. D., Vasquez, J., Long, A., Albert, S., and Brennan, M. (1973). A human cell line from a pleural effusion derived from a breast carcinoma. *J. Natl. Cancer Inst.* **51,** 1409–1413.

Soule, H. D., Maloney, T., Wolman, S., Peterson, W., Brentz, R., McGrath, C. M., Russo, J., Jones, R., and Brooks, S. C. (1990). Isolation and characterization of a spontaneously immortalized human breast epitheliod cell line, MCF-10. *Cancer Res.* **50,** 6075–6086.

Tait, L., Soule, H. D., and Russo, J. (1990). Ultrastructural and immunocytochemical characterization of an immortalized human breast epithelial cell line, MCF-10. *Cancer Res.* **50,** 6087–6094.

Tibbetts, L. M., Poisson, M. H., Tibbetts, L. L., and Cummings, F. J. (1988). A human breast stromal sarcoma cell line with features of malignant cystosarcoma phyllodes. *Cancer* 2176–2182.

Yamane, M., Nishiki, M., Kataoka, T., Kishi, N., Amano, K., Nakagawa, K., Okumichi, T., Naito, M., Ito, A., and Ezaki, H. (1984). Establishment and characterization of a new cell line (YMB-1) derived from a human breast carcinoma. *Hiroshima J. Med. Sci.* **33,** 715–720.

Vandewalle, B., d'Hooghe, M. C., Savary, J. B., Vilain, M. O., Peyrat, J. P., Deminatti, M., Delobelle-Deroide, A., and Lefebvre, J. (1987). Establishment and characterization of a new cell line (VHB-1) derived from a primary breast carcinoma. *J. Cancer Res. Clin. Oncol.* **113,** 550–558.

Whitehead, R. H., Bertoncello, J., Webber, L. M., and Pedersen, J. S. (1983). A new breast carcinoma cell line (PMC42) with stem cell characteristics. I. Morphological characterization. *J. Natl. Cancer Inst.* **70,** 649–661.

Young, R. K., Cailleau, R. M., Mackay, B., and Reeves, W. J., Jr. (1974). Establishment of MDA-MB-157 from metastatic pleural effusion of human breast carcinoma. *In Vitro* **9,** 239–245.

Hepatocellular Carcinomas

Masahiro Miyazaki and Masayoshi Namba
Department of Cell Biology
Institute of Molecular and Cellular Biology
Okayama University Medical School
Okayama 700, Japan

I. Introduction 185
II. Undifferentiated Hepatocellular Carcinoma Cell Line HLE 186
III. Differentiated Hepatocellular Carcinoma Cell Lines 189
   A. HuH-7 189
   B. PLC/PRF/5 193
C. Hep 3B 196
IV. Differentiated Hepatoblastoma Cell Lines 199
   A. HuH-6 Clone 5 199
   B. Hep G2 201
V. Conclusions 205
References 205

I. Introduction

Primary liver cancers involve two types of carcinomas, that is, hepatocellular carcinoma (HCC) and cholangioma, which originate from liver parenchymal cells (hepatocytes) or intrahepatic biliary epithelial cells, respectively. HCC is one of the major malignant diseases of the world, being responsible for approximately one million deaths per year. Currently, no satisfactory treatment for HCC exists, so the worldwide incidence rate nearly equals the mortality rate. As such, HCC remains one of the most challenging areas of study in medical oncology.

The causative factors associated with HCC development remain elusive. Epidemiological and experimental data suggest an association between the incidence of HCC and chronic hepatitis virus infection and mycotoxin exposure. The role of cellular oncogenes in the sequence of cellular events leading to HCC has become an area of great interest. Quantitative or qualitative activation or inactivation of certain cellular oncogenes or tumor suppressor genes in the development of HCC has been demonstrated using various animal models, animal or human hepatoma cell lines, or primary human tumors. However, the interaction, cooperativity, and exact functional role of the oncogenes in HCC development have not yet been defined.

Establishment of *in vitro* cell lines derived from HCC is considered a logical approach to the laboratory investigation of the tumor. Permanent cell lines derived from HCCs would broaden the scope of investigations on the tumor, since pure cell cultures could be used for a variety of studies that cannot be carried out on short-lived biopsy or postmortem tissue. Chen established the first hepatoma cell line in 1964. Since then, many laboratories including this one have reported establishment of HCC cell lines that express various liver functions and/or produce hepatitis B surface antigen (HBsAg) (Table I). However, reports on establishment of HCC cell lines remarkably decreased after 1985. One of the reasons for this phenomenon is the difficulty in obtaining liver cancer tissues suitable for isolation and cultivation of intact viable cells, because transcatheter arterial embolization with lipiodol and chemotherapeutic agents have been developed for treatment of the disease. Actually, only in 2 cases of 20 trials over 3 years did we succeed in obtaining cell lines from human primary liver cancer tissues. The two cell lines derived from the liver cancer tissues were from patients who received no treatment with transcatheter arterial embolization.

In this chapter, we describe characteristics of HCC (HLE and HuH-7) and hepatoblastoma (HuH-6 Clone 5) cell lines that were established in our laboratory. Characteristics of other HCC cell lines (PLC/PRF/5 and Hep 3B) and a hepatoblastoma cell line (Hep G2) are described also.

II. Undifferentiated Hepatocellular Carcinoma Cell Line HLE

The cell line HLE was derived from an autopsy hepatoma tissue of a 68-year-old Japanese male patient (Doi *et al.*, 1975). The hepatoma was diagnosed histologically as undifferentiated HCC. α-Fetoprotein (AFP) was detected in the serum of the patient by the double immunodiffusion test.

The hepatoma tissue was washed thoroughly with phosphate buffered saline (PBS) containing penicillin (100 U/ml) and streptomycin (100 μg/ml), minced with a pair of scalpels, and digested with 0.2% trypsin to isolate cells. The isolated cells were inoculated at a cell number of 5×10^5/ml into test tubes (16 $\times$ 150 mm) containing Eagle's minimal essential medium (MEM) supplemented with heat-inactivated bovine serum (20%), penicillin (100 U/ml), and streptomycin (100 μg/ml). Cultures were incubated at 37°C under closed conditions and the medium was renewed twice a week (Doi *et al.*, 1975).

About 3 months after the initiation of primary culture, the growth of two colonies was detected in the same tube. One colony consisted of epithelial cells and the other of fibroblast-like cells. On culture day 114, the two colonies

Human Hepatocellular Carcinoma and Hepatoblastoma Cell Lines

Cell line	Origin			Characteristics				Reference
	Diagnosis[a]	Age/sex[b]	Morphology[c]	Mode of chromosome number	Doubling time (hr)	Tumorigenicity	Protein production[d]	
HLE	HCC	68y/M	Epi	68	40–45	−	AFP(+) → (−)	Doi et al. (1975)
HLF	HCC	68y/M	Fibro	66	53	+		Doi et al. (1975)
HuH-6 Clone 5	HB	1y/M	Epi	48	99	+	AFP, ALB	Doi (1976)
PLC/PRF/5	HCC	24y/M	Epi	104		+	AFP, HBsAg	Alexander et al. (1976)
Hep G2	HB	15y/M	Epi	55		−	AFP, ALB	Aden et al. (1979)
Hep 3B	HCC	8y/M	Epi	60		+	AFP, ALB, HBsAG	Aden et al. (1979)
DELSH-5	HCC	50y/M	Epi	61	30–40		AFP, ALB, HBsAG	Das et al. (1980)
huH-1	HCC	53y/M	Epi	69		+	TAT, HBsAg	Huh and Utakoji (1981)
huH-4	HCC	51y/M	Epi	70			HBsAg	Huh and Utakoji (1981)
HuH-7	HCC	57y/M	Epi	55–65	36	+	AFP, ALB, PRE, AAT, CP, FIB, FN, HP, HX, BLP, AMG, BMG, TF, C3, C4, CEA	Nakabayashi et al. (1982)
c-HB-3	HB	1y/M	Epi			−	AFP	Hata et al. (1982)
KG-55T	HCC	53y/M	Epi	69	50–60		ALB, AAT, AMG, TAT	Matsuura (1983)
HCC-M	HCC	34y/M	Epi	63	24	+	ALB(+) → (−)	Watanabe et al. (1983)
FOCUS	HCC	63y/M	Epi	61–70	42–48	+	AFP, AAT, FIN, AATase, G6Pase, CEA	He et al. (1984)
KYN-1	HCC	58y/M	Epi	61–74	31	+	AFP, ALB, AAT, FER, BMG, CEA	Yano et al. (1986)
HCC-T	HCC	69y/M	Epi	64	24	+	AFP	Saito et al. (1989)

[a] HCC, hepatocellular carcinoma; HB, hepatoblastoma.
[b] y, year; M, male.
[c] Epi, epithelial; Fibro, fibroblast-like.
[d] AFP, α-fetoprotein; ALB, albumin; PRE, prealbumin; AAT, α_1-antitrypsin; CP, ceruloplasmin; FER, ferritin; FIB, fibrinogen; FN, fibronectin; HP, haptoglobin; HX, hemopexin; BLP, β-lipoprotein; AMG, α_2-macroglobulin; BMG, β_2-microglobulin; TF, transferrin; C3 and C4, Complement component 3 and 4; AATase, aspartate aminotransferase; G6Pase, glucose 6-phosphatase; TAT, tyrosine aminotransferase; CEA, carcinoembryonic antigen; HBsAg, hepatitis B surface antigen.

were separated by trypsinization; the culture consisting of epithelial cells was designated the HLE line. Currently, HLE cells are maintained in the same medium in an ordinary humidified atmosphere of 95% air and 5% CO_2 at 37°C (Doi *et al.*, 1975).

The morphological features of HLE cells are typically epithelial with rough granules in the cytoplasm. The cytoplasm varies in size and the cells have prominent rounded or oval-shaped vesicular nuclei containing one or more nucleoli (Fig. 1). The doubling time of the population of HLE cells was 40–45 hr during culture days 170 to 387. HLE cells were heteroploid with a modal number of 68 on culture day 238. In rotation culture, the cells form aggregates that resemble the original hepatoma tissue histologically. The cells are nontumorigenic when transplanted into the cheek pouch of adult hamsters with cortisone acetate (Doi *et al.*, 1975).

Staining of HLE cells with periodic acid–Schiff (PAS) reagent reveals strongly positive, large cytoplasmic granules (glycogen granules) that are digested by treatment with diastase. The cell line produced AFP during the first 6 months, after which this function ceased (Doi *et al.*, 1975). Other than AFP, no other plasma proteins were detected in the spent medium of HLE cell culture (Nakabayashi *et al.*, 1982).

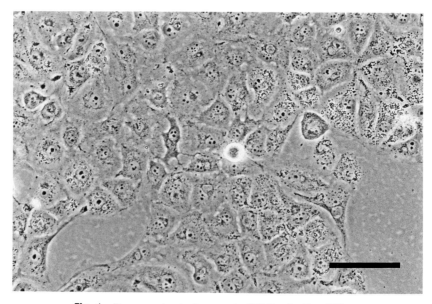

Fig. 1. Phase-contrast micrograph of HLE cells. Bar, 100 μm.

III. Differentiated Hepatocellular Carcinoma Cell Lines

A. HuH-7

Cell line HuH-7 was derived from a hepatoma tissue removed during surgery from a 57-year-old Japanese male patient (Nakabayashi et al., 1982). The hepatoma was diagnosed histologically as well-differentiated HCC. AFP and carcinoembryonic antigen (CEA) were detected in the serum of the patient at concentrations of 16.0 μg/ml and 1.6 ng/ml, respectively, but HBsAg was not detected.

The hepatoma tissue was minced and cultivated in RPMI 1640 medium supplemented with 20% heat-inactivated bovine serum and 0.4% lactalbumin hydrolysate (LAH). An epithelial cell colony isolated from primary culture on day 28 was designated HuH-7. On day 198 (10 passages) in culture with serum, the growth medium for HuH-7 cells was changed to serum-free RPMI 1640 supplemented with 0.4% LAH. The cells in the serum-free medium grew at a more rapid rate than did those in serum-containing medium. The population doubling times of HuH-7 cells at passage 15 in serum-containing and serum-free medium were 126.5 and 78.9 hr, respectively. Thereafter, as the growth rate increased further, the population doubling time decreased to 35.8 hr at passage 19 (Nakabayashi et al., 1982).

Because HuH-7 cells grew even better in the serum-free medium supplemented with LAH, an additional attempt was made to eliminate LAH from the medium. For growth of HuH-7 cells, selenium as disodium selenite (Na_2SeO_3, 1×10^{-8}–3×10^{-7} M) could replace LAH, which gives an equivalent amount of selenium at the added concentration of 0.4%. Thus, HuH-7 cells were maintained in RPMI 1640 supplemented with 3×10^{-8} M disodium selenite from day 406 (23 passages). The population doubling time of HuH-7 cells in the chemically defined medium was 44–56 hr (Nakabayashi et al., 1982). For further improvement of the chemically defined medium, various organic compounds and inorganic trace elements that stimulated growth of HuH-7 cells were added to the selenium-supplemented RPMI 1640 containing 1.3 g/L sodium bicarbonate: linoleic acid, 3×10^{-9} M; oleic acid, 3×10^{-9} M; $FeSO_4 \cdot 7H_2O$, 1×10^{-7} M; $MnCl_2 \cdot 4H_2O$, 3×10^{-10} M; $(NH_4)_6Mo_7O_{24} \cdot 4H_2O$, 3×10^{-9} M; NH_4VO_3, 1×10^{-8} M, and HEPES, 5×10^{-3} M. This improved chemically defined medium was designated IS-RPMI. From day 463, HuH-7 cells have been maintained in IS-RPMI. Since then, the cells have been replicating with doubling times of 47–51 hr (Nakabayashi et al., 1984).

HuH-7 cells exhibit a typical epithelial feature with pavement-like cell arrangement. The cells appear flattened and polygonal, having fine granules in the cytoplasm (Fig. 2). The PAS-stained cells have strongly positive, large cytoplasmic granules that are digested almost completely by treatment with 0.1% diastase solution. The chromosome numbers of HuH-7 cells at various

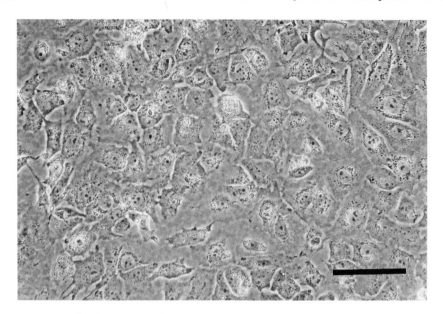

Fig. 2. Phase-contrast micrograph of HuH-7 cells. Bar, 100 μm.

passages are distributed between 55 and 65. HuH-7 cells produce tumors in athymic nude mice when inoculated subcutaneously into the mice at doses of 8×10^6–3×10^7 cells/mouse. The tumors produced in nude mice resemble the original hepatoma tissue of the patient histologically (Nakabayashi et al., 1984).

As just described, HuH-7 cells grow autonomously in the serum-free chemically defined medium IS-RPMI. HuH-7 cells express mRNAs of platelet-derived growth factor (PDGF), insulin-like growth factor I (IGF-I), and their respective receptors (Tsai et al., 1988). The reverse transcription–polymerase chain reaction (RT–PCR) reveals that HuH-7 cells contain both IGF-IA mRNA (representing exons I, II, III, and V) and IGF-IB mRNA (representing exons I, II, III, and IV). IGF-IA mRNA is 10-fold more abundant than IGF-IB mRNA in the cells (Nagaoka et al., 1991). Further, HuH-7 cells produce a novel growth factor with potent growth-promoting activity for the cell themselves; this factor is designated human hepatoma-derived growth factor (HuHGF). HuHGF, with the relatively large molecular size of 64 kDa, is acid and heat labile, and sensitive to reduction and trypsin digestion (Nakamura et al., 1989). Thus, the autonomous growth of HuH-7 cells appears to be supported by these growth factors.

The pattern of carbohydrate-metabolizing enzymes in HuH-7 cells is characterized by the resemblance to the original tumor enzyme pattern (Nakabayashi et al., 1982). Small activities of the liver-specific enzymes glucose 6-

phosphatase (2.6 units/g protein) and fructose 1,6-diphosphatase (5.7 units/g protein) are present in the cells.

HuH-7 cells synthesize and secrete 15 different plasma proteins, including AFP, albumin, prealbumin, α_1-antitrypsin, ceruloplasmin, fibrinogen, fibronectin, haptoglobin, hemopexin, β-lipoprotein, α_2-macroglobulin, β_2-microglobulin, transferrin, complement C3, and complement C4. In addition to these plasma proteins, the cells secrete CEA into the culture medium (Nakabayashi *et al.*, 1982). HuH-7 cells secrete high and stable levels of AFP, in the range of 53–110 ng/day/1 $\times$ 10^4 cells. The level of albumin secreted by 1 $\times$ 10^7 cells in 2 days is approximately 1/500 the level of albumin in serum. Production of these plasma proteins by HuH-7 cells was maintained stably over 3 yr in the chemically defined medium IS-RPMI (Nakabayashi *et al.*, 1984).

Further, HuH-7 cells secrete vitamin D-binding protein (Egawa *et al.*, 1987), apolipoproteins (A-I, A-II, B, and E) (Yamamoto *et al.*, 1987), and two novel lipoproteins—apolipoprotein B-containing high density lipoprotein and apolipoprotein A-I-containing lipoprotein (Yamamoto *et al.*, 1990). In addition, pro-α 1(I), pro-α2(I), pro-α1(IV), and pro-α2(V) collagen mRNAs also are detected in HuH-7 cells by Northern blot hybridization analysis (Inagaki *et al.*, 1987).

As described earlier, the cell line HuH-7 is one of the hepatoma cell lines producing the highest amount of AFP. Dexamethasone (3 $\times$ 10^{-7} to 3 $\times$ 10^{-6} M) causes 1.3- to 1.4-fold stimulation of AFP secretion in HuH-7 cells in the serum-free chemically defined medium IS-RPMI. Triiodothyronine (3 $\times$ 10^{-9} M) also increases AFP secretion in HuH-7 cells 1.2-fold. Further, dibutyryl cyclic AMP at 0.25–1.0 mM also slightly increases AFP secretion (about 1.2-fold over the control). However, the addition of insulin at concentrations greater than 5 $\times$ 10^{-8} M partially inhibits AFP secretion. Dexamethasone and dibutyryl cyclic AMP have the greatest additive effect on stimulation of AFP secretion in HuH-7 cells. On the other hand, when dexamethasone and insulin are added simultaneously, the dexamethasone-mediated stimulation of AFP secretion is diminished (Nakabayashi *et al.*, 1985).

Since the level of AFP mRNA in HuH-7 cells is elevated by the addition of dexamethasone, the glucocorticoid hormone appears to regulate the AFP gene primarily at the level of transcription. Further, dexamethasone stimulation is suppressed effectively by the glucocorticoid antagonist RU 486, indicating that this effect is mediated by glucocorticoid receptors. The transcriptional regulation by the steroid hormones is thought to be mediated by initial binding of the steroid to its receptor, followed by specific interaction of the hormone–receptor complex with a DNA element termed the glucocorticoid responsive element (GRE). Actually, the location of AFP GRE is identified to be the 71-bp region from 98 to 169 bp upstream of the human AFP gene (Nakabayashi *et al.*, 1989).

The AFP and albumin genes are similar in structure, are arranged tandemly, and are believed to be derived from a common ancestral gene. The effects of transfection of the normal c-Ha-*ras* gene, *ras*$^{Gly\ 12}$, and its oncogenic mutant, *ras*$^{Val\ 12}$ on expression of the AFP and albumin genes have been examined using HuH-7 cells. The mutant and, to a lesser extent, the normal *ras* gene cause reduction of the AFP mRNA but not of the albumin mRNA level in the transfected HuH-7 cells. Co-transfection experiments with a *ras*$^{Val\ 12}$ expression plasmid and a chloramphenicol acetyltransferase reporter gene fused to AFP regulatory sequences show that *ras*$^{Val\ 12}$ suppresses the activity of enhancer and promoter regions containing AT-rich sequences (AT motif). In contrast, *ras*$^{Val\ 12}$ does not affect the promoter activity of the albumin gene although the promoter contains homologous AT-rich elements (Nakao *et al.*, 1990).

Southern blot analysis reveals the presence of an aberrant albumin gene as well as a normal one in HuH-7 cells (Urano *et al.*, 1991). The aberrant albumin gene is formed by a linkage of albumin and nonalbumin sequences in intron 11 of the albumin gene. The nonalbumin sequence is assigned to chromosome 4q12-q13 by *in situ* hybridization. Since the human albumin gene is mapped at 4q11-q13, these results indicate an interstitial deletion of a chromosomal segment within 4q11-q13 in HuH-7 cells. The truncated albumin gene was detected in an early passage of HuH-7 cells and has been maintained stably in cell culture. These findings suggest that the deletion of the albumin locus took place *in vivo*. Deletions of DNAs in chromosomes 4q, 11p, 13q, 16q, and 17p also have been shown in many human HCCs. From these findings, one of the tumor suppressor genes might be present near the albumin gene. Therefore, the genomic clone carrying the rearranged albumin sequence may prove useful in identifying the suppressor oncogene.

Hepatitis B virus (HBV) is a causative agent for hepatitis. The viral genome can be integrated into patient liver cell chromosomes; a significant fraction of such patients later develop HCC. Therefore, studies with this virus on the mechanism of regulation of gene expression and replication are of substantial importance in virology as well as in public health. To elucidate a detailed mechanism of HBV propagation in hepatocytes, a cell culture system is needed that allows replication of this virus under controlled experimental conditions. The production of particles resembling hepatitis B virions (Dane particles) has been achieved by transfection of HuH-7 cells with cloned HBV DNA (Chang *et al.*, 1987; Yaginuma *et al.*, 1987). By *in vitro* mutagenesis assay of the template HBV DNA, the *P* gene as well as the *C* gene product has been demonstrated to be essential to the production of HBV particles (Yaginuma *et al.*, 1987). Further, a detailed mutational analysis of the regulatory DNA sequence elements that control expression of the HBV major surface antigen gene was performed in the cell line HuH-7 using transient transfection assay (Raney *et al.*, 1991). Seven regions of the major surface antigen promoter, located within 200 nucleotides of the RNA initiation site, have been identified that influence the level of transcription from this promoter. The level of expression of the major surface antigen gene appears to be controlled by the complex interplay between a minimum of six transcription factors that activate transcription from this gene and one transcription factor that represses this activity.

B. PLC/PRF/5

Cell line PLC/PRF/5 was derived from an autopsy tumor tissue of a 24-year-old Shangaan male patient with a primary liver carcinoma (Alexander *et al.*, 1976). AFP and HBsAg were present in patient serum. Histological examination of postmortem liver tissue showed cirrhosis and malignant neoplasia with features indistinguishable from hepatoma. Ultrastructural examination of the original tumor tissue showed cells with large nuclei and lysosome-like structures but no discernible virus particles.

The autopsy specimen, approximately 1 cm in size, obtained from the secondary tumor mass in the liver was minced with scissors into 0.1- to 1-mm^3 pieces and treated for 5 min with 0.25% trypsin in PBS. The trypsinized tissue fragments were planted into 25-cm^2 tissue culture flasks containing 1.5 ml growth medium consisting of MEM with nonessential amino acids, 10% heat-inactivated fetal bovine serum, 100 U/ml penicillin, and 100 μg/ml streptomycin. After 2 days to allow for cellular attachment, the growth medium was replaced with 5 ml fresh medium. Within 5 days, the attached clumps were surrounded by areas of epithelial cell outgrowth. Cell division continued, but death of cells was pronounced; no overall increase in number of cells occurred for about 5 weeks. Thereafter, the death rate declined and the cells increased in number. Contaminating fibroblasts were removed initially by mechanical scraping around the epithelial cell islands and later by differential trypsinization with mechanical scraping. The tumor cells first were subcultured after 4 months *in vitro*. Since passage 4, the cultures have been completely free of fibroblasts. Thus, the cell line was established after 18 months in culture and designated PLC/PRF/5. At approximately passage 40, PLC/PRF/5 cell culture could be passaged at 1 : 6 dilution, the plating efficiency was 40–50%, and the cell doubling time was 35–40 hr. Cell division did not stop in confluent cultures and, although the cells did not pile up, large numbers detached from the surface (Alexander *et al.*, 1976).

PLC/PRF/5 cells are polygonal in shape with well-defined borders (Fig. 3). Ultrastructurally, the cells have a dense cytoplasmic matrix with few elements of the endoplasmic reticulum, abundant polysomes, and mitochondria with a dense matrix. Lysosome-like structures containing membrane fragments and other electron-dense material are also present in the cells. Sparsely distributed glycogen granules are evident and some areas of the cell surface are particularly rich in microvilli. Bundles of filaments are found occasionally in the cytoplasmic matrix. The nuclei are characterized by the absence of condensed chromatin. Most cells have more than one nucleolus. Virus particles were not detected in the cells by electron microscopy (Alexander *et al.*, 1976).

The karyology of PLC/PRF/5 cells is male and human. The number of chro-

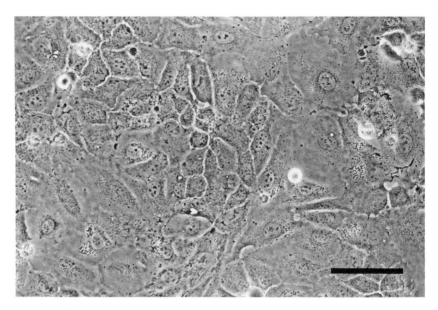

Fig. 3. Phase-contrast micrograph of PLC/PRF/5 cells. Bar, 100 μm.

mosomes distributes between 48 and 61, with a mean number of 56 (Alexander *et al.*, 1976). Northern blot analyses show that PLC/PRF/5 cells express six kinds of cellular oncogenes: c-*abl*, c-*fes*, c-*fms*, c-*myc*, c-Ha-*ras*, and c-*sis* (Motoo *et al.*, 1986). Further, immunoprecipitation after [^{35}S]methionine labeling of PLC/PRF/5 cells demonstrates that p53 protein, a tumor suppressor, is reduced in concentration greatly in these cells (Bressac *et al.*, 1990).

Subcutaneous injection of PLC/PRF/5 cells into athymic BALB/c nude mice results in the growth of well-circumscribed, moderately differentiated HCC. The intervals until tumor appearance and tumor take rates are dependent on inoculum dose (Shouval *et al.*, 1981). Immunosuppression of athymic nude mice treated with antilymphocyte serum (ALS) or irradiation increases tumor incidence in mice, reduces the latency period, and produces local invasiveness. Similarly, treatment of mice with anti-mouse interferon (IFN) globulin also increases the tumor incidence and reduces the latency period. On the other hand, co-cultivation of nude mouse spleen cells with PLC/PRF/5 cells induces secretion of murine IFNα. Treatment of mice with ALS, irradiation, or anti-IFN globulin abolishes natural killer (NK) cell activity against PLC/PRF/5 cells. These findings, therefore, suggest that the IFN/NK cell system may play a role in limiting tumorigenicity and invasiveness of HBV-infected human HCC cells (Shouval *et al.*, 1983).

PLC/PRF/5 cells grow at high density in the absence of exogenous growth factors (Nakabayashi *et al.*, 1982; Hoshi and McKeehan, 1985). PLC/PRF/5 cells also express mRNAs for PDGF and IGF-I (Tsai *et al.*, 1988). Moreover, the cells secrete a growth factor that has physicochemical properties similar to those of a novel growth factor extracted from brain tissue (Hoshi and McKeehan, 1985). The autogenous secretion of these growth factors may support the autonomous growth of the cells at high density.

In PLC/PRF/5 cell-conditioned medium, 14 different plasma proteins are detected: AFP, albumin, transferrin, β-lipoprotein, fibronectin, α_2-macroglobulin, β_2-microglobulin, α_2-antitrypsin, ceruloplasmin, plasminogen, fibrinogen, α_1-acid glycoprotein, complement C3, and complement C4 (Knowles *et al.*, 1980; Nakabayashi *et al.*, 1982; Nakagawa *et al.*, 1985). In addition, the cells secrete CEA (Nakabayashi *et al.*, 1982), heparin cofactor II (Jaffe *et al.*, 1985); factor XIIIa (Nagy *et al.*, 1986); and α_1-microglobulin (Vincent *et al.*, 1987).

Secretion of AFP by PLC/PRF/5 cells is reduced by sodium *n*-butyrate (0.4−1 mM), which causes a marked reduction in growth rate, colony-forming efficiency in soft agar, and *de novo* synthesis of DNA, as well as remarkable morphological changes (Nakagawa *et al.*, 1985). However, insulin (1 × 10^{-7} M), which stimulates DNA synthesis and growth of the cells, enhances AFP secretion of PLC/PRF/5 cells (Motoo *et al.*, 1988). On the other hand, albumin production by PLC/PRF/5 cells is enhanced by treatment with sodium *n*-butyrate (Nakagawa *et al.*, 1985).

Although AFP is recognized to be a secretory-type protein, monoclonal antibody (19F12) against human AFP, which is obtained by the hybridoma technique, detects AFP over the surface of PLC/PRF/5 cells with uniform distribution (Hosokawa *et al.*, 1989). The monoclonal antibody 19F12 specifically binds to the epitope of AFP present on the membrane surface of hepatoma cells. Therefore, this antibody may be a suitable one for immunotoxin therapy or imaging of AFP-producing cancer cells.

PLC/PRF/5 cells respond to a factor produced by endotoxin-stimulated human monocytes by synthesizing C-reactive protein (CRP), a prototype acute-phase reactant (Goldman and Liu, 1987). Also, the cells produce CRP in response to conditioned medium from lipopolysaccharide-stimulated human monocytes. This CRP induction by the conditioned medium is potentiated 180-fold by 2 mM caffeine (Ganapathi *et al.*, 1990). The induction of CRP in PLC/PRF/5 cells by the conditioned medium is inhibited by an antibody raised against *Escherichia coli*-derived recombinant human interleukin 6 (IL-6) (Ganapathi *et al.*, 1988). Thus IL-6 appears to play a central role in induction of CRP in PLC/PRF/5 cells.

The cell line PLC/PRF/5 contains four to seven copies per cell of HBV DNA that is integrated into chromosomes (Brechot *et al.*, 1980; Chakraborty *et al.*, 1980; Edman *et al.*, 1980; Marion *et al.*, 1980; Ziemer *et al.*, 1985). The cells produce HBsAg but no other known HBV proteins, such as hepatitis B core antigen (HBcAg) and hepatitis B envelope antigen (HBeAg), and contain no viral DNA polymerase. Thus, the surface antigen gene is the only intact HBV transcription unit present in these integrated sequences (Marion et al., 1979; Skelly et al., 1979). In addition, nonintegrated HBV DNA also exists in the cells. This DNA is found in the cell cytoplasm as a DNA–protein complex (Marquardt et al., 1982).

Titers of up to 1.3 μg/ml HBsAg are obtained by incubating confluent cultures of PLC/PRF/5 cells in maintenance medium for 7 days. The predominant form of HBsAg secreted by the cells is the 20- to 25-nm spherical particle. The buoyant density is 1.19 g/ml. Thus, the HBsAg produced by the PLC/PRF/5 line is very similar in size, morphology, and buoyant density to the spherical particles of circulating serum HBsAg (Skelly et al., 1979). Histochemically, HBsAg is shown to be localized in the cytoplasm and on the surface of PLC/PRF/5 cells (Gerber et al., 1981).

Secretion of HBsAg by PLC/PRF/5 cells is stimulated by dexamethasone (1×10^{-6} M), which suppresses growth of the cells (Oefinger et al., 1981). Hydrocortisone (1×10^{-3} M) also stimulates antigen production, but does not exhibit any effect on cell growth. Treatment of the cells with a combination of hydrocortisone (1×10^{-3} M) and adenine arabinoside (1×10^{-5} M) greatly stimulates HBsAg production, suggesting that the inhibition of DNA synthesis by adenine arabinoside amplifies the effect of hydrocortisone (Clementi et al., 1983). On the other hand, insulin (100 U/ml), which has no effect on growth of PLC/PRF/5 cells, shows an inhibitory effect on HBsAg production by the cells (Clementi et al., 1984). Recombinant human IFN-α2a (1200 IU/ml), as well as natural human IFNα (608 IU/ml) and IFNβ (367 IU/ml), also suppresses both HBsAg production and proliferation of the cells (Yamashita et al., 1988).

A vaccine against HBV has been prepared using HBsAg derived from PLC/PRF/5 cells grown in the interstices of a Diaflo hollow filter unit. This cell culture vaccine was as potent as a human plasma-derived vaccine and proved safe in tests in chimpanzees and in human subjects who were in the late stages of cancer of the central nervous system and receiving therapy (McAleer et al., 1984).

C. Hep 3B

The cell line Hep 3B originated from a liver tumor biopsy obtained during extended lobectomy of an 8-year-old black male from the United States. The histology of the biopsy specimen revealed well-differentiated HCC with a trabecular pattern (Aden et al., 1979).

Primary culture of tissue minces of an HCC biopsy was initiated on feeder layers of the irradiated mouse cell line STO in Williams E medium supplemented with 10% fetal bovine serum. This method promoted the growth of cells that had fastidious requirements but prevented fibroblast overgrowth. After an initial period of apparent cell proliferation, growth was restricted to a single colony that was maintained in a flask for several months. Cells from the flask containing the single large colony were dissociated by trypsinization and transferred onto a new feeder layer, eventually producing the proliferating cell line Hep 3B. Thereafter, Hep 3B cells became able to grow without feeder layers (Aden et al., 1979; Knowles et al., 1980). Currently the cells are maintained in 10% fetal bovine serum-supplemented Eagle's MEM, that contains nonessential amino acids, sodium pyruvate, and Earle's balanced salt solution (BSS).

Hep 3B cells can grow in a simple defined medium consisting of 75% MEM, 25% MAB (Waymouth's medium) 87/3, and $3 \times 10^{-8} M$ selenium (Darlington et al., 1987). The autonomous growth of the cells may be supported by their autocrine growth factors such as PDGF and IGF-I (Tsai et al., 1988).

Hep 3B cells are epithelial in shape. Their morphological characteristics are compatible with those of liver parenchymal cells (Fig. 4). Ultrastructurally, the cells contain prominent cisterns of granular endoplasmic reticulum and a moderate number of mitochondria. Stacks of annulate lamellae and aggregates of glycogen are found in the cytoplasm (Aden et al., 1979).

Hep 3B cells are chromosomally abnormal, with a modal number of 60, and contain distinctive rearrangements of chromosome 1 (Aden et al., 1979). In Hep 3B cells, a major portion of the p53 gene is deleted. This deletion is accompanied by the absence of p53 transcripts and p53-encoded protein in this cell line. Also, the loss of p53 expression or the presence of abnormal forms of the protein frequently is associated with other HCC cell lines. Therefore, these observations suggest that alterations in p53 may be important

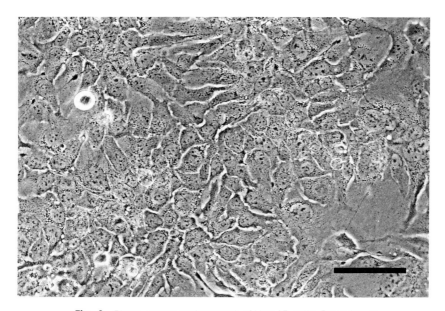

Fig. 4. Phase-contrast micrograph of Hep 3B cells. Bar, 100 μm.

events in the transformation of hepatocytes to the malignant phenotype (Bressac *et al.*, 1990).

Hep 3B cells are tumorigenic in athymic nude mice when injected under the kidney capsule. The resulting tumors are histologically similar to the hepatoma from which the cells were derived (Knowles *et al.*, 1980).

The cell line Hep 3B has been shown to have one to two copies of HBV DNA integrated into the host genome (Twist *et al.*, 1981). Using probes that cover different regions of the HBV genome, five species of RNA are observed, of sizes 4.0, 3.3, 2.9, 2.6, and 2.2 kb. The RNAs include surface antigen gene, pre-S and X regions. None have a core antigen sequence (Su *et al.*, 1986). Hep 3B cultures containing 1×10^7 cells are capable of producing an average of 220 ng HBsAg within a 24-hr period (Aden *et al.*, 1979). The production of HBsAg is suppressed by addition of low concentrations (0.1–1 nM) of insulin to serum-free medium. In addition, the suppression of HBsAg production by insulin is paralleled by a decrease in HBsAg mRNA abundance. Thus, Hep 3B cells carry a speciifc receptor with high affinity for insulin ($k_d = 1.8$ nM). The receptor shows an insulin-dependent protein tyrosine kinase activity. These findings suggest that insulin may act through its receptor binding to suppress HBsAg expression in Hep 3B cells (Chou *et al.*, 1989).

Knowles *et al.* (1980) demonstrated that the cell line Hep 3B synthesizes and secretes 17 of the major human plasma proteins. The detected proteins are AFP, albumin, α_2-macroglobulin, α_1-antitrypsin, α_1-antichymotrypsin, transferrin, haptoglobin, ceruloplasmin, plasminogen, Gc-globulin, complement C3, C3 activator, α_1-acid glycoprotein, fibrinogen, α_2-HS-glycoprotein, β-lipoprotein, and retinol-binding protein. In addition, the cells produce apolipoproteins (A-I, A-II, B, C-II, C-III, and E; Zannis *et al.*, 1981), α_1-plasmin inhibitor (Saito *et al.*, 1982), IGF carrier protein (Moses *et al.*, 1983), vitamin D-binding protein (Haddad *et al.*, 1983), α_1-microglobulin (Vincent *et al.*, 1987), vitamin D_3 25-hydroxylase (Tam *et al.*, 1988), and ferritin (Costanzo *et al.*, 1989).

Hep 3B cells also respond to the culture fluid of endotoxin-stimulated human monocytes by synthesizing the prototype acute phase reactant CRP (Goldman and Liu, 1987). Similarly, the synthesis of CRP and serum amyloid A is increased after exposure to conditioned medium from lipopolysaccharide-activated monocytes. The induction of CRP and serum amyloid A by monocyte-conditioned medium is inhibited effectively by antibody against human IL-6. However, IL-6 alone has no discernible effect on the induction of CRP and serum amyloid A in the cells, but is capable of causing increased synthesis of α_1-protease inhibitor and fibrinogen and reduced synthesis of albumin. The addition of IL-1 to IL-6 leads to induction of both CRP and serum amyloid A in Hep 3B cells (Ganapathi *et al.*, 1988). The induction of CRP by these cytokines in Hep 3B cells is potentiated extensively by caffeine, although the mechanism of this interaction is presently unclear (Ganapathi *et al.*, 1990).

Hep 3B cells respond to heat-shock treatment by expressing mRNA for microsomal heme oxygenase, which catalyzes the rate-limiting step in the oxidative metabolism of heme (Mitani et al., 1989). The maximum induction of heme oxygenase mRNA (5- to 7-fold) is observed with treatment of the cells at 43.5°C for 60 min. The heat-mediated induction of heme oxygenase mRNA is blocked by simultaneous treatment of the cells with actinomycin D or cycloheximide.

IV. Differentiated Hepatoblastoma Cell Lines

A. HuH-6 Clone 5

Cell line HuH-6 Clone 5 was derived from a tumor biopsy from the right lobe of the liver of a 1-year-old Japanese boy (Doi, 1976). The histology of the tumor showed a well-differentiated massive nodular pattern consisting of tumor cells, vascular lakes, extramedullary hematopoiesis, mixed components of squamous cells, and focal chondro-osteogenic tissues. No cirrhosis was found. The tumor was diagnosed as hepatoblastoma of the well-differentiated type. AFP was detected in the serum and ascites, but HBsAg was not detected.

Cells isolated from the tumor tissue by treatment with 0.1% trypsin were inoculated at 1×10^6 cells per 35-mm Falcon plastic dish containing 3 ml LD medium (0.5% LAH + saline D; Takaoka, 1958) supplemented with 20% bovine serum, and cultured in a humidified atmosphere of 95% air and 5% CO_2. Fibroblast growth was suppressed effectively under these culture conditions. About 4 weeks after initiation of the culture, HuH-6 Clone 5 cells were obtained from an epithelial cell colony using 0.1% trypsin-saturated filter paper. At this time, the culture medium was switched to RPMI 1640 supplemented with 0.4% LAH and 20% bovine serum. Thereafter, the concentrations of LAH and bovine serum in the medium were reduced gradually to 0.2% and 2.5%, respectively (Doi, 1976).

Morphological features of HuH-6 Clone 5 cells are typically epithelial with pavement-like cell arrangements (Fig. 5). Their cytoplasm has abundant granules. The ratio of nucleus to cytoplasm is approximately 1:1 to 1:2; the cytoplasm differs in size depending on the center or periphery of the colony. Nuclei have 1–2 prominent nucleoli. PAS staining of the cells shows strongly positive, large cytoplasmic granules that are digested to a large extent by treatment with 0.1% diastase. Electron microscopically, HuH-6 Clone 5 cells are epithelial with numerous desmosomes connecting bundles of tonofibrils. Many glycogen granules are seen in the cytoplasm. The nucleoli are large in proportion to the condensed chromatin, suggesting differentiating nuclei. The cells tend to grow in multiple layers. The multiple-layered cells form round or ballooned aggregates and, after floating in the medium, they adhere to another

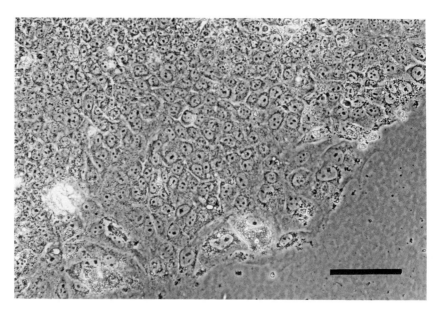

Fig. 5. Phase-contrast micrograph of HuH-6 Clone 5 cells. Bar, 100 μm.

surface of the culture dish and begin to grow. Their growth is relatively slow; the doubling time is 99 hr. The cells are hyperdiploid with a modal number of 48 (68%), and are transplantable into the cheek pouch of young hamsters treated with antithymocyte serum. Histologically, these tumors show ductule formation consisting of cultured cells in vascular lakes (Doi, 1976).

HuH-6 Clone 5 cells can grow in a chemically defined serum-free medium, RPMI 1640, supplemented with disodium selenite at $3 \times 10^{-8} M$ (Nakabayashi et al., 1982). The cells express mRNAs for PDGF, IGF-I, and their receptors (Tsai et al., 1988). Thus, the autonomous growth of HuH-6 Clone 5 cells may be supported by these autocrine growth factors.

In the spent serum-free medium of HuH-6 Clone 5 cells, 15 different plasma proteins are detected by double immunodiffusion with individual antibodies. The cells secrete AFP, albumin, prealbumin, α_1-antitrypsin, ceruloplasmin, fibrinogen, fibronectin, haptoglobin, hemopexin, β-lipoprotein, α_2-macroglobulin, β_2-microglobulin, transferrin, complement C3, and complement C4 (Nakabayashi et al., 1982). In addition to these plasma proteins, HuH-6 Clone 5 cells produce CEA. Fibronectin synthesized by the cells is secreted into the medium and distributed on the cell surfaces. This protein has been shown to be identical to human plasma fibronectin with reference to molecular weight, electrophoretic mobility, and antigenicity (Tanaka et al., 1983). Laminin and fibronectin used as culture substrates enhance production of AFP and albumin in the cells, suggesting that genomic expression of some differentiated functions can be modulated by extracellular matrix substances (Tokiwa *et al.*, 1988). On the other hand, treatment of HuH-6 Clone 5 cells with either dibutyryl cyclic AMP (0.5–1 m*M*) or retinoic acid (0.5–1 μg/ml) inhibits the expression of AFP mRNA but enhances the expression of albumin mRNA (Sato and Enomoto, 1991).

Bile acid synthesis is one of many liver-specific functions. HuH-6 Clone 5 cells synthesize bile acids and secrete them into the culture medium (Amuro *et al.*, 1982). Cholic, chenodeoxycholic and lithocholic acids are found in the culture medium; a portion of the chenodeoxycholic acid and all the lithocholic acid are sulfated. Chenodeoxycholic acid is the main bile acid in the medium. Lithocholic acid usually is formed from chenodeoxycholic acid by intestinal bacteria in humans. To our knowledge, this is the first report of lithocholic acid synthesis and sulfation in a cell line derived from human hepatocytes.

Pancreatic secretory trypsin inhibitor (PSTI) is known to be a specific trypsin inhibitor in the pancreas and pancreatic juice. PSTI is synthesized in HuH-6 Clone 5 cells and released on stimulation with various cytokines: IL-1, tumor necrosis factor (TNF), and IFN-β (Murata *et al.*, 1988). Further, HuH-6 Clone 5 cells produce intestinal-type alkaline phosphatase (Yamamoto *et al.*, 1984). Thus, the HuH-6 Clone 5 cell line is a useful *in vitro* model to study the regulatory mechanism for phenotypic expression of intestinal-type alkaline phosphatase isoenzymes in liver cancer cells.

HuH-6 Clone 5 cells have been transfected with a recombinant HBV DNA (Tsurimoto *et al.*, 1987). One clone thus obtained (HB611) produces and releases surface antigen and envelope antigen into medium at a high level and accumulates core particles intracellularly. This clone has a chromosomally integrated set of the original recombinant DNA and produces a 3.5-kb transcript corresponding to the pre-genome RNA as well as HBV DNAs in an extrachromosomal form. Most of these DNAs in single-stranded or partially double-stranded form are packaged in the intracellular core particles. The particles containing HBV DNA are present in the medium and are morphologically indistinguishable from Dane particles. Thus, the cells can be used as a model system for analysis of gene expression and DNA replication of HBV in human hepatocytes.

B. Hep G2

The cell line Hep G2 was derived from a liver tumor biopsy obtained during extended lobectomy of a 15-year-old Caucasian male from Argentina (Aden *et al.*, 1979). The liver tumor was diagnosed histologically as a well-differentiated hepatoblastoma (Knowles *et al.*, 1980).

Tumor minces were cultured on irradiated mouse cell layers (STO) in Williams E medium supplemented with 10% fetal bovine serum, in the same manner as cell line Hep 3B. The cell line Hep G2 also was established after the biopsy specimen had been cultured for several months. The cell line Hep G2 does not produce HBsAg (Aden *et al.*, 1979; Knowles *et al.*, 1980). Currently the cells are maintained in 10% fetal bovine serum-supplemented Eagle's MEM that contains nonessential amino acids, sodium pyruvate, and Earle's BSS.

Even under serum-free conditions, Hep G2 cells proliferate in the presence of 5–10 μg/ml insulin (Hoshi and McKeehan, 1985; Darlington *et al.*, 1987). Further, Hep G2 cells can grow in the simple defined medium consisting of 75% MEM, 25% MAB (Waymouth's medium) 87/3, and 3×10^{-8} *M* selenium (Darlington *et al.*, 1987), suggesting that the cells are producing factors required for their proliferation. Actually, Hep G2 cells also express mRNAs for PDGF, IGF-I, and IGF-I receptor (Tsai *et al.*, 1988), and secrete a growth factor that has physicochemical properties similar to those of a novel growth factor from brain tissue (Hoshi and McKeehan, 1985).

Hep G2 cells are epithelial in shape and resemble liver parenchymal cells morphologically (Fig. 6). Electron microscopy of the cells reveals desmosomes between adjacent cells (Bouma *et al.*, 1989). Bile canaliculus-like structures and Golgi apparatus complexes are particularly developed. Rough endoplasmic reticulum is abundant, slightly dilated, and contains an electron-dense material. Smooth endoplasmic reticulum is poorly developed. Mitochondria are round or oval in shape. Free cytoplasmic ribosomes are particularly numerous. Glycogen particles are detected essentially in the cells maintained in glucose-enriched medium.

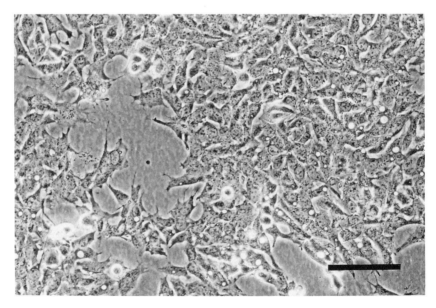

Fig. 6. Phase-contrast micrograph of Hep G2 cells. Bar, 100 μm.

Hep G2 cells are chromosomally abnormal, with a modal number of 55, and contain distinctive rearrangements of chromosome 1 (Knowles et al., 1980). The expression of the c-myc gene is elevated and deregulated in Hep G2 cells (Huber and Thorgeirsson, 1987). Futher, the Hep G2 N-ras gene also is activated to a dominant-acting transforming gene by a missense mutation in codon 61 (Richards et al., 1990).

Hep G2 cells originally were described as nontumorigenic when injected under the kidney capsule of athymic nude mice (Knowles et al., 1980). However, later Hep G2 cells and a single-cell clonal Hep G2 line, HLD_2-6 were found to be equally tumorigenic when injected subcutaneously into athymic nude mice (Huber et al., 1985). Tumors are nonencapsulated, highly invasive adenocarcinomas and are positive for γ-glutamyltranspeptidase activity and bile production. Plasma from tumor-bearing mice is positive for human AFP, but negative for HBsAg.

The cell line Hep G2 has been examined extensively for liver functions. Knowles et al. (1980) reported that Hep G2 cells synthesize and secrete 17 human plasma proteins: AFP, albumin, α_2-macroglobulin, α_1-antitrypsin, α_1-antichymotrypsin, transferrin, haptoglobin, ceruloplasmin, plasminogen, complement C3, complement C4, C3 activator, α_1-acid glycoprotein, fibrinogen, α_2-HS-glycoprotein, β-lipoprotein, and retinol-binding protein. In addition to these proteins, the cells secrete testosterone–estradiol-binding globulin (Khan et al., 1981); α_2-plasmin inhibitor (Saito et al., 1982); plasminogen activator (Levin et al., 1983); thyroxine-binding globulin (Bartalena et al., 1984); coagulation factor V (Wilson et al., 1984); human serum spreading factor (Barnes and Reing, 1985); somatomedin-binding protein (Povoa et al., 1985); factor VII, protein C, protein S, protein C inhibitor (Fair and Marlar, 1986); bile acids (Everson and Polokoff, 1986); plasminogen activator inhibitor (Wun and Kretzmer, 1987); α_1-microglobulin (Vincent et al., 1987); erythropoietin (Ueno et al., 1989); ferritin (Lescoat et al., 1989); and IGF-binding protein (Conover and Lee, 1990). Also, Hep G2 cells have various liver-specific enzyme activities: major enzymes of intra- and extracellular cholesterol metabolism (lecithin-cholesterol acyltransferase and cholesterol 7α-hydroxylase; Chen et al., 1986; Erickson and Fielding, 1986); drug-metabolizing enzymes (cytochrome P450, mixed function oxidase, and UDP–glucuronyl transferase; Sassa et al., 1987; Doostdar et al., 1988; Grant et al., 1988); carboxypeptidase N (Grimwood et al., 1988); vitamin D_3 25-hydroxylase (Tam et al., 1988); glutathione transferase isoenzymes (Dierickx, 1989); and hepatic triglyceride lipase (Busch et al., 1989a,b). Glucose 6-phosphatase activity also has been demonstrated cytochemically in the cells (Bouma et al., 1989). Further, the cells express receptors for asialoglycoprotein (Schwartz et al., 1981); transferrin (Ciechanover et al., 1983); heparin-binding growth factor (DiSorbo et al., 1988); insulin (Briata et al., 1989; Hatada et al., 1989); and epidermal growth factor (Clementi et al., 1989).

The accumulations in the medium of albumin, AFP, tranferrin, and α_1-antitrypsin are 13.2 ± 1.9, 10.7 ±1.7, 3.2 ± 0.4, and 4.9 ± 1.5 $\mu g/10^6$ cells/24 hr, respectively (Bouma et al., 1989). Treatment of Hep G2 cells with triiodothyronine (1 × 10^{-8} M) causes 4- and 2.5-fold increases in secretion of albumin and α_1-acid glycoprotein, whereas it causes 8- and 2-fold inhibition of AFP and thyroxine-binding protein secretion, respectively (Darlington et al., 1987; Kobayashi et al., 1988; Conti et al., 1989). On the other hand, insulin (10 $\mu g/ml$) reduces the level of total secreted protein by 30% (Darlington et al., 1987). Individually, secretion of albumin, AFP, and α_1-antitrypsin declines 4-, 3.6-, and 2.6-fold, respectively.

Hep G2 cells respond to conditioned media of human squamous carcinoma (COLO-16) cells and lipopolysaccharide-stimulated human monocytes by increasing the synthesis of α_1-acid glycoprotein, haptoglobin, complement C3, α_1-antichymotrypsin, α_1-antitrypsin, and fibrinogen while decreasing the synthesis of albumin. The regulation of the acute-phase proteins is mediated by hepatocyte-stimulating factors (HSF) and IL-1 present in the conditioned medium. Dexamethasone alone has no effect on acute-phase protein synthesis but enhances the response to various HSF several-fold (Baumann et al., 1987). Phorbol ester [12-O-tetradecanoyl·phorbol-13-acetate (TPA, 0.5 μM)] partially mimics the stimulatory effect of IL-6 (100–250 U/ml) on the synthesis of α_1-acid glycoprotein, haptoglobin, and α_1-antichymotrypsin. TPA and IL-6 act synergistically. However, prolonged pretreatment of Hep G2 cells with TPA results in a drastically reduced response to the cytokine (Baumann et al., 1988). On the other hand, synthesis of α_2-HS-glycoprotein in Hep G2 cells is down-regulated by the conditioned medium from lipopolysaccharide-activated human monocytes, recombinant human IL-6, and IL-1β (Daveau et al., 1988).

Hep G2 cells secrete lipids (triglycerides and cholesterol), apolipoproteins (A-I, A-II, B, C-II, C-III, and E); very low density lipoproteins (VLDL); low density lipoproteins (LDL); and high density lipoproteins (HDL) (Zannis et al., 1981; Archer et al., 1985; Wettesten et al., 1985; Dashti and Wolfbauer, 1987; Hoeg et al., 1988; Ranganathan and Kottke, 1990; Sato et al., 1990). Secretion of two apolipoproteins (A-I and C-II) increases 2.0- and 2.5-fold in response to 20 nM estrogen, respectively (Tam et al., 1985). Further, exposure of Hep G2 cells to cholesterol (100 $\mu g/ml$) and oleic acid (1.12 mM), which elevates intracellular cholesterol levels, stimulates apolipoprotein B secretion and reduces receptor-mediated uptake of LDL (Fuki et al., 1989).

Hep G2 cells express receptors for LDL (Dashti et al., 1984) and HDL (Dashti et al., 1985; Kambouris et al., 1988). Incubation of Hep G2 cells with physiological concentrations of HDL with a density between 1.16 and 1.20 g/ml (heavy HDL) results in an approximately 7-fold increase in LDL receptor acitivity, whereas incubation with physiological concentrations of LDL results in a maximum 2-fold decrease in LDL receptor activity (Havekes et al., 1986). Addition of estradiol (10 $\mu g/ml$) to Hep G2 cells growing in lipoprotein-deficient medium also increases cell surface LDL receptor activity by 141% (Semenkovich and Ostlund, 1987). Similarly, inclusion of insulin (100 mU/ml) in the incuba-tion medium containing lipoprotein-deficient serum results in an approxi-mately 2-fold increase in the amounts of LDL receptor mRNA and LDL receptor protein (Wade et al., 1988,1989).

The cell line Hep G2 has been transfected with a plasmid carrying four tandem copies of the HBV genome. Thus, two clones (Hep G2T2.2.15 and Hep G2T8) have been obtained. Hep G2T2.2.15 secretes complete HBV particles as well as HBsAg, HBeAg, and HBcAg, whereas Hep G2T8 does not secrete HBV particles but does secrete high levels of HBsAg and HBeAg (Sells et al., 1987; Roingeard et al., 1990a,b). No detectable HBcAg is present in the culture medium of Hep G2T8 cells, but HBcAg is accumulated in the intracellular compartment. Immunofluorescent staining at confluence shows a localization pattern of strong nuclear HBcAg in Hep G2T8 cells but mostly cytoplasmic and membranous HBcAg in Hep G2T2.2.15 cells. In the Hep G2T8 clone, signifi-cant cell death is observed 4 days after cell confluency, corresponding with noticeable HBcAg release into the culture medium. These findings suggest a cytopathic effect associated with HBcAg accumulation in the HBV nonpro-ducer clone, and suggest that viral factors as well as the host's immune response may be considered in explaining liver injury that occurrs in hepatitis B (Roingeard et al., 1990b).

V. Conclusions

We have described characteristics of HCC cell lines (HLE, HuH-7, PLC/PRF/ 5, and Hep 3B) and hepatoblastoma cell lines (HuH-6 Clone 5 and Hep G2). These HCC and hepatoblastoma cell lines offer good models not only for the investigation of hepatocarcinogenesis but also for the elucidation of regulatory mechanisms of liver-specific gene expression. Further, an in vitro system for the production of HBV particles has been established by transfection of HBV DNA, using the HCC and hepatoblastoma cell lines as recipients. These systems provide a new approach to unanswered questions regarding HBV-related gene functions and HBV replication. Through these studies, new treat-ment for and prevention of liver cancers may be developed in the near future.

References

Aden, D. P., Fogel, A., Plotkin, S., Damjanov, I., and Knowles, B. B. (1979). Controlled synthesis of HBsAg in a differentiated human liver carcinoma-derived cell line. Nature (London) **282,** 615–616.

Alexander, J. J., Bey, E. M., Geddes, E. W., and Lecatsas, G. (1976). Establishment of a continu-ously growing cell line from primary carcinoma of the liver. S. Afr. Med. J. **50,** 2124–2128.

Amuro, Y., Tanaka, M., Higashino, K., Hayashi, E., Endo, T., Kishimoto, S., Nakabayashi, H., and Sato, J. (1982). Bile acid synthesis by long-term cultured cell line established from a human hepatoblastoma. J. Clin. Invest. **70,** 1128–1130.

Archer, T. K., Tam, S. P., Deugau, K. V., and Deeley, R. G. (1985). Apolipoprotein C-II mRNA levels in primate liver. *J. Biol. Chem.* **260,** 1676–1681.

Barnes, D. W., and Reing, J. (1985). Human spreading factor: Synthesis and response by Hep G2 hepatoma cells in culture. *J. Cell. Physiol.* **125,** 207–214.

Bartalena, L., Tata, J. R., and Robbins, J. (1984). Characterization of nascent and secreted thyroxine-binding globulin in cultured human hepatoma (Hep G2) cells. *J. Biol. Chem.* **259,** 13605–13609.

Baumann, H., Richards, C., and Gauldie, J. (1987). Interaction among hepatocyte-stimulating factors, interleukin 1, and glucocorticoids for regulation of acute phase plasma proteins in human hepatoma (Hep G2) cells. *J. Immunol.* **139,** 4122–4128.

Baumann, H., Isseroff, H., Latimer, J. J., and Jahreis, G. P. (1988). Phorbol ester modulates interleukin 6- and interleukin 1-regulated expression of acute phase plasma proteins in hepatoma cells. *J. Biol. Chem.* **263,** 17390–17396.

Bouma, M. E., Rogier, E., Verthier, N., Labarre, C., and Feldmann, G. (1989). Further cellular investigation of the human hepatoblastoma-derived cell line Hep G2: Morphology and immuno-cytochemical studies of hepatic-secreted proteins. *In Vitro Cell. Dev. Biol.* **25,** 267–275.

Brechot, C., Pourcel, C., Louise, A., Rain, B., and Tiollasis, P. (1980). Presence of integrated hepatitis B virus DNA sequences in cellular DNA of human hepatocellular carcinoma. *Nature (London)* **286,** 533–535.

Bressac, B., Galvin, K. M., Liang, T. J., Isselbacher, K. J., Wands, J. R., and Ozturk, M. (1990). Abnormal structure and expression of p53 gene in human hepatocellular carcinoma. *Proc. Natl. Acad. Sci. U.S.A.* **87,** 1973–1977.

Briata, P., Gherzi, R., Adezati, L., and Cordera, R. (1989). Effect of two different glucose concentrations on insulin receptor mRNA levels in human hepatoma Hep G2 cells. *Biochem. Biophys, Res. Commun.* **160,** 1415–1420.

Busch, S. J., Krstenansky, J. L., Owen, T. J., and Jackson, R. L. (1989a). Human hepatoma (Hep G2) cells secrete a single 65K dalton triglyceride lipase immunologically identical to postheparin plasma hepatic lipase. *Life Sci.* **45,** 615–622.

Busch, S. J., Martin, G. A., Barnhart, R. L., and Jackson, R. L. (1989b). Heparin induces the expression of hepatic triglyceride lipase in a human hepatoma (Hep G2) cell line. *J. Biol. Chem.* **264,** 9527–9532.

Chakraborty, P. R., Ruiz-Opazo, N., Shouval, D., and Shafritz, D. A. (1980). Identification of integrated hepatitis B virus DNA and expression of viral RNA in an HBsAg-producing human hepatocellular carcinoma cell line. *Nature (London)* **286,** 531–533.

Chang, C., Jeng, K., Hu, C., Lo, S. J., Su, T., Ting, L. P., Chou, C. K., Han, S., Pfaff, E., Salfeld, J., and Schaller, H. (1987). Production of hepatitis B virus in vitro by transient expression of cloned HBV DNA in a hepatoma cell line. *EMBO J.* **6,** 675–680.

Chen, J. M. (1964). Establishment *in vitro* and some preliminary observations on a strain of human liver cell carcinoma. *Acta Un. Int. Cancer* **20,** 1314–1315.

Chen, C. H., Forte, T. H., Cahoon, B. E., Thrift, R. N., and Albers, J. J. (1986). Synthesis and secretion of lecithin-cholesterol acyltransferase by the human hepatoma cell line Hep G2. *Biochim. Biophys. Acta* **877,** 433–439.

Chou, C. K., Su, T. S., Chang, C., Hu, C., Huang, M. Y., Suen, C. S., Chou, N. W., and Ting, L. P. (1989). Insulin suppresses hepatitis B surface antigen expression in human hepatoma cells. *J. Biol. Chem.* **264,** 15304–15308.

Ciechanover, A., Schwartz, A. L., Dautry-Varsat, A., and Lodish, H. F. (1983). Kinetics of internalization and recycling of transferrin and the transferrin receptor in a human hepatoma cell line. *J. Biol. Chem.* **258,** 9681–9689.

Clementi, M., Bagnarelli, P., and Pauri, P. (1983). Effect of steroids and adenine-arabinoside (araA) on growth and HBsAg production of a human hepatoma cell line. *Arch. Virol.* **75,** 137–141.

Clementi, M., Testa, I., Bagnarelli, P., Festa, A., Pauri, P., Brugia, M., Calegari, L., and Martinis, C.

(1984). Insulin reduces HBsAg production by PLC/PRF/5 human hepatoma cell line. *Arch. Virol.* **81**, 177–184.

Clementi, M., Festa, A., Testa, I., Bagnarelli, P., Devescovi, G., and Carloni, G. (1989). Expression of high- and low-affinity epidermal growth factor receptors in human hepatoma cell lines. *FEBS Lett.* **249**, 297–301.

Conover, C. A., and Lee, P. D. K. (1990). Insulin regulation of insulin-like growth factor-binding protein production in cultured Hep G2 cells. *J. Clin. Endocrinol. Metab.* **70**, 1062–1067.

Conti, R., Ceccarini, C., and Tecce, M. F. (1989). Thyroid hormone effect on α-fetoprotein and albumin coordinate expression by a human hepatoma cell line. *Biochim. Biophys. Acta* **1008**, 315–321.

Costanzo, F., Bevilacqua, M. A., Giordano, M., and Cimino, F. (1989). Expression of genes of ferritin subunits in human hepatoma cell lines. *Biochem. Biophys, Res. Commun.* **161**, 902–909.

Darlington, G. J., Kelly, J. H., and Buffone, G. J. (1987). Growth and hepatospecific gene expression of human hepatoma cells in a defined medium. *In Vitro Cell. Dev. Biol.* **23**, 349–354.

Das, P. K., Nayak, N. C., Tsiquaye, K. N., and Zuckerman, A. J. (1980). Establishment of a human hepatocellular carcinoma cell line releasing hepatitis B virus surface antigen. *Br. J. Exp. Pathol.* **61**, 648–654.

Dashti, N., and Wolfbauer, G. (1987). Secretion of lipids, apolipoproteins, and lipoproteins by human hepatoma cell line, Hep G2: Effects of oleic acid and insulin. *J. Lipid Res.* **28**, 423–436.

Dashti, N., Wolfbauer, G., Koren, E., Knowles, B., and Alaupovic, P. (1984). Catabolism of human low density lipoproteins by human hepatoma cell line Hep G2. *Biochim. Biophys, Acta* **794**, 373–384.

Dashti, M., Wolfbauer, G., and Alaupovic, P. (1985). Binding and degradation of human high-density lipoproteins by human hepatoma cell line Hep G2. *Biochim. Biophys. Acta* **833**, 100–110.

Daveau, M., Davrinche, C., Julen, N., Hiron, M., Arnaud, P., and Lebreton, J. P. (1988). The synthesis of human α-2-HS glycoprotein is down-regulated by cytokines in hepatoma Hep G2 cells. *FEBS Lett.* **241**, 191–194.

Dierickx, P. J. (1989). Partial purification and characterization of the soluble glutathione transferase isoenzymes from cultured Hep G2 cells. *Cell Biol. Int. Rep.* **13**, 585–593.

DiSorbo, D., Shi, E., and McKeehan, W. L. (1988). Purification from hepatoma cells of a 130-kDa membrane glycoprotein with properties of the heparin-binding growth factor receptor. *Biochem. Biophys. Res. Commun.* **157**, 1007–1014.

Doi, I. (1976). Establishment of a cell line and its clonal sublines from a patient with hepatoblastoma. *Gann* **67**, 1–10.

Doi, I., Namba, M., and Sato, J. (1975). Establishment and some characteristics of human hepatoma cell lines. *Gann.* **66**, 385–392.

Doostdar, H., Duthie, S. J., Burke, M. D., Melvin, W. T., and Grant, M. H. (1988). The influence of culture medium composition on drug metabolising enzyme activities of the human liver derived Hep G2 cell line. *FEBS Lett.* **241**, 15–18.

Edman, J. C., Gray, P., Valenzuela, P., Rall, L. B., and Rutter, W. J. (1980). Integration of hepatitis B virus sequences and their expression in a human hepatoma cell. *Nature (London)* **286**, 535–538.

Egawa, T., Yamaoto, H., and Kishimoto, S. (1987). Vitamin D-binding protein synthesized by a human hepatocellular carcinoma cell line. *Clin. Chim. Acta* **167**, 273–284.

Erickson, S. K., and Fielding, P. E. (1986). Parameters of choresterol metabolism in the human hepatoma cell line, Hep G2. *J. Lipid Res.* **27**, 875–883.

Everson, G. T., and Polokoff, M. A. (1986). A human hepatoblastoma cell line exhibiting defects in bile acid synthesis and conjugation. *J. Biol. Chem.* **261**, 2197–2201.

Fair, D. S., and Marlar, R. A. (1986). Biosynthesis and secretion of factor VII, protein C, protein S, and the protein C inhibitor from a human hepatoma cell line. *Blood* **67**, 64–70.

Fuki, I. V., Preobrazhensky, S. N., Misharin, A. Y., Bushmakina, N. G., Menschikov, G. B., Repin,

V. S., and Karpov, R. S. (1989). Effect of cell cholesterol content on apolipoprotein B secretion and LDL receptor activity in the human hepatoma cell line, Hep G2. *Biochim. Biophys. Acta* **1001,** 235–238.

Ganapathi, M. K., May, L. T., Schultz, D., Brabenec, A., Weinstein, J., Sehgal, P. B., and Kushner, I. (1988). Role of interleukin-6 in regulating synthesis of C-reactive protein and serum amyloid A in human hepatoma cell lines. *Biochem. Biophys. Res. Commun.* **157,** 271–277.

Ganapathi, M. K., Mackiewicz, A., Samols, D., Brabenec, A., Kushner, I., Schultz, D., and Hu, S. I. (1990). Induction of C-reactive protein by cytokines in human hepatoma cell lines is potentiated by caffeine. *Biochem. J.* **269,** 41–46.

Gerber, M. A., Garfinkel, E., Hirschman, S. Z., Thung, S. N., and Panagiotatos, T. (1981). Immune and enzyme histochemical studies of a human hepatocellular carcinoma cell line producing hepatitis B surface antigen. *J. Immunol.* **126,** 1085–1089.

Goldman, N. D., and Liu, T. Y. (1987). Biosynthesis of human C-reactive protein in cultured hepatoma cells is induced by a monocyte factor(s) other than interleukin-1. *J. Biol. Chem.* **262,** 2363–2368.

Grant, M. H., Duthie, S. J., Gray, A. G., and Burke, M. D. (1988). Mixed function oxidase and UDP-glucuronyltansferase activities in the human Hep G2 hepatoma cell line. *Biochem. Pharmacol.* **37,** 4111–4116.

Grimwood, B. G., Plummer, T. H., Jr., and Tarentino, A. L. (1988). Characterization of the carboxy-peptidase N secreted by Hep G2 cells. *J. Biol. Chem.* **263,** 14397–14401.

Haddad, J. G., Aden, D. P., and Kowalski, M. A. (1983). Characterization of the human plasma binding protein for vitamin D and its metabolites synthesized by the human hepatoma-derived cell line, Hep 3B. *J. Biol. Chem.* **258,** 6850–6854.

Hata, Y., Uchino, J., Sato. K., Sasaki, F., Une, Y., Naito, H., Manabe, K., Kuwahara, T., and Kasai, Y. (1982). Establishment of an experimental model of human hepatoblastoma. *Cancer* **50,** 97–101.

Hatada, E. N., McClain, D. A., Potter, E., Ullrich, A., and Olefsky, J. M. (1989). Effects of growth and insulin treatment on the levels of insulin receptors and mRNA in Hep G2 cells. *J. Biol. Chem.* **264,** 6741–6747.

Havekes, L. M., Schouten, D., Wit, E. C. M., Cohen, L. H., Griffioen, M., Hinsbergh, V. W. M., and Princen, H. M. G. (1986). Stimulation of the LDL receptor activity in the human hepatoma cell line Hep G2 by high-density serum fractions. *Biochim. Biophys. Acta* **875,** 236–246.

He, L., Isselbacher, K. J., Wands, J. R., Goodman, H. M., Shih, C., and Quaroni, A. (1984). Establishment and characterization of a new human hepatocellular carcinoma cell line. *In Vitro* **20,** 493–504.

Hoeg, J. M., Meng, M. S., Ronan, R., Demosky, S. J., Jr., Fairwell, T., and Brewer, H. B., Jr. (1988). Apolipoprotein B synthesized by Hep G2 cells undergoes fatty acid acylation. *J. Lipid Res.* **29,** 1215–1220.

Hoshi, H., and McKeehan, W. L. (1985). Production of an auto-stimulatory growth factor by human hepatoma cells abrogates requirement for a brain-derived factor. *In Vitro Cell. Dev. Biol.* **21,** 125–128.

Hosokawa, S., Muramatsu, M., and Nagaike, K. (1989). Detection of membrane-bound α-fetoprotein in human hepatoma cell lines by monoclonal antibody 19F12. *Cancer Res.* **49,** 361–366.

Huber, B. E., and Thorgeirsson, S. S. (1987). Analysis of c-*myc* expression in a human hepatoma cell line. *Cancer Res.* **47,** 3414–3420.

Huber, B. E., Dearfield, K. L., Williams, J. R., Heilman, C. A., and Thorgeirsson, S. S. (1985). Tumorigenicity and transcriptional modulation of c-*myc* and N-*ras* oncogenes in a human hepatoma cell line. *Cancer Res.* **45,** 4322–4329.

Huh, N., and Utakoji, T. (1981). Production of HBs-antigen by two new human hepatoma cell lines and its enhancement by dexamethasone. *Gann* **72,** 178–179.

Inagaki, Y., Tsunokawa, Y., Sakamoto, H., Hirohashi, S., Kobayashi, K., Hattori, N., Ramirez, F., Terada, M., and Sugimura, T. (1987). Presence of different types of procollagen messenger RNAs in human hepatoma cell lines. *Biochem. Biophys. Res. Commun.* **148,** 869–875.

Jaffe, E. A., Armellino, D., and Tollefsen, D. M. (1985). Biosynthesis of functionally active heparin cofactor II by a human hepatoma-derived cell line. *Biochem. Biophys. Res. Commun.* **132,** 368–374.

Kambouris, A. M., Roach, P. D., and Nestel, P. J. (1988). Demonstration of a high density lipoprotein (HDL)-binding protein in Hep G2 cells using colloidal gold-HDL conjugates. *FEBS Lett.* **230,** 176–180.

Khan, M. S., Knowles, B. B., Aden, D. P., and Rosner, W. (1981). Secretion of testosterone-estradiol-binding globulin by a human hepatoma-derived cell line. *J. Clin. Endocrinol. Metab.* **53,** 448–449.

Knowles, B. B., Howe, C. C., and Aden, D. P. (1980). Human hepatocellular carcinoma cell lines secrete the major plasma proteins and hepatitis B surface antigen. *Science* **209,** 497–499.

Kobayashi, M., Horiuchi, R., Hachisu, T., and Takikawa, H. (1988). Dualistic effects of thyroid hormone on a human hepatoma cell line: Inhibition of thyroxine-binding globulin synthesis and stimulation of α_1-acid glycoprotein. *Endocrinology* **123,** 631–640.

Lescoat, G., Loreal, O., Moirand, R., Dezier, J. F., Pasdeloup, N., Deugnier, Y., and Brissot, P. (1989). Iron induction of ferritin synthesis and secretion in human hepatoma cell (Hep G2) cultures. *Liver* **9,** 179–185.

Levin, E. G., Fair, D. S., and Loskutoff, D. J. (1983). Human heptoma cell line plasminogen activator. *J. Lab. Clin. Med.* **102,** 500–508.

McAleer, W. J., Markus, H. Z., Wampler, D. E., Buynak, E. B., Miller, W. J., Weibel, R. T., McLean, A. A., and Hilleman, M. R. (1984). Vaccine against human hepatitis B virus prepared from antigen derived from human hepatoma cells in culture (41801). *Proc. Soc. Exp. Biol. Med.* **175,** 314–319.

Marion, P. L. Salazar, F. H., Alexander, J. J., and Robinson, W. S. (1979). Polypeptides of hepatitis B virus surface antigen produced by a hepatoma cell line. *J. Virol.* **32,** 796–802.

Marion, P. L., Salazar, F. H., Alexander, J. J., and Robinson, W. S. (1980). State of hepatitis B viral DNA in a human hepatoma cell line. *J. Virol.* **33,** 795–806.

Marquardt, O., Zaslavsky, V., and Hofschneider, P. H. (1982). Evidence for nonchromosomal hepatitis B virus surface (HBsAg)- and core antigen (HBcAg)-specific DNA sequences in a hepatoma cell line. *J. Gen. Virol.* **61,** 105–109.

Matsuura, H. (1983). Primary cultured cells and an established cell line of human hepatocellular carcinomas. *Acta Med. Okayama* **37,** 341–352.

Mitani, K., Fujita, H., Sassa, S., and Kappas, A. (1989). Heat shock induction of heme oxygenase mRNA in human Hep 3B hepatoma cells. *Biochem. Biophys. Res. Commun.* **165,** 437–441.

Moses, A. C., Freinkel, A. J., Knowles, B. B., and Aden, D. P. (1983). Demonstration that a human hepatoma cell line produces a specific insulin-like growth factor carrier protein. *J. Clin. Endocrinol. Metab.* **56,** 1003–1008.

Motoo, Y., Kobayashi, K., and Hattori, N. (1988). Effect of insulin on the growth of a human hepatoma cell line PLC/PRF/5: A possible role of insulin receptor. *Tohoku J. Exp. Med.* **156,** 351–357.

Motoo, Y., Mahmoudi, M., Osther, K., and Bollon, A. P. (1986). Oncogene expression in human hepatoma cells PLC/PRF/5. *Biochem. Biophys. Res. Commun.* **135,** 262–268.

Murata, A., Ogawa, M., Uda, K., Matsuura, N., Watanabe, Y., Baba, T., and Mori, T. (1988). Release of pancreatic secretory trypsin inhibitor from human hepatoblastoma cells on stimulation with cytokines. *Life Sci.* **43,** 1233–1240.

Nagaoka, I., Someya, A., Iwabuchi, K., and Yamashita, T. (1991). Expression of insulin-like growth factor-IA and factor-IB mRNA in human liver, hepatoma cells, macrophage-like cells and fibroblasts. *FEBS Lett.* **280,** 79–83.

Nagy, J. A., Henriksson, P., and McDonagh, J. (1986). Biosynthesis of factor XIII$_B$ subunit by human hepatoma cell lines. *Blood* **68,** 1272–1279.

Nakabayashi, H., Taketa, K., Miyano, K., Yamane, T., and Sato, J. (1982). Growth of human hepatoma cell lines with differentiated functions in chemically defined medium. *Cancer Res.* **42,** 3858–3863.

Nakabayashi, H., Taketa, K., Yamane, T., Miyazaki, M., Miyano, K., and Sato, J. (1984). Phenotypical stability of a human hepatoma cell line, HuH-7, in long-term culture with chemically defined medium. *Gann* **75,** 151–158.

Nakabayashi, H., Taketa, K., Yamane, T., Oda, M., and Sato, J. (1985). Hormonal control of α-fetoprotein secretion in human hepatoma cell lines proliferating in chemically defined medium. *Cancer Res.* **45,** 6379–6383.

Nakabayashi, H., Watanabe, K., Saito, A., Otsuru, A., Sawadaishi, K., and Tamaoki, T. (1989). Transcriptional regulation of α-fetoprotein expression by dexamethasone in human hepatoma cells. *J. Biol. Chem.* **264,** 266–271.

Nakagawa, T., Nakao, Y., Matsui, T., Koizumi, T., Matsuda, S., Maeda, S., and Fujita, T. (1985). Effects of sodium n-butyrate on alpha-fetoprotein and albumin secretion in the human hepatoma cell line PLC/PRF/5. *Br. J. Cancer* **51,** 357–363.

Nakamura, H., Kambe, H., Egawa, T., Kimura, Y., Ito, H., Hayashi, E., Yamamoto, H., Sato, J., and Kishimoto, S. (1989). Partial purification and characterization of human hepatoma-derived growth factor. *Clin. Chim. Acta* **183,** 273–284.

Nakao, K., Lawless, D., Ohe, Y., Miyao, Y., Nakabayashi, H., Kamiya, H., Miura, K., Ohtsuka, E., and Tamaoki, T. (1990). c-Ha-*ras* down regulates the α-fetoprotein gene but not the albumin gene in human hepatoma cells. *Mol. Cell. Biol.* **10,** 1461–1469.

Oefinger, P. E., Bronson, D. L., and Dreesman, G. R. (1981). Induction of hepatitis B surface antigen in human hepatoma-derived cell lines. *J. Gen. Virol.* **53,** 105–113.

Povoa, G., Isaksson, M., Jornvall, H., and Hall, K. (1985). The somatomedin-binding protein isolated from a human hepatoma cell line is identical to the human amniotic fluid somatomedin-binding protein. *Biochem. Biophys. Res. Commun.* **128,** 1071–1078.

Raney, A. K., Milich, D. R., and McLachlan, A. (1991). Complex regulation of transcription from the hepatitis B virus major surface antigen promoter in human hepatoma cell lines. *J. Virol.* **65,** 4805–4811.

Ranganathan, S., and Kottke, B. A. (1990). Rapid regulation of apolipoprotein A-I secretion in Hep G2 cells by a factor associated with bovine high-density lipoproteins. *Biochim. Biophys. Acta* **1046,** 223–228.

Richards, C. A., Short, S. A., Thorgeirsson, S. S., and Huber, B. E. (1990). Characterization of a transforming N-*ras* gene in the human hepatoma cell line Hep G2: Additional evidence for the importance of c-*myc* and *ras* cooperation in hepatocarcinogenesis. *Cancer Res.* **50,** 1521–1527.

Roingeard, P., Lu, S., Sureau, C., Freschlin, M., Arbeille, B., Essex, M., and Romet-Lemonne, J. L. (1990a). Immunocytochemical and electron microscopic study of hepatitis B virus (HBV) antigen and complete particle production in HBV DNA transfected cells. *Hepatology* **11,** 277–285.

Roingeard, P., Romet-Lemonne, J. L., Leturcq, D., Goudeau, A., and Essex, M. (1990b). Hepatitis B virus core antigen (HBcAg) accumulation in an HBV nonproducer clone of HepG2-transfected cells is associated with cytophatic effect. *Virology* **179,** 113–120.

Saito, H., Goodnough, L. T., Knowles, B. B., and Aden, D. P. (1982). Synthesis and secretion of α$_2$-plasmin inhibitor by established human liver cell lines. *Proc. Natl. Acad. Sci. U.S.A.* **79,** 5684–5687.

Saito, H., Morizane, T., Watanabe, T., Kagawa, T., Iwabuchi, N., Kumagai, N., Inagaki, Y., Tsuchimoto, K., and Tsuchiya, M. (1989). Establishment of a human cell line (HCC-T) from a patient with hepatoma bearing no evidence of hepatitis B or A virus infection. *Cancer* **64,** 1054–1060.

Sassa, S., Sugita, O., Galbraith, R. A., and Kappas, A. (1987). Drug metabolism by the human hepatoma cell. Hep G2. *Biochem. Biophys. Res. Commun.* **143,** 52–57.

Sato, R., Imanaka, T., and Takano, T. (1990). The effect of HMG-CoA reductase inhibitor (CS-514) on the synthesis and secretion of apolipoprotein B and A-I in the human hepatoblastoma Hep G2. *Biochim. Biophys. Acta* **1042,** 36–41.

Satoh, M., and Enomoto, K. (1991). Study of differentiation in human hepatoblastoma cells:

Changes in cell properties by differentiation-inducing agents and co-culture with fibroblastic cells. *Sapporo-Igaku-Zasshi* **60**, 173–182 (*in Japanese*).

Schwartz, A. L., Fridovich, S. E., Knowles, B. B., and Lodish, H. F. (1981). Characterization of the asialoglycoprotein receptor in a continuous hepatoma line. *J. Biol. Chem.* **256**, 8878–8881.

Sells, M. A., Chen, M. L., and Acs, G. (1987). Production of hepatitis B virus particles in Hep G2 cells transfected with cloned hepatitis B virus DNA. *Proc. Natl. Acad. Sci. U.S.A.* **84**, 1005–1009.

Semenkovich, C. F., and Ostlund, R. E., Jr. (1987). Estrogens induce low-density lipoprotein receptor activity and decrease intracellular cholesterol in human hepatoma cell line Hep G2. *Biochemistry* **26**, 4987–4992.

Shouval, D., Reid, L. M., Chakraborty, P. R., Ruiz-Opazo, N., Morecki, R., Gerber, M. A., Thung, S. N., and Shafritz, D. A. (1981). Tumorigenicity in nude mice of a human hepatoma cell line containing hepatitis B virus DNA. *Cancer Res.* **41**, 1342–1350.

Shouval, D., Rager-Zisman, B., Quan, P., Shafritz, D. A., Bloom, B. R., and Reid, L. M. (1983). Role in nude mice of interferon and natural killer cells in inhibiting the tumorigenicity of human hepatocellular carcinoma cells infected with hepatitis B virus. *J. Clin. Invest.* **72**, 707–717.

Skelly, J., Copeland, J. A., Howard, C. R., and Zuckerman, A. J. (1979). Hepatitis B surface antigen produced by a human hepatoma cell line. *Nature (London)* **282**, 617–618.

Su, T. S., Lin, L. H., Chou, C. K., Chang, C., Ting, L. P., Hu, C., and Han, S. H. (1986). Hepatitis B virus transcripts in a human hepatoma cell line, Hep3B. *Biochem. Biophys. Res. Commun.* **138**, 131–138.

Takaoka, T. (1958). Fluid-suspension culture of rat ascites hepatoma cells and tissue culture strain cells. *Jpn. J. Exp. Med.* **28**, 381–393.

Tam, S. P., Archer, T. K., and Deeley, R. G. (1985). Effects of estrogen on apolipoprotein secretion by the human hepatocarcinoma cell line, HepG2. *J. Biol. Chem.* **260**, 1670–1675.

Tam, S. P., Strugnell, S., Deeley, R. G., and Jones, G. (1988). 25-Hydroxylation of vitamin D_3 in the human hepatoma cell lines Hep G2 and Hep 3B. *J. Lipid Res.* **29**, 1637–1642.

Tanaka, M., Kawamura, K., Fang, M., Higashino, K., Kishimoto, S., Nakabayashi, H., and Sato, J. (1983). Production of fibronectin by HuH6 c15 cell line established from a human hepatoblastoma. *Biochem. Biophys. Res. Commun.* **110**, 837–841.

Tokiwa, T., Miyagiwa, M., Kusaka, Y., Muraoka, A., and Sato, J. (1988). Effects of various substrates on human hepatoblastoma and hepatoma cell culture. *Cell Biol. Int. Rep.* **12**, 131–142.

Tsai, T. F., Yauk, Y. K., Chou, C. K., Ting, L. P., Chang, C., Hu, C., Han, S. H., and Su, T. S. (1988). Evidence of autocrine regulation in human hepatoma cell lines. *Biochem. Biophys. Res. Commun.* **153**, 39–45.

Tsurimoto, T., Fujiyama, A., and Matsubara, K. (1987). Stable expression and replication of hepatitis B virus genome in an integrated state in a human hepatoma cell line transfected with the cloned viral DNA. *Proc. Natl. Acad. Sci. U.S.A.* **84**, 444–448.

Twist, E. M., Clark, H. F., Aden, D. P., Knowles, B. B., and Plotkin, S. A. (1981). Integration pattern of hepatitis B virus DNA sequences in human hepatoma cell lines. *J. Virol.* **37**, 239–243.

Ueno, M., Seferynska, I., Beckman, B., Brookins, J., Nakashima, J., and Fisher, J. W. (1989). Enhanced erythropoietin secretion in hepatoblastoma cells in response to hypoxia. *Am. J. Physiol.* **257**, C743–C749.

Urano, Y., Watanabe, K., Lin, C. C., Hino, O., and Tamaoki, T. (1991). Interstitial chromosomal deletion within 4q11-q13 in a human hepatoma cell line. *Cancer Res.* **51**, 1460–1464.

Vincent, C., Marceau, M., Blangarin, P., Bouic, P., Madjar, J. J., and Revillard, J. P. (1987). Purification of α_1-microglobulin produced by human hepatoma cell lines. Biochemical characterization and comparison with α_1-microglobulin synthesized by human hepatocytes. *Eur. J. Biochem.* **165**, 699–704.

Wade, D. P., Knight, B. L., and Soutar, A. K. (1988). Hormonal regulation of low-density lipoprotein (LDL) receptor activity in human hepatoma Hep G2 cells. Insulin increases LDL receptor activity and diminishes its suppression by exogenous LDL. *Eur. J. Biochem.* **174**, 213–218.

Wade, D. P., Knight, B. L., and Soutar, A. K. (1989). Regulation of low-density-lipoprotein-receptor mRNA by insulin in human hepatoma Hep G2 cells. *Eur. J. Biochem.* **181,** 727–731.

Watanabe, T., Morizane, T., Tsuchimoto, K., Inagaki, Y., Munakata, Y., Nakamura, T., Kumagai, N., and Tsuchiya, M. (1983). Establishment of a cell line (HCC-M) from a human hepatocellular carcinoma. *Int. J. Cancer* **32,** 141–146.

Wettesten, M., Bostrom, K., Bondjers, G., Jarfeldt, M., Norfeldt, P. I., Carrella, M., Wiklund, O., and Olofsson, S. O. (1985). Pulse-chase studies of the synthesis of apolipoprotein B in a human hepatoma cell line, Hep G2. *Eur. J. Biochem.* **149,** 461–466.

Wilson, D. B., Salem, H. H., Mruk, J. S., Maruyama, I., and Majerus, P. W. (1984). Biosynthesis of coagulation factor V by a human hepatocellular carcinoma cell line. *J. Clin. Invest.* **73,** 654–658.

Wun, T. C., and Kretzmer, K. K. (1987). cDNA cloning and expression in *E. coli* of a plasminogen activator inhibitor (PAI) related to a PAI produced by HepG2 hepatoma cells. *FEBS Lett.* **210,** 11–16.

Yaginuma, K., Shirakata, Y., Kobayashi, M., and Koike, K. (1987). Hepatitis B virus (HBV) particles are produced in a cell culture system by transient expression of transfected HBV DNA. *Proc. Natl. Acad. Sci. U.S.A* **84,** 2678–2682.

Yamamoto, H., Tanaka, M., Nakabayashi, H., Sato, J., Okochi, T., and Kishimoto, S. (1984). Intestinal-type alkaline phosphatase produced by human hepatoblastoma cell line HuH-6 Clone 5. *Cancer Res.* **44,** 339–344.

Yamamoto, T., Takahashi, S., Moriwaki, Y., Hada, T., and Higashino, K. (1987). A newly discovered apolipoprotein B-containing high-density lipoprotein produced by human hepatoma cells. *Biochim. Biophys. Acta* **922,** 177–183.

Yamamoto, T., Suda, M., Moriwaki, Y., Takahashi, S., and Higashino, K. (1990). Novel apoprotein A-I-containing lipoprotein produced by a human hepatoma-derived cell line HuH-7. *Biochim. Biophys. Acta* **1047,** 49–56.

Yamashita, Y., Koike, K., Takaoki, M., and Matsuda, S. (1988). Suppression of HBsAg production in PLC/PRF/5 human hepatoma cell line by interferons. *Microbiol. Immunol.* **32,** 1119–1126.

Yano, H., Kojiro, M., and Nakashima, T. (1986). A new human hepatocellular carcinoma cell line (KYN-1) with a transformation to adenocarcinoma. *In Vitro Cell. Dev. Biol.* **22,** 637–646.

Zannis, V. I., Breslow, J. L., SanGiacomo, T. R., Aden, D. P., and Knowles, B. B. (1981). Characterization of the major apolipoproteins secreted by two human hepatoma cell lines. *Biochemistry* **20,** 7089–7096.

Ziemer, M., Garcia, P., Shaul, Y., and Rutter, W. J. (1985). Sequence of hepatitis B virus DNA incorporated into the genome of a human hepatoma cell line. *J. Virol.* **53,** 885–892.

Hematopoietic Cell Lines

8

Hans G. Drexler and Suzanne M. Gignac
Department of Human and Animal Cell Cultures
German Collection of Microorganisms and Cell Cultures
D-38124 Braunschweig, Germany

Jun Minowada
Fujisaki Cell Center
Hayashibara Biochemical Laboratories, Inc.
Okayama 702, Japan

I. Introduction 213

II. Culture of Hematopoietic Cells 214

III. Leukemia–Lymphoma Cell Lines 217

IV. Characterization of Leukemia Cell Lines 219
A. Origin of Cells 220
B. *In Vitro* Culture 221
C. Morphology 222
D. Cytochemistry 225
E. Immunophenotyping 225
F. Genotyping 236
G. Cytogenetics 237
H. Cytokines 239

I. Functional Features 239

V. Growth Factor-Dependent Leukemia Cell Lines 242

VI. Hodgkin's Disease-Derived Cell Lines 243

VII. Future Prospects 244
A. Data Bank 244
B. Collection of Cell Lines 247
C. Establishment of New Cell Lines 247
D. Characterization Program, Quality and Identity Control 248

VIII. Lists of Leukemia Cell Lines 249

References 249

I. Introduction

Most mature hematopoietic cells are short lived and are replaced continuously throughout life (Metcalf, 1989). Blood cells originate from a self-renewing pool of multipotential stem cells that generate in the bone marrow, progenitor cells that are committed irreversibly to one of the various hematopoietic cell lineages (Table I). A pattern of maturation running parallel to final clonal

Table I

Hierarchy of Hematopoietic Proliferation and Differentiation[a]

Class of cells	Proliferation/differentiation
Stem cells	Self-renewal and proliferation; differentiation commitment
Commited progenitor cells	Proliferation; further differentiation commitment
Morphologically identifiable immature cells	Proliferation; maturation
Mature cells	Postmitotic; differentiated

[a] Adapted from Metcalf (1989).

proliferation is common to all blood cell lineages (Table II). The mature cells are fully differentiated and, except for lymphoid cells, are incapable of further division (Metcalf, 1989).

Regulation of hematopoiesis is achieved by interacting systems: specialized stromal cells in the bone marrow and lymphoid organs (e.g., lymph nodes, thymus) and regulatory molecules (called hematopoietic growth factors or cytokines) control and stimulate the proliferation of stem and progenitor cells and their progeny and initiate the maturation process to produce fully mature cells (Metcalf, 1989).

A significant handicap for studies on hematopoietic cells is that nearly all types of normal precursor and immature cells are present only at very low quantities in the respective organs and require elaborate purification steps. The culturing of hematopoietic cells has opened new avenues of research.

II. Culture of Hematopoietic Cells

Two methods commonly are used to propagate hematopoietic cells *in vitro* in short-term, long-term, or continuous culture: semi-solid media (e.g., soft agar, methylcellulose) and suspension cultures. The first technique is used mainly for short-term culture (for days and weeks) of normal or leukemic cells, for example, colony formation assays of progenitor or immature cells. Suspension culture usually is employed for long-term culturing (for weeks and months) and for attempts to establish continuous cell lines.

Most normal and malignant hematopoietic cells have a very short survival time *in vitro,* ranging from a few days to weeks. If the cells begin to divide, they occasionally can survive for several weeks or even months. Rarely, these cells form established cell lines. Continuous cell lines have been defined as cultures that apparently are capable of an unlimited number of population dou-

Table II

Hematopoietic Stem, Progenitor, and Mature Cells[a]

Pluripotent stem cell	→	Lymphoid progenitor (lymphopoiesis)		
Pluripotent stem cell	→	Mixed myeloid progenitor (granulo-, erythro-, megakaryo-, monopoiesis)		
Lymphoid progenitor	→	B lymphocyte		
	→	T lymphocyte		
Mixed myeloid progenitor	→	Erythroid progenitor	→	Erythrocyte
	→	Megakaryocyte progenitor	→	Platelet
	→	Granulocyte/monocyte progenitor	→	Monocyte[b]
			→	Granulocyte → Neutrophil
				Basophil[c]
				Eosinophil

[a] The eight main types of blood cells are underlined.
[b] Derivates of monocytes in tissues are macrophages and histiocytes.
[c] Derivates of basophils are mast cells.

blings (immortalization); an immortalized cell is not necessarily one that is neoplastically or malignantly transformed (Schaeffer, 1990).

So-called "normal" hematopoietic cell lines must be discerned from malignant hematopoietic cell lines, that is, from leukemia–lymphoma cell lines. The former lines require transformation by a virus, for example, Epstein–Barr virus (EBV)-transformed B-lymphoblastoid cell lines (B-LCL), for immortalization. Leukemia–lymphoma cell lines are immortalized by virus infection with EBV or human T-cell leukemia virus (HTLV) or arise spontaneously. For instance, specific chromosomal translocations can activate proto-oncogenes, leading to continuous autocrine growth stimulation. Previously normal T or B lymphocytes grow readily when transforming virus such as EBV, HTLV-I, and HLTV-II or other retroviruses are present in the cultures (Lange, 1989). Nonneoplastic cells are inevitably present as residual cells in any hematopoietic explant (e.g., peripheral blood, bone marrow, lymph node, tonsil). Since lymphoblastoid lines—mostly EBV$^+$ B-LCL—may grow out from this normal population, a cell line from a human malignancy is not automatically of neoplastic origin (Table III; Nilsson and Pontén, 1975). In fact, the odds of establishing an EBV$^+$ B-LCL from a leukemic peripheral blood sample are certainly 10- to 100-fold better than those of establishing a leukemia cell line. Few systematic attempts to

Table III

Differences between EBV$^+$ Lymphoblastoid Cell Lines and Leukemia Cell Lines[a]

	EBV$^+$ Lymphoblastoid cell lines	Leukemia cell lines
Source of material	Normal or malignant tissue	Malignant tissue
Efficiency of establishment	High	Low
Population doubling time (hr)	24–48	12–120
Growth in suspension	Large dense clumps	All variations possible: adherence, single cells, small/large clumps
Morphological diversity		
between cell lines	None	Present
within cell lines	Heterogeneous	Homogenous[b]
Karyotype	Diploid	Aneuploid[c]
Clonality	Polyclonal (initially)[d]	Monoclonal[e]
EBV genome	Always present	Present or absent

[a] Modified from Nilsson and Pontén (1975).
[b] With some notable exceptions (e.g., Hodgkin's cell lines).
[c] Cells usually show numerical (polyploid) and distinct structural chromosome abnormalities.
[d] In long-term culture, a predominant monoclonal population will grow out.
[e] In theory, a leukemia cell line is established from a single cell.

develop continuous leukemia–lymphoma cell lines have been reported, with an average success rate of 1–5%.

Difficulties in establishing continuous human leukemia–lymphoma cell lines may originate not only from the inappropriate selection of nutrients and growth factors for these cells, but also from some of the biological characteristics of the *in vivo* tumor mass. Based on *in vivo* growth kinetics studies, the malignant cells do not necessarily proliferate faster than their normal counterparts; in fact, the proliferation time seems to be significantly longer. Many neoplastic cells might not be capable of indefinite proliferation per se. Judging empirically from the number of established cell lines (assuming that workers make equal numbers of attempts to produce cell lines from the various cell lineages), the success rate is certainly much higher for more immature than for more mature cells, and for lymphoid (T or B cell) than for myeloid (including monocytic and erythroid) cells.

III. Leukemia–Lymphoma Cell Lines

Distinguishing cell lines established from the "liquid phase" of human neoplasms of hematopoiesis (i.e., leukemias spread throughout the blood and bone marrow) from those derived from the "solid phase" of these diseases (i.e., lymphomas, primarily localized to lymphatic tissues) is virtually impossible. For instance, a T cell from a sample of T-cell leukemia is often very difficult to discern from a T cell lymphoblastic lymphoma. Therefore, we use the term "leukemia cell lines" here in lieu of the more correct "leukemia–lymphoma cell lines."

A relatively large number of human leukemia cell lines has been established in the past 10–20 yrs, most of them in the 1980s (Ferrero and Rovera, 1984). The variety of available cell lines includes all major categories of hematopoietic malignancies (the various cell lineages are outlined in Table II). Recently, attention has shifted to the establishment of cell lines with specific properties, for example, dependency on externally added growth factors, multi- or pluripotential stem cell lines, and so on. However, the establishment of most of these cell lines has been purely accidental (Ferrero and Rovera, 1984). As mentioned earlier, most leukemic cells do not grow well in tissue culture despite their proliferative capacity *in vivo*. A second drawback is the fact that some leukemias (usually myelomonocytic, but also other cell lineages) only survive several passages before dying out and, thus, cannot be regarded as established cell lines. In contrast to routine immortalization of lymphoid cells by co-cultivation with B-lymphotrophic EBV or T-lymphotrophic human HTLV-I, no such procedure is available for the establishment of leukemia cell lines. In most instances, these cell lines are independent of known growth factors, but usually are established with fetal bovine serum (FBS). Different additives have been employed but no single protocol has been shown to be more efficient than others. Also, the source of leukemic cells (e.g., peripheral blood, bone marrow) does not seem to influence the efficiency of immortalization (Lübbert and Koeffler, 1988). The successful establishment of a leukemia cell line was not correlated with the specific type of leukemia or its clinical aggressiveness. However, most cell lines contain cells with multiple karyotypic abnormalities. Overall, the mechanisms leading to immortalization of fresh leukemic cells after prolonged culturing remain unclear, making establishment of these lines largely fortuitous (Lübbert *et al.*, 1991).

Burkitt's lymphoma cell lines are comparatively easy to establish and were the first of these cell lines generated. Several milestones in the study of human leukemia cell lines are highlighted in Table IV. After establishment of the first human leukemia cell lines, many more were developed, so now a wide array of lines representing different stages of the various cell lineages is available. These lines serve as models for their respective normal cell types and for the

Table IV

Significant Milestones in the Study of Human Leukemia Cell Lines[a]

Cell line	Approximate year	Significance[b]
RAJI	1964	Burkitt lymphoma (B cells)
RPMI-8226	1966	Multiple myeloma (plasma cells)
RPMI-7206	1966	"Normal" EBV-transformed B cells
K-562	1970	CML in blast crisis[c]
MOLT 1-4	1971	T-cell ALL (T cells)
U-698-M	1973	Non-Burkitt malignant lymphoma (B cells)
REH	1973	Common ALL (pre-B cells)
U-937	1974	Histiocytic lymphoma (monocytic cells)
NALM-1	1975	CML in lymphoid blast crisis (pre-B cells)
BALM-1,-2	1976	B-cell ALL (B cells)
HL-60	1977	AML (myeloid cells)
L-428	1978	Hodgkin's cell line (lymphoid cell?)
MT-1	1978	Adult T-cell leukemia (mature T cells)
HEL	1980	Erythroleukemia (erythroid cells)
JOK-1	1981	Hairy cell leukemia (B cells)
MEG-01	1983	CML in megakaryoblastic crisis (megakaryocytic cells)

[a] Modified from Minowada (1982).

[b] Abbreviations: ALL, acute lymphoblastic leukemia; AML, acute myeloid leukemia; CML, chronic myeloid leukemia.

[c] K-562 cells later were found to have pluripotential differentiation abilities and to synthesize hemoglobin.

type of leukemia from which they were derived originally (Lübbert and Koeffler, 1988). However, the comparability of the primary neoplastic clone and the established cell line has limitations: cells in culture may lose some of their original characteristics and develop additional chromosomal abnormalities; they may change their patterns of gene expression, presumably because of constant selection for *in vitro* growth advantage. These alterations necessitate that observations derived from cell lines be correlated with properties of normal and primary leukemia cells (Lübbert and Koeffler, 1988).

IV. Characterization of Leukemia Cell Lines

In most cases, the cell culturist cannot discriminate continuous cell lines established from patients with leukemia from those cell lines originating from patients with lymphoma on the basis of the morphological appearance of the cells. Also, cell lines originating from different cell lineages are often impossible to distinguish by morphology alone. Since leukemia cell lines commonly grow as suspension cultures, the cells can be subjected easily to various examinations for the purpose of characterization and classification. Although some types of leukemia cell line grow in dense clusters or adhere loosely to the flask, single cell populations can be prepared readily for subsequent analysis. Table V lists a variety of parameters useful for the description of the cells and a panel of possible tests applicable to the phenotypic characterization of most cell lines.

Leukemic cells (and, by analogy, cell lines derived from these primary cells) are thought to be malignant counterparts of normal immature cells. The neoplastic cells are arrested in their differentiation and are "frozen" at a particular point of maturation and differentiation. Whereas the corresponding normal counterparts develop further to fully mature, nonproliferating cells (terminal differentiation), uncoupling of proliferation and differentiation and stabilization of a normally transitory phenotype occur in leukemic cells (Greaves, 1986). Leukemic phenotypes, however, are not perfect replicas of normal ones. Usually specific chromosomal changes, with consequent dysregulation of particular genes, are involved in the malignant process. Leukemic cells might show some asynchrony of phenotypic expression in comparison with their normal equivalent cells (Greaves, 1986). Additionally, cells could display truly aberrant gene expression, especially cell lines in which specific clones with dominant *in vitro* growth advantages might have overgrown other populations.

However, most cell lines preserve a pattern of gene expression that reflects the developmental level of clonal expansion and maturation arrest. The necessary multiparameter examination of the cellular phenotype provides important information on the likely cell of origin, the variable stringency of maturation arrest, and the predominantly normal pattern of gene expression. In the following discussion, we highlight some of the features of the phenotypic profiles of

Table V
Analytical Characterization of Leukemia Cell Lines

Parameter	Details and examples
Origin	Patient's data
In vitro culture	Growth kinetics, proliferative characteristics
Morphology	*In situ* (flask, plate) under inverted microscope
	Light microscopy (May–Grünwald–Giemsa-stained)
	Electron microscopy (transmission and scanning)
Cytochemistry	Acid phosphatase, α-naphthyl acetate esterase, and others
Immunophenotyping	Surface marker antigens (fluorescence microscopy, flow cytometry)
	Intracytoplasmic and nuclear antigens (immunoenzymatic staining)
Genotyping	Southern blot analysis of T-cell receptor (TCR) and immunoglobulin (Ig) heavy and light chain gene rearrangements
	Northern analysis of expression of TCR and Ig transcripts
Cytogenetics	Structural and numerical abnormalities
	Specific chromosomal markers
Cytokines	Production of cytokines
	Expression of cytokine receptors
	Response to cytokines, dependency on cytokines
Function aspects/specific features	Phagocytosis
	Antigen presentation
	Immunoglobulin production and secretion
	Hemoglobin synthesis
	Inducibility of differentiation
	Positivity for EBV, HTLV-I, or other viruses
	Heterotransplantability into nude or SCID mice
	Production/secretion of specific proteins
	Natural killer cell activity

leukemic cell lines most often studied that are essential for their description and classification.

A. Origin of Cells

The origin of an established cell line must be documented sufficiently. A minimum set of data will insure the authenticity and uniqueness of each cell line (Table VI). Note that each cell line is unique, consisting of cells arrested at a specific point in the differentiation continuum along a particular pathway. The cells of a given cell line express particular phenotypic features that are specific to this type of cell or are associated with the malignant process.

Table VI
Minimum Data Set Essential for the Identification and Description of a Leukemia Cell Line

Parameter	Example[a]
Name of cell line	HL-60
Cell phenotype	Myeloid
Original disease of patient	Acute myeloid leukemia (FAB M2)
Patient data (age, sex, race)	35-yr-old Caucasian woman
Source of material	Peripheral blood
Year of establishment	1976
Literature reference	Collins et al. (1977)
Culture medium	RPMI 1640 + 10% FBS
Subcultivation routine	Maintain at 0.1–0.5 × 10^6 cells/ml, split ratio 1 : 5 to 1 : 10, every 2–3 days
Minimum cell density	0.5–1 × 10^5 cells/ml
Maximum cell density	1–1.5 × 10^6 cells/ml
Doubling time	24–36 hr
Cell storage conditions	RPMI 1640 + 20% FBS + 10% DMSO
In situ morphology	Round single cells in suspension

[a] Abbreviations: RPMI, Roswell Park Memorial Institute; FBS, fetal bovine serum; DMSO, dimethyl sulfoxide.

Permanently cultured cell lines will, sooner or later, show changes in their expression of distinct characteristics, albeit to various extents, since many cell lines appear to be cytogenetically unstable.

In many cases, a sample from one patient with leukemia or lymphoma was split into several cultures, so different subclones with, in some instances, quite explicit differences began to grow from the outset. Another type of subclone generation is the accidentally emerging or deliberately produced subclone with features different from those of the parental cell line. For instance, duplicates of the progenitor cell line K-562 grown in different laboratories express various types of hemoglobin. Another example consists of Jurkat/JM cell lines maintained in different laboratories, which demonstrate differences in interleukin 2 (IL-2)-producing ability.

B. In Vitro *Culture*

The standard culture medium for human leukemia cell lines is RPMI 1640. Although many cell lines were established in other media such as Iscove's modified Dulbecco's medium, alpha-modified Eagle's medium, McCoy's 5A, Ham's F-10, and others, nearly all cell lines can be adapted to RPMI 1640 medium. Cell lines are cultured most often in the presence of fetal or newborn bovine serum (5–20%). Although all cell lines established during the first 20 years of leukemia cell culture work grow autonomously, recent efforts are directed toward establishing cell lines that are dependent on exogenous growth factors. These cell lines will not survive in the absence of these growth factors. However, growth factor-independent clones might often outgrow the originally dependent cells during prolonged culture. Cells are frozen in 70% medium, 20% FBS, and 10% dimethylsulfoxide (DMSO) for cryopreservation in liquid nitrogen at $-196°C$.

The density of the cells in culture is, for the most part, in the range of $0.1–2.0 \times 10^6$ cells/ml. Some pre-B acute lymphoblastic leukemia (ALL) cell lines can reach maximal cell densities of $4–6 \times 10^6$ cells/ml. Cultures containing large or giant cells (e.g., erythroleukemia and Hodgkin's cell lines) will not exceed a concentration of $0.3–0.5 \times 10^6$ cell/ml. Most cell lines require high cell densities for continuous growth; the cells may secrete growth factors that are active in an autocrine fashion. Although cloning leukemia cell lines is possible, the technique is often tedious and difficult, resulting in low clonogenic yields. A wide variety in doubling times ranges from 12 hr to 5 or more days. In general, most mycoplasma-free, optimally cultured cell lines have doubling times of 24–48 hr.

C. Morphology

The morphology of leukemia cell lines can be appreciated at three different levels: morphology of the cells in liquid culture under the inverted microscope; morphology of single cells on cytospin slide or smear preparations stained with May–Grünwald–Giemsa; and morphology of cells observed by transmission or scanning electron microscopy. The electron microscopic examinations of cell lines will not be discussed in this chapter since they require special skills and equipment and have, apart from studies done in the 1970s, rarely been performed on leukemia cell lines.

1. *In Situ* Morphology

In general, human leukemia cell lines grow in suspension in culture flasks or on microtiter plates and are not adherent (Fig. 1). Occasionally, some cells might adhere loosely to the plastic (for example, cells from monocytic leukemia cell lines or mature B cells from plasma cell leukemia or myeloma cell lines); however, the adherence is significantly less pronounced than in fibroblast- or epithelial-like cell lines. Commonly, the cells can be detached easily by shaking the flask or by pipetting the cultured cells vigorously (Fig. 2). After treatment with biomodulators, cells might adhere more strongly (Fig. 3). Cells grow as single cells or clustered in small or large, loose or tight clumps. No conclusion about the cell lineage can be drawn from the type of growth pattern. Nevertheless, EBV$^+$ cell lines are always, and mature T or B cell lines

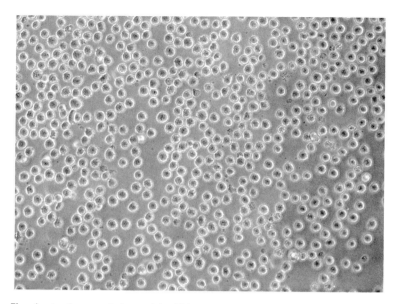

Fig. 1. *In vitro* morphology of the EM-2 cell line (chronic myeloid leukemia in blast crisis). In liquid medium, these cells grow as a nonadherent single-cell suspension. (Inverted phase-contrast microscopy.)

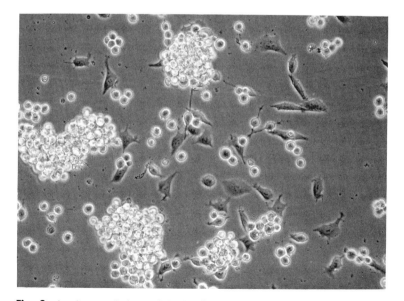

Fig. 2. *In situ* morphology of the MEG-01 cell line (acute megakaryoblastic leukemia). These cells grow partially adherent and partially in dense clumps in suspension. No trypsin is required to dislodge cells during subculturing. (Inverted phase-contrast microscopy.)

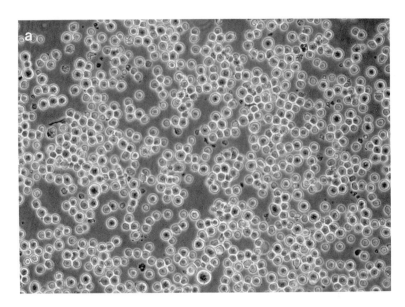

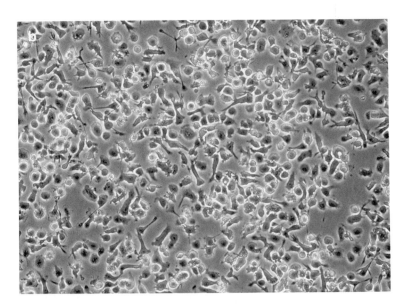

Fig. 3. *In situ* morphology of HL-60 cells prior to (a) and following (b) treatment with the phorbol ester TPA (10^{-8} M TPA). (a) Untreated HL-60 cells (acute myeloblastic leukemia M2) grow as single non-adherent cells. (b) After induction of differentiation to monocyte–macrophage-type cells, the majority of HL-60 cells are adherent with cytoplasmic extensions and pseudopodia. (Inverted phase-contrast microscopy.)

(including Burkitt's lymphoma) are often, clustered in large aggregates containing 10–100 or more cells (Fig. 4). Some small clumps of 5–10 cells occur to some extent in all types of cell lines.

2. Cellular Morphology

Cultures of leukemia cells propagated *in vitro* display a wide variety of morphological appearances, ranging from the expected classical type of cell resembling their normal counterparts entirely or partly to the unexpected morphological appearance incongruous with the stage of differentiation or the assignment to a particular cell lineage (Figs. 5–19).

D. Cytochemistry

Cytochemical stainings are standard methods in the classification of leukemias. Cytochemistry combined with classical morphological criteria (Romanovsky or May–Grünwald–Giemsa staining) is still a useful parameter in the characterization of leukemia cell lines, providing valuable information about lineage and differentiation of the cells.

The most common stains are: acid phosphatase, alkaline phosphatase, periodic acid–Schiff (PAS), myeloperoxidase, Sudan Black, esterases (naphthol AS-D-acetate esterase, α-naphthyl acetate/butyrate esterase, chloroacetate esterase), and β-glucuronidase (Flandrin and Daniel, 1981). In addition, numerous special stains are indicative of particular cell types, for example, toluidine blue for basophils, Oil Red O for Burkitt cells, methyl green pyronine for plasma cells, and muraminidase for monocytes (Scott, 1989).

E. Immunophenotyping

One of the most important characterization tools of human leukemia cell lines is immunophenotyping the cells, that is, immunofluorescently or immunoenzymatically demonstrating cell lineage-specific or -associated proteins (antigens) on the surface, in the cytoplasm, or in the nucleus of the cells (Minowada, 1982; Figs. 20,21). This method can assign any given cell line to a specific cell lineage (T cell, B cell, myeloid, monocytic, erythroid, megakarocytic) and to the stage of differentiation (e.g., pre-B cell, mature B cell, plasma cell) (Matsuo and Minowada, 1988). The nomenclature for these antigens has been elaborated in several workshops defining leukocyte antigen expression in "clusters of differentiation" (CD). Examples of CD and other markers for the immunological classification of human leukemia cell lines are summarized in Table VII.

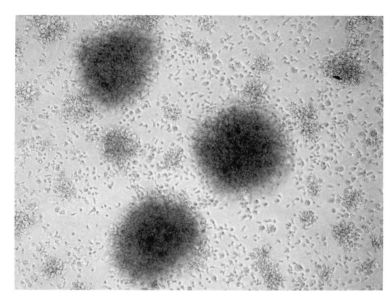

Fig. 4. *In situ* morphology of an EBV-transformed B-lymphoblastoid cell line. Cells grow clumped, to a large extent in very dense clusters of several hundred cells. (Inverted phase-contrast microscopy.)

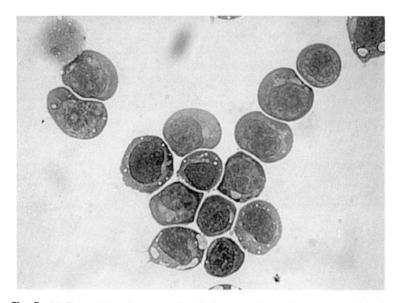

Fig. 5. MKB-1 cell line (immature T cells from a patient with acute myeloblastic leukemia). These cells are large, round or oval, and have a high nucleus:cytoplasm ratio, basophilic cytoplasm with occasional vacuoles, a round or kidney-shaped nucleus with fine chromatin, and 1–3 prominent nucleoli. Cytoplasmic protrusions can occur. Cells have a diameter of 12–15 μm, and grow as single cells in suspension in RPMI 1640 medium up to 1.0–1.5 × 10^6 cells/ml. (Figs. 5–19,24,25: cytospin slide preparations; May–Grünwald–Giemsa staining.)

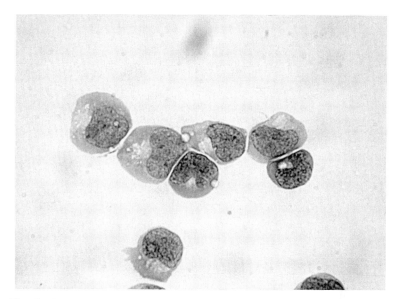

Fig. 6. CML-T1 cell line (immature T cells from a patient with chronic myeloid leukemia in blast crisis). These cells are 10–20 μm in diameter. The sometimes scanty cytoplasm is lightly basophilic without any granules. Many cells have a perinuclear clear or finely vacuolated zone. Many cells show nuclear irregularity with clefts and indentations or lobulations (often seen in T-cell leukemias). Chromatin is fine to moderately clumped, with the presence of one or more nucleoli in some cells. The line grows as single cells in suspension in RPMI 1640 medium, up to 2.5–3.0 × 10⁶ cells/ml.

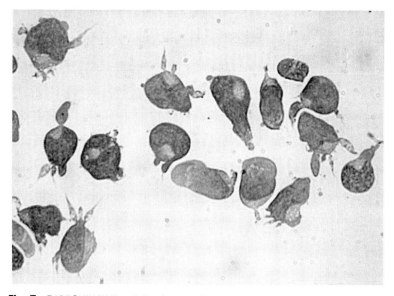

Fig. 7. P12/ICHIKAWA cell line (mature T cells from a patient with acute lymphoblastic leukemia). This cell line is characterized by irregular and bizarre shapes, often with cytoplasmic extensions. The cytoplasm is basophilic and scanty with small vacuoles. The nuclei are polymorphic (round, elongated, lobulated, clefted, indented), containing fine chromatin. Growth is as clumped cells in suspension in RPMI 1640 medium, up to 1.5 × 10⁶ cells/ml.

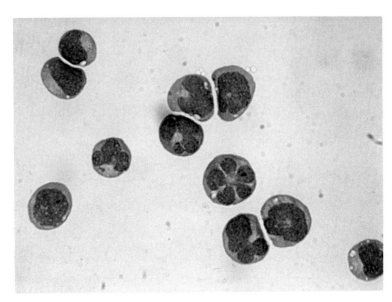

Fig. 8. MOLT-3 cell line (mature T cells from a patient with acute lymphoblastic leukemia). These cells are round to ovoid, with only scanty basophilic cytoplasm. The nuclei are markedly irregular and lobulated with fine or condensed chromatin. Growth is as single cells in suspension in RPMI 1640 medium, up to 1.0×10^6 cells/ml.

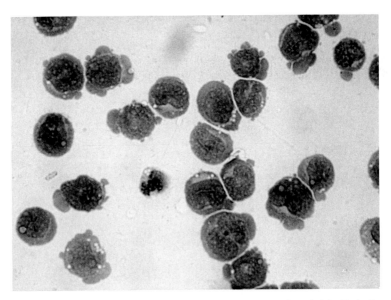

Fig. 9. 697 cell line (pre-B cells from a patient with acute lymphoblastic leukemia). These cells have a blast-like appearance with a high nucleus : cytoplasm ratio. The rim of cytoplasm is basophilic, has no granules, but is sometimes vacuolated. The nuclei frequently are indented and contain a fine chromatin network. The cellular form is polymorphic. Growth is as single cells in suspension in RPMI 1640 medium, up to $2.0–3.0 \times 10^6$ cells/ml.

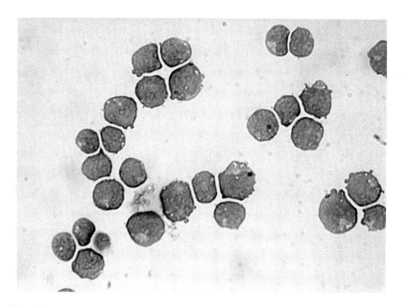

Fig. 10. 380 cell line (pre-B cells from a patient with acute lymphoblastic leukemia). These cells are significantly smaller than the 697 cells (Fig. 9). The cells display only a small rim of basophilic cytoplasm underlining the blast morphology. Growth is as single cells in suspension in RPMI 1640 medium, up to $4-6 \times 10^6$ cells/ml.

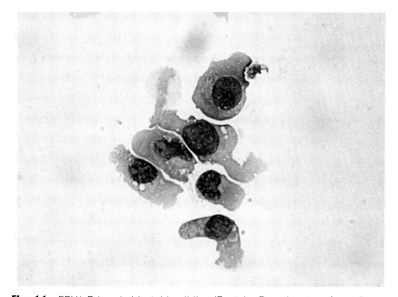

Fig. 11. EBV$^+$ B-lymphoblastoid cell line (Epstein–Barr virus-transformed mature B cells from a normal individual). The cells have abundant cytoplasm with variable vacuolation and exhibit irregularity of the surface membrane with blebs and projections. Their nuclei are usually round with coarse or fine chromatin and large nucleoli. Growth is as large, densely packed clumps of several hundred cells in suspension (see Fig. 4).

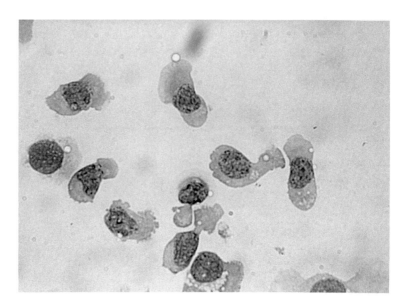

Fig. 12. JVM-2 cell line (mature B cells from a patient with B-prolymphocytic leukemia). The morphological appearance of this cell line is similar to that of "normal" EBV-transformed lymphoblastoid cell lines (Fig. 11). The JVM-2 cells are also EBV⁺ but are derived from the leukemic clone. Growth is as single cells in suspension or clustered in floating clumps in RPMI 1640 medium, with a maximal density of $0.5-1.0 \times 10^6$ cells/ml.

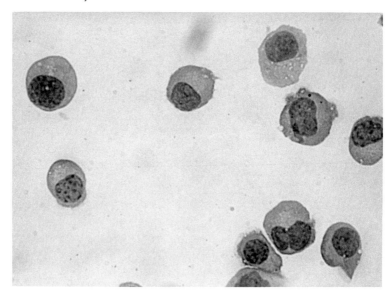

Fig. 13. U-266 cell line (plasma cells from a patient with myeloma). These cells have an eccentrically located nucleus with coarse, or even clumped, chromatin and no demonstrable, or indistinct, nucleoli. The cytoplasm is basophilic and has a juxtanuclear clear area. Some cells are larger and binucleated. The centrally placed nucleus has finely granular chromatin and distinct nucleoli. Growth is as single or clustered cells in suspension in RPMI 1640 medium; some cells are loosely adherent.

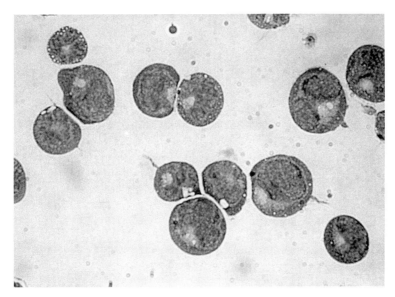

Fig. 14. RC-2A cell line (myeloid cells from a patient with acute myelomonocytic leukemia). These cells are round or oval and large, ranging from 10 to 20 μm in diameter with a preponderance of larger cells. The smaller cells have a small amount of basophilic cytoplasm and a round or kidney-shaped nucleus. The larger cells have irregularly shaped, indented and lobulated nuclei with fine chromatin and numerous cytoplasmic vacuoles. Growth is as single cells in suspension in RPMI 1640 medium, up to 1.0×10^6 cells/ml.

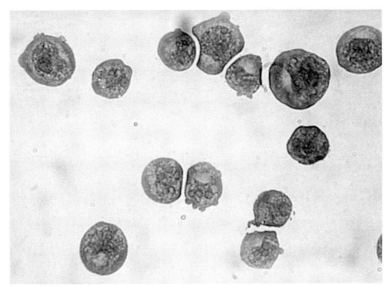

Fig. 15. HL-60 cell line (myeloid cells from a patient with acute myeloblastic leukemia). These cells are round to oval with pale or basophilic cytoplasm containing fine granules. The nuclei are round, oval, or convoluted with prominent nucleoli and coarse chromatin. Some cells have blunt pseudopods. Growth is as single cells in suspension in RPMI 1640, up to $1-2 \times 10^6$ cells/ml.

231

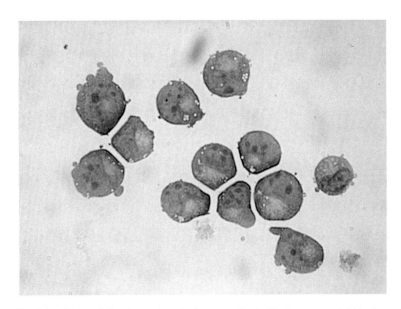

Fig. 16. EM-2 cell line (myeloid cells from a patient with chronic myeloid leukemia in blast crisis). These cells have polygonal forms with a marked variation in the nucleus : cytoplasm ratio. The cytoplasm is deeply basophilic and contains fine vacuoles and eosinophilic granules. The nuclei are pleomorphic with a fine chromatin pattern and two or more prominent nucleoli. Growth is as single cells in suspension in RPMI 1640 medium, up to $1.0-1.5 \times 10^6$ cells/ml.

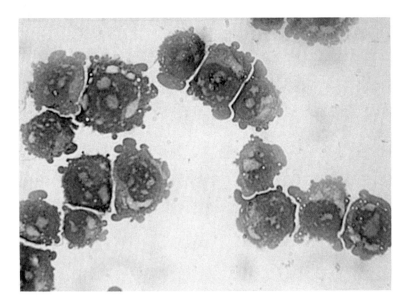

Fig. 17. JOSK-I cell line (monocytic cells from a patient with acute myelomonocytic leukemia). These cells are large and round to polygonal in shape. All cells show bleb formations. The cytoplasm of the cells is basophilic with fine granules and small vacuoles. The nuclei are round or indented with a fine chromatin structure and one to three large prominent nucleoli. Growth is as single cells in suspension in RPMI 1640 medium, up to $1.5-2.0 \times 10^6$ cells/ml.

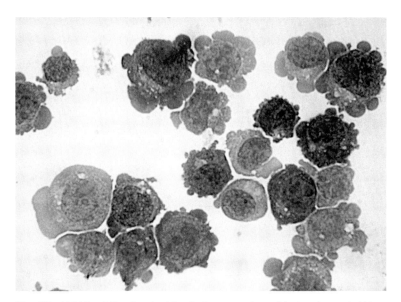

Fig. 18. K-562 cell line (erythroid cells from a patient with chronic myeloid leukemia in blast crisis). The cells are very large, and round to oval in shape. The cytoplasm is slightly or deeply basophilic, having no granules but occasionally some vacuoles. The nucleus is round and spongy, with one or several large distinct nucleoli. The cytoplasm has irregular forms, often with extensive blebs. Growth is as single cells in suspension in RPMI 1640 medium, up to 0.5×10^6 cells/ml.

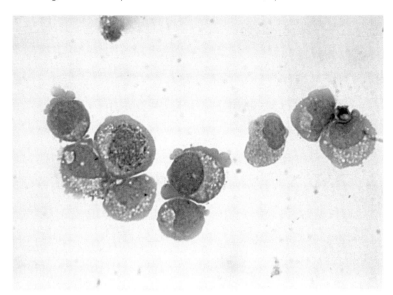

Fig. 19. M-07e cell line (megakaryocytic cells from a patient with acute megakaryoblastic leukemia). These cells have large irregular and lobulated nuclei with delicate chromatin and no evident nucleoli. The cytoplasm is abundant, basophilic, vacuolated, and granulated with occasional buds. Growth is as single cells in suspension in Iscove's modified Dulbecco's medium, up to 0.5×10^6 cells/ml. Cells are dependent on interleukin 3 (IL-3) or granulocyte–macrophage colony stimulating factor (GM–CSF).

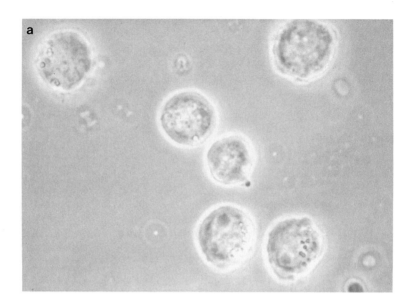

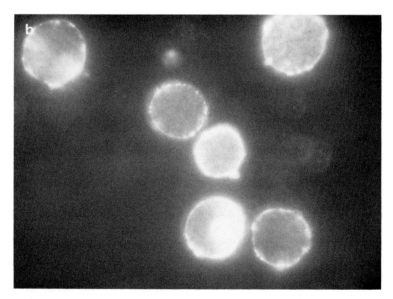

Fig. 20. Immunofluorescence staining of surface antigens on the Hodgkin's cell line KM-H2. The same field is shown in phase contrast (a) and under fluorescent light (b). All cells are positive for expression of the CD30 antigen. (Fluorescence microscopy.)

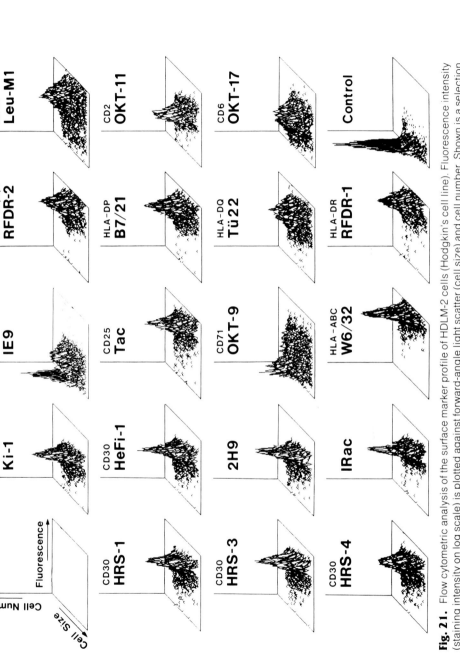

Fig. 21. Flow cytometric analysis of the surface marker profile of HDLM-2 cells (Hodgkin's cell line). Fluorescence intensity (staining intensity on log scale) is plotted against forward-angle light scatter (cell size) and cell number. Shown is a selection of positive cell surface markers (see also Table VII).

Table VII

Immunological Subclassification of Human Leukemia Cell Lines by Immunophenotyping[a,b]

T Cell	B Cell	Myeloid	Monocytic	Erythroid	Megakaryocytic
CD1	CD10	CD13	CD11c	CD13	CD13
CD2	CD11c	CD15	CD13	CD15	CD33
CD3	**CD19**	CD33	**CD14**	CD33	CD34
CD4	**CD20**	CD34	CD15	CD34	**CD41**
CD5	**CD21**	CDw65	CD33	**GlyA**	**CD42**
CD7	**CD22**	HLA-DR	CD34	**H antigen**	**CD61**
CD8	CD34	**MPO**	CDw65	HLA-DR	HLA-DR
CD25	**CD37**		**CD68**		
CD34	CD38		HLA-DR		
HLA-DR	**FMC7**		**MPO**		
TCR α/β	HLA-DR				
TCR γ/δ	**Ig(s/cy)**				
TdT	TdT				

[a] Immunological markers (CD codes according to international workshops for the definition of leukocyte antigens) can be used to immunophenotype cells belonging to the various cell lineages and stages of differentiation (Van Dongen *et al.*, 1988; Drexler and Scott, 1989). Antigens characteristically associated with a cell lineage are boldfaced.

[b] Abbreviations: CD, cluster of differentiation; GlyA, glycophorin A; HLA-DR, MHC Class II antigen; Ig(s/cy), immunoglobulin (surface/cytoplasmic); MPO, myeloperoxidase; TCR, T-cell receptor complex; TdT, terminal deoxynucleotidyl transferase.

F. Genotyping

The normal maturation process in the lymphoid system is associated with somatic gene rearrangements of immunoglobulin (Ig) heavy and light chain gene complexes in B cells and of T cell receptor (TCR) gene complexes in T cells (Naiem *et al.*, 1990). Mature cells will express these receptors (Ig on B cells and TCRα/β or TCRγ/δ on T cells) on their surfaces. Gene rearrangements are random processes, so the final structure of the rearranged genes varies from one cell to the next. Clonally expanded cells all have identical rearrangements. The Southern blotting technique allows, in principle, a detailed analysis of clonally rearranged Ig heavy and light chain genes and clonally rearranged TCRα, TCRβ, TCRγ, or TCRδ genes (Van Dongen and Wolvers-Tettero, 1991; Fig. 22).

Although these rearrangements are very informative with respect to detection of monoclonality, they are not specific for B or T cells, respectively. Although most B and T cell-derived cell lines will have one of the Ig and TCR

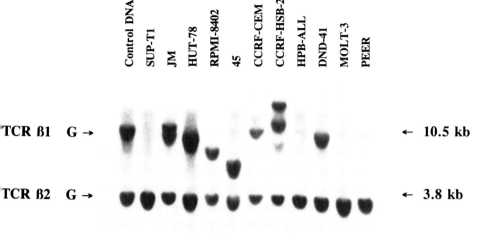

Fig. 22. Example of a Southern blot analysis of T-cell receptor gene rearrangement in a series of T-cell leukemia lines. This autoradiogram depicts a Southern blot of *Eco*RI-digested DNA to detect rearrangements in the TCR β1 region, using a Cβ probe. The two germ-line bands (G) of about 10.5 kb (Cβ1) and 3.8 kb (Cβ2) are indicated. In all cell lines, the TCRβ1 genes appear to be rearranged clonally or deleted. Digestion with other restriction enzymes and probing with additional cDNAs provide complementary information. (Autoradiogram kindly provided by Dr. T. Hansen-Hagge, University of Ulm, Germany.)

genes rearranged, respectively, cross-lineage rearrangements may occur: T-cell lines with Ig gene rearrangements and B-cell lines with TCR gene rearrangements have been detected. Rearranged Ig and TCR genes have been found even in myeloid cell lines. Therefore, Ig and TCR rearrangements cannot be used as absolute indicators of cell lineage. However, the combined use of immunophenotyping and genotyping can provide sufficient information in most cases about the lineage derivation of a cell line.

Other rearrangements detectable by Southern blotting are associated with specific chromosome translocations (see also, Section IV,G): the *bcl-2* gene in t(14;18), the c-*abl* gene in t(9;22), and others.

G. Cytogenetics

Chromosome banding analysis has shown that most human hematological malignancies have chromosomal abnormalities, most of them nonrandomly associated with a particular disease (Fig. 23). For some types of leukemias, specific chromosomal markers could be defined. Numerical (aneuploidy) and structural (translocations, deletions, inversions, and so on) abnormalities also have been examined in human leukemia cell lines. Table VIII lists some of the cytogenetic changes most commonly seen in cell lines.

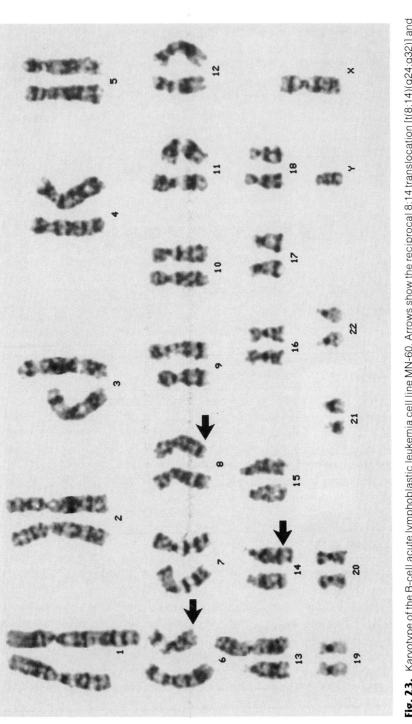

Fig. 23. Karyotype of the B-cell acute lymphoblastic leukemia cell line MN-60. Arrows show the reciprocal 8;14 translocation [t(8:14)(q24;q32)] and a 6q— deletion. Other secondary chromosome rearrangements are not described. (Karyotype kindly provided by Dr. R. A. F. MacLeod, Deutsche Sammlung von Mikroorganismen und Zellkulturen, Braunschweig, Germany.)

A problem often encountered is that the karyotype published does not correspond to the actual karyotype of a given cell line. Likely causes for the loss or gain of aberrations are (1) an inherent instability of karyotypes of continuous cell lines, (2) mistakes in the original report, and (3) mislabeling of cell lines.

H. Cytokines

Normal hematopoiesis is under the control of a group of glycoproteins called hematopoietic growth factors or colony-stimulating factors (CSF) (Metcalf, 1989). Depending on the stimulation with one or a combination of these growth factors, precursor cells differentiate into several mature cell types (see Section I and Table II). With respect to defined biological and biochemical characteristics, five hematopoietic growth factors have been described to date: granulocyte CSF (G-CSF), granulocyte–macrophage CSF (GM-CSF), macrophage CSF (M-CSF), interleukin 3 (IL-3), and erythropoietin.

A vast range of other polypeptide products (in conjunction with the growth factors, these proteins are collectively termed cytokines) participates in a variety of cellular responses including the regulation of the immune system: the interleukins (1–13, at the latest count), interferons (IFNα, -β, -γ), tumor necrosis factors (TNFα, -β), transforming growth factors (TGFα, -β), leukemia inhibitory factor (LIF), and others.

Four points are of interest regarding cytokines. (1) Some cell lines are dependent on exogenous cytokines (see Section V). (2) Cytokines can have stimulatory or inhibitory effects on the growth or differentiation of cell lines; cell lines that produce their own growth factors can be inhibited by antibodies against the growth factor or against the cellular receptor for the factor. (3) Cell lines can express the gene encoding a cytokine at the RNA level and even produce and secrete the protein. (4) Cell lines can express cytokine receptors, which are subject to external modulation.

I. Functional Features

Most cell lines are characterized by specific features relating to certain functional aspects or other physiological or pathophysiological parameters. Here, we can mention only a few representative highlights: phagocytic activity by monocytes (e.g., cell line THP-1); antigen presentation (e.g., Hodgkin's cell line L-428); immunoglobulin production and secretion (many B chronic lymphocytic leukemia, B prolymphocytic leukemia, and myeloma cell lines); synthesis of mostly fetal and embryonic hemoglobins (e.g., erythroleukemia cell line HEL); inducibility of differentiation by pharmacological and physiological agents (e.g., the acute myeloid leukemia cell line HL-60 along the granulocytic or monocytic pathway); heterotransplantability into mice (e.g., the pre-B

Table VIII

Cytogenetic Abnormalities in Human Leukemia Cell Lines[a,b]

Chromosomal abnormality	Disease	Example of cell line (origin)	Gene affected
Associated with B cells			
t(1;19)(q23;p13)	pre-B ALL	697 (pre-B ALL)	*PBX, E2A*
t(2;8)(p12;q24)	B-ALL, B-cell lymphoma	ROS-17 (Burkitt lymphoma)	*c-myc*
t(4;11)(q21;q23)	pre-B ALL, mixed acute leukemia	RS4;11 (mixed acute leukemia)	
		B1 (pre-B ALL)	
		KARPAS-422 (B-cell NHL)	
		MN-60 (B-ALL)	
del(6q)	pre-B ALL	TC-78 (pre-B ALL)	
t(8;14)(q24;q32)	B-ALL, B-cell lymphoma	380 (pre-B ALL)	*c-myc*
		ROS-1 (Burkitt lymphoma)	
t(8;22)(q24;q11)	B-ALL, B-cell lymphoma	KAL-1 (B-cell NHL)	*c-myc*
		NAB-2 (Burkitt lymphoma)	
t(11;14)(q13;q32)	B-cell lymphoma	JVM-2 (B-PLL)	
		HBL-2 (B-cell NHL)	
t(11;19)(q23;p13)	ALL	BS (pre-B ALL)	
del(12p)	pre-B ALL	TC-78 (pre-B ALL)	
+12	B-CLL, B-cell lymphoma	BONNA-12 (HCL)	
		FL-18 (B-cell NHL)	
14q+	Mature B-cell diseases	L-363 (plasma cell leukemia)	
		FRAVEL (myeloma)	
t(14;18)(q32;q21)	B-CLL, B-cell lymphoma	380 (pre-B ALL)	*bcl-2*
		KARPAS-422 (B-cell lymphoma)	

Associated with T cells			
t(2;5)(p23;q35)	Lymphoma	KARPAS-299 (T-cell lymphoma)	
t(8;14)(q24;q11)	T-ALL	TALL-104 (T-ALL)	*c-myc*
t(10;14)(q24;q11)	T-ALL	MKB-1 (T-cell line)	*HOX-11*
t(11;14)(p13;q11)	T-ALL	TALL-104 (T-ALL)	*Rhombotin2/Ttg-2*
Associated with myeloid cells			
+4	AML M2 and M4	KU-812 (CML-myeloid BC)	
		EOL-3 (AML)	
+8	AML	GDM-1 (CML-myeloid BC)	
		NKM-1 (AML M2)	
t(8;21)(q22;q22)	AML M2	KASUMI-1 (AML M2)	
t(9;22)(q34;q11)	AML, pre-B ALL, CML	TOM-1 (pre-B ALL)	*c-abl/bcr*
		EM-2 (CML-myeloid BC)	
		NALM-1 (CML-lymphoid BC)	
		MEG-01 (AML M7)	
t(15;17)(q22;q11-12)	AML M3	NB-4 (AML M3)	*PML/RAR*
inv(16)	AML M4	ME-1 (AML M4)	
i(17q)	AML	KU-812 (CML-myeloid BC)	
−Y	AML	KASUMI-1 (AML M2)	

[a] Adapted from Heim and Mitelman (1987), Naiem *et al.* (1990), and Rabbitts (1991).

[b] Abbreviations: ALL, acute lymphoblastic leukemia; AML M2, acute myeloblastic leukemia; AML M3, acute promyelocytic leukemia; AML M4, acute myelomonocytic leukemia; AML M7, acute megakaryoblastic leukemia; CLL, chronic lymphocytic leukemia; CML(-BC), chronic myeloid leukemia (in blast crisis); HCL, hairy cell leukemia; NHL, non-Hodgkin's lymphoma; PLL, prolymphocytic leukemia.

ALL cell line A1); production and secretion of particular proteins (e.g., amylase by the myeloma cell line KMS-12 and histamine by the myeloid cell line KU-812); oncogene expression (e.g., c-fgr, c-fms, c-myb, c-myc, and c-pim by the histiocytic cell line DEL). These unique features render the respective cell lines particularly interesting.

V. Growth Factor-Dependent Leukemia Cell Lines

The autonomous growth of malignant cells frequently is driven by the same factors that stimulate normal cell proliferation. Cancer cell proliferation may occur in the context of autocrine loops in which neoplastic cells produce their own growth factors for proliferation. Growth also can take place in response to paracrine circuits in which the neoplastic cells divide in response to growth factors released by surrounding cells. With respect to leukemia cell lines, the autocrine scenario refers to independently growing cell lines whereas growth factor-dependent cell lines require paracrine or external stimulation.

Until recently, all human leukemia cell lines were grown in medium supplemented with FBS and were not dependent on exogenous hematopoietic growth factors (also known as cytokines). Although cell lines dependent on IL-2 have been established, this success was restricted to a particular category of T-cell lines, namely, adult T-cell leukemia lines. Early attempts to support myeloid leukemia cell growth *in vitro* with conditioned media from stimulated lymphocytes succeeded in sustaining growth only for a limited time period of several weeks. Thus, growth factors are at least permissive for expansion of leukemia clones.

The availability of purified or recombinant cytokines permitted further progress in this area. A bank of factor-dependent cell lines was established from patients with various types of leukemia, all of which are absolutely dependent on addition of growth factors to the medium for proliferation and survival (Oval and Taetle, 1990). A selection of representative cytokine-dependent leukemia cell lines is listed in Table IX. The cytokines involved include some of the interleukins (e.g., IL-2, IL-3, IL-6) and the so-called colony-stimulating factors (GM-CSF, G-CSF, M-CSF) (Oval and Taetle, 1990). Many of these cell lines are dependent on either IL-3 or GM-CSF, reflecting the ability of these cytokines to stimulate early progenitor and immature cells.

Preliminary studies also suggest that non-lineage-specific hormones such as insulin and insulin-like growth factor (IGF-I) can stimulate growth of factor-dependent leukemia cell lines synergistically. Further studies combining hormones such as IGF-I and cytokines will provide new insights into the significance of these findings. Thus, establishing factor-dependent cell lines may allow for the continuous growth of leukemic cells that previously were eliminated during selection in standard culture medium.

Table IX

Growth Factor-Dependent Leukemia Cell Lines

Cell line	Cell phenotype	Original disease[a]	Growth factor(s) on which dependent[b]
ATL cell lines	T cell	ATL	IL-2
AML-193	Myeloid	AML M5	IL-3 or GM-CSF
F-36P	Erythroid	MDS	IL-3 or GM-CSF
GM/SO	Myeloid	CML-BC	GM-CSF
HMS-2	Plasma cell	B-ALL	IL-6
ILKM-1	Plasma cell	Myeloma	IL-6
JJN-2	Plasma cell	PCL	IL-6
KIT-225	T cell	T-CLL	IL-2
KT-3	T cell	T-cell NHL	IL-6
M-07e	Megakaryocytic	AML M7	IL-3 or GM-CSF
MB-02	Erythroid/megakaryocytic	AML M7	GM-CSF
MM-A1	Plasma cell	Myeloma	IL-6
MT-ALL	Myelomonocytic	T-ALL	IL-2, IL-3, or GM-CSF
MV4-11	Myelomonocytic	AML/B-ALL	IL-3 or GM-CSF
OCI-AML1	Myelomonocytic	AML M4	IL-3 or GM-CSF
TALL-101	Myeloid	T-ALL	IL-3 or GM-CSF
TALL-104	T cell	T-ALL	IL-2
TF-1	Erythroid	AML M6	IL-3, GM-CSF, or Epo
TMD2	B cell	B-CLL	IL-3
U-1996	Plasma cell	Myeloma	IL-6
UT-7	Megakaryocytic	AML M7	IL-3, GM-CSF, or Epo

[a] ALL, acute lymphoblastic leukemia; AML M4, acute myelomonocytic leukemia; AML M5, acute monocytic leukemia; AML M6, erythroleukemia; AML M7, acute megakaryoblastic leukemia; ATL, adult T-cell leukemia; CLL, chronic lymphocytic leukemia; CML-BC, chronic myeloid leukemia in blast crisis; MDS, myelodysplastic syndrome; NHL, non-Hodgkin's lymphoma; PCL, plasma cell leukemia.

[b] Epo, erythropoietin; GM-CSF, granulocyte–macrophage colony stimulating factor; IL, interleukin.

VI. Hodgkin's Disease-Derived Cell Lines

The origin of Hodgkin and Reed–Sternberg cells, the cells thought to represent the malignant population in Hodgkin's disease, remains a highly controversial issue. Studies on noncultured, freshly disaggregated biopsy material have not definitely established the lineage of Hodgkin cells, although evidence has been gathered that clearly favors one or the other hematopoietic cell type (Drexler et al., 1987).

Over the last 15 years, numerous cell lines have been established from patients with Hodgkin's disease, all of which are claimed to represent *in vitro* counterparts of Hodgkin and Reed–Sternberg cells (Figs. 24–26). These cell lines have geno- and immunophenotypic and other characteristics (cytochemistry, cytokines, oncogenes, karyotypes, functional features) that are reminiscent of lymphoid cells (T or B cell) or compatible with a monocyte/macrophage origin (Table X; Drexler, 1992; Drexler and Minowada, 1992). The results from the studies on Hodgkin's cell lines suggest the following conclusions: (1) the Hodgkin's cell lines fulfill most of the criteria for neoplastic cells *in vitro*. (2) The established Hodgkins's cell lines are unique compared with the bank of well-characterized, permanently established leukemia–lymphoma cell lines of lymphoid, myelomonocytic, megakaryocytic, and erythroid origin. (3) The Hodgkin's cell lines are not identical or even similar to one another; they show common features, but also show a number of important differences in phenotype that might reflect the *in vivo* heterogeneity of this disease. (4) Intensive investigations on these cell lines have failed to yield definitive evidence concerning the origin of Hodgkin and Reed–Sternberg cells to date. (5) Results obtained through examination of the *in vitro* cell lines cannot necessarily be extrapolated to the *in vivo* situation. The establishment of the cell lines was difficult to achieve in each case; most cell lines stem from biological conditions rather similar to suspension cultures, namely, pleural effusion and bone marrow. (6) Despite these reservations about whether certain types of cells were selected in the artificial *in vitro* system and whether the cultured cells do, indeed, represent their *in vivo* counterparts, the availability of Hodgkin's cell lines has had a tremendous impact on studies in an area in which research on fresh tumor biopsy material is hampered severely by the fact that the neoplastic cells constitute only a fraction (0.1–1%) of the total population (Drexler *et al.*, 1987).

VII. Future Prospects

A. Data Bank

To our knowledge, no comprehensive listing of human leukemia cell lines is currently available. The most extensive index of cell lines has been assembled at the Fujisaki Cell Center and covers some 130 lines. At the Deutsche Sammlung von Mikroorganismen und Zellkulturen (DSM), we are in the process of accumulating information from the published literature on established leukemia cell lines for a data bank. We estimate that at least 500 cell lines (not including the more easily established Burkitt's cell lines) have been established over the last 25 years. One serious drawback is our inability to predict, in many cases, whether the cell lines described are truly continuously proliferating lines and whether these cell lines do, in fact, still exist.

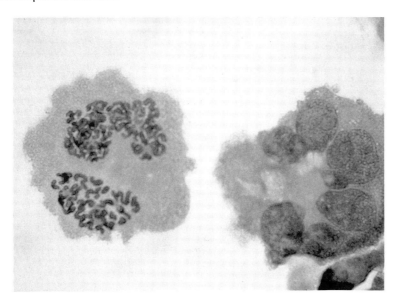

Fig. 24. Two giant cells of the Hodgkin's cell line HDLM-2. The cell on the right is multinucleated with six nuclei. The cell on the left shows a tri-star mitosis, attesting to endomitosis and disturbed cytokinesis.

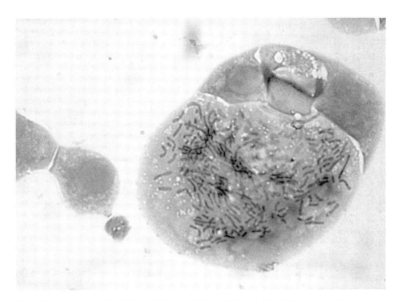

Fig. 25. A giant HDLM-2 cell (Hodgkin's cell line) 80–100 μm in diameter in which several nuclei appear to be in simultaneous mitosis.

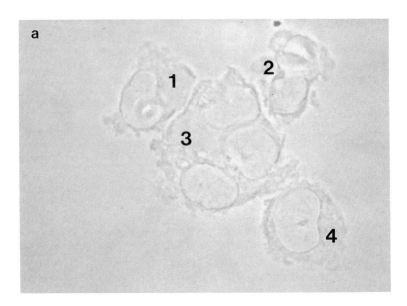

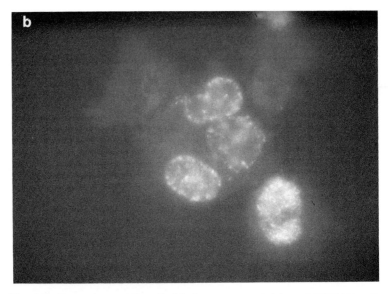

Fig. 26. Bromodeoxyuridine (BrdU) fluorescence staining of HDLM-2 cells (Hodgkin's cell line). BrdU positivity indicates cells in S phase. The same field is shown in phase contrast (a) and under fluorescence (b). Cells 1 and 2 are negative; cell 3 (with three nuclei) and cell 4 (binucleated) are positive. (Fluorescence microscopy.)

Table X
Hodgkin's Disease-Derived Cell Lines[a]

Cell line	Histological subtype[b]	Cell type
CO	NS	Lymphoid (T)
DEV	NS	Lymphoid (B)
HD-70	NS	Lymphoid (B)
HDLM-1/-2/-3	NS	Lymphoid (T)
HO	NS	Lymphoid (T)
HUT-11	MC	Monocyte/macrophage
KM-H2	MC	Lymphoid (B)
L-428	NS	Lymphoid
L-540	NS	Lymphoid (T)
SU/RH-HD-1	NS	Monocyte/macrophage
SUP-HD1	NS	Lymphoid (B)
ZO	NS	Lymphoid (B)

[a] Adapted from Drexler et al. (1987) and Drexler and Minowada (1992).
[b] Abbreviations: MC, mixed cellularity; NS, nodular sclerosing.

B. Collection of Cell Lines

Many of the most often used and best-known cell lines are available as authentic cell lines at the major culture collections (American Type Culture Collection, European Collection of Animal Cell Cultures, Japanese Cancer Research Resources Bank). At the DSM, we intended to collect as many such cell lines as possible. However, these attempts are hampered severely by the unwillingness of researchers to provide their cell lines to cell banks: we requested more than 200 cell lines published in the scientific literature, but received less than 20%, half of which could not be propagated. Thus, the awareness of scientists of the benefits of institutionalized cell culture collections must be heightened.

C. Establishment of New Cell Lines

As stated earlier, the deliberate establishment of cell lines remains a random process. The use of cytokines permits the long-term culture of many fresh leukemia cells. Subsequently, in some cases, continuous cell lines might evolve. Future technical innovations, for example, transfection of the target cells with oncogenes or viral DNA and selective activation of genes, might improve the success rate of establishing continuous cell lines. Concerted efforts to analyze the various parameters of leukemia cell culture certainly will

Table XI

Representative Human Leukemia Cell Lines Belonging to Different Hematopoietic Cell Lineages[a,b]

T cell		B cell		
Immature T cell	Mature T cell	Pre-B cell	B cell	Plasma cell
CCRF-CEM (ALL)[c]	HH (NHL)	697 (ALL)	DAUDI (Burkitt)	ARH-77 (PCL)
DND-41 (ALL)	HUT-78 (SS)	BLIN (ALL)	EB-1 (Burkitt)	FR (myeloma)
HPB-ALL (ALL)	HUT-102 (ATL)	BV-173 (CML-BC)	JOK-1 (HCL)	IM-9 (myeloma)
JM (ALL)	KARPAS-299 (NHL)	HAL-01 (ALL)	JVM-2 (CLL)	KARPAS-707 (myeloma)
KE-37 (ALL)	KIT-225[d] (CLL)	MR-87 (ALL)	KAL-1 (NHL)	KMM-1 (myeloma)
MOLT-3 (ALL)	MT-1 (ATL)	NALM-1 (CML-BC)	KARPAS-422 (NHL)	L-363 (PCL)
MOLT-13 (ALL)	SKW-3 (CLL)	NALM-6 (ALL)	NAMALWA (Burkitt)	LP-1 (myeloma)
PEER (ALL)	ST-4 (NHL)	REH (ALL)	RAJI (Burkitt)	OPM-2 (myeloma)
RPMI-8402 (ALL)	SUP-T1 (NHL)	RS4;11 (ALL)	TMD2[d] (CLL)	RPMI-8226 (myeloma)
TALL 104[d] (ALL)	SZ-4 (MF)	TOM-1 (ALL)	U-698-M (NHL)	U-266 (myeloma)

[a] Cell lines were assigned to one of the indicated cell lineages on the basis of their immunophenotypic, cytochemical, genotypic, cytogenetic, functional, and other features.

[b] Abbreviations: ALL, acute lymphoblastic leukemia; AML/AML M2, acute myeloblastic leukemia; AML M3, acute promyelocytic leukemia; AML M4, acute myelomonocytic leukemia; AML M5, acute monocytic leukemia; AML M6, erythroleukemia; AML M7, acute megakaryoblastic leukemia; ATL, adult T-cell leukemia; CLL, chronic lymphocytic leukemia; CML(−BC), chronic myeloid

lead to significant improvements in the efficiency of immortalization. Ultimately, to be able to generate a cell line from any given sample of *ex vivo* leukemia cells is desirable.

D. Characterization Program, Quality and Identity Control

Prior to the publication of newly established leukemia cell lines, the cells should be analyzed according to a required standardized characterization program. The preparation of guidelines for such a characterization program is underway. At the DSM, we found that 30–40% of cultures obtained from the original investigators were contaminated with mycoplasma. Further, about 10% of the cell lines contained cells different from those expected, probably because of misidentification or cross-examination. Also, karyotypes published from several cell lines were incorrect. Thus, a stringent quality and identity control, in conjunction with a characterization program for leukemia cell lines, is of great importance.

Table XI

Continued

Myeloid cell	Monocytic cell	Erythroid cell	Megakaryocytic cell
AML-193[d] (AML M5)	CTV-1 (AML M5)	F-36P[d] (AML M6)	CMK (AML M7)
EM-2 (CML-BC)	DEL (MH)	HEL (AML M6)	DAMI (AML M7)
EOL-1 (AML eosino)	JOSK-I (AML M4)	JK-1 (CML-BC)	M-07e[d] (AML M7)
HL-60 (AML M2)	JOSK-K (AML M5)	K-562 (CML-BC)	MB-02[d] (AML M7)
HMC-1 (AML baso)	ML-2 (AML M4)	KMOE (AML M6)	MEG-01 (AML M7)
KASUMI-1 (AML 2)	MONO-MAC-6 (AML M5)	LAMA-84 (CML-BC)	UT-7[d] (AML M7)
KG-1 (AML M2)	PLB-985 (AML M4)	MB-03 (AML M6)	
KU-812 (CML-BC)	THP-1 (AML M5)	OCI-M1 (AML M6)	
MV4-11[d] (AML)	U-937 (HL)	RM-10 (AML M6)	
NB-4 (AML M3)	YK-M2 (AML M5)	TF-1[d] (AML M6)	

leukemia (in blast crisis); HCL, hairy cell leukemia; HL, histiocytic lymphoma; MF, mycosis fungoides; MH, malignant histiocytosis; NHL, non-Hodgkin's lymphoma; PCL, plasma cell leukemia; SS, Sézary syndrome.
[c] Name of cell line as published; original diagnosis of the patient from whom the cell line was established is given in parentheses.
[d] These cell lines are dependent on externally added cytokines (e.g., GM-CSF, IL-3),

VIII. Lists of Leukemia Cell Lines

Listings of human leukemia cell lines are shown in Table XI (see also Tables IV, VIII, and IX, in which other cell lines are mentioned). These listings are not intended to be exhaustive, since several hundred cell lines have been described in the literature. The cell lines are subdivided according to the normal cell lineage to which they apparently belong, based on their immunophenotypic, cytochemical, genotypic, and functional features. Hodgkin's disease-derived cell lines do not fit into any classification and are listed separately (Table X).

References

Collins, S. J., Gallo, R. C., and Gallagher, R. E. (1977). Continuous growth and differentiation of human myeloid leukaemic cells in suspension culture. *Nature (London)* **270,** 347–349.

Drexler, H. G. (1993). Recent results on the biology of Hodgkin and Reed–Sternberg cells. II. Continuous cell lines. *Leuk. Lymph.* **9,**1–26.

Drexler, H. G., and Minowada, J. (1992). Hodgkin's disease derived cell lines. A review. *Hum. Cell* **5,** 42–53.

Drexler, H. G., and Scott, C. S. (1989). Morphological and immunological aspects of leukaemia diagnosis. *In* "Leukaemia Cytochemistry" (C. S. Scott, ed.), pp. 13–67. Ellis Horwood, Chichester.

Drexler, H. G., Amlot, P. L., and Minowada, J. (1987). Hodgkin's disease-derived cell lines— Conflicting clues for the origin of Hodgkin's disease. *Leukemia* **1,** 629–637.

Ferrero, D., and Rovera, G. (1984). Human leukaemic cell lines. *Clin. Haematol.* **13,** 461–487.

Flandrin, G., and Daniel, M. T. (1981). Cytochemistry in the classification of leukemias. *In* "The Leukemic Cell" (D. Catovsky, ed.), pp. 29–48. Churchill-Livingstone, Edinburgh.

Greaves, M. F. (1986). Differentiation-linked leukemogenesis in lymphocytes. *Science* **234,** 697–704.

Heim, S., and Mitelman, F. (eds.) (1987). "Cancer Cytogenetics." Liss, New York.

Lange, B. J. (1989). Growth of human leukaemia cells in vitro. *In* "Cell Growth and Division" (R. Baserga, ed.), pp. 61–79. IRL Press, Oxford.

Lübbert, M., and Koeffler, H. P. (1988). Leukemic cell lines as a paradigm for myeloid differentiation. *Cancer Rev.* **10,** 33–62.

Lübbert, M., Herrmann, F., and Koeffler, H. P. (1991). Expression and regulation of myeloid-specific genes in normal and leukemic myeloid cells. *Blood* **77,** 909–924.

Matsuo, Y., and Minowada, J. (1988). Human leukemia cell lines. Clinical and theoretical significances. *Hum. Cell* **1,** 263–274.

Metcalf, D. (1989). The molecular control of cell division, differentiation commitment and maturation in haemopoietic cells. *Nature (London)* **339,** 27–30.

Minowada, J. (1982). Immunology of leukemic cells. *In* "Leukemia" (F. W. Gunz and E. S. Henderson, eds.), 4th Ed., pp. 119–139. Grune & Stratton, New York.

Naiem, F., Gatti, R. A., and Yunis, J. J. (1990). Recent advances in diagnosis and classification of leukemias and lymphomas. *Dis. Markers* **8,** 231–264.

Nilsson, K., and Pontén, J. (1975). Classification and biological nature of established human hematopoietic cell lines. *Int. J. Cancer* **15,** 321–341.

Oval, J., and Taetle, R. (1990). Factor-dependent human leukemia cell lines: New models for regulation of acute non-lymphocytic leukemia cell growth and differentiation. *Blood Rev.* **4,** 270–279.

Rabbitts, T. H. (1991). Translocations, master genes, and differences between the origins of acute and chronic leukemias. *Cell* **67,** 641–644.

Schaeffer, W. I. (1990). Terminology associated with cell, tissue and organ culture, molecular biology and molecular genetics. *In Vitro Cell. Dev. Biol.* **26,** 97–101.

Scott, C. S. (1989). Cytochemical and immunocytochemical techniques. *In* "Leukaemia Cytochemistry" (C. S. Scott, ed.), pp. 362–379. Ellis Horwood, Chichester.

Van Dongen, J. J. M., and Wolvers-Tettero, I. L. M. (1991). Analysis of immunoglobulin and T cell receptor genes. *Clin. Chim. Acta* **198,** 1–174.

Van Dongen, J. J. M., Adriaansen, H. J., and Hooijkaas, H. (1988). Immunophenotyping of leukemias and non-Hodgkin lymphomas: Immunological markers and their CD codes. *Neth. J. Med.* **33,** 298–314.

Human Sarcoma Cells in Culture

Richard B. Womer and Albert E. Wilson

Division of Oncology, Children's Hospital of Philadelphia and
Department of Pediatrics, University of Pennsylvania School of Medicine
Philadelphia, Pennsylvania 19104

I. Introduction 251

II. Methods of Establishment and Maintenance 252
A. Obtaining Fresh Tissue 252
B. Confirming the Diagnosis 252
C. Technique of Explantation 253
D. Collagenase 253
E. Selection of Media 253
F. Supplements 254

G. Propagation 254
H. Distinguishing Normal Cell from Tumor Cell Cultures 254

III. Specific Cell Line Characteristics 255
A. Osteosarcomas 256
B. Rhabdomyosarcomas 260
C. Other Sarcomas 264
References 265

I. Introduction

Establishing and maintaining human sarcoma cells in culture is a difficult undertaking that challenges even accomplished cell biologists. The tumors are infiltrated heavily with normal stromal cells and the tumor cells *in vitro* are often slow growing, so overgrowth of cultures with fibroblasts is common. The tumor cells themselves, being derived from connective tissue, are often similar to normal fibroblasts in morphology, growth requirements, and biological characteristics, which makes it difficult or impossible to find conditions that will separate sarcoma cells from fibroblasts. Sarcoma cells are also very fastidious, with unpredictable requirements or preferences for different media and supplements.

The result is that fewer than one-fourth of human sarcomas can be propagated in culture to produce permanent cell lines. Success is most common with osteosarcomas, intermediate for Ewing's sarcomas, and rarest in rhabdomyosarcomas. Basic biological differences are likely to exist between the sarcomas that generate cell lines and the majority that do not. Tissue culture is a highly stressful and selective environment that is very different from the

Atlas of Human Tumor Cell Lines

human body. Different lines from the same diagnostic category (for example, two different osteosarcoma cell lines) may differ greatly in important properties. Samples of the same cell line maintained in different laboratories may diverge in important properties over time. Thus, generalizing from a single osteosarcoma cell line to all osteosarcoma cell lines, let alone to osteosarcomas *in vivo,* is dangerous. One must maintain a healthy skepticism in the interpretation of results obtained with any of these lines.

Another difficulty is that no published controlled studies are available that compare the efficiencies of different methods of tumor explantation and propagation. Each laboratory, and each worker, develops methods that seem to work by individual processes of trial and error. The procedures described in this chapter are those most often reported in the literature, and those we have found most reliable at The Children's Hospital of Philadelphia.

II. Methods of Establishment and Maintenance

A. Obtaining Fresh Tissue

The first step is crucial to success in cultivating these difficult specimens. A close relationship with a surgeon and pathologist who will provide a fresh generous specimen promptly is indispensable. A specimen that has rested on a gauze pad on the operating room instrument table for 1 hr will be too desiccated to be viable, and one handled without sterile technique in the surgical pathology suite will be contaminated with bacteria and yeast.

Large sarcomas frequently have areas of necrosis and dense connective tissue, neither of which is suitable for explantation. A pathologist can identify areas of viable tumor, with minimal stroma, that offer the best chances for successful culture.

Patient care often requires evaluation of the surgical margins to insure that they are free of tumor. The traditional technique is to coat the outside of the tumor mass with India ink, the particles of which are visible under the microscope. This procedure can make maintenance of sterility difficult. Alternatives are the use of sterilized ink or inking the specimen, incising it deeply with a sterile scalpel blade, and obtaining tissue from deep in the mass using another sterile blade. Occasionally the surgeon can submit biopsies of the tissue outside the surgical margins for analysis, and make inking unnecessary.

B. Confirming the Diagnosis

Since sarcomas (especially rhabdomyosarcomas) are often difficult to diagnose, the final histological diagnosis may not be the same as the clinical diagnosis or the frozen section diagnosis. Electron microscopy and immunohistochemistry may take several days to complete, and solid tumor cytogenetics and molecular analyses may lead to changes in diagnosis weeks after the initial histological diagnosis. Thus, confirming the diagnosis with the pathologist after special studies are completed is essential.

C. Technique of Explantation

All human cell culture procedures are carried out in a certified biological safety cabinet that is capable of containing tissue-borne pathogens, as well as protecting the specimens from airborne bacteria and fungal spores. As is the case whenever fresh human tissues are handled, gloves and protective clothing and training in biological safety techniques are important.

Our procedure is to mince the tissue into small fragments (2 mm or less on a side) using two scalpels in a 60-mm tissue culture dish with a few drops of medium. The fragments are drawn into a sterile pipette and are distributed among 2–4 additional 60-mm dishes; enough medium is added to barely cover the bottom of each dish. If too much medium is added, the tissue fragments float and cell adhesion to the dish does not occur. The plates are incubated in a humidified atmosphere, and 5 ml medium are added the next day.

An important part of the explantation technique is avoidance of desiccation, which occurs very rapidly in the strong currents of a laminar air-flow tissue culture hood. Tissue fragments always should be covered with medium, and their container should be covered whenever possible.

Some laboratories use various methods of mechanical disaggregation, such as repeated passage through syringe needles of decreasing size or straining through tissue culture seives, to achieve unicellular (or nearly unicellular) suspensions. We have found these techniques to be cumbersome, and find that they increase the chances of bacterial and fungal contamination more than they increase the chances for successful culture.

D. Collagenase

Many laboratories incubate minced tissue specimens for a few hours or overnight in medium supplemented with collagenase in an attempt to digest the acellular stroma investing the tumor cells and to release more of them into the medium. The technique is described in detail by Limon et al. (1986). We have not been impressed by any increased number of successful cultures using collagenase, and have found that its use often leaves a thick film of debris on the surface of the plates that may impede attachment of floating cells.

E. Selection of Media

Almost any standard tissue culture medium supplemented with 10–20% bovine serum (heat inactivated) should provide satisfactory results. We generally use the high-glucose formulation of Dulbecco's modified Eagle's medium (DMEM) with 10% heat-inactivated bovine serum (Hyclone). Each lot of serum is tested before use for its ability to sustain the growth of MG-63 human osteosarcoma cells; occasionally a batch of medium does not support cell growth, or even kills the cells.

Many laboratories use a mixture of Ham's F-12 medium and DMEM. We explanted 18 consecutive sarcoma specimens into the two media and examined the morphology of the cells twice weekly. Although some explants grew better in one medium or the other, no consistent pattern was seen in the results, that is, some explants grew better in Ham's/DMEM whereas others grew better in DMEM. We obtained similar results when comparing DMEM with MCDB-104, a medium specifically designed for human cells.

F. Supplements

A mixture of insulin, selenium, and transferrin (ITS) frequently is added to media to increase culture efficiency and the vigor of cell growth. Again, no controlled data are available that demonstrate its efficacy for human sarcoma cell isolation, and its cost is substantial.

G. Propagation

We generally trypsinize the original culture and transfer the cells to a single 60-mm plate once substantial cell outgrowth is observed (usually 7–14 days). Thereafter, we split the cultures (1 : 3) whenever the cells achieve confluence. We have tried keeping the cells superconfluent, on the theory that this condition would encourage the growth of anchorage-independent tumor cells. We also have tried keeping the cells from becoming confluent, thinking that this condition would prevent fibroblast crowding of tumor cells. However, neither of these approaches makes a consistent difference in outcome.

H. Distinguishing Normal Cell from Tumor Cell Cultures

The most vexing problem in sarcoma cell culture is distinguishing tumor cells from normal cells. The issue can be approached on two levels: methods to try to suppress the growth of fibroblasts and encourage the growth of tumor cells and methods to confirm that the resulting lines are actually tumor cells.

1. Selection for Tumor Cells

Probably the most reliable technique is to inoculate an immunodeficient mouse (nude or SCID) with cells from an early-passage culture and to harvest the lesions that result. These lesions should be histologically identical to the original tumor. Unfortunately, few cultures will lead to nodule growth, and even well-established well-characterized lines sometimes fail to grow. Other approaches are picking tumor cell colonies off early passage plates with wire loops, fine pipettes, or cloning cylinders. Careful trypsinization of cultures should dislodge tumor cells or fibroblasts selectively (Smith *et al.*, 1976). No method works reliably or consistently.

2. Confirmation of Tumor Origin of Cultivated Cells

The techniques used to confirm that cell lines are of tumor origin depend on the characteristics of the tumors.

a. Karyotyping. Karyotyping is the "gold standard" technique for tumors that, at diagnosis, had a characteristic chromosomal abnormality. Alveolar rhabdomyosarcomas, for example, have a t(2;13) translocation; synovial sarcomas have a t(X;18) abnormality (Reeves *et al.*, 1989); osteosarcomas frequently have a bizarre karyotype with many marker chromosomes (Biegel *et al.*, 1989). Cultures derived from such tumors should, of course, have the same karyotypic abnormality.

b. Immunohistochemistry. Immunohistochemical staining reactions developed for surgical pathology can be adapted to tissue culture specimens, although considerable trial-and-error is necessary since cell monolayers are thinner than histological sections. Cells can be grown on slides with removable wells (Nunc, Inc., Napierville, IL), so several cultures can be checked in one set of reactions. Rhabdomyosarcoma cells can be identified by reactivity with antibodies against the intermediate filament protein desmin; synovial sarcomas react with antikeratin antibodies. No immunohistochemical stain for osteosarcoma or fibrosarcoma is in general use.

c. Cytochemistry. Some osteosarcoma cell lines express alkaline phosphatase, which can be detected cytochemically with commercially available kits (Sigma, St. Louis, Missouri, and others).

d. Xenografting. Some cell lines will form tumors of the appropriate histological type when xenografted into nude or SCID mice. Although this event is reassuring when it occurs, such a result is not a *sine qua non* for malignancy; several well-established, well-characterized, aneuploid lines will not form tumors in mice.

III. Specific Cell Line Characteristics

We have selected eight commonly used cell lines from the American Type Culture Collection (ATCC) for discussion, representing osteosarcomas, rhabdomyosarcomas, and a fibrosarcoma. Although many other lines are maintained by the ATCC (Table I) and individual laboratories, they are not as widely used.

Table I
Human Sarcoma Cell Lines in ATCC

Description	Cell line	ATCC number
Chondrosarcoma	SW 1353	HTB 94
Fibrosarcoma	HT-1080	CCL 121
Fibrosarcoma	SW 684	HTB 91
Fibrosarcoma, metastatic to lung	Hs 913T	HTB 152
Leiomyosarcoma, vulva, primary	SK-LMS-1	HTB 88
Liposarcoma	SW872	HTB 92
Osteogenic sarcoma, bone, primary	U2-OS	HTB 96
Osteogenic sarcoma, TE 85, clone F-5	HOS	CRL 1543
Osteogenic sarcoma, chemically transformed	MNNG/HOS	CRL 1547
Osteosarcoma, HOS (Kirsten transformed)	HOS/NP	CRL 1544
Osteosarcoma, HOS (revertant)	KHOS-240S	CRL 1545
Osteosarcoma, HOS (revertant)	KHOS-312H	CRL 1546
Osteosarcoma versus Ewing sarcoma, bone	SK-ES-1	HTB 86
Osteosarcoma	G292, clone A141B1	CRL 1423
Osteosarcoma	MG-63	CRL 1427
Osteosarcoma, primary	Saos-2	HTB 85
Osteosarcoma, TK$^-$	143B	CRL 8303
Osteosarcoma	143B PML BK TK	CRL 8304
Rhabdomyosarcoma	A-204	HTB 82
Rhabdomyosarcoma	A-673	CRL 1598
Rhabdomyosarcoma	RD	CCL 136
Rhabdomyosarcoma, left leg	Hs 729	HTB 153
Synovial sarcoma	SW982	HTB 93
Uterine, mixed mesodermal tumor, consistent with leiomyosarcoma grade II	SK-UT-1	HTB 114
Uterine, mixed mesodermal tumor, consistent with leiomyosarcoma grade II	SK-UT-1B	HTB 115

A. Osteosarcomas

Osteosarcoma cell lines abound: 12 are listed in the ATCC catalog and many others exist in individual laboratories. We selected the four most widely used lines for discussion. Osteosarcoma cell lines generally share several characteristics: they are relatively fast-growing, with doubling times of 24–48 hr in 10–15% serum; they have highly variable, hyperploid karyotypes (often 60–100 chromosomes per cell) with many marker chromosomes; and they lack functional *p53* and *RB* genes. Many of these lines express alkaline phosphatase activity and have vitamin D receptors, which are osteoblast markers *in vivo*. Many well-established lines also produce histological osteosarcomas in xenografts.

1. U2-OS

U2-OS was established in 1964 and is widely used (Pontén and Saksdela, 1967). In culture, U2-OS grows rapidly to a high density, with a mixture of spindle-shaped and cuboidal cells (Fig. 1). Serum starvation does not cause growth arrest, but does cause cell death once nutrients are depleted. The karyotype is typical of osteosarcoma: inconsistent, aneuploid, and bizarre, with varying numbers of chromosomes and many markers.

U2-OS has been shown to produce finctional platelet-derived growth factor (PDGF) molecules, which appear to be A-chain dimers (Heldin *et al.*, 1986). Betsholtz and colleagues (1984) demonstrated autocrine growth stimulation; the secreted PDGF acts on cellular receptors, causing their phosphorylation, although the association of growth factor and receptor may occur below the cell surface (Richter and Graves, 1988). This line is one of the few osteosarcoma cell lines shown to express the retinoblastoma (*RB*) gene, and it continues to grow after transfection with an *RB* expression vector (Huang *et al.*, 1988). Conflicting information exists regarding the *p53* tumor suppressor gene. One group found it to be present in one copy, but without mRNA or protein expression (Masuda *et al.*, 1987). Others found two normal copies of the *p53* gene and apparently normal *p53*-encoded protein at the same levels found in human fibroblasts (Diller *et al.*, 1990).

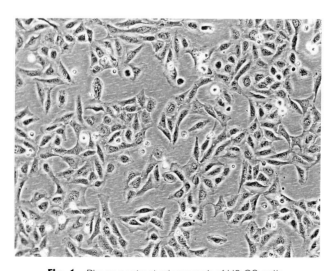

Fig. 1. Phase-contrast micrograph of U2-OS cells.

2. HOS

HOS was reported first in 1971 (McAllister *et al.*, 1971). This line initially was named MT (the patient's initials) and others have referred to it as T-85 (Rhim *et al.*, 1975a,b). HOS is relatively slow growing and becomes growth arrested at a low density; the cells are usually polygonal or stellate with large nuclei and multiple nucleoli, with occasional spindle-shaped cells and rare multinucleated giant cells (Fig. 2). The line is generally nontumorigenic in nude mice; this characteristic and its low saturation density have made it useful for studies of transformation, since chemical carcinogens and activated *ras* genes induce tumorigenicity (Tainsky *et al.*, 1987). This line has a bizzarre polyploid karyotype with variable markers (McAllister *et al.*, 1971).

Contradictory information exists on *p53* gene expression. A codon 156 mutation was found in several HOS-derived lines and no wild-type *p53* mRNA was detectable (Romano *et al.*, 1989). Although Masuda found no *p53* mutations and found both the mRNA and the protein expressed at elevated levels (Masuda *et al.*, 1987), this protein is apparently functionally abnormal since it binds the heat-shock protein hsc70 (Diller *et al.*, 1990). No published information is available on the status of the *RB* gene in HOS.

3. MG-63

MG-63 was established in 1974 from a tumor in a 14-year-old boy (Heremans, 1978). The isolation technique was unusual: the tumor was explanted into tissue culture medium over suspended lens paper, and only cells capable of traversing the lens paper were able to adhere to the dish and propagate.

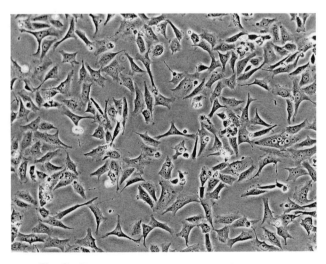

Fig. 2. Phase-contrast micrograph of HOS cells.

These cells initially were exploited for the study of interferon (β_2 type), which they produce in large quantities (Billiau *et al.*, 1977; Wantanabe, 1983).

MG-63 cells grow rapidly (1–3 day doubling time in 10% bovine serum) to a high cell density (10^5 cells or more per cm^2). The morphology varies from plump spindle-shaped cells to stellate cells at low density, although cells are usually cuboidal at high density (Fig. 3). Occasionally these cells form clumps. Serum starvation (0.5% for 5–7 days or MCDB-104 medium without serum) causes growth arrest, which is reversible by addition of insulin, insulin-like growth factor I (IGF-I), or serum (Womer *et al.*, 1987; Furlanetto and Womer, 1989). Either insulin or IGF-I can sustain growth in serum-free medium, acting through both the insulin and the Type I IGF receptor (Furlanetto and Womer, 1989). The cells have PDGF receptors, but do not respond mitogenically to this growth factor (Womer *et al.*, 1987), nor do they express c-*sis* mRNA or secrete a PDGF-like mitogen (Graves *et al.*, 1984).

The karyotype is bizarre, with a variable number of chromosomes (hypertriploid, modal chromosome number 66) and many markers. The cells have an abnormal *p53* DNA pattern as revealed by Southern blotting and contain no detectable *p53*-encoded protein by immunoblot analysis (Masuda *et al.*, 1987).

4. Saos-2

Saos-2 was reported first in 1975. This line grows as plump spindle-shaped or polygonal cells (Fig. 4), is nontumorigenic in nude mice, and has the characteristic bizarre aneuploid karyotype of osteosarcoma lines, with several markers (Hay *et al.*, 1992). A very complete biochemical characterization

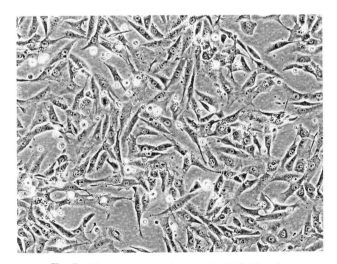

Fig. 3. Phase-contrast micrograph of MG-63 cells.

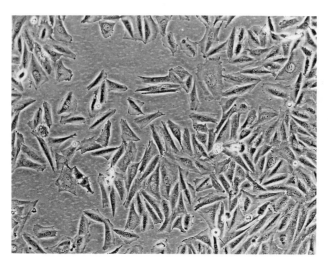

Fig. 4. Phase-contrast micrograph of Saos-2 cells.

demonstrated a moderate growth rate and final cell density (37-hr doubling time and $300,000/cm^2$ saturation density in 10% serum), high levels of alkaline phosphatase activity which varied with the phase of growth, parathormone-modulated adenylate cyclase activity, and apparently normal vitamin D receptors (Rodan *et al.*, 1987).

Saos-2 expresses a mutant *RB* gene that encodes a truncated 95-kDa protein lacking T-antigen binding ability (Shew *et al.*, 1990). Introduction of a wild-type *RB* gene in an expression vector suppresses growth (Huang *et al.*, 1988) and microinjection of functional RB protein causes growth arrest late in G_1 (Goodrich *et al.*, 1991). A similar situation exists for *p53*. Saos-2 cells have a deletion in the coding region of the *p53* gene and express neither mRNA nor protein (Masuda *et al.*, 1987). Introduction of a wild-type *p53* gene in an expression vector profoundly inhibits growth (Chen *et al.*, 1990).

B. Rhabdomyosarcomas

Rhabdomyosarcomas are exceedingly difficult to cultivate; thus, only one genuine rhabdomyosarcoma cell line is listed in the ATCC catalog. There are two main histological types of rhabdomyosarcoma: embryonal and alveolar. Several alveolar rhabdomyosarcoma cell lines exist in individual laboratories, but embryonal lines are exceedingly rare. Rhabdomyosarcoma explants generally are overrun with fibroblasts. On the rare occasions that a line emerges, it tends to undergo terminal differentiation into myotubes after 20–30 passages. This behavior has made cell biological research on rhabdomyosarcomas difficult; any conclusions based on limited numbers of cell lines should be generalized only with great caution.

The identity of alveolar rhabdomyosarcoma cell lines can be confirmed by the presence of the characteristic chromosomal translocation t(2;13) (Douglass *et al.,* 1987; Wang-Wuu *et al.,* 1988). Embryonal rhabdomyosarcomas often are reduced to homozygosity for markers on chromosome 11 when their genomic DNA is analyzed (Scrable *et al.,* 1989). The n-*myc* proto-oncogene apparently is amplified in alveolar, but not in embryonal, tumors (Dias *et al.,* 1990a).

For both histological types, expression of the muscle differentiation gene *myo*D1 is the definitive characteristic for histological identity: *myo*D1 mRNA and protein are the earliest known markers of muscle differentiation (Scrable *et al.,* 1989; Dias *et al.,* 1990b,c). Excellent evidence indicates that this gene is the "master switch" for myogenesis (Davis *et al.,* 1987).

Autocrine growth stimulation occurs in rhabdomyosarcoma cells in culture, through production of IGF-II and stimulation of the Type I IGF receptor (El Badry *et al.,* 1990). The *p53* tumor suppressor gene is mutated frequently (Felix *et al.,* 1992).

1. RD

The only true rhabdomyosarcoma in the ATCC catalog, and the only embryonal rhabdomyosarcoma cell line in general use is RD. Characterization of this line was reported first in 1969 (McAllister *et al.,* 1969). Interestingly, the patient from whom the cells were explanted had been treated with cyclophosphamide, and perhaps radiation, before the biopsies were taken; the mutagenic effect of these agents may have contributed to successful cultivation.

In culture, RD grows as a mixed population of spindle-shaped and stellate cells, with occasional multinucleated giant cells (Fig. 5). Although RD generally lacks light or electron microscopic evidence of muscle differentiation, the cells express *myo*D1 and myosin (Hiti *et al.,* 1989), and can be induced to fuse into myotubes by cultivation in low-serum medium or treatment with phorbol esters (Aguanno *et al.,* 1990) or cytarabine (Crouch *et al.,* 1991). These cells are tumorigenic in nude mice; different clonal isolates have differing tendencies to metastasize (Lollini *et al.,* 1991).

The cells have a hyperdiploid karyotype with markers, and lack the t(2;13) of alveolar rhabdomyosarcoma. There is apparent loss of heterozygosity for tumor suppressor loci on 11p15 since fusion of chromosome 11p-containing microcells with RD cells causes marked inhibition of growth (Loh *et al.,* 1992). RD cells have an unusual activating n-*ras* mutation at codon 61 (Chardin *et al.,* 1985) and a codon 248 mutation in the *p53* tumor suppressor gene (Felix *et al.,* 1992). No published information is available on the status of the *RB* gene.

Helman and colleagues have demonstrated autocrine growth stimulation in RD cells; secreted IGF-II stimulates mitosis through the Type I IGF receptor (El Badry *et al.,* 1990). IGF-II also stimulates RD cell motility, acting through the Type II IGF/mannose 6-phosphate receptor (Minniti *et al.,* 1992).

Stratton and co-workers (1989) have demonstrated convincingly that the

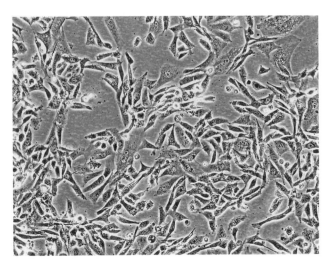

Fig. 5. Phase-contrast micrograph of RD cells.

TE671 cell line, widely reputed to be a medulloblastoma, actually is derived from RD.

2. A-204

A-204 generally is considered a rhabdomyosarcoma. This line was established from the tumor of a 1-year-old girl and first was reported in 1973 (Giard *et al.*, 1973). However, the line does not express *myo*D1 or myosin heavy chain and does not undergo muscle differentiation in culture (Hiti *et al.*, 1989), so its histological identity is questionable. In culture, the line grows as a mixed monolayer of spindle-shaped cells and plump stellate cells, some with very long cytoplasmic processes (Fig. 6). The tumorigenicity of A-204 in nude mice is suppressed by introduction of a normal chromosome 11, despite a normal karyotype (46;XX) (Oshimura *et al.*, 1990).

3. A-673

Like A-204, the putative rhabdomyosarcoma A-673 cell line does not express *myo*D1 (Hiti *et al.*, 1989). Chen cites information that the tumor was derived from a peripheral neuroepithelioma (Chen *et al.*, 1991). The morphology of this line in culture is consistent with that diagnosis: the cells grow in widely spaced clumps of round-to-stellate cells (Fig. 7). Although the cells were aneuploid with marker chromosomes when originally characterized (Giard *et al.*, 1973), no karyotype has been published since; whether this line has the t(11;22) characteristic of neuroepitheliomas is unknown.

Although A-673 has two apparently normal *p53* alleles, it does not express the protein; introduction of a wild-type *p53* gene in an expression vector

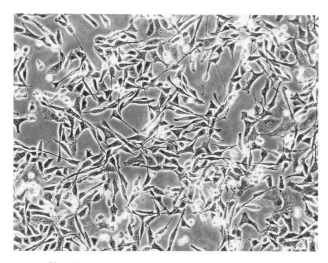

Fig. 6. Phase-contrast micrograph of A-204 cells.

suppresses tumorigenicity in nude mice, although it has little or no effect on growth rate *in vitro* or on soft-agar colony formation (Chen *et al.,* 1991). The cells produce a 25-kDa protein that is, apparently, transforming growth factor beta (TGFβ) (Romeo and Mizel, 1989), and produce a TGFα epidermal growth factor (EGF)-related mRNA as well (Jakowlew *et al.,* 1988). In addition, several high molecular weight (15–48 kDa) proteins that bind to the EGF receptor

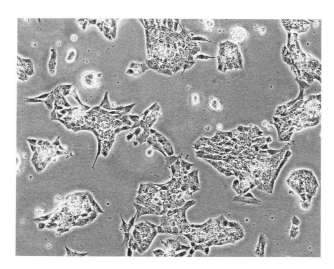

Fig. 7. Phase-contrast micrograph of A-673 cells.

(Stromberg *et al.*, 1986) are produced. The cells also secrete a 34-kDa IGF-binding protein normally found in cerebrospinal fluid (Romanus *et al.*, 1989).

C. Other Sarcomas

Although a variety of other sarcoma cell lines exists in the ATCC catalog and and in individual laboratories, only the HT-1080 fibrosarcoma cell line has been widely used.

1. HT-1080

This cell line from a poorly differentiated fibrosarcoma was established in 1972 (Rasheed *et al.*, 1974) and has been very widely used. Morphologically, the cells in culture vary from round to spindle shaped, with occasional multinucleated giant cells (Fig. 8). The karyotype is pseudodiploid, with occasional markers (Rasheed *et al.*, 1974). The cells grow in serum-free medium and the growth rate is not increased with the addition of PDGF or EGF (McCormick *et al.*, 1987). The cells have been shown to express the c-*sis* proto-oncogene and to produce large quantities of both PDGF-like and EGF-like molecules (Pantotis *et al.*, 1985; Jakowlew *et al.*, 1988). Evidence also suggests that these cells produce the hematopoietic growth factor *kit* ligand/stem-cell growth factor (Toyota *et al.*, 1992), a remarkable characteristic for a connective tissue cell line.

HT-1080 cells have an activating codon 61 mutation of n-*ras*. Paterson and co-workers (1987) mutagenized HT-1080 cells and isolated slow-growing "re-

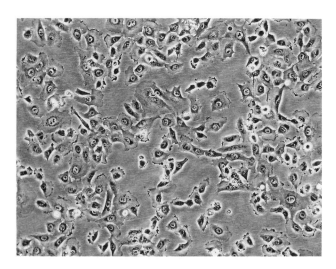

Fig. 8. Phase-contrast micrograph of HT-1080 cells.

vertants" with repeated FUdR–Ara-C selection. These revertant cells had an orderly "flat" morphology, grew slowly to a lower density, formed few colonies in soft agar, had diminished tumorigenicity in nude mice, and were pseudotetraploid. These properties were shown to be related to a reduced "dose" of mutated n-*ras* DNA and protein in the revertants. Transfection with a mutated *ras* gene restored the fully transformed phenotype.

References

Aguanno, S., Bouche, M., Adamo, S., and Molinaro, M. (1990). 12-O-Tetradecanoylphorbol-13-acetate-induced differentiation of a human rhabdomyosarcoma cell line. *Cancer Res.* **50(11),** 3377–3382.

Betsholtz, C., Westermark, B., Ek, B., and Heldin, C.-H. (1984). Coexpression of a PDGF-like growth factor and PDGF receptors in a human osteosarcoma cell line: Implications for autocrine receptor activation. *Cell* **39,** 447–457.

Biegel, J. A., Womer, R. B., and Emanuel, B. S. (1989). Complex karyotypes in a series of pediatric osteosarcomas. *Cancer Genet. Cytogenet.* **38(1),** 89–100.

Billiau, A., Edy, V. G., Hubertine, H., Van Damme, J., Desmyter, J., Georgiades, J. A., and De Somer, P. (1977). Human interferon: Mass production in a newly established cell line, MG-63. *Antimicrob. Agents Chemother.* **12(1),** 11–15.

Chardin, P., Yeramian, P., Madaule, P., and Tavitian, A. (1985). N-*ras* gene activation in the RD human rhabdomyosarcoma cell line. *Int. J. Cancer* **(35(5),** 647–652.

Chen, P. L., Chen, Y. M., Bookstein, R., and Lee, W. H. (1990). Genetic mechanisms of tumor suppression by the human p53 gene. *Science* **250(4987),** 1576–1580.

Chen, Y. M., Chen, P. L., Arnaiz, N., Goodrich, D., and Lee, W. H. (1991). Expression of wild-type p53 in human A673 cells suppresses tumorigenicity. *Oncogene* **6(10),** 1799–1805.

Crouch, G., Kalebic, T., Tsokos, M., and Helman, L. (1991). Ara-c inhibits the growth and induces differentiation of a human rhabdomyosarcoma (rms) cell line. *Proc. Ann. Meet. Am. Soc. Clin. Oncol.* **10,** 310.

Davis, R. L., Weintraub, H., and Lassar, A. B. (1987). Expression of a single transfected cDNA converts fibroblasts to myoblasts. *Cell* **51,** 987–1000.

Dias, P., Kumar, P., Marsden, H. B., Gattmeneni, H. R., Heighway, J., and Kumar, S. (1990a). N-*myc* gene is amplified in alveolar rhabdomyosarcomas (rms) but not in embryonal rms. *Int. J. Cancer* **45(4),** 593–596.

Dias, P., Parham, D. M., Shapiro, D. N., Webber, B. L., and Houghton, P. J. (1990b). Myogenic regulatory protein (MyoD1) expression in childhood solid tumors: Diagnostic utility in rhabdomyosarcoma. *Am. J. Pathol.* **137(6),** 1283–1291.

Diller, L., Kassel, J., Nelson, C. E., Gryka, M. A., Litwak, G., Gerhardt, M., Bressac, B., Ozturk, M., Baker, S. J., Vogelstein, B., and Friend, S. H. (1990). p53 functions as a cell cycle control protein in osteosarcomas. *Mol. Cell. Biol.* **10(11),** 5772–5781.

Douglass, E. C., Valentine, M., Etchubanas, E., Parham, D., Webber, B. L., Houghton, P. J., and Green, A. A. (1987). A specific chromosomal abnormality in rhabdomyosarcoma. *Cytogenet. Cell Genet.* **45,** 148–155.

El Badry, O. M., Minniti, C., Kohn, E. C., Houghton, P. J., Daughaday, W. H., and Helman, L. J. (1990). Insulin-like growth factor II acts as an autocrine growth and motility factor in human rhabdomyosarcoma tumors. *Cell Growth Diff.* **1(7),** 325–331.

Felix, C. A., Kappel, C. C., Mitsundomi, T., Nau, M. N., Toskos, M., Crouch, G. D., Nisen, P. D., Winick, N. J., and Helman, L. J. (1992). Frequency and diversity of p53 mutations in childhood rhabdomyosarcoma. *Cancer Res.* **52,** 2243–2247.

Furlanetto, R. W., and Womer, R. B. (1989). Analysis of IGF action using MG-63 human osteosarcoma cells. *J. Cell Biochem.* **13B,** 149.

Giard, D. J., Aaronson, S. A., Todaro, G. J., Arnstein, P., Kersey, J. H., Dosik, H., and Parks, W. P. (1973). In vitro cultivation of human tumors: Establishment of cell lines derived from a series of solid tumors. *J. Natl. Cancer Inst.* **51(5),** 1417–1423.

Goodrich, D. W., Wang, N. P., Qian, Y. W., Lee, E. Y., and Lee, W. H. (1991). The retinoblastoma gene product regulates progression through the G_1 phase of the cell cycle. *Cell* **67(2),** 293–302.

Graves, D. T., Antoniades, H. N., Williams, S. R., and Owen, A. J. (1984). Evidence for funtional platelet-derived growth factor receptors on MG-63 human osteosarcoma cells. *Cancer Res.* **44,** 2966–2970.

Hay, R. J., Caputo, J., Chen, T. R., Macy, M., McClintock, P., and Reid, Y. (1992). "American Type Culture Collection Catalogue of Cell Lines and Hybridomas." American Type Culture Collection, Rockville, Maryland.

Heldin, C. H., Johnsson, A., Wennergren, S., Wernstedt, C., Betsholtz, C., and Westermark, B. (1986). A human osteosarcoma cell line secretes a growth factor structurally related to a homodimer of PDGF A-chains. *Nature (London)* **319(6),** 511–514.

Heremans, H., Billian, A., Cassiman, J. J., Mulier, J. C., and deSomer, P. (1978). In vitro cultivation of human tumor tissues. II. Morphological and virological characterization of three cell lines. *Oncology* **35,** 246–252.

Hiti, A. L., Bogenmann, E., Gonzales, F., and Jones, P. A. (1989). Expression of the MyoD1 muscle determination gene defines differentiation capability but not tumorigenicity of human rhabdomyosarcomas. *Mol. Cell Biol.* **9(11),** 4722–4730.

Huang, H. J. S., Yee, J. K., Shew, J. Y., Chen, P. L., Bookstein, R., Friedman, T., Lee, Y. H. P., and Lee, W. H. (1988). Suppression of the neoplastic phenotype by replacement of the RB gene in human cancer cells. *Science* **242,** 1563–1566.

Jakowlew, S. B., Kondaiah, P., Dillard, P. J., Sporn, M. B., and Roberts, A. B. (1988). A novel low molecular weight ribonucleic acid (RNA) related to transforming growth factor alpha messenger RNA. *Mol. Endocrinol.* **2(11),** 1056–1063.

Limon, J., Dal Cin, P., and Sandberg, A. A. (1986). Application of long-term collagenase disaggregation for the cytogenetic analysis of human solid tumors. *Cancer Genet. Cytogenet.* **23,** 305–313.

Loh, W. J., Scrable, H. J., Livanos, E., Arboleda, M. J., Cavenee, W. K., Oshimura, M., and Weissman, B. E. (1992). Human chromosome 11 contains two different growth suppressor genes for embryonal rhabdomyosarcoma. *Proc. Natl. Acad. Sci. U.S.A.* **89(5),** 1755–1759.

Lollini, P. L., De, G. C., Landuzzi, L., Nicolletti, G., Scotlandi, K., and Nanni, P. (1991). Reduced metastatic ability of in vitro differentiated human rhabdomyosarcoma cells. *Invasion Metastasis* **11(2),** 116–124.

McAllister, R. M., Melnyk, J., Finkelstein, J. Z., Adams, E. C., Jr., and Gardner, M. B. (1969). Cultivation in vitro of cells derived from a human rhabdomyosarcoma. *Cancer* **24(3),** 520–526.

McAllister, R. M., Gardner, M. B., Greene, A. E., Bradt, C., Nichols, W. W., and Landing, B. H. (1971). Cultivation *in vitro* of cells derived from a human osteosarcoma. *Cancer* **27,** 397–402.

McCormick, J. J., Schilz, R., Ryan, P. A., and Maher, V. M. (1987). Growth factor requirements of normal, transformed, and fibrosarcoma-derived human fibroblasts. *Proc. Annu. Meet. Am. Assoc. Cancer Res.* **28,** 56.

Masuda, H., Miller, C., Koeffler, H. P., Battifora, H., and Cline, M. J. (1987). Rearrangement of the p53 gene in human osteogenic sarcomas. *Proc. Natl. Acad. Sci. U.S.A.* **83(21),** 7716–7719.

Minniti, C. P., Kohn, E. C., Grubb, J. H., Sly, W. S., Oh, Y., Muller, H. L., Rosenfeld, R. G., and Helman, L. J. (1992). The insulin-like growth factor II (IGF-II)/mannose 6-phosphate receptor mediates IGF II-induced motility in human rhabdomyosarcoma cells. *J. Biol. Chem.* **267(13),** 9000–9004.

Oshimura, M., Kugoh, H., Koi, M., Shimizu, M., Yamada, H., Satoh, H., and Barrett, J. C. (1990). Transfer of a normal human chromosome 11 suppresses tumorigenicity of some but not all tumor cell lines. *J. Cell Biochem.* **42(3),** 135–142.

Pantotis, P., Pelicci, P. G., Dalla-Favera, R., and Antoniades, H. N. (1985). Synthesis and secretion of proteins resembling platelet-derived growth factor by human glioblastoma and fibrosarcoma cells in culture. *Proc. Natl. Acad. Sci. U.S.A.* **82**, 2404–2408.

Paterson, H., Reeves, B., Brown, R., Hall, A., Furth, M., Bos, J., Jones, P., and Marshall, C. (1987). Activated N-*ras* controls the transformed phenotype of HT1080 human fibrosarcoma cells. *Cell* **51**, 803–812.

Pontén, J., and Saksela, E. (1967). Two established *in vitro* cell lines from human mesenchymal tumours. *Int. J. Cancer* **2**, 434–447.

Rasheed, S., Nelson-Rees, W. A., Toth, E. M., Arnstein, P., and Gardner, M. B. (1974). Characterization of a newly derived human sarcoma cell line (HT-1080). *Cancer* **33**, 1027–1033.

Reeves, B. R., Smith, S., Fisher, C., Warren, W., Knight, J., Martin, C., Chan, A. M., Gusterson, B. A., Westbury, G., and Cooper, C. S. (1989). Analysis of a specific chromosomal translocation, t(x;18), found in human synovial sarcomas. *Cancer Cells* **7**, 69–73.

Rhim, J. S., Cho, H. Y., Vernon, M. L., Arnstein, P., Huebner, R. J., Gilden, R. V., and Nelson-Rees, W. A. (1975a). Characterization of non-producer human cells induced by Kirsten sarcoma virus. *Int. J. Cancer* **16**, 840–849.

Rhim, J. S., Han, Y. C., and Huebner, R. J. (1975b). Non-producer human cells induced by murine sarcoma virus. *Int. J. Cancer* **15**, 23–29.

Richter, M. R., and Graves, D. T. (1988). DNA synthesis in U-2 OS human osteosarcoma cells is independent of PDGF binding to functional cell surface receptors. *J. Cell Physiol.* **135(3)**, 474–480.

Rodan, S. B., Imai, Y., Thiede, M. A., Wesolowski, G., Thompson, D., Z., Bar-Shavit, Shull, S., Mann, K., and Rodan, G. A. (1987). Characterization of a human osteosarcoma cell line (Saos-2) with osteoblastic properties. *Cancer Res.* **47(18)**, 4961–4966.

Romano, J. W., Ehrhart, J. C., Duthu, A., Kim, C. M., Appella, E., and May, P. (1989). Identification and characterization of a p53 gene mutation in a human osteosarcoma cell line. *Oncogene* **4(12)**, 1483–1488.

Romanus, J. A., Tseng, L. Y., Yang, Y. W., and Rechler, M. M. (1989). The 34 kilodalton insulin-like growth factor binding proteins in human cerebrospinal fluid and the A673 rhabdomyosarcoma cell line are human homologues of the rat brl-3a binding protein. *Biochem. Biophys. Res. Commun.* **163(2)**, 875–881.

Romeo, D. S., and Mizel, S. B. (1989). Partial purification of an immunosuppressive protein from a human tumor cell line and analysis of its relationship to transforming growth factor beta. *Cell. Immunol.* **122(2)**, 483–492.

Scrable, H., Witte, D., Shimada, H., Seemayer, T., Wang-Wuu, S., Soukup, S., Koufos, A., Houghton, P., Lampkin, B., and Cavenee, W. (1989). Molecular differential pathology of rhabdomyosarcoma. *Genes Chrom. Cancer* **1**, 23–25.

Shew, J. Y., Lin, B. T., Chen, P. L., Tseng, B. Y., Yang, F. T., and Lee, W. H. (1990). C-Terminal truncation of the retinoblastoma gene product leads to functional inactivation. *Proc. Natl. Acad. Sci. U.S.A.* **87(1)**, 6–10.

Smith, H. S., Owens, R. B., Hiller, A. J., Nelson-Rees, W., and Johnston, J. O. (1976). The biology of human cells in tissue culture. I. Characterization of cells derived from osteogenic sarcoma. *Int. J. Cancer* **17**, 219–234.

Stratton, M. R., Darling, J., Pilkington, G. J., Lantos, P. L., Reeves, B. R., and Cooper, C. S. (1989). Characterization of the human cell line TE671. *Carcinogenesis* **10(5)**, 905.

Stromberg, K., Hudgins, W. R., Fryling, C. M., Hazarika, P., Dedman, J. R., Pardue, R. L., Hargreaves, W. R., and Orth, D. N. (1986). Human A673 cells secrete high molecular weight EGF-receptor binding growth factors that appear to be immunologically unrelated to EGF or TGF-alpha. *J. Biol. Chem.* **32(4)**, 247–259.

Tainsky, M. A., Shamanski, F. L., Blair, D., and Van de Woude, G. (1987). Human recipient cell for oncogene transfection studies. *Mol. Cell. Biol.* **7(3)**, 1280–1284.

Toyota, M., Hinoda, Y., Itoh, F., Tsujisaki, M., Imai, K., and Yachi, A. (1992). Expression of two types of *kit* ligand mRNAS in human tumor cells. *Int. J. Hematol.* **55(3),** 301–304.

Wang-Wuu, S. S., Ballard, E., Gotwals, B., and Lampkin, B. (1988). Chromosomal analysis of sixteen rhabdomysarcomas. *Cancer Res.* **48,** 983–987.

Wantanabe, Y. (1983). Production and purification of human MG63 cell interferon. *Microbiol. Immunol.* **27(5),** 433–444.

Womer, R. B., Frick, K., Mitchell, C. D., Ross, A. H., Bishayee, S., and Scher, C. D. (1987). PDGF induces c-*myc* mRNA expression in MG-63 human osteosarcoma cells but does not stimulate cell replication. *J. Cell Physiol.* **132(1),** 65–72.

Cell Lines from Esophageal Tumors

Tetsuro Nishihira, Masafumi Katayama, Yuh Hashimoto, and
Takashi Akaishi
Second Department of Surgery
Tohoku University School of Medicine
Sendai 980, Japan

I. Introduction 269

II. Methods of Establishment
and Maintenance 270
A. Establishment of the TE Series 270
B. Brief Review of Major Papers 273

III. Morphology 273
A. Morphological Studies by Phase-
Contrast Microscopy 273

B. Ultrastructural Features and
Immunocytochemical Studies 277

IV. Other Characteristics 277
A. Chromosome Analysis 277
B. Tumorigenicity in Nude Mice 278
C. Growth and Differentiation 279

V. Discussion 279

VI. Future Prospects 282
References 282

I. Introduction

Esophageal cancer occurs worldwide, although large variations in incidence occur even within the same country (Banks-Schlegel and Quintero, 1986a). The disease is extremely frequent in certain parts of South Africa, Russia, Iran, Iraq, and China, as well as in Japan, Puerto Rico, and the Dutch West Indies (Day, 1975; Adkins, 1981).

Epidemiological studies of esophageal cancer indicate that nutritional and environmental factors are important causes of the disease (Stoner et al., 1985). An increased risk of esophageal cancer has been correlated with tobacco smoking, the consumption of alcoholic beverages, and the use of certain plant products (Stoner et al., 1985). The most common symptom, dysphagia, develops when the primary tumor is already an advanced lesion. Further, esophageal cancer often metastasizes to regional lymph nodes and distant organs. Despite advances in surgery, radiation therapy, and chemotherapy, the 5-yr survival rate for patients with esophageal cancer continues to be less than 20%

Atlas of Human Tumor Cell Lines

in Japan (Japanese Society for Esophageal Disease, personal communication) and less than 10% in the United States (Silverberg *et al.*, 1990).

The establishment of continuously growing cell lines from human esophageal cancer was reported in 1976 (Bey *et al.*, 1976). Since then, the other cell lines listed in Table I have been reported. In general, these cell lines have been established in countries marked by a high incidence of esophageal cancers, namely, China, South Africa, and Japan. All cell lines but one (TE-7) were derived from squamous cell carcinoma of the esophagus, since this type of cancer is by far the most common neoplasm of the esophagus.

We have established 15 cell lines of esophageal cancer (Table II; Nishihira *et al.*, 1979,1985; Akaishi *et al.*, 1986). Three cell lines were derived from poorly differentiated squamous cell carcinoma, four were from moderately differentiated squamous cell carcinoma, and seven were from well-differentiated squamous cell carcinoma. TE-7 was derived from adenocarcinoma of the esophagus. Of these cell lines, 13 were derived from primary lesions, TE-3 was from a subcutaneous lymph node metastasis in the chest, and TE-9 was from a pleural effusion. These lines are known as the TE series and have been used worldwide as models for investigating the biological behavior of these types of tumors *in vitro*.

II. Methods of Establishment and Maintenance

A. Establishment of the TE Series

Establishment of the TE series is discussed in detail by Nishihira *et al.* (1979,1985) and by Akaishi *et al.* (1986). Most of the tumor samples were obtained from resected primary lesions of esophageal cancer. TE-3 was derived from a metastatic lymph node and TE-9 was derived from a pleural effusion. TE-4 was established from a transplanted tumor in a nude mouse. Specimens were collected in medium and transported to the laboratory. Some specimens were diced using scissors and were applied to explant cultures (TE-1, -2, -3, -4, and -5); the others were minced with knives, digested wtih collagenase, and inoculated directly into the medium. RPMI 1640, Eagle's minimum essential medium (MEM), or DM-160 (Katsuta *et al.*, 1976) with 6, 10, or 20% fetal bovine serum (FBS) was used for the culture of esophageal cancer cells. Antibiotics [5 μg/ml gentamycin, 5 μg/ml amphotericin B, and 100 μg/ml sodium piperacillin (Toyama Chemical Co., Tokyo, Japan) and/or 1 mg/ml kanamycin sulfate] were used for primary culture. Fibroblasts were removed by differential trypsinization or mechanical scraping. Cells were passaged by trypsinization. All TE series cell lines are now maintained in DM-160 with 10% FBS.

Human Esophageal Cancer Cell Lines

Cell line	Originator (date) and country	Reference
SN	E. Bey (1976), South Africa	Rabin et al. (1982)
ECa109	Department of Cell Biology, CICAM (1976), China	Banks-Schlegal and Quintero (1986a); Pan et al. (1989)
CaEs-17	Laboratory Section, Department of Surgery, People's Hospital, Peking Medical College (1976), China	Rabin et al. (1982); Banks-Schlegel and Quintero (1986a); Pan et al. (1989)
Ec56	S. Li (1979), China	Pan et al. (1989)
TE series (TE-1,2,3,4,5,6,7,8,9, 10,11,12,13,14,15)	T. Nishihira (1979), Japan	Rabin et al. (1982); Nishihira et al. (1985); Akaishi et al. (1986); Banks-Schlegel and Quintero (1986b); Kamata et al. (1986); Yamamoto et al. (1986); Chida et al. (1988); Hiraizumi et al. (1990); Takano et al. (1990a,b); Whang-Peng et al. (1990); Yoshida et al. (1990); Katayama and Kan (1991)
Hcu series (Hcu-10,13,18, 33,35,37,39,57)	K. M. Robinson (1980), South Africa	Rabin et al. (1982); Robinson et al. (1982); Robinson and Maistry (1983); Angorn et al. (1985); Banks-Schlegel and Quintero (1986b); Botha et al. (1986); van Helden et al. (1988); Whang-Peng et al. (1990)
SGF series (SGF-3,5)	T. Shinbo (1981), Japan	Saito et al. (1990)
B series	K. M. Robinson (1983), South Africa	Rabin et al. (1982); Robinson et al. (1986)
CE series (CE-48,69,81T/VGH)	C. Hu (1984), Taiwan	Pan et al. (1989)
HCE series (HCE-1,3,4,5,6,7, 8,9)	S. P. Banks-Schlegel (1986), United States	Banks-Schlegel and Quintero (1986b); Whang-Peng et al. (1990)
EC/CUHK 1	C. H. Mok (1987), Hong Kong	Hollstein et al. (1990)
EC-GI	K. Sato (1987a), Japan	Sato et al. (1987,1988)
KSE series	H. Matsuoka (1987), Japan	Matsuoka et al. (1989,1990a,b,1991); Ueo et al. (1990)
KYSE series (KYSE-30,50,70, 110,140,150,170,180,190, 200,220,240,270,280,350, 360,390,410,450,510,520)	Y. Shimada (1992), Japan	Naito et al. (1990)

Table II

TE Series Cell Lines and Clinical Characteristics of Patients[a]

Cell line	Age/sex[b]	Histology (differentiation)[c]	Tumor location[d]	Stage[e]	Source for cell culture
TE-1	58/M	WD SCC	Im,Ei	Stage II	Primary lesion
TE-2	56/M	PD SCC	Im	Stage IV	Primary lesion
TE-3	48/M	WD SCC	Iu,Im	Stage IV	Subcutaneous lymph node metastasis
TE-4	48/F	WD SCC	Ei,Im	Stage III	Primary lesion
TE-5	73/F	PD SCC	Im,Ei	Stage IV	Primary lesion
TE-6	71/M	WD SCC	Ce,Ph	Stage IV	Primary lesion
TE-7	72/M	Adenocarcinoma	Ei,Ea	Stage II	Primary lesion
TE-8	63/M	MD SCC	Ei,Im	Stage III	Primary lesion
TE-9	48/M	PD SCC	Im	Stage IV	Pleural effusion
TE-10	58/M	WD SCC	Im,Iu	Stage IV	Primary lesion
TE-11	58/M	MD SCC	Ce,Iu	Stage IV	Primary lesion
TE-12	54/M	MD SCC	Ei,Im	Stage III	Primary lesion
TE-13	65/F	PD SCC	Ce,Iu	Stage IV	Primary lesion
TE-14	57/M	MD SCC	Im,Ei	Stage IV	Primary lesion
TE-15	58/F	WD SCC	Ei,Ea	Stage IV	Primary lesion

[a] References: Nishihira et al. (1979, 1985), Rabin et al. (1982), Akaishi et al. (1986), Banks-Schlegel and Quintero (1986b), Kamata et al. (1986), Yamamoto et al. (1986), Chida et al. (1988), Hiraizumi et al. (1990), Takano et al. (1990a,b), Whang-Peng et al. (1990), Yoshida et al. (1990), Katayama and Kan (1991).

[b] Age in years. M, male; F, female.

[c] WD, well differentiated; MD, moderately differentiated; PD, poorly differentiated; SCC, squamous cell carcinoma.

[d] Iu, upper intrathoracic esophagus; Im, middle intrathoracic esophagus; Ei, lower intrathoracic esophagus; Ea, abdominal esophagus; Ce, cervical esophagus; Ph, pharynx. (These abbreviations are usually used in Japan to show the location of esophageal cancer.)

[e] Japanese staging of esophageal cancer (stage I–Stage IV).

B. Brief Review of Major Papers

Tumor samples were obtained from resected tumors at the time of surgery (Nishihira *et al.*, 1979; Robinson *et al.*, 1980,1982; Hu *et al.*, 1984; Banks-Schlegel and Quintero, 1986a; Shimada *et al.*, 1992), from biopsy specimens at the time of diagnostic esophagoscopy (Bey *et al.*, 1976; Robinson *et al.*, 1980,1982), or from metastatic lymph nodes (Matsuoka *et al.*, 1987,1991; Sato *et al.*, 1987a). Tumors in nude mice also were used to establish cell lines (Sato *et al.*, 1987a; Shimada *et al.*, 1992). Specimens were cut into small fragments and used for explant cultures (Nishihira *et al.*, 1979; Robinson *et al.*, 1980,1982; Hu *et al.*, 1984). Robinson *et al.* (1980) reported that a dry explant technique for the initiation of cultures proved most successful. Bey *et al.* (1976) treated their materials with trypsin. Single-cell suspensions of lymph nodes were prepared by mechanical pipetting (Matsuoka *et al.*, 1991). Shimada *et al.* (1992) made a crude cell suspension by pressing minced tissue through stainless steel mesh.

Eagle's MEM (Bey *et al.*, 1976; Robinson *et al.*, 1980,1982; Matsuoka *et al.*, 1987), Dulbecco's modified Eagle's medium (DMEM) (Hu *et al.*, 1984; Matsuoka *et al.*, 1991), Medium 199 (Banks-Schlegel and Quintero, 1986a), or Ham's F-12 medium (Sato *et al.*, 1987a) supplemented with 10% FBS was used for the culture of esophageal cancer cells. Shimada *et al.* (1992) used Ham's F-12/RPMI 1640 medium containing 2–20% FBS. Antibiotics such as penicillin, streptomycin, amphotericin B, mycostatin, gentamycin, and sodium piperacillin were used for primary cultures (Bey *et al.*, 1976; Robinson *et al.*, 1980; Hu *et al.*, 1984; Matsuoka *et al.*, 1987,1991; Shimada *et al.*, 1992). Fibroblasts were removed by differential trypsinization or mechanical scraping (Bey *et al.*, 1976; Nishihira *et al.*, 1979; Robinson *et al.*, 1980; Hu *et al.*, 1984; Shimada *et al.*, 1992). Matsuoka *et al.* (1990b,1991) reported that lethally irradiated confluent monolayers of BALB/c 3T3 cells provided a favorable culture condition for initiating primary cultures.

III. Morphology

A. Morphological Studies by Phase-Contrast Microscopy

Cultures of normal human esophageal epithelial cells and esophageal carcinoma cell lines are morphologically distinguishable when examined by phase-contrast microscopy (Banks-Schlegel and Quintero, 1986a). Normal human epithelial cells are uniform in appearance and polygonal, whereas esophageal carcinoma cells are pleomorphic, varying greatly in size and shape, and demonstrate the tendency to pile up in an organized fashion (Banks-Schlegel and Quintero, 1986a). The morphology of cultured normal human esophageal epithelial cells (Katayama *et al.*, 1984) and that of esophageal carcinoma cell lines (TE-1, -2, -3, -4, -5) (Nishihira *et al.*, 1979,1985; Akaishi *et al.*, 1986) is shown in Fig. 1, in which the different features of these cell lines can be observed. Cells that were less epithelioid and more spindle shaped in appearance were observed in some cell lines (Banks-Schlegel and Quintero, 1986a; Matsuoko *et al.*, 1991). Dispase-detached sheets of carcinoma cells were reported to reveal a more limited and less orderly stratification process than normal esophageal keratinocytes (Banks-Schlegel and Quintero, 1986a). Morphological characteristics of esophageal cancer cells at the late log phase of growth (KYSE-30, -50, -200, and -520 cell lines; Shimada *et al.*, 1992) are shown in Fig. 2 as examples of the typical basic structure of esophageal cancer cell lines.

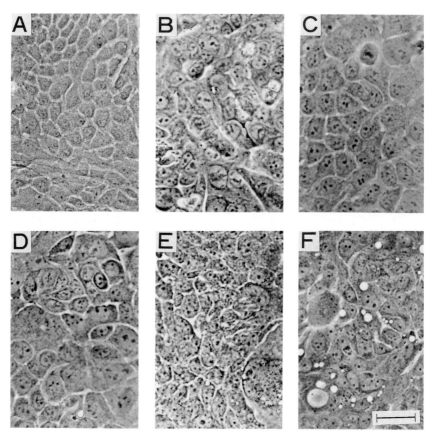

Fig. 1. Morphological examination of cell cultures of normal human esophageal epithelial cells and esophageal carcinoma cell lines by phase-contrast microscopy. (A) Normal human esophageal epithelial cells. (B) TE-1. (C) TE-2. (D) TE-3. (E) TE-4. (F) TE-5. Bar, 50 μm.

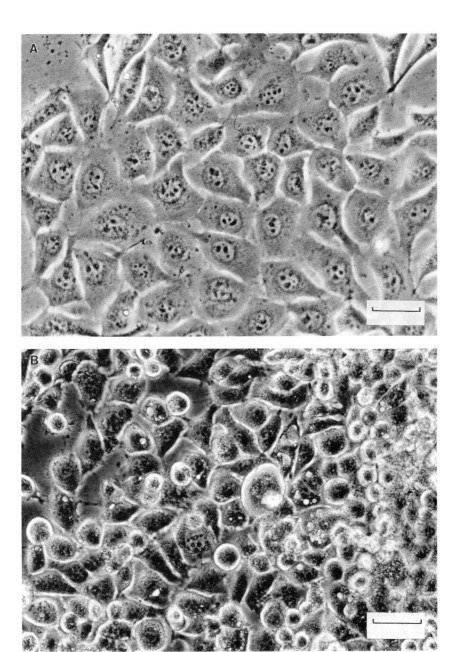

Fig. 2. Morphological characteristics of esophageal cancer cells at late log phase of growth as shown by phase-contrast microscopy (original magnification, 200x; Bar, 50 μm). The cells are uniform in appearance and polygonal in shape, and demonstrate, in some cases, the tendency to pile up in a disorganized fashion, varying greatly in size and shape. The KYSE-30, -50, -200, and -520 cell lines were chosen because they best demonstrate the typical basic structure of esophageal cancer cell lines. (A) The KYSE-30 cell line demonstrates the largest cells in which the nuclei and nucleolus are easily visible. (B) The KYSE-50 cell line demonstrates a basic tendency to bunch together in a grape-like manner.

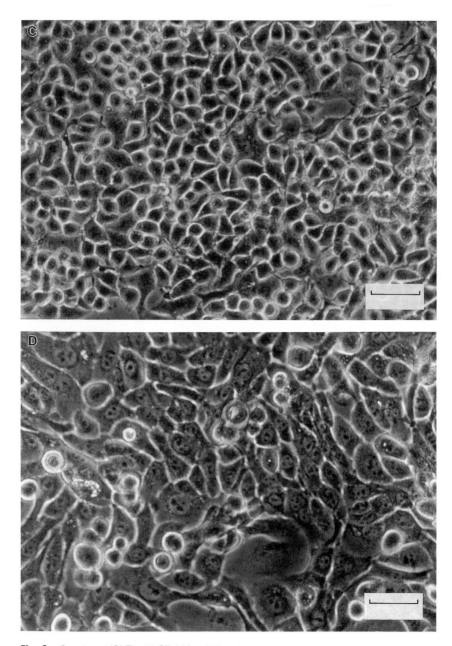

Fig. 2. *Continued* (C) The KYSE-200 cell line demonstrates small cells of uniform size and a tendency toward spherical shape. (D) The KYSE-520 cell line exhibits the most typical demonstration of polygonal cells, which interlock like a jigsaw puzzle. Note that the variance in phase density is due to photographic film only and is not related to "lighter" or "darker" cells.

B. Ultrastructural Features and Immunocytochemical Studies

Transmission electron microscopy of cultured esophageal cancer cells showed the presence of epithelial markers such as desmosomes (Nishihira et al., 1979; Rabin et al., 1982; Mok et al., 1987; Shimada et al., 1992) and tonofilaments (Bey et al., 1976; Rabin et al., 1982; Mok et al., 1987). We reported the presence of abundant cell organelles in TE-1 and a few organelles in the scanty cytoplasm TE-2 (Nishihira et al., 1979). Immunocytochemical studies showed the presence of keratin in cultured esophageal cancer cells (Banks-Schlegel and Quintero, 1986a; Matsuoka et al., 1987,1991). In general, the esophageal carcinoma cells were stained less intensely and were not characterized by a predominantly perinuclear staining pattern compared with normal esophageal keratinocytes (Banks-Schlegel and Quintero, 1986a). The pattern of keratin staining was found to vary markedly among the different carcinoma cell lines, exhibiting a fine lacy network in some and a thicker cobweb-like appearance in others (Banks-Schlegel and Quintero, 1986a).

IV. Other Characteristics

A. Chromosome Analysis

Chromosome analysis of cancer cell lines provides important information about possible genetic alterations underlying the development of cancer. Detailed analyses of the TE series, the HCE series, the HCU series, and Chinese esophageal cancer cells were reported by Su et al. (1988) and Whang-Peng et al. (1990). Represenatative karyotypic aberrations of TE-4, -5, -6, -7, -8, and -10 are listed in Table III. Esophageal cancer cells, like other types of cancer, have aneuploid chromosome numbers that vary according to cell lines. The chromosomes involved most frequently in the structural abnormalities were reported to be chromosomes 1, 9, and 11 (Whang-Peng et al., 1990). Loss of heterozygosity at a variety of chromosomal sites now has been described in many different types of tumors (Ponder, 1988). Wagata et al. (1991) analyzed allelic deletions at 23 loci on 18 different chromosomes in 35 esophageal squamous cell carcinoma tissues using restriction fragment length polymorphism (RFLP) markers. Loss of heterozygosity was detected on chromosomes 2, 3, 6, 7, 11–14, 16–18, 21, and 22, whereas no loss was detected on chromosomes 1, 4, and 8–10. Only the loss of chromosome 17p was detected with high frequency (45%); losses on other chromosomes had frequencies of less than 22% (Wagata et al., 1991).

Table III

Chromosomal Aberrations

Cell line	Chromosomal analysis
TE-4	51,−X,−X,−1,−3,−4,−9,−11,−12,−12,−13,−14,−14,−15,+2,+16,+18,+18,+der (X)t(X;?)(q26;?),der(1)t(1;?)(q32;?),del(1)(p32),+del(7)(q21),del(7)(q32),+der(9) t(9;?)(p11,?),+der(12)t(12;?)(q22;?),+der(13)t(13)t(13;?)(q32;?),+der(16)t(16;?) (q22;?),+7mar
TE-5	61,XX,−3,−4,−4,−5,−8,−8,−11,−13,−13,−14,−15,−15,−17,−20,+del(1)(q21q32), del(2)(p12p14),+der(4)t(4;?)(q31;?),+der(5)t(5;?)(q22;?),+del(6)(p23),del(7)(p13), del(7)(p13),del(9)(q32),+der(9)t(9;?)(p11;?),+der(10)t(10;?)(q22;?),+der(17)t (17;?)(q25;?),+22mar
TE-6	55,−X,−X,−2,−12,−13,−13,−14,−14,−15,−15,−18,−21,−21,−22,+1,+5,+5, +9,+11,+16,+16,+der(X)t(X;?)(q?1;?),del(3)(q12),+del(8)(q22),+der(12)t(12;?) (p13;?),+der(14)t(14;?)(p11;?),+der(14)t(14;?)(p11;?),+der(14)t(14;?)(p11;?),+der (16)t(16;?)(q24;?),+der(18)t(18;?)(q23;?),+der(18)t(18;?)(q23;?),+i(21q),+mar1, +2mar2,+mar3,+mar4,+mar5
TE-7	61,XX,−3,−4,−4,−5,−8,−8,−11,−13,−13,−14,−15,−15,−17,−20,+del(1) (q21q32),del(2) (p12p14),+der(4)t(4;?)(q31;?),+der(5)t(5;?)(q22;?),+del(6)(p23), del(7)(p13),del(7)(p13),del(9)(q32),+der(9)t(9;?)(p11;?),+der(10)t(10;?) (q22;?),+der(17)t(17;?)(q25;?),+mar22
TE-8	56,X,t(8;X)(p23;q13),+i(1q),−3,−3,+der(3)t(3;?)(p11;?),+der(3)t(3;?)(p11;?),der(4) (p14),+5,−6,−6,−7,−7,+i(7q),+der(9)t(9;?)(p13;?),+9,−10,−10,+der(10)t(10;?) (p11;?),+der(10)t(10;?)(p11;?),+del(11)(p14),+der(11)t(11;?)(q23;?),+12,+14, +16,+der(17)t(17;?)(p11;?),+2mar1,+mar2
TE-10	62,XY,+Y,+10,+12,+16,+16,−1,−1,−5,−7,−11,−13,−14,−17,+der(1)t(1;?) (q32;?),+der(1)t(1;?)(q11;?),+der(1)t(1;?)(q32;?),+der(6)t(6;?)(q21;?),+der(7)t (7;?)(p12;?),+der(17)t(17;?)(p11;?),+2mar1,+2mar2,+mar3,+mar4,+mar5, +mar6,+mar7,+mar8,+mar9,+mar10,+mar11

B. Tumorigenicity in Nude Mice

Tumorigenicity testing in nude mice is a useful method of screening cell lines for malignant potential. Robinson and Maistry (1983) extensively ana-lyzed the tumorigenicity of 10 human esophageal carcinoma cell lines. Al-though the latent period and tumor take rate varied depending on the cell line inoculated, distinct tumor masses grew in nude mice as a direct result of the introduction of human esophageal carcinoma cells from each of 10 lines; 56 tumors developed in 53 of the 239 mice inoculated, giving an overall take rate of 21.9% (a range of 5–90%) (Robinson and Maistry, 1983; Angorn *et al.*, 1985). No obvious correlations existed among take rates, the latent periods before tumor appearance, the sex of the mice, and the degree of histological differen-tiation of the human carcinomas from which the inoculated lines were derived

(Robinson and Maistry, 1983; Angorn *et al.*, 1985). Banks-Schlegel and Quintero (1986a) reported that tumorigenic potential was augmented greatly by the use of cell lines derived from selected clones grown in soft agarose.

C. Growth and Differentiation

Human esophageal carcinoma cell lines differ from their normal counterparts in terms of their growth properties and the expression of certain differentiated functions, namely, keratin proteins and cross-linked envelopes (Banks-Schlegel and Quintero, 1986a). When grown under optimal growth conditions, the carcinoma cells generally grow to a higher saturation density than their nontransformed counterparts. The generation times of these cells are variable (Banks-Schlegel and Quintero, 1986a). Transformed cells grow better under stringent growth conditions than do nontransformed human esophageal keratinocytes, and grow in an anchorage-independent fashion in soft agarose, although the colony-forming efficiency and the size of the colonies vary among the different cell lines (Banks-Schlegel and Quintero, 1986a).

Altered expression of keratin proteins and cross-linked envelopes were observed in the carcinoma cell lines and generally reflected those changes seen in primary esophageal carcinomas (Grace *et al.*, 1985; Banks-Schlegel and Quintero, 1986a). Growth characteristics in tissue culture may reflect the malignant potential of the tumors from which the cells were derived. Robinson *et al.* (1980,1982) found that adaptation to *in vitro* culture conditions correlated well with invasiveness *in vivo* and with prognosis at 6 months after surgery.

V. Discussion

Many cell lines of human esophageal cancer have been established to develop methods for studying tumor biology and for testing the therapeutic sensitivity of individual tumors.

Cancer is thought to appear as a result of multiple genetic alterations (Wagata *et al.*, 1991). Advances in molecular biology have clarified that oncogenes and tumor suppressor genes are involved in the development of human malignancies to varying degrees and in various combinations, depending on the type of tumor. *ras* (Hollstein *et al.*, 1988; Meltzer *et al.*, 1990), c-*myc* (Lu *et al.*, 1988), *hst*-1 (Tsuda *et al.*, 1989), *int*-2 (Tsuda *et al.*, 1989; Wagata *et al.*, 1991), and c-*erbB2* (Hollstein *et al.*, 1988) oncogenes were analyzed using tumor samples of esophageal cancer. Amplification of the c-*myc* gene and coamplification of the *hst*-1 and *int*-2 genes were detected in esophageal cancer. However, esophageal cancer seems to be an exception with respect to activation of the *ras* oncogenes (Hollstein *et al.*, 1988; Meltzer *et al.*, 1990). The *p53* tumor suppressor gene apparently is involved in the development of many cancers. Hollstein *et al.* (1990) demonstrated the presence of *p53* mutations, not only in esophageal cancer cell lines but also in tumor samples of this cancer.

Growth factors and their receptors, as well as various factors involved in signal transduction, regulate the growth and differentiation of cells and are closely related to oncogenes. Epidermal growth factor (EGF) is a potent mitogen for a wide variety of cultured cells (Carpenter and Cohen, 1979). Many cell types, including epithelial cells, express the EGF homolog transforming growth factor α (TGFα) (Coffey *et al.*, 1987). Expression of mRNA for EGF and TGFα was detected in the TE series (Yoshida *et al.*, 1990). Banks-Schlegel and Quintero (1986b) found that esophageal carcinoma cell lines (HCU, HCE, and TE series) contained lower quantities of surface EGF receptors (2- to 100-fold) and that the affinity of the EGF receptor was increased (6- to 100-fold) compared with normal esophageal epithelial cells. Growth inhibitory effects of EGF on the TE series were reported by Kamata *et al.* (1986). Amplification of the EGF receptor gene was found in the TE series (Yamamoto *et al.*, 1986) as well as in primary tumors (Hollstein *et al.*, 1988; Lu *et al.*, 1988). These data suggest that changes in EGF receptors may play an important role in the pathogenesis of esophageal cancer. The heparin-binding (fibroblast) growth factor family also contains a wide spectrum of mitogens that support epithelial cells in culture (Burgess and Maciag, 1989). We demonstrated heparin-binding growth-promoting activity in cell extracts of the TE series (Katayama and Kan, 1991). Protein kinase C (PKC) is considered to play a crucial role in the signal transduction in the cell membrane that is mediated by hormones, growth factors, neurotransmitters, and tumor promoters. Chida *et al.* (1988) screened 41 cell lines for PKC activity and found that cancer cells, including the TE series, generally showed much lower activity than their normal counterparts.

Altered glycosylation of cell surface glycoproteins occurs in association with malignant transformation (Warren *et al.*, 1978). We previously demonstrated altered glycosylation of human esophageal cancer cell products (Hiraizumi *et al.*, 1990) and suggested that terminal residues of N-glycosidic carbohydrates of the TE series might contribute to the recognition sites of macrophages (Takano *et al.*, 1990a). Using TE-1 cells and esophageal cancer specimens, we demonstrated that the increase in β1-6-branched oligosaccharides in esophageal carcinomas was an important trait of the tumors with respect to invasion into the surrounding tissue (Takano *et al.*, 1990a,b).

Humoral hypercalcemia associated with solid tumors is caused by a variety of factors such as a parathormone (PTH)-like factor, TGFs, and interleukin 1α (IL-1α)-like factors (Sato *et al.*, 1987a,b,1988). Sato *et al.* (1987a) established a cell line designated EC-GI from a patient with esophageal carcinoma who developed humoral hypercalcemia, and showed the production of IL-1α and a PTH-like factor by this cell line (Sato *et al.*, 1987b,1988).

The elevated level of plasminogen activator content (Angorn *et al.*, 1985; Robinson, 1986) and prostaglandin production (Botha *et al.*, 1986) in the HCU series seems to correlate with the invasive and metastastic potential of these cells. Markers related to carcinomas, for example, carcinoembryonic antigen (CEA) (Nishihira *et al.*, 1979; Hu *et al.*, 1984) and glutathione S-transferase π (GST-π) (Tsuchida *et al.*, 1989), were measured in the TE series (CEA and GST-π) and the CE series (CEA). We observed the elevation of GST-π in the TE series as well as in esophageal cancer tissues (Tsuchida *et al.*, 1989). A monoclonal antibody against KYSE-50 cells was also reported (Naito *et al.*, 1990).

Esophageal carcinoma occurs more frequently in males, and the postoperative course is generaly more favorable in females (van Andel *et al.*, 1979). These findings suggest that the growth of esophageal cancer cells and the prognosis of the patient might be influenced by hormones such as testosterone or estrogen. Matsuoka *et al.* (1987,1990a) demonstrated an inhibitory effect of estrogen and a stimulatory effect of testosterone on the growth of KSE-1 cells. These cells have a binding content of 4.2 fmol/mg protein for the estrogen receptor and 2.2 fmol/mg protein for the testosterone receptor (Matsuoka *et al.*, 1987). However, Ueo *et al.* (1990) found that KSE-2 cells were unaffected by these sex hormones and that the heterotransplanted tumors of KSE-2 cells in nude mice had neither estrogen receptors nor androgen receptors. Yamashita *et al.* (1989) analyzed androgen receptors using tumor specimens of human esophageal cancer and xenografts implanted into nude mice, and reported that the administration of estrogen did not inhibit tumor growth.

Human tumor cell lines provide a useful model for the study of therapeutic response. We investigated the sensitivities of the TE series to anticancer drugs (Nishihira *et al.*, 1985; Akaishi *et al.*, 1986). The responses of the cell lines to drugs seem to be "individual," that is, no trends or rules were detected. Similar variability was found also with xenografts implanted into nude mice (Nishihira *et al.*, 1985). The development of simultaneous resistance to multiple structurally unrelated drugs is a major impediment to cancer chemotherapy. Shen *et al.* (1986) demonstrated that multidrug resistance in human carcinoma cell lines correlated with the amplification of the *mdr*-1 gene. We analyzed multidrug resistance in the TE series using Southern blot analysis. The amplification of the *mdr*-1 gene was detected in only one cell line (TE-5) of the 13 cell lines of the TE series. Hyperthermia has been used for therapeutic treatment of solid tumors. Matsuoka *et al.* (1989) evaluated the effect of the combined application of hyperthermia, chemotherapy, and irradiation on KSE-1 cells. Saito *et al.* (1990) found that esophageal cancer cells (SGF-3 and SGF-5) were sensitive to hyperthermia at 42° for 72 hr.

Contamination of cell lines by other cells is a phenomenon not unknown to scientists. Such events have been estimated to occur quite frequently (van Helden *et al.*, 1988). Using DNA fingerprint analysis, van Helden *et al.* (1988) surveyed the cross-contamination of HCU cell lines and found that lines HCU 10, 18, 33, 37, and 39 were genetically identical. Shimada *et al.* (1992) demonstrated that all their original 19 cell lines were genetically distinct using DNA fingerprint analysis.

VI. Future Prospects

Alterations generating dominantly acting oncogenes and inactivating suppressor genes are involved as genetic events in multistage carcinogenesis (Marshall, 1991). The multiplicity of genetic lesions found in most cancers may necessitate complex therapeutic regimens. Human tumor cell lines could provide information useful to understanding such complexity. However, cellular and molecular biological studies on esophageal cancer began only relatively recently, and data are insufficient to understand this cancer completely. Considering the multiplicity of genetic alterations in cancer cells, analyzing individual types of cancers using cell lines seems very important. Findings thus obtained will lead to development of more efficient therapies for different types of esophageal cancer in the future.

Acknowledgment

The authors thank Yutaka Yamada, Faculty of Medicine, Kyoto University, Kyoto, Japan, for kindly providing us with the photographs presented in Fig. 2.

References

Adkins P. C. (1981). Tumors of the esophagus. *In* "Davis–Christopher Textbook of Surgery" (D. C. Sabiston, Jr., ed.) pp. 841–855. Saunders, Philadelphia.

Akaishi, T., Sekine, Y., Sanekata, K., Nishihira, T., and Kasai, M. (1986). Characteristics of growth and effectiveness of anti-cancer drugs on tissue cultured human esophageal cancer cells. *In* "Esophageal Cancer" (M. Kasai, ed.), pp 35–38. Excerpta Medica, Tokyo.

Angorn, I. B., Robinson, K. M., and Haffejee, A. A. (1985). An *In vitro* human cell system for carcinoma of the esophagus. *In* "*In Vitro* Models of Cancer Research" (M. M. Webber, ed.), Vol. I, pp. 39–63. CRC Press, Boca Raton, Florida.

Banks-Schlegel, S. P., and Quintero, J. (1986a). Growth and differentiation of human esophageal carcinoma cell lines. *Cancer Res.* **46,** 250–258.

Banks-Schlegel, S. P., and Quintero, J. (1986b). Human esophageal carcinoma cells have fewer, but higher affinity epidermal growth factor receptors. *J. Biol. Chem.* **261,** 4359–4362.

Bey, E., Alexander, J., Whitcutt, J. M., Hunt, J. A., and Gear, J. H. S. (1976). Carcinoma of the esophagus in Africans: Establishment of a continuously growing cell line from a tumor specimen. *In Vitro* **12,** 107–114.

Botha, J. H., Robinson, K. M., Ramchurren, N., Reddi, K., and Norman, R. J. (1986). Human esophageal carcinoma cell lines: Prostaglandin production, biological properties, and behavior in nude mice. *J. Natl. Cancer Inst.* **76,** 1053–1056.

Burgess, W. H., and Maciag, T. (1989). The heparin-binding (fibroblast) growth factor family of proteins. *Ann. Rev. Biochem.* **58,** 575–606.

Carpenter, G., and Cohen, S. (1979). Epidermal growth factor. *Ann. Rev. Biochem.* **48,** 193–216.

Chida, K., Kato, N., Yamada, S., and Kuroki, T. (1988). Protein kinase C activities and bindings of a phorbol ester tumor promoter in 41 cell lines. *Biochem. Biophys. Res. Commun.* **157,** 1–8.

Coffey, R. J., Jr., Derynck, R., Wilcox, J. N., Bringman, T. S., Goustin, A. S., Moses, H. L., and Pittelkow, M. R. (1987). Production and auto-induction of transforming growth factor-α in human keratinocytes. *Nature* (*London*) **328,** 817–820.

Day, N. E. (1975). Some aspects of the epidemiology of esophageal cancer. *Cancer Res.* **35,** 3304–3307.

Department of Cell Biology of CICAMS (1976). Establishment of an epithelial cell line from human esophageal carcinoma. *Chin. Med. J. (Engl. Ed.)* **56**, 412–415.

Grace, M. P., Kim, K. H., True, L. D., and Fuchs, E. (1985). Keratin expression in normal esophageal epithelium and squamous cell carcinoma of the esophagus. *Cancer Res.* **45**, 841–846.

Hiraizumi, S., Takasaki, S., Nishihira, T., Mori, S., and Kobata, A. (1990). Comparative study on the N-linked oligosaccharides released from normal human esophageal epithelium and esophageal squamous carcinoma. *Jpn. J. Cancer Res.* **81**, 363–371.

Hollstein, M. C., Smits, A. M., Galiana, C., Yamasaki, H., Bos, J. L., Mandard, A., Partensky, C., and Montesano, T. (1988). Amplification of epidermal growth factor receptor gene but no evidence of *ras* mutation in primary human esophageal cancers. *Cancer Res.* **48**, 5119–5123.

Hollstein, M. C., Metcalf, R. A., Welsh, J. A., Montesano, R., and Harris, C. C. (1990). Frequent mutation of the p53 gene in human esophageal cancer. *Proc. Natl. Acad. Sci. U.S.A.* **87**, 9958–9961.

Hu, C., Hsieh, H., Chien, K., Wang, P., Wang, C., Chen, C., Lo, S. J., Wuu, K., and Chang, C. (1984). Biologic properties of three newly established human esophageal carcinoma cell lines. *J. Natl. Cancer Inst.* **72**, 577–588.

Kamata, N., Chida, K., Rikimaru, K., Horikowhi, M., Enomoto, S., and Kuroki, T. (1986). Growth-inhibitory effects of epidermal growth factor and overexpression of its receptors on human squamous cell carcinomas in culture. *Cancer Res.* **46**, 1648–1653.

Katayama, M., and Kan, M. (1991). Heparin-binding (fibroblast) growth factors are potential autocrine regulators of esophageal epithelial cell proliferation. *In Vitro Cell. Dev. Biol.* **27A**, 533–541.

Katayama, M., Akaishi, T., Nishihira, T., Kasai, M., Kan, M., and Yamane, I. (1984). Primary culture of human esophageal epithelial cells. *Tohoku J. Exp. Med.* **143**, 129–140.

Katsuta, H., and Takaoka, T. (1976). Improved synthetic media suitable for tissue culture of various mammalian cells. *In* "Methods in Cell Biology" (D. M. Prescott, ed.), Vol. XIV, pp. 145–158. Academic Press, New York.

Laboratory Section, Department of Surgery, People's Hospital, Peking Medical College (1976). Establishment of a cell line from human esophageal carcinoma. *Chin. Med. J. (Engl. Ed.)* **2**, 357–364.

Li, S., *et al.* (1979). The establishment and characterization of a human esophageal cancer epithelial cell line and its epithelioid and fusiform clonal lines. *Acta Zool. Sin.* **25**, 297.

Lu, S., Hsieh, L., Luo, F., and Weinstein, I. B. (1988). Amplification of the EGF receptor and c-*myc* genes in human esophageal cancers. *Int. J. Cancer* **42**, 502–505.

Marshall, C. J. (1991). Tumor suppressor genes. *Cell* **64**, 313–326.

Matsuoka, H., Sugimachi, K., Ueo, H., Kuwano, H., Nakano, S., and Nakayama, M. (1987). Sex hormone response of a newly established squamous cell line derived from clinical esophageal carcinoma. *Cancer Res.* **47**, 4134–4140.

Matsuoka, H., Sugimachi, K., Mori, M., Kuwano, H., Ohno, S., and Nakano, S. (1989). Effects of hyperthermochemoradiotherapy on KSE-1 cells, a newly established human squamous cell line derived from esophageal carcinoma. *Eur. Surg. Res.* **21**, 49–59.

Matsuoka, H., Ueo, H., Yano, K., Kido, Y., Shirabe, K., Mitsudomi, T., and Sugimachi, K. (1990a). Estradiol sensitivity test using contact-sensitive plates of confluent BALB/c 3T3 cell monolayers. *Cancer Res.* **50**, 2113–2118.

Matsuoka, H., Mori, M., Ueo, H., and Sugimachi, K. (1990b). Role of confluent monolayer surfaces on the growth of a newly established human esophageal carcinoma cell line. *In Vitro Cell. Dev. Biol.* **26**, 741–742.

Matsuoka, H., Mori, M., Ueo, H., Sugimachi, K., and Urabe, A. (1991). Characterization of human esophageal carcinoma cell line established on confluent monolayer, and advantage of confluent monolayer surface structure for attachment and growth. *Pathobiology* **59**, 76–84.

Meltzer, S. J., Mane, S. M., Wood, P. K., Resau, J. H., Newkirk, C., Terzakis, J. A., Korelitz, B. I., Weinstein, W. M., and Needleman, S. W. (1990). Activation of c-Ki-*ras* in human gastrointestinal dysplasias determined by direct sequencing of polymerase chain reaction products. *Cancer Res.* **50**, 3627–3630.

Mok, C. H., Chew, E. C., Riches, D. J., Lee, J. C. K., Huang, D. P., Hadgis, C., and Crofts, T. J. (1987). Biological characteristics of a newly established human oesophageal carcinoma cell line. *Anticancer Res.* **7**, 409–416.

Naito, M., Imamura, M., Kannagi, R., and Tobe, T. (1990). Production of monoclonal antibody to human esophageal cancer cell line. *Jpn. J. Surg.* **20**, 170–179.

Nishihira, T., Kasai, M., Mori, S., Watanabe, T., Kuriya, Y., Suda, M., Kitamura, M, Hirayama, K., Akaishi, T., and Sasaki, T. (1979). Characteristics of two cell lines (TE-1 and TE-2) derived from human squamous cell carcinoma of the esophagus. *Gann* **70**, 575–584.

Nishihira, T., Kasai, M., Kitamura, H., Hirayama, K., Akaishi, T., and Sekine, Y. (1985). Biological characteristics of cultured cell lines of human esophageal carcinomas and tumors transplantable to nude mice originating from human esophageal carcinomas and their clinical application. *In* "*In Vitro* Models of Cancer Research" (M. M. Webber, ed.), Vol. I, pp. 65–79. CRC Press, Boca Raton, Florida.

Pan, Q. (1989). Studies of esophageal cancer cells *in vitro*. *Proceedings of the Chinese Academy of Science and the Peking Union Medical College* **4**, 52–57.

Ponder, B. (1988). Gene losses in human tumors. *Nature (London)* **335**, 400–402.

Rabin, H., Gonda, M. A., Benton, C. V., and Robinson, K. M. (1982). Cell culture studies of simian and human esophageal carcinoma. *In* "Cancer of the Esophagus" (C. J. Pfeiffer, ed.), Vol. II, pp. 99–111. CRC Press, Boca Raton, Florida.

Robinson, K. M. (1986). Evaluation of the biological properties of continuous human esophageal carcinoma cell lines *in vitro* and *in vivo* in the nude mouse. *In* "Esophageal Cancer" (M. Kasai, ed.), pp. 39–42. Excerpta Medica, Tokyo.

Robinson, K. M., and Maistry, L. (1983). Tumorigenicity and other properties of cells from ten continuous human esophageal carcinoma cell lines in nude mice. *J. Natl. Cancer Inst.* **70**, 89–93.

Robinson, K. M., Haffejee, A. A., and Angorn, I. B. (1980). Tissue culture and prognosis in carcinoma of the esophagus. *Clin. Oncol.* **6**, 125–136.

Robinson, K. M., Haffejee, A. A., and Angorn, I. B. (1982). Tissue culture of tumor biopsies in oesophageal carcinoma and correlations with prognosis. *S. Afr. J Surg.* **20**, 245–252.

Saito, M., Shinbo, T., Saito, T., Kato, H., Otagiri, H., Karaki, Y., Tazawa, K., and Fujimaki, M. (1990). Temperature sensitivity on proliferation and morphologic alteration of human esophageal carcinoma cells in culture. *In Vitro Cell. Dev. Biol.* **26**, 181–186.

Sato, K., Kasono, K., Ohba, Y., Yoshiro, T., Fujii, Y., Yoshida, M., Tsushima, T., and Shizume, K. (1987a). Establishment of a parathyroid hormone-like factor-producing esophageal carcinoma cell line (EC-GI). *Jpn. J. Cancer Res. (Gann)* **78**, 1044–1048.

Sato, K., Han, D. C., Ozawa, M., Fujii, Y., Tsushima, T., and Shizume, K. (1987b). A highly sensitive bioassay for PTH using ROS 17/2.8 subclonal cells. *Acta Endocrinol.* **116**, 113–120.

Sato, K., Fujii, Y., Kasono, K., Tsushima, T., and Shizume, K. (1988). Production of interleukin 1 and a parathyroid hormone-like factor by a squamous cell carcinoma of the esophagus (EC-GI) derived from a patient with hypercalcemia. *J. Clin. Endocrinol. Metab.* **67**, 592–601.

Shen, D.-W., Fojo, A., Chin, J. E., Roninson, I. B., Richert, N., Pastan, I., and Gottesman, M. M. (1986). Multidrug-resistant cell lines: Increased *mdr*-1 expression can precede gene amplification. *Science* **232**, 643–645.

Shimada, Y., Imamura, M., Wagata, T., Yamaguchi, N., and Tobe, T. (1992). Characterization of 21 newly established esophageal cancer cell lines. *Cancer* **69**, 277–284.

Shinbo, T., Otagiri, H., Karaki, Y., Saitoh, J., and Fujimaki, M. (1981). Characterization of an established cell line (SGF-3) derived from a human esophageal carcinoma. *Jpn. J. Gastroenterol. Surg.* **14**, 1646.

Silverberg, E., Boring, C. C., and Squires, T. S. (1990). Cancer Statistics 1990. *Ca—A Cancer J. Clin.* **40**, 9–26.

Stoner, G. D., Babcock, M. S., Scaramuzzino, D. A., and Gunning, W. T. (1985). Cultured rat esophageal epithelial cells for studies on differentiation and carcinogenesis. *In* "*In Vitro* Models of Cancer Research" (M. M. Webber, ed.), Vol. I, pp. 81–96. CRC Press, Boca Raton, Florida.

Su, Y., Wang, X., Hu, N., Pei, X., and Wu, M. (1988). G-banded chromosome analyses of mucosal epithelium adjacent to esophageal cancer (EC)—Some consistent chromosomal changes. *Sci. Sin.* (*B*) **31**, 710–718.

Takano, R., Nose, M., Kanno, H., Nishihira, T., Hiraizumi, S., Kobata, A., and Kyogoku, M. (1990a). Recognition of N-glycosidic carbohydrates on esophageal carcinoma cells by macrophage cell line THP-1. *Am. J. Pathol.* **137**, 139–401.

Takano, R., Nose, M., Nishihira, T., and Kyogoku, M. (1990b). Increase of β1-6-branched oligosaccharides in human esophageal carcinomas invasive against surrounding tissue *in vivo* and *in vitro*. *Am. J. Pathol.* **137**, 1007–1011.

Tsuchida, S., Sekine, Y., Shineha, R., Nishihira, T., and Sato, K. (1989). Elevation of the placental glutathione *S*-transferase form (GST-π) in tumor tissues and the levels in sera of patients with cancer. *Cancer Res.* **49**, 5225–5229.

Tsuda, T., Tahara, E., Kajiyama, G., Sakamoto, H., Terada, M., and Sugimura, T. (1989). High incidence of coamplification of *hst-1* and *int-2* genes in human esophageal carcinomas. *Cancer Res.* **49**, 5505–5508.

Ueo, H., Matsuoka, H., Sugimachi, K., Kuwano, H., Mori, M., and Akiyoshi, T. (1990). Inhibitory effects of estrogen on the growth of a human esophageal carcinoma cell line. *Cancer Res.* **50**, 7212–7215.

van Andel, J. G., Dees, J., Dijkhuis, C. M., Fokkens, W., van Houten, H., de Jong, P. C., and van Woerkom-Eykenboom, W. M. H. (1979). Carcinoma of the esophagus. *Ann. Surg.* **190**, 684–689.

van Helden, P. D., Wiid, I. J. F., Albrecht, C. F., Theron, E., Thornley, A. L., and Hoal-van Helden, E. G. (1988). Cross-contamination of human esophageal squamous carcinoma cell lines detected by DNA fingerprint analysis. *Cancer Res.* **48**, 5660–5662.

Wagata, T., Ishizaki, K., Imamura, M., Shimada, Y., Ikenaga, I., and Tobe, T. (1991). Deletion of 17p and amplification of the *int-2* gene in esophageal carcinomas. *Cancer Res.* **51**, 2113–2117.

Warren, L., Buck, C. A., and Tuszynski, G. P. (1978). Glycopeptide changes and malignant transformation, a possible role for carbohydrate in malignant behavior. *Biochim. Biophys. Acta* **516**, 97–127.

Whang-Peng, J., Banks-Schlegel, S. P., and Lee, E. C. (1990). Cytogenetic studies of esophageal carcinoma cell lines. *Cancer Genet. Cytogenet.* **45**, 101–120.

Yamamoto, T., Kamata, N., Kawano, H., Shimizu, S., Kuroki, T., Toyoshima, K., Rikimaru, K., Nomura, N., Ishizaki, R., Pastan, I., Gamou, S., and Shimizu, N. (1986). High incidence of amplification of the epidermal growth factor receptor gene in human squamous carcinoma cell lines. *Cancer Res.* **46**, 414–416.

Yamashita, Y., Hirai, T., Mukaida, H., Kawano, K., Toge, T., Niimoto, M., and Hattori, T. (1989). Detection of androgen receptors in human esophageal cancer. *Jpn. J. Surg.* **19**, 195–202.

Yoshida, K., Kyo, E., Tsuda, T., Tsujino, T., Ito, M., Niimoto, M., and Tahara, E. (1990). EGF and TGF-α, the ligands of hyperproduced EFGR in human esophageal carcinoma cells, act as autocrine growth factors. *Int. J. Cancer* **45**, 131–135.

Gastric Tumor Cell Lines

M. Sekiguchi
Department of Surgery I, Saitama Medical School
Saitama 350-04, Japan

T. Suzuki
Department of Pathology II, Fukushima Medical College
Fukushima 950-12, Japan

I. Introduction 287

II. Methods of Establishment
and Maintenance 288

III. Morphology 292
A. General Morphological
Features 292
B. Morphology of Representative Cell
Lines 293

IV. Other Characteristics 309

V. Future Prospectives 312
A. Need for Culture of Normal
Gastrointestinal Epithelial
Cells 312
B. Development of Serum-Free Culture
Media 312
C. Oncogenes in Gastric Carcinoma
Cell Lines 313

References 314

I. Introduction

Gastric carcinoma is the most prevalent form of cancer in east Asia, including Korea and Japan (where the disease has been the leading cause of death) and is moderately prevalent in certain countries in Europe [Segi, 1981; Ministry of Health and Social Affairs, Korea (MHSAK), 1987]. The cure rate is still relatively low, despite the extensive use of radical surgery combined with other modes of treatment, although the mortality rate fortunately has been declining gradually of late (Davis *et al.*, 1990).

Cultured cell lines derived from human gastric carcinoma may provide very useful tools with which to study the cell biology and histogenesis of this disease, which shows one of the most variable histological pictures among carcinomas, and to develop new therapeutic approaches to the disease, which has considerable cellular heterogeneity. In general, gastric carcinoma has been more difficult to establish in long-term culture (Sekiguchi and Suzuki,

1980; Sekiguchi, 1989) than esophageal carcinoma (Sekiguchi, 1989) or colorectal carcinoma (Park et al., 1987). A modest number of gastric carcinoma cell lines has been established, primarily in east Asian countries (Sekiguchi and Suzuki, 1980; Lin et al., 1984; Sekiguchi, 1989; Park et al., 1990) and sporadically in Europe (Dobrynin, 1963; Wolff and Wolff, 1966; Maunoury, 1977; Laboisse et al., 1982; Dippold et al., 1987; Whelan et al., 1988) and in the United States (Barranco et al., 1983; Park et al., 1990). (See Table I for a list of reported and existing cell lines.)

Several reports (Ming et al., 1967; Sasano et al., 1969; Nevalainen and Järvi, 1977) on the histogenesis of human gastric cancer have indicated that two major types can be identified, according to cancer morphology and histochemistry. One type is tubular adenocarcinoma, well to moderately differentiated, which originates from intestinal metaplasia of gastric mucosa. The term intestinal metaplasia implies morphological and histochemical transformation of gastric mucosa to intestinal mucosa. Accordingly, absorptive cells, goblet cells, and Paneth cells can be seen in intestinal metaplastic glands. Ultrastructurally, well-developed microvilli (brush borders) on the luminal surfaces of absorptive cells as well as mucus in the goblet cells are observed. Many similarities are demonstrated between intestinal metaplasia and intestinal-type well-differentiated gastric carcinomas.

The other type is poorly differentiated adenocarcinoma which derives directly from gastric glands, although the frequency is far lower than that of intestinal-type carcinoma. This type of gastric carcinoma exhibits morphological and histochemical characteristics that are compatible with the gastric gland cells from which these carcinoma cells are derived. These features include parietal cells, mucus neck cells, foveolar cells, and pyloric gland cells.

Small subpopulations of gastric cancers demonstrate endocrine cell differentiation or lack of differentiation.

In this chapter, we describe the morphological and ultrastructural features of representative gastric cancer cell lines that have a wide variety of histological features, ranging from well-differentiated to poorly differentiated adenocarcinoma and signet ring cell carcinoma, and we correlate the in vivo growth pattern with histogenesis.

II. Methods of Establishment and Maintenance

Most of the continuous cell lines have been established from malignant ascitic or pleural effusions (Sekiguchi, 1989), because cancer cells in effusions are abundant, floating free, and primed for in vitro growth. However, effusions usually contain mesothelial cells, which may survive more than 1 yr in vitro and are sometimes indistinguishable from malignant cells (Whitehead and Hughes, 1975). Another group of cell lines has originated from metastatic

Table I

Reported and Existing Cell Lines

Cell line	Histological diagnosis	Source	Originator(s)	Date of initial culture	Chromosomes	Tumor markers and receptors[a]	Other features	Reference
KATO-III	Signet ring cell carcinoma	Pleural effusion	Sekiguchi, M.	Apr. 1974	mode 88	CEA	Plemorphic round cell, floating free	Sekiguchi et al. (1978)
HPE-GAC-T	Carcinoma simplex	Ascites	Morikawa, S.	Mar. 1974		CEA	Signet ring cell-like, floating free	Kanazawa et al. (1982)
HPE-GAC-2	Adenocarcinoma, partially poorly differentiated	Ascites	Morikawa, S.	Apr. 1975				Kanazawa et al. (1982)
HPE-GAC-3	Adenocarcinoma	Ascites	Morikawa, S.	Dec. 1975		CEA	Adherent, partially floating	Kanazawa et al. (1982)
HPE-GAC-4	Adenocarcinoma	Ascites	Morikawa, S.	Jul. 1978		CEA, EGFR	Floating free, partially adherent	Taniguchi et al. (1986)
HLN-GAC-5	Poorly differentiated adenocarcinoma	Lymph node metastasis	Morikawa, S.	Jul. 1987			Spindle cells, adherent	Morikawa et al. (1988)
MKN-1	Adenosquamous carcinoma	Primary tumor	Hojo, H., and Suzuki, T.	Jun. 1975	mode 39		Adherent	Hojo (1977)
MKN-7	Well-differentiated adenocarcinoma	Lymph node metastasis	Hojo, H., and Suzuki, T.	Jul. 1975	mode 88–91		c-erb B2 DNA amplification	Hojo (1977)
MKN-28	Well-differentiated adenocarcinoma	Lymph node metastasis	Hojo, H., and Suzuki, T.	Aug. 1975	59–80		Adherent	Hojo (1977)
MKN-45	Poorly differentiated adenocarcinoma	Liver metastasis	Hojo, H., and Suzuki, T.	Sep. 1975	mode 45–52	CEA	Adherent and floating	Hojo (1977)
MKN-74	Moderately differentiated adenocarcinoma	Liver metastasis	Hojo, H., and Suzuki, T.	Jul. 1976	mode 36		Adherent	Motoyama et al. (1986)
NC1-N87	Well-differentiated adenocarcinoma	Liver metastasis	Park, J.-G.	Aug. 1976	38–45 mode 43	CEA, CA19-9	Adherent	Park et al. (1990)

(continues)

Table I
Continued

Cell line	Histological diagnosis	Source	Originator(s)	Date of initial culture	Chromosomes	Tumor markers and receptors[a]	Other features	Reference
SNU-1	Poorly differentiated adenocarcinoma	Primary tumor	Park, J.-G.	Apr. 1984	mode 47	CEA, CA19-9	Adherent and floating	Park et al. (1990)
SNU-5	Poorly differentiated adenocarcinoma	Ascites	Park, J.-G.	Jun. 1987	85–98 mode 89	CEA, CA19-9	Adherent and floating	Park et al. (1990)
SNU-16	Poorly differentiated adenocarcinoma	Ascites	Park, J.-G.	Jul. 1987	81–94 mode 92	CEA	c-myc, c-erb B-2-RNA, adherent and floating	Park et al. (1990)
SGC-7901	Adenocarcinoma	Lymph node metastasis	Lin, C.-H.	Aug. 1979	mode 67		Adherent	Lin et al. (1984)
HGT-1	Poorly differentiated adenocarcinoma	Primary tumor	Laboisse, C. L.		55–60 mode 57	H_2-R	Adherent	Laboisse et al. (1982)
AGS	Moderately–poorly differentiated adenocarcinoma	Primary tumor	Barranco, S. C.		39–92 mode 47		Adherent	Barranco et al. (1983)
TE-7	Poorly differentiated adenocarcinoma	Primary tumor	Akaishi, T.	Aug. 1981				Akaishi et al. (1986)
HOGT	Tridermal mature teratoma	Primary tumor	Ishiwata, I.	Jun. 1982	mode 46	CEA, AFP	Spindle-shape, fibroblast-like	Ishiwata et al. (1985)
KWS	Poorly differentiated adenocarcinoma	Ascites	Sekiguchi, M.	Oct. 1982	mode 43		Adherent and floating	Sekiguchi (1989)
Takigawa	Moderately differentiated adenocarcinoma	Lymph node metastasis	Sekiguchi, M.	Dec. 1982			Adherent	Sekiguchi (1989)
TMK-1	Poorly differentiated adenocarcinoma	Lymph node metastasis	Ochiai, A.		mode 63–65	CEA, EGFR	Adherent, piling up	Ochiai et al. (1985)
HC-154	Poorly differentiated adenocarcinoma	Lymph node metastasis	Usugane, M.	Feb. 1983	mode 97	CEA	Cell aggregate	Usugane et al. (1986)

Cell line	Histology	Source	Investigator	Date	Mode	Markers	Morphology	Reference
MKO	Poorly differentiated adenocarcinoma	Primary tumor	Usugane, M.	May 1984	mode 99	CEA	Cell aggregate	Usugane et al. (1986)
KS-1	Poorly differentiated adenocarcinoma	Ovarian metastasis	Whelan, R.	Apr. 1984	48–77 mode 62–65		Pleomorphic round cell, floating free	Whelan et al. (1988)
MA IV	Signet ring cell carcinoma	Pleural effusion	Kuki, K.			CEA	Floating	Kuki et al. (1985)
IT-25	Well-differentiated adenocarcinoma	Primary tumor	Ide, K., and Iwamura, T.	Oct. 1985		CEA, CA19-9	Adherent	Ide et al. (1988)
MZ-sto-1	Moderately differentiated adenocarcinoma	Ascites	Dippold, W. G.		72–74	CEA, Lewis	Adherent	Dippold et al. (1987)
HuG-1	Moderately differentiated, adenocarcinoma	Ascites	Imanishi, H.	Oct. 1986	mode 60	CEA	Floating	Imanishi et al. (1988)
OCUM-1	Poorly differentiated adenocarcinoma	Pleural effusion	Kubo, T.	Sep. 1987	mode 50	CEA, CA19-9, Span-1	Lymphoid-like	Kubo (1991)
JR-St	Poorly differentiated adenocarcinoma	Spinal fluid	Terano, A.	Oct. 1987	mode 59	CEA, CA19-9, TPA, Procollagen III	Adherent	Terano et al (1991)
GCIV	Poorly differentiated adenocarcinoma	Ascites	Nosue, M.	Mar. 1988	mode 57	CEA, CA19-9, CA125	Adherent, mucin producing	Nosue et al. (1989)
Nakajima	Adenocarcinoma, scirrhous carcinoma	Pleural effusion	Yanoma, S.	May 1988		CA19-9	Adherent, piling up	Yanoma and Tsukuda (1989)
TSG-6	Signet ring cell carcinoma	Lymph node metastasis	Aizawa, K.	Nov. 1988			Floating	Aizawa et al. (1990)
KKLS	Poorly differentiated adenocarcinoma	Primary tumor	Sawaguchi, K.	Nov. 1988	mode 84		Adherent	Sawaguchi et al. (1989)
HNGA	Poorly differentiated adenocarcinoma	Ascites	Ishiwata, I., and Soma, M.	Dec. 1988		TPA	Adherent and floating	Ishiwata et al. (1989)

[a] Abbreviations: CEA, carcinoembryonic antigen; EGFR, epidermal growth factor receptor; AFP, α-fetoprotein; TPA, 12-O-tetradecanoyl phorbol-13-acetate.

tissues, predominantly regional lymph nodes (Sekiguchi, 1989). Epithelial types of cells growing out from lymph nodes can be identified as cancer cells because an epithelial element does not exist in lymphoid tissues. Cell lines also have been established from distant metastatic sites such as liver (Hojo, 1977; Park et al., 1990) or brain (Maunoury, 1977). The third and smallest group of cell lines has been isolated from primary tumors (Sekiguchi, 1989), which are the most difficult to cultivate. The pimary tumor tissue must be disaggregated using enzymes, which may damage harvested cells and reduce the viable yield. Moreover, contaminating stromal fibroblast-like cells often overgrow the cancer cells, particularly during the early phases of cultivation. In addition, a small number of cell lines has been derived from xenografts of gastric carcinoma established in athymic nude mice (Sekiguchi, 1989).

Usually, dispersed tumor cells are plated in plastic culture flasks with a basal medium such as RPMI 1640, Dulbecco's minimal essential medium (MEM), or Ham's F-10, supplemented with 10–20% heat-inactivated fetal bovine serum (FBS). Vigorously growing cultures are passaged by trypsinization or mechanical dispersion at appropriate intervals and split ratios. Only the KS-1 cell line was established by a soft agarose culture method (Whelan et al., 1988).

III. Morphology

A. General Morphological Features

The varying histological types of gastric carcinoma are closely related to histogenesis. Well-differentiated adenocarcinoma is thought to arise from foci of intestinal metaplasia; the cancer cells show features of intestinal epithelium by mucus stains, enzyme immunohistochemistry, and electron microscopy. Poorly differentiated adenocarcinoma, however, is believed to originate from the ordinary gastric mucosa; the cancer cells exhibit features of the gastric epithelium per se.

Morphological characteristics of cultured cells derived from well-differentiated adenocarcinoma include pavement-like arrangements of continuously expanding colonies and tubular structures with cellular polarity. The free surfaces of the cells have microvilli with core filaments resembling those of the intestinal mucosa and have well-developed junctional complexes. Cultured cells from poorly differentiated carcinomas proliferate in a scattered fashion in vitro and have ultrastructural features of gastric parietal cells with intracellular canaliculi. Occasionally, they also may express some features of intestinal epithelial cells. Cultured cells from signet ring cell carcinomas grow as free floating suspension cells. Ultrastructurally, these cells have microvilli with no central core filaments and mucous granules with highly electron-dense central cores. These properties are depicted in Fig. 1.

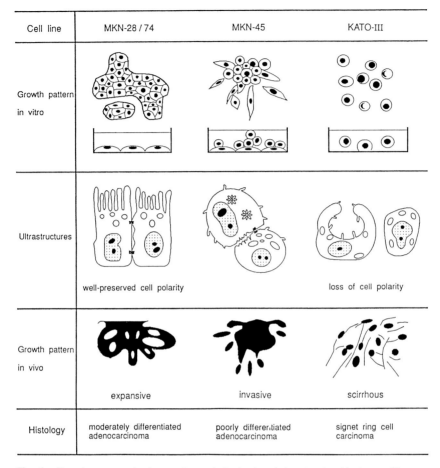

Cell line	MKN-28 / 74	MKN-45	KATO-III
Growth pattern in vitro			
Ultrastructures	well-preserved cell polarity		loss of cell polarity
Growth pattern in vivo	expansive	invasive	scirrhous
Histology	moderately differentiated adenocarcinoma	poorly differentiated adenocarcinoma	signet ring cell carcinoma

Fig. 1. Growth patterns *in vitro*, and morphological and ultrastructural features of human gastric cancers and cell lines.

B. Morphology of Representative Cell Lines

1. MKN-28, MKN-74, and MKN-7

a. *Morphology.*   The MKN-28 and MKN-74 cell lines were derived from moderately differentiated tubular adenocarcinomas that had metastasized to a regional lymph node (MKN-28; Hojo, 1977) or to liver (MKN-74; Motoyama *et al.*, 1986). Both cell lines grow adherently in a pavement-like arrangement. Before confluency, the cells form anastomosing, thick trabeculae (Fig. 2). At conflluency, both cell lines form highly piled-up bridge-like structures that sometimes show short branchings (Fig. 3). Dome-like structures also are visible. At early passage, both cell lines were positive for periodic acid–Schiff

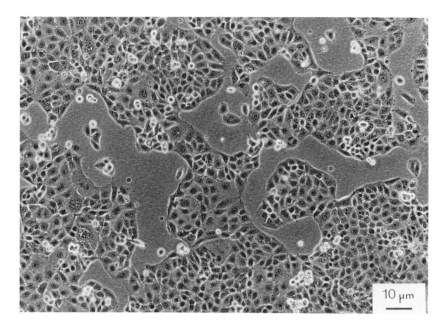

Fig. 2. Phase-contrast micrograph of MKN-28 cells reveals tightly adherent, cohesive growth forming cellular branching trabeculae.

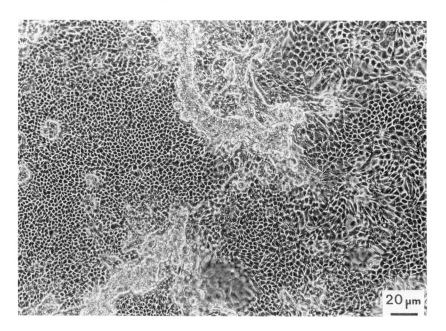

Fig. 3. In stationary phase, the MKN-28 cell line forms columnar piled-up structures with short branchings.

(PAS) and alcian blue staining, but at later passage they were negative for mucus staining. Moderate activity of alkaline phosphatase (the heat-labile type) has been detected in these two cell lines (Aizawa, 1988). Nude mouse tumors formed by these cell lines have histologies simiilar to those of their corresponding primary tumors. The xenografts show expansive growth and form tubules occasionally (Fig. 4). Intraperitoneal inoculation into nude mice also gives rise to moderately to poorly differentiated adenocarcinomas. Mucus production in these xenograft tumors is not histochemically evident.

The MKN-7 cell line, derived from a nodal metastasis of a well-differentiated tubular adenocarcinoma of the stomach, has an *in vitro* morphology similar to

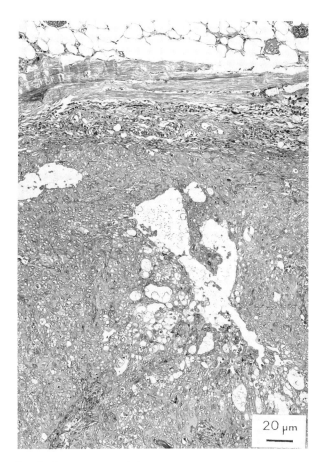

Fig. 4.   Subcutaneous MKN-28 tumor in an athymic nude mouse is an adenocarcinoma, mainly poorly differentiated and partly moderately differentiated, with an occasional tubular structure. Invasive growth is not observed.

that of the MKN-28 and -74 lines prior to confluency (Fig. 5) MKN-7 cells are flat and show pleomorphism during the exponential growth phase (Fig. 5). However, at confluency, MKN-7 cells become more uniform in shape but do not pile up (Fig. 6).

b. Ultrastructure. Scanning electron microscopy reveals rather flat cells with numerous microvilli on the cell surface and long thread-like cytoplasmic processes at the cell margins (Fig. 7). Adjacent cells are tightly attached (Fig. 8). Transmission electron microscopy reveals cells with moderately developed cell organelles and well-developed junctional complexes between neighboring cells. Microvilli with glycocalyx and rootlets were detected at early passage (Hojo, 1977), suggesting an origin from intestinal metaplastic cells of the stomach. At later passage, these specific microvilli are rare, but moderately developed microvilli can be observed even in intercellular spaces. Mucous granules, however, seldom are observed (Fig. 9). Vertical section of the mono-layered cells discloses filamentous bundles at the cell base and several annulate lamellae. A basement membrane is not formed (Fig. 10).

2. MKN-45

a. Morphology. The MKN-45 cell line was established from a poorly differentiated adenocarcinoma that was metastatic to the liver (Hojo, 1977). MKN-45 cells grow partly as adherent epithelial cells and partly as piled-up cells. The

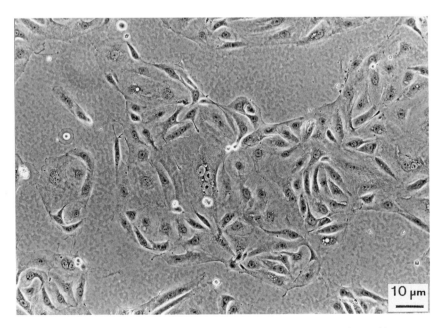

Fig. 5. Exponential growth phase of the MKN-7 cell line discloses the polymorphic appearances of the tumor cells.

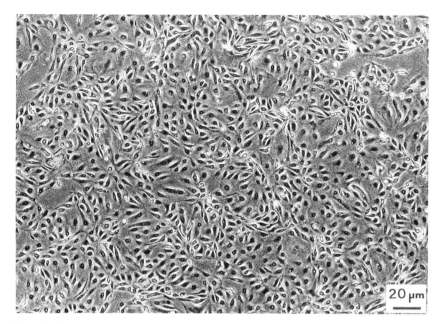

Fig. 6. Confluent MKN-7 cells are rather uniform in cell shape and more spindle shaped. The cells do not pile up, even during the growth phase.

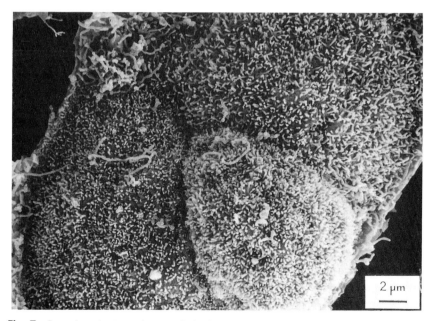

Fig. 7. Scanning electron micrograph depicts coherent tumor cells with numerous short microvilli on the cell surface and long, slender microvilli at the cell margins (MKN28).

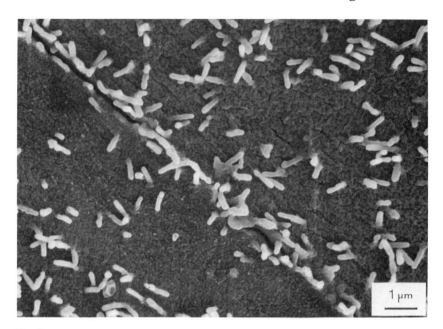

Fig. 8. Two adjacent flat cells face tightly at the cell margin, with partial overlapping of their cytoplasms (MKN-28).

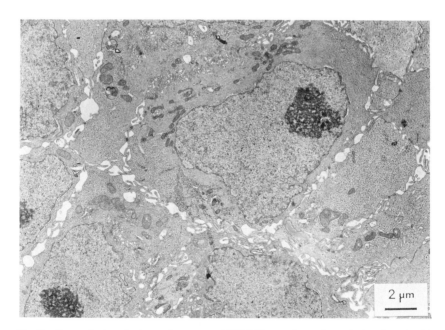

Fig. 9. Transmission electron microscopy of MKN-28 reveals moderately developed cell organelles and abundant free ribosomes. Narrow intercellular spaces contain varying numbers of microvilli.

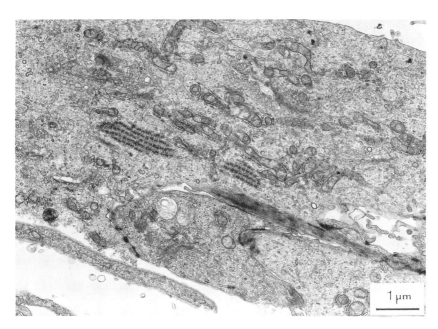

Fig. 10. Vertical section of an MKN-74 cell discloses well-developed filamentous bundles at the cell base and several annulate lamellae. Basal laminae are not observed.

cell arrangement is irregular, and growing cell margins are poorly defined (Fig. 11). Most cells are negative for mucus stains including PAS and alcian blue (Hojo, 1977). Moderate reactivity for alkaline phoshatase (the heat-labile type) is present in this cell line (Aizawa, 1988). Subcutaneous heterotransplantation into nude mice forms a tumor that frequently ulcerates. MKN-45 xenografts show invasive growth into the subcutaneous muscle of the host animal. The tumor histology is of a medullary, poorly differentiated adenocarcinoma without any tubular structures and mucus production (Fig. 12).

b. Ultrastucture. Scanning electron microscopy discloses round cells, as well as piled-up and adherent fusiform cells. Moderately developed microvilli on the cell surface and filopodia-like fine cytoplasmic processes are also present (Fig. 13). Several adhesions among cytoplasmic processes of the neighboring round cells are observed (Fig. 14). Transmission electron microscopy shows intracytoplasmic canaliculi (Fig. 15) that are characteristic of gastric parietal cells. These structures occasionally fuse together to form well-developed intercellular canals (Fig. 16). Accumulated microvilli at the apical poles and immature mucus granules also are encountered (Fig. 17).

3. KATO-III and TSG-6

a. Morphology. The KATO-III cell line was derived from a signet ring cell carcinoma (Fig. 18) and grows *in vitro* as free floating cells. The cells are rounded and highly pleomorphic (15–44 μm in diameter), and proliferate as

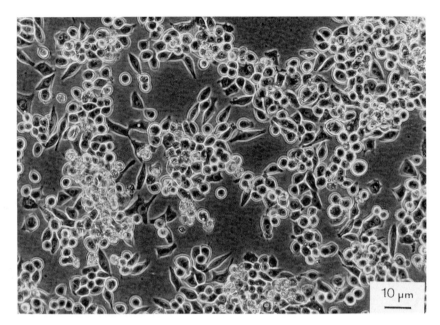

Fig. 11. Phase-contrast micrograph of the MKN-45 cell line demonstrates an irregular cellular arrangement, including focal piled-up and multidirectional cell growth.

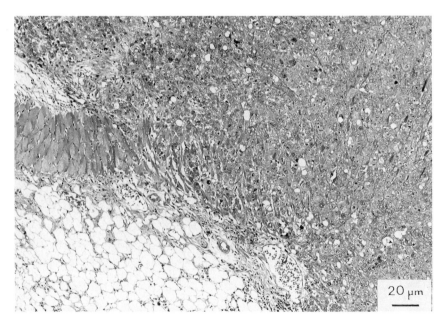

Fig. 12. Subcutaneous MKN-45 tumor in an athymic nude mouse, demonstrating an undifferentiated adenocarcinoma, vigorously invading the surrounding fat and muscle tissues.

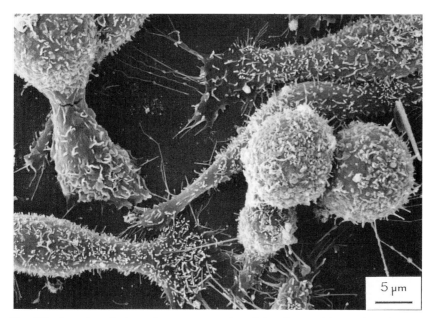

Fig. 13. Scanning electron microscopy reveals cellular pleomorphism and moderately developed microvilli on the cell surface. Filopodia-like fine cytoplasmic processes are also observed (MKN-45).

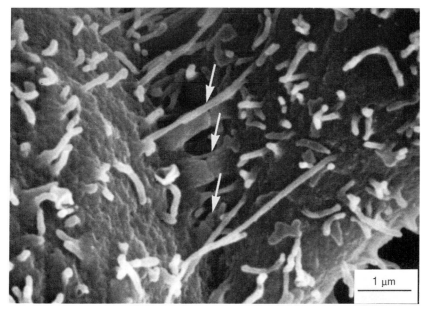

Fig. 14. Adhesions between cytoplasmic processes, suggestive of a desmosomal cell junction (MKN-45).

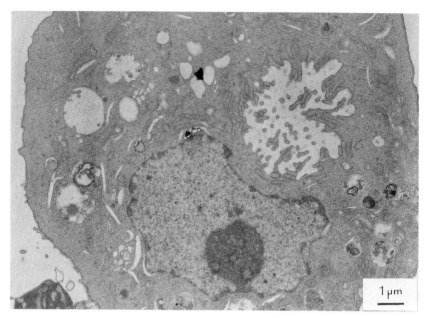

Fig. 15. An intracytoplasmic canaliculus in an MKN-45 cell.

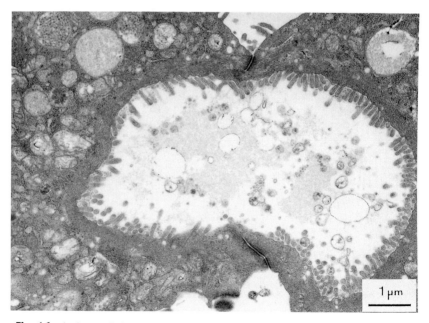

Fig. 16. An intercellular canal formed by fusion of intracellular canaliculi (MKN-45).

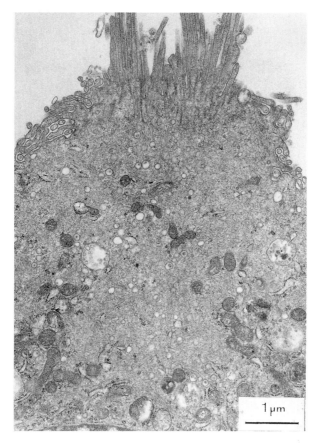

Fig. 17. A group of long microvilli at the apical portion and numerous small vesicles with or without cores in an MKN-45 cell.

separate cells that never aggregate or adhere to the substrate (Fig. 19). By phase contrast microscopy, intracellular structures cannot be discerned. Giemsa stain shows that most of the cells have relatively large round or oval nuclei with scant cytoplasm. Less than 10% of the cells have the eccentric nuclei characteristic of signet ring cells, which were observed primarily in the original primary tumor and the malignant effusion. PAS- and mucicarmine-positive granules are present in the cytoplasm. Production of alkaline phosphatase is not observed (Aizawa, 1988).

Xenogenic transplantation of the cells into the cheek pouches of Syrian golden hamsters that were immunosuppressed with antithymocyte serum developed tumors with an inoculum of 10^7 cells, but not of 2×10^6 cells. Histological appearance of the tumors was consistent with poorly differentiated

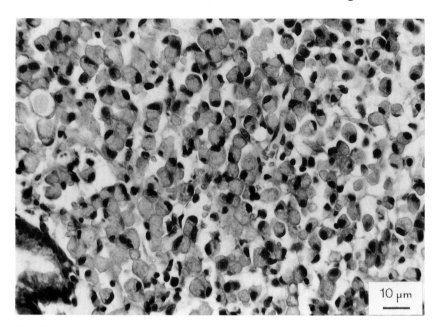

Fig. 18. Histology of the original signet ring cell carcinoma from which the KATO-III cell line was established.

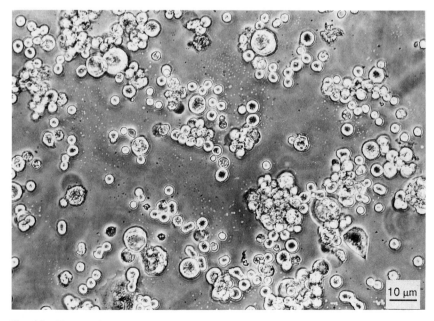

Fig. 19. Phase contrast micrograph of the KATO-III cell line. The cells are round, of various sizes, and grow freely in suspension.

adenocarcinoma with acinar structures, and differed from that of the original tumor (Sekiguchi *et al.*, 1978; Fig. 20).

The cells (10^7 cells) did not take in untreated athymic nude mice. However, cells (2×10^7 cells) transplanted in 3 of 16 nude mice irradiated with 5 Gy γ rays. The histological picture showed undifferentiated carcinoma without signet ring cells (Motoyama *et al.*, 1986). The differences in histological patterns are not contradictory, because a variety of histological patterns was present in the primary tumor.

As a floating cell line, TSG-6 was established from nodal metastasis of a scirrhous gastric carcinoma (Aizawa *et al.*, 1990). This line grows as floating single cells or cell clusters (Fig. 21). Signet ring-like structures are also apparent in some cells (Fig. 21). PAS and alcian blue stain and carcinoembryonic antigen (CEA) immunostaining are positive.

b. Ultrastructure.   Scanning electron microscopy of the KATO-III cells shows that the cultured cells are covered with numerous microvilli with rounded tips and additional thinner and shorter microvilli. Occasionally, the cells have cave-like depressions on the cell wall (Fig. 22). Such membranous depressions are sometimes irregular and may extend deeply and resemble

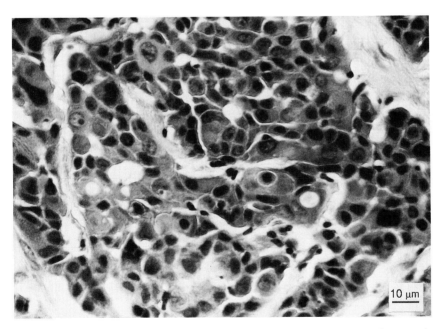

Fig. 20. A KATO-III tumor in the cheek pouch of a Syrian golden hamster is a poorly differentiated medullary carcinoma. (Reproduced from Sekiguchi *et al.*, 1978, with permission.)

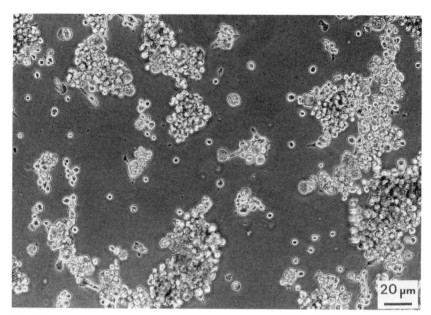

Fig. 21. The floating cell line TSG-6 grows as single cells and in cell clusters of varying size. Several signet ring-shaped tumor cells can be seen.

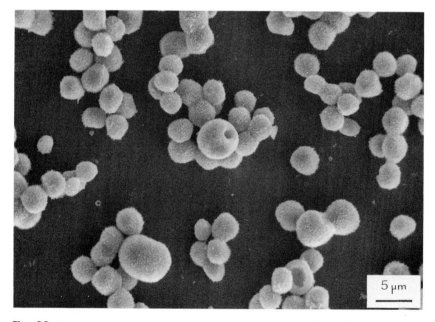

Fig. 22. Low power scanning electron microscopic view of KATO-III cells, showing rounded, variedly sized cells, loosely attached to each other and invested with numerous microvilli.

volcanos. Few of the cells have multiple depressions, both large and small. The concave surface also is covered with a few microvilli (Fig. 23).

Transmission electron microscopy reveals that the cells have nuclei of irregular shape as well as moderately developed intracellular organelles. Vacuoles containing an electron-lucent substance were observed occasionally in the cytoplasm. The mononucleated cells usually had a small number of microvilli and little glycocalyx. The large multinucleated cells, encountered rarely, contained well-developed microvilli. Cyst-like structures, found in the cytoplasm, also had microvilli on their inner surfaces (Fig. 24). Mucus granules with electron-dense cores, 0.5–1.0 μm in diameter, were found infrequently and resembled those of the gastric mucus neck cells or the pyloric glandular cells, but were more irregular in shape (Fig. 25). The ultrastructural findings, both scanning and transmission electron microscopically, suggest that the cyst-like organization generates the deep cavity or the depression of the cell surface. Cells with eccentric nuclei, observed by light microscopy, corresponded to those containing side-wall cavities by scanning electron microscopy.

4. MKN-1

a. Morphology. The MKN-1 cell line was established from a nodal metastasis of a gastric carcinoma of the adenosquamous type (Hojo, 1977). MKN-1 cells grow adherently without conspicuous piling up and adhere tightly to each

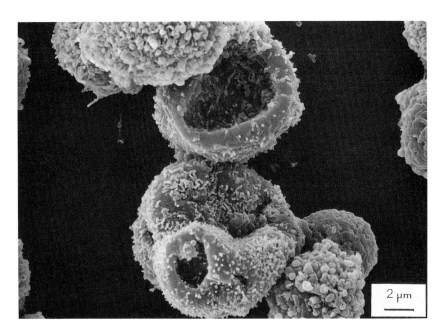

Fig. 23. Indented or dimpled cells of the KATO-III line, as seen by scanning electron microscopy.

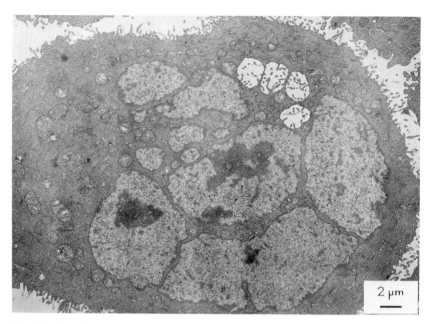

Fig. 24. A multinucleated giant cell of the KATO-III line has short microvilli. Small "cysts" with microvilli (which reflect cytoplasmic indentations) can be seen in the upper right corner.

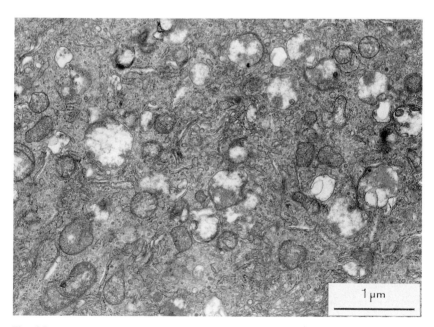

Fig. 25. Several round cytoplasmic structures resemble mucin granules of the gastric mucus neck cells (KATO-III).

other. Most of cells are distinctly polygonal and are intermingled with large round cells that have ample cytoplasm (Fig. 26). All cells are negative for mucus stains such as PAS and alcian blue (Hojo, 1977). High activity of alkaline phosphatase has been detected in this cell line (Aizawa, 1988). Subcutaneous heterotransplantation of MKN-1 into athymic nude mice forms poorly differentiated squamous cell carcinomas (Fig. 27). Glandular structures are absent. These observations confirm that MKN-1 is a squamous cell carcinoma cell line.

 b. Ultrastructure.   Transmission electron microscopy reveals abundant microfilaments, as shown in Fig. 28. Desmosomes, and tight junctions are also well developed between cell membranes. Mucus granules and/or secretory granules are not observed.

IV. Other Characteristics

Most cultured cells of stomach cancer origin are tumorigenic in athymic nude mice, irrespective of their degree of differentiation (Hojo, 1977; Motoyama *et al.,* 1986; Sekiguchi, 1989; Park *et al.,* 1990). Of interest, however, is that cell lines derived from poorly differentiated adenocarcinoma and signet ring cell carcinoma are not always tumorigenic (Sekiguchi, 1989).

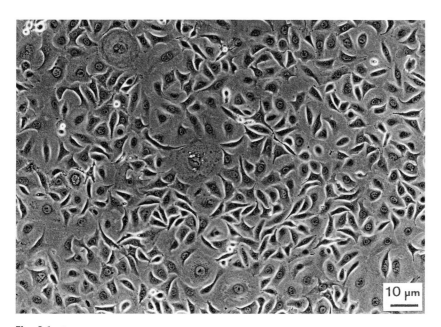

Fig. 26. Phase-contrast micrograph of the MKN-1 cell line reveals polygonal and large, round tumor cells tightly attached to each other.

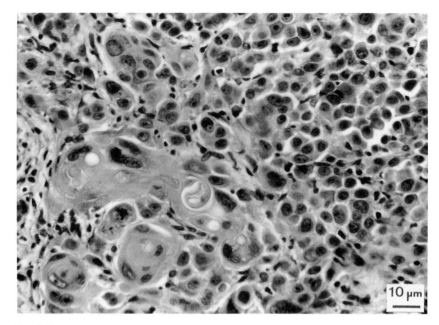

Fig. 27. Subcutaneous MKN-1 tumor in an athymic nude mouse is a squamous cell carcinoma of the poorly differentiated type, with no glandular structures.

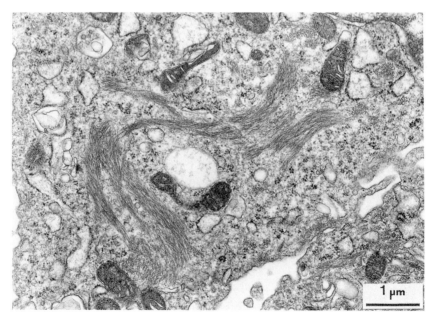

Fig. 28. Transmission electron micrograph of the MKN-1 cell line shows well-developed microfilamentous bundles in the cytoplasm.

Many of the gastric carcinoma cell lines produce glycoprotein antigens associated with the gastrointestinal epithelium, for example, CEA and CA19-9 (Laboisse et al., 1982; Motoyama and Watanabe, 1983; Dippold et al., 1987; Sekiguchi, 1989; Park et al., 1990).

Prominent collagen formation is a characteristic feature of some gastric carcinomas, especially of the scirrhous type, and usually is attributed to the accompanying stromal cells. However, production of collagen is not restricted to mesenchymal cells and may be a function of epithelial cells, as was discovered for hepatic parenchymal cells (Sakakibara et al., 1976). These authors demonstrated formation of collagen by two cloned gastric carcinoma cell lines (Sakakibara et al., 1982). Both MKN-28 and MKV-74 cloned lines synthesized type I collagen in vitro; the cancer cells stained with specific antibodies against this type of collagen in vitro and in xenografts. These findings strongly suggest an important role for gastric carcinoma cells in the formation of stromal collagen.

On the other hand, augmented production of collagen by fibroblasts cultured with gastric carcinoma cells was elucidated by direct co-cultivation and parabiotic cultivation (Naito et al., 1984) and by measurement of synthesized glycosaminoglycan and dermatan sulfate in vitro (Tsuji, 1982). A possible stimulating factor for fibroblasts from gastric cancer cells, in conjunction with the autoproduction mechanism, may account for the formation of stromal collagen in gastric cancer, especially in the scirrhous type of gastric cancer.

The expression of major histocompatibility complex (MHC) Class II antigens was examined with normal gastric epithelia, primary gastric carcinoma tissues, and gastric carcinoma cell lines (Sakai et al., 1987). DR and DP antigens but not DQ antigens were found in normal fundic epithelia; DR antigen was expressed in 11 of 15 gastric carcinoma specimens tested. Among 3 carcinoma cell lines, only the KATO-III cell line expressed DR antigen. However, on treatment with interferon γ (IFNγ), the KATO-III line enhanced expression of DR antigen and induced DQ and DP antigens and the MKN-45 line induced expression of DR antigen alone.

Proliferation and differentiation of a variety of carcinoma cells is known to be regulated in vitro and in vivo by epidermal growth factor (EGF). Sakai et al. (1986) reported that EGF receptor was expressed in more than 50% of gastric carcinomas, irrespective of histological types and positive lymph node metastases. Yasui et al. (1988) reported, however, that EGF was found in none of 26 early gastric carcinomas and in 38 of 130 advanced gastric carcinomas, and EGF receptor was expressed in 1 of the 26 early carcinomas and in 44 of the 130 advanced carcinomas. Moreover, patients whose tumors expressed EGF and its receptor were found to have a poorer prognosis than those lacking these features. Ochiai et al. (1988) detected EGF receptor in the TMK-1, MKN-1, MKN-7, MKN-28, and MKN-45 cell lines but not in the MKN-74 cell line. Growth and DNA synthesis in vitro of these cell lines expressing EGF receptor were stimulated by EGF, irrespective of histological type of the cell line.

Gastrin is a well-known gastrointestinal hormone that stimulates acid secretion from gastric parietal cells and has trophic effects on gastric mucosa. Weinstock and Baldwin (1988) reported that cultured human gastric cancer cells (Okajima, MKN-1, MKN-28, MKN-45, and MKN-74) had an ability to bind gastrin. This property is, however, common to other human carcinoma cell lines. Moyer et al. (1986) found that gastrin as well as glutamine stimulated growth of normal and malignant gastric mucosal cells in vitro. Somatostatin also stimulated the growth of gastric cancer cells, but inhibited the growth of normal gastric cells.

Prostaglandins (PGs) are known to have multifaceted biological functions. PGD inhibited proliferation of gastric carcinoma cell lines (KATO-III, MKN45, and HGC-11) in vitro (Sakai et al., 1984). PGE_2 and $PGF_{2\alpha}$ also affected growth of KATO-III cells in vitro by increasing the cellular level of cyclic AMP (Nakamura et al., 1989).

V. Future Prospectives

A. Need for Culture of Normal Gastrointestinal Epithelial Cells

Long-term cultivation of normal gastrointestinal epithelial cells has been almost unsuccessful to date. However, comparative studies with paired normal and malignant gastrointestinal epithelial cells, particularly from the same organ of the same donor, are strongly required for many aspects of cancer research (Moore, 1990). Moyer (1983) succeeded in cultivating normal gastrointestinal epithelial cells (the stomach, the small intestine, and the colon) for several months. The successful method consisted of mechanical harvesting of mucosa with a blunt-ended instrument, followed by mechanical dissociation of the tissue; use of conditioned medium and basal medium supplemented with many growth-promoting substances in addition to 1–10% FBS; and coating of substrates. Using modified Ham's F-12 medium, Terano et al. (1983) maintained normal human gastric mucosal cells obtained by endoscopic biopsy in vitro for 2 wk without overgrowth by fibroblasts. These pioneer studies will lead to development of new culture media and new culture methods suitable for normal gastrointestinal epithelial cells.

B. Development of Serum-Free Culture Media

Serum-free chemically defined synthetic media are very useful in analyzing low levels of biologically active substances synthesized or secreted by culture cells derived from human tumors, as well as in examining effects of certain substances on cultured cells (Katsuta and Takaoka, 1973; Yamaguchi et al., 1990). Many synthetic media with different compositions have been developed (Katsuta and Takaoka, 1973) and many cultured cells that had been established in serum-containing media have adapted successfully to serum-free chemically defined media. For example, Yamaguchi *et al.* (1990) successfully adapted 10 human pancreatic carcinoma cell lines to a serum-free chemically defined medium, examined morphological changes, and detected small quantities of biological substances secreted from the cell lines. Unfortuantely, such studies have not been successful with cultured cells of gastric carcinoma origin. Another advantage of serum-free chemically defined medium was recognized when it was used to establish cell lines derived from human neoplasms. Takahashi *et al.* (1982) developed a new serum-free chemically defined medium, RITC80-7, based on modified MEM and supplemented with EGF, insulin, transferrin, triiodothyronine, hydrocortisone, and 0.5% bovine serum albumin. These investigators successfully maintained 7 of 10 primary carcinoma cultures for 6–30 months. One of the cultures became a permanent cell line. The new medium suppressed overgrowth by fibroblasts in the early stage of the primary culture and promoted the desirable growth of cancer cells.

C. Oncogenes in Gastric Carcinoma Cell Lines

Chromosomal abnormalities are observed in every type of gastric carcinoma cell line without specific patterns (Motoyama *et al.,* 1986; Sekiguchi, 1989). These changes already have occurrerd *in vivo* during tumor initiation and progression, and are accelerated further by long-term cultivation *in vitro* (Hojo, 1977). In addition, Park *et al.* (1990) reported a high incidence of double minute chromosomes. Karyotype analysis of cloned KATO-III cells reveals a homogeneously staining region (HSR) in chromosome 10, suggesting that amplified DNA segments are present in the cells (Matsuzawa *et al.,* 1988). In fact, Nakatani *et al.* (1986) demonstrated DNA fragments that were amplified several hundred times in the KATO-III and MKN-7 cell lines by a DNA renaturation method. In this study, an HSR was detected on an abnormal chromosome 11 and was thought to contain the amplified DNA fragments. The researchers later identified a new gene, designated K-*sam,* in these fragments. This gene, having tyrosine kinase activity, was found to be expressed frequently in poorly differentiated gastric carcinomas but not in well-differentiated ones (Yoshida *et al.,* 1989). Conversely, c-*erb*B2 generally was amplified in the well-differentiated tumors (Hattori *et al.,* 1990), suggesting that the degree of differentiation in gastric carcinomas may reflect the genetic background of the tumor. Of interest, K-*sam* also is amplified in the SNU-16 cell line Park *et al.,* 1990; Mor *et al.,* 1991), but, in this line, the amplified sequences are located on double minute chromosomes that also contain amplified c-*myc* sequences (Bar-Am *et al.,* 1992).

Oncogenes such as c-*myc* (Koda *et al.,* 1985), c-*yes*-1 (Seki *et al.,* 1985), and c-*erb*B2 (Yokota *et al.,* 1986) were reported to be amplified in primary gastric cancers., Studies of the relationship between oncogenes and gastric cancer are still in the initial stages and require further research.

References

Aizawa, K. (1988). Heat-stable alkaline phosphatase in cultured human gastric cancer cells. *Niigata-Igakkai-Zasshi* **102**, 8–24 (*in Japanese*).

Aizawa, K., Muto, I., Katayanagi, N., Suzuki, T., Tanaka, O., and Muto, T. (1990). A newly established cell line from scirrhous gastric carcinoma (TSG6): Its characteristics and sensitivity to various anticancer agents. *Proc. Jpn. Cancer Assoc.* **49**, 220 (*in Japanese*).

Akaishi, T., *et al.* (1986). In "Esophageal Cancer" (M. Kasai, ed.), p. 35. Excerpta Medica, Tokyo.

Bar-Am, I., Mor, O., Yeger, H., Shiloh, Y., and Avivi, L. (1992). Detection of amplified DNA sequences in human tumor cell lines by fluorescence *in situ* hybridization. *Genes Chrom. Cancer* **4**, 314–320.

Barranco, S. C., Townsend, C. M., Jr., Casartelli, C., Macik, B. G., Broerwinkle, W. R., and Gourley, W. K. (1983). Establishment and characterization of an *in vitro* model system for human adenocarcinoma of the stomach. *Cancer Res.* **43**, 1703–1709.

Davis, O. L., Hoel, D., Fox, J., and Lopez, A. (1990). International trends in cancer mortality in France, West Germany, Italy, Japan, England and Wales, and the United States. *Ann. N.Y. Acad. Sci.* **609**, 5–48.

Dippold, W. G., *et al.* (1987). Signet ring stomach cancer: Morphological characterization and antigenic profile of a newly established cell line (Mz-Sto-1). *Eur. J. Cancer Clin. Oncol.* **23**, 697–706.

Dobrynin, Y. V. (1963). Establishment and characteristics of cell strains from some epithelial tumors of human origin. *J. Natl. Cancer Inst.* **31**, 1193–1195.

Hojo, H. (1977). Establishment of cultured cell lines of human stomach cancer origin and their morphological characteristics. *Niigata-Igakkai-Zasshi* **91**, 737–752 (*in Japanese*).

Ide, K., *et al.* (1988). Establishment and characterization of human colonic and gastric cancer cell lines. *Hum. Cell* (*Tokyo*) **1**, 245–249 (*in Japanese*).

Imanishi, H., *et al.* (1988). Effect of sodium butyrate on morphological features and several tumor markers (alkaline phosphatase, CA19-9, TPA, CEA) in a human gastric cancer cell line (HuG-1). *Proc. Jpn. Cancer Assoc.* **47**, 396.

Ishiwata, I., *et al.* (1985). Establishment and characterization of a new clonal strain from human benign gastric teratoma. *Exp. Pathol.* **27**, 143.

Ishiwata, I., and Soma, M. (1989). A cultured human gastric carcinoma cell line, HNGA. *Hum. Cell* (*Tokyo*) **2**, 335 (*in Japanese*).

Kanazawa, S., *et al.* (1982). Carcinoembryonic antigen producing cultured cell lines enable detection of autoantibodies in sera from patients with gastrointestinal cancer. *Cancer* **50**, 107.

Katsuta, H., and Takaoka, T. (1973). Cultivation of cells in protein- and lipid-free synthetic media. In "Methods in Cell Biology" (D. M. Prescott, ed.), Vol. 6, pp. 1–42. Academic Press, New York.

Koda, T., Matsushima, S., Sasaki, A., Danjo, Y., and Kakinuma, M. (1985). c-*myc* gene amplification in primary stomach cancer. *Jpn. J. Cancer Res. (Gann)* **76**, 551–554.

Kubo, T. (1991). Establishment and characterization of a new gastric cancer cell line (OCUM-1), derived from Borrmann type IV tumor. *J. Jpn. Surg. Soc.* **92**, 1451–1460 (*in Japanese*).

Kuki, K., *et al.* (1985). Growth appearance of cancer cells derived from human gastric carcinoma with special reference to morphological features in vitro. *Nihon-Ikadaigaku-Zasshi* **52**, 485–492 (*in Japanese*).

Laboisse, C. L., Augeron, C., Couturier-Turpin, M.-H., Gespach, C., Cheret, A.-M., and Potet, F. (1982). Characterization of a newly established human gastric cancer cell line HGT-1 bearing histamine H_2-receptors. *Cancer Res.* **42**, 1541–1548.

Lin, C.-H., Fu, Z.-M., Liu, Y.-L., Yang, J.-L., Xu, J.-F., Chen, Q.-S., and Chen, H.-M. (1984). Investigation of SGC-7901 cell line established from human gastric carcinoma cells. *Chin. Med. J.* **97**, 831–834.

Matsuzawa, Y., Hayata, I., Ichikawa, T., Ichikawa, Y., and Toida, T. (1988). Cytogenetic study on a subline of KATO-III with HSR, a cell line derived from human gastric carcinoma. *Proc. Jpn. Acad.* **64B**, 57–60.

Maunoury, R. (1977). Establissement et caractérisation de 3 lignées cellulaires humaines derivées de tumeurs métastatiques intracélébrales. *C. R. Acad. Sci. Paris* **D284,** 991–994 (*in French*).

Min, S.-C., Goldman, H., and Freiman, D. G. (1967). Intestinal metaplasia and histogenesis of carcinoma in human stomach. Light and electron microscopic study. *Cancer* **20,** 1418–1429.

Ministry of Health and Social Affairs (1987). One year's report for cancer registry programme in the Republic of Korea July 1, 1985–June 30, 1986. *J. Korean Cancer Assoc.* **19,** 131–258 (*in Korean*).

Moore, G. E. (1990). The need for a human cell organization. *Cancer* **66,** 617–618.

Mor, O., *et al.* (1991). Novel DNA sequences at chromosome 10q26 are amplified in human gastric carcinoma cell lines: Molecular cloning by competitive DNA reassociation. *Nucleic Acids Res.* **19,** 117–123.

Morikawa, S., *et al.* (1988). Establishment of a novel human gastric cancer cell line expressing both cytokeratin and vimentin in their cytoskeleton. *Proc. Jpn. Cancer, Assoc.* **47,** 294 (*in Japanese*).

Motoyama, T., and Watanabe, H. (1983). Carcinoembryonic antigen production in human gastric cancer cell lines *in vitro* and in nude mice. *Gann* **74,** 679–686.

Motoyama, T., Hojo, H., and Watanabe, H. (1986). Comparison of cell lines derived from human gastric carcinomas. *Acta Pathol. Jpn.* **36,** 65–83.

Moyer, M. P. (1983). Culture of human gastrointestinal epithelial cells. *Proc. Soc. Exp. Biol. Med.* **174,** 12–15.

Moyer, M. P., Armstrong, A., Aust, J. B., Levine, B. A., and Sirinek, K. R. (1986). Effect of gastrin, glutamine, and somatostatin on the *in vitro* growth of normal and malignant human gastric mucosal cells. *Arch. Surg.* **121,** 285–288.

Naito, Y., Kino, I., Horiuchi, K., and Fujimoto, D. (1984). Promotion of collagen production by human fibroblasts with gastric cancer cells *in vitro*. *Virchows Arch. Cell Pathol.* **46,** 145–154.

Nakamura, A., Chiba, T., Yamatani, T., Yamaguchi, A., Inui, T., Morishita, T., Kadowaki, S., and Fujita, T. (1989). Prostglandin E_2 and $F_{2\alpha}$ inhibit growth of human gastric carcinoma cell line KATO-III with simultaneous stimulation of cyclic AMP production. *Life Sci.* **44,** 75–80.

Nakatani, H., *et al.* (1986). Detection of amplified DNA sequences in gastric cancer by a DNA renaturation method in gel. *Jpn. J. Cancer Res. (Gann)* **77,** 849–853.

Nevalainen, T. J., and Järvi, O. (1977). Ultrastructure of intestinal and diffuse type gastric carcinoma. *J. Pathol.* **122,** 129–136.

Nosue, M., *et al.* (1989). Establishment and characterization of the gastric cancer cell line (GCIY) producing CA19-9. *Hum. Cell (Tokyo) (Suppl.)* **2(3),** 89–90 (*in Japanese*).

Ochiai, A., *et al.* (1985). Growth promoting effect of gastrin in human gastric carcinoma cell line TMK-1. *Jpn. J. Cancer Res (Gann)* **76,** 1064–1071.

Ochiai, A., Takanashi, A., Takekura, N., Yoshida, K., Miyamori, S., Harada, T., and Tahara, E. (1988). Effect of human epidermal growth factor on cell growth and its receptor in human gastric carcinoma cell lines. *Jpn. J. Clin. Oncol.* **18,** 15–25.

Park, J.-G., Oie, H. K., Sugarbaker, P. H., Henslee, J. G., Chen, T.-R., Johnson, B. E., and Gazdar, A. F. (1987). Characteristics of cell lines established from human colorectal carcinoma. *Cancer Res.* **47,** 6710–6718.

Park, J.-G., *et al.* (1990). Characteristics of cell lines established from human gastric carcinoma. *Cancer Res.* **50,** 2773–2780.

Sakai, K., Mori, S., Kamamoto, T., Taniguchi, S., Kobori, O., Morioka, Y., Kuroki, T., and Kano, K. (1986). Expression of epidermal growth factor receptors on normal human gastric epithelia and gastric carcinoma. *J. Natl. Cancer Inst.* **77,** 477–483.

Sakai, K., Takiguchi, M., Mori, S., Kobori, O., Inoko, H., Sekiguchi, M., and Kano, K. (1987). Expression and function of class II antigens on gastric carcinoma cells and gastric epithelia: Differential expression of DR, DQ and DP antigens. *J. Natl. Cancer Inst.* **79,** 923–932.

Sakai, T., Yamaguchi, N., Shiroko, Y., Sekiguchi, M., Fujii, G., and Nishino, H. (1984). Prostaglandin D_2 inhibits the proliferation of human malignant tumor cells. *Prostaglandins* **27,** 17–26.

Sakakibara, K., Saito, M., Umeda, K., and Tsukada, Y. (1976). Native collagen formation by liver parenchymal cells in culture. *Nature (London)* **262,** 316–318.

Sakakibara, K., *et al.* (1982). Biosynthesis of an interstitial type of collagen by cloned human gastric carcinoma cells. *Cancer Res.* **42**, 2019–2027.

Sasno, N., Nakamura, K., Arai, M., and Akazaki, K. (1969). Ultrastructural cell patterns in human gastric carcinoma compared with non-neoplastic gastric mucosa-histogenetic analysis fo carcinoma by mucin histochemistry. *J. Natl. Cancer Inst.* **43**, 783–802.

Sawaguchi, K., *et al.* (1989). Establishment and characterization of human gastric cancer cell line (KKLS). *Proc. Jpn. Cancer Assoc.* **48**, 178 (*in Japanese*).

Segi, M., Aoki, K., and Kurihara, M. (1981). World cancer mortality. *Gann Monogr.* **26**, 121–250.

Seki, T., Fujii, G., Mori, S., Tamaoki, N., and Shibuya, M. (1985). Amplification of c-*yes*-1 proto-oncogene in a primary human gastric cancer. *Jpn. J. Cancer Res. (Gann)* **76**, 907–910.

Sekiguchi, M. (1989). A list of cultured human cell lines derived from the gastrointestinal tracts. *Hum. Cell (Tokyo)* **2**, 310–340, 439–463 (*in Japanese*).

Sekiguchi, M., and Suzuki, T. (1980). List of human cell lines established and maintained in Japan. *Tissue Cult. (Tokyo)* **6**, 527–548 (*in Japanese*).

Sekiguchi, M., Sakakibara, K., and Fujii, G. (1978). Establishment of cultured cell lines derived from a human gastric carcinoma. *Jpn. J. Exp. Med.* **48**, 61–68.

Takahashi, M., Matsuno, S., Sato, T., Saze, K., Kudo, T., and Tachibana, T. (1982). Efficacy of a serum-free medium (RITC80-7), for the establishment of human cancer cell lines. *Kosankinbyo-kenkyujo-Zasshi* **39**, 217–227 (*in Japanese*).

Taniguchi, M., *et al.* (1986). Epidermal growth factor-sensitive human gastric carcinoma cell lines. *Proc. Jpn. Cancer Assoc.* **45**, 198 (*in Japanese*).

Terano, A., Mach, T., Stachura, J., Sekhon, S., Tarnawski, A., and Ivey, K. J. (1983). A monolayer culture of human gastric epithelial cells. *Digest. Dis. Sci.* 28, 595–603.

Terano, A., *et al.* (1989). Characterization of a newly established cell line (JR-St) derived from gastric signet ring cell cancer, producing tumor markers. *Gastroenterol. Jpn.* **26**, 7–13.

Tsuji, K. (1982). Studies on the characteristic changes in human fibroblasts induced by the proliferation of gastric cancer cells *in vitro*. *Jpn. J. Gastroenterol.* **79**, 931–939 (*in Japanese*).

Usugane, M., *et al.* (1986). Establishment of human gastric, colonic and pancreatic cancer cell lines. *Proc. Jpn. Cancer Assoc.* **45**, 240 (*in Japanese*).

Weinstock, J., and Baldwin, G. S. (1988). Binding of gastrin$_{17}$ to human gastric cell lines. *Cancer Res.* **48**, 932–937.

Whelan, R., Gibby, E., Sheer, D., Povey, S., and Hill, B. T. (1988). Characterization of a continuous cell line in culture established from a Krukenberg tumour of the ovary arising from a primary gastric adenocarcinoma. *Eur. J. Cancer Clin. Oncol.* **24**, 1397–1408.

Whitehead, R. H., and Hughes, L. E. (1975). Tissue culture studies of malignant effusions. *Br. J. Cancer* **3**, 512–518.

Wolff, E. T., and Wolff, E. M. (1966). Cultures organotypiques de longue durée de deux tumeurs humaines du tube digestif. *Eur. J. Cancer* **2**, 93–103 (*in French*).

Yamaguchi, N., Yamamura, Y., Koyama, K., Ohtsuji, E., Imanishi, J., and Ashihara, T. (1990). Characterization of new human pancreatic cell lines which propagate in a protein-free chemically defined medium. *Cancer Res.* **50**, 7008–7014.

Yanoma, S., and Tsukuda, M. (1989). Establishment and characterization of CA19-9 producing human gastric cancer cell line (Nakajima). *Hum. Cell (Tokyo) (Suppl.)* **2(3)**, 93 (*in Japanese*).

Yasui, W., Hata, J., Yokozaki, H., Nakatani, H., Ochiai, A., Ito, H., and Tahara, E. (1988). Interaction between epidermal growth factor and its receptor in progression of human gastric carcinoma. *Int. J. Cancer* **41**, 211–217.

Yokota, J., *et al.* (1986). Amplification of c-*erb* B-2 oncogene in human adenocarcinoma *in vivo*. *Lancet* **i**, 765–766.

Yoshida, T., Hattori, Y., Sakamoto, H., Odagiri, H., Miyagawa, K., Nakatani, H., Tahara, E., Sugimura, T., and Terada, M. (1989). Amplified gene in gastric cancer, *sam*, codes for a novel transforming receptor kinase. *Proc. Am. Assoc. Cancer Res.* **30**, 783.

Colorectal Cancer Cell Lines

12

Jae-Gahb Park and Han-Kwang Yang
Korean Cell Line Bank, Cancer Research Institute
Seoul National University College of Medicine
Seoul 110-744, Korea

Robert J. Hay
Cell Culture Department, American Type Culture Collection
Rockville, Maryland 20852

Adi Gazdar
Department of Pathology, Simmons Cancer Center
Southwestern Medical College, Dallas, Texas 75235

I. Introduction 317

II. Establishment of Cell Lines 320

III. Culture Characteristics 324

IV. Morphological Characteristics 325

A. Well Differentiated 328
B. Moderately Differentiated 328
C. Poorly Differentiated 330
D. Mucinous 331

V. Other Characteristics 334

References 338

I. Introduction

Colorectal cancer is one of the most common solid tumors; its incidence is second only to lung cancer in the United States. An estimated 155,000 new cases will be diagnosed in 1990; of the afflicted, 60,900 will die of their disease (Silverberg *et al.*, 1990). Cell lines established from human colorectal carcinomas may provide useful tools to study the biology of this disease and to develop and test new therapeutic approaches.

Any in-depth study requires a comprehensive bank of well-characterized cell lines. Although over 80 lines exist (Leibovitz *et al.*, 1976; Brattain *et al.*, 1982; McBain *et al.*, 1984; Park *et al.*, 1987; Table I), these lines represent the

Table I

Reported Cell Lines from Primary and Metastatic Human Colorectal
Adenocarcinoma and Adenoma[a]

Cell line	Origin	Reference
HCT116	Colon	Brattain *et al.* (1982,1983)
HCT116a	Colon	
HCT116b	Colon	
HCTC	Colon	
HCTC Col	Colon	
Co-115	Colon	Carrel *et al.* (1979)
DLD-1	Colon	Dexter et al. (1979,1981)
HCT-15	Colon	
LoVo	Supraclavicular node	Drewinko *et al.* (1976)
C168	Colon	Durrant *et al.* (1986,1987)
C168Node	Liver metastasis	
C170	Colon	
C146	Colon	
SK-CO-1	Ascites	Fogh and Trempe (1975)
Caco-2	Colon	Fogh et al. (1977)
HT-29	Colon	
GW39	Colon	Goldenberg *et al.* (1966)
HCA-2	Colon	Kirkland and Bailey (1986)
HCA-7	Colon	
HCA-24	Colon	
HCA-46	Colon	
HRA-16	Rectum	
HRA-19	Rectum	
SW-48	Colon	Leibovitz *et al.* (1976).
SW-403	Colon	
SW-480	Colon	
SW-620	Lymph node	
SW-707	Rectum	
SW-742	Colon	
SW-802	Colon	
SW-837	Rectum	
SW-948	Colon	
SW-1083	Rectum	
SW-1116	Colon	
SW-1222	Colon	
SW-1345	Rectum	
SW-1398	Colon	
SW-1417	Colon	
SW-1463	Rectum	
VACO 1	Liver metastasis	McBain *et al.* (1984)
VACO 3	Mesentery node	
VACO 4A	Rectum	
VACO 4S	Rectum	
VACO 9P	Rectum	
VACO 9M	Liver metastasis	

Table I

Continued

Cell line	Origin	Reference
VACO 10MS	Omentum metastasis	
VACO 5	Cecum	
VACO 5A	Cecum	
VACO 6	Cecum	
VACO 8	Cecum	
VACO 10P	Cecum	
WiDr	Colon	Noguchi *et al.* (1986)
NCI-H498	Peritoneum (cecum)	Park *et al.* (1987)
NCI-H508	Abdominal wall (cecum)	
NCI-H548	Colon	
NCI-H630	Liver metastasis (rectum)	
NCI-H684	Liver metastasis (colon)	
NCI-H716	Ascites (cecum)	
NCI-H742	Colon	
NCI-H747	Node metastasis (colon)	
NCI-H768	Cecum	
NCI-H958	Liver metastasis (colon)	
SNU-C1	Peritoneum (colon)	
SNU-C2A	Xenograft (cecum)	
SNU-C4	Xenograft (colon)	
SNU-C5	Xenograft (cecum)	
COLO320	Colon	Quinn *et al.* (1979)
COLO320DM	Colon	
COLO320HSR	Colon	
COLO321	Colon	
T84	Lung metastasis	Reid *et al.* (1978)
LS123	Colon	Rutzky *et al.* (1983)
COLO 201	Ascites	Semple *et al.* (1978)
COLO 205	Ascites	
COLO 206	Ascites	
LS174T	Colon	Tom *et al.* (1976)
LS180	Colon	
HCT-8	Cecum	Tompkins *et al.* (1974)
HRT-18	Rectum	
HT-55	Rectum	Watkins and Sanger (1977)
LIM1215	Omentum	Whitehead *et al.* (1985,1987)
LIM1863	Cecum	
VACO 206	Colon	Wilson *et al.* (1987)
VACO 241	Liver metastasis	
VACO 235	Colon, villous polyp	
VACO 330	Colon, villous polyp	
HuCCL-14	Colon	Yaniv *et al.* (1978)

[a] See Fig. 1 and Hay *et al.* (1992) for additional details.

culmination of numerous attempts over many years in several different laboratories. Whether these lines constitute a representative bank or whether they form a minor subgroup is unknown. A large bank of well-characterized cell lines may reflect the diversity of tumor phenotypes and may provide adequate models for studying tumor heterogeneity. (See Fig. 1 for morphologies of six reference cell lines.)

II. Establishment of Cell Lines

Some human tumors, such as melanomas, are relatively easy to culture. Others, such as pheochromocytomas and islet cell carcinomas, are exceedingly difficult to culture; no established cell lines currently exist for these tumors. Colorectal carcinoma culture appears to be of intermediate difficulty. A variety of culture techniques has been employed including the use of feeder layers (Brattain *et al.*, 1982), enriched serum-containing media (McBain *et al.*, 1984), and collagen gels (Wilson, 1987).

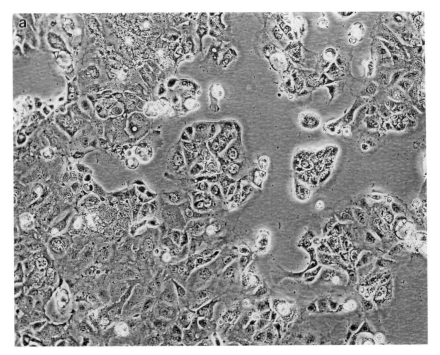

Fig. 1. Typical morphologies of six reference colon cancer lines in wide use. (a) Caco-2 (ATCC.HTB 37) was derived from the primary tumor of a 72-year-old white male. This line forms a monolayer of squamous epithelia.

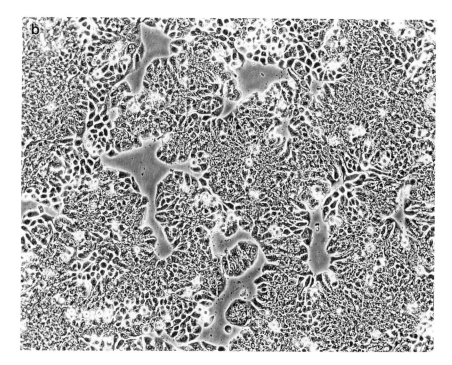

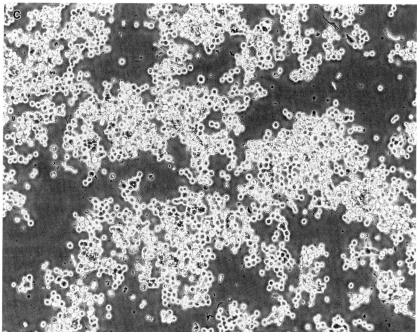

Fig. 1. *Continued* (b) HT-29 (ATCC.HTB 38) is also from a primary adenocarcinoma of a 44-year-old white female. Tightly packed cuboidal epithelia are apparent in monolayer, with groups of highly refractile cells present especially at the periphery of the expanding cell sheet. (c) COLO205 (ATCC.CCL 222) was developed from ascitic fluid from a 70-year-old white male with carcinoma of the colon. Epithelial-like cells are apparent that proliferate in large, undifferentiated clusters loosely attached to the substrate.

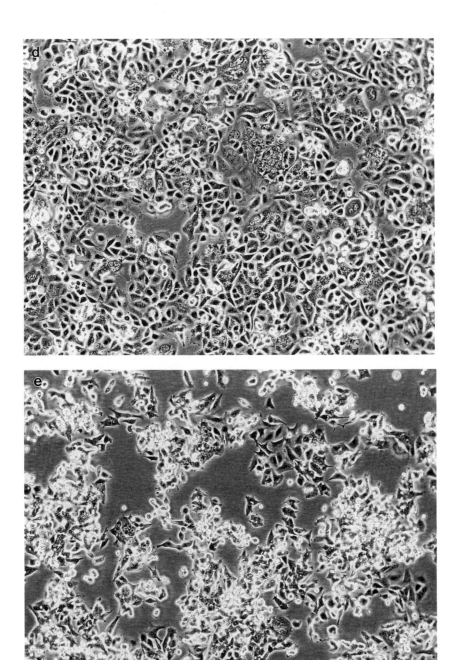

Fig. 1. *Continued* (d) SW480 (CCL 228) was isolated from a primary adenocarcinoma of a 51-year-old white male. Islands of epithelia are predominant, but bipolar cells are also apparent. (e) LoVo (CCL 229) was initiated from a metastatic nodule in the supraclavicular region of a 56-year-old white male. The line is polymorphic with epithelial-like, bi- and multipolar cells that proliferate predominantly as loosely attached cells in large groups and clusters to form a dense layer. Acinar structures and signet ring cells have been noted.

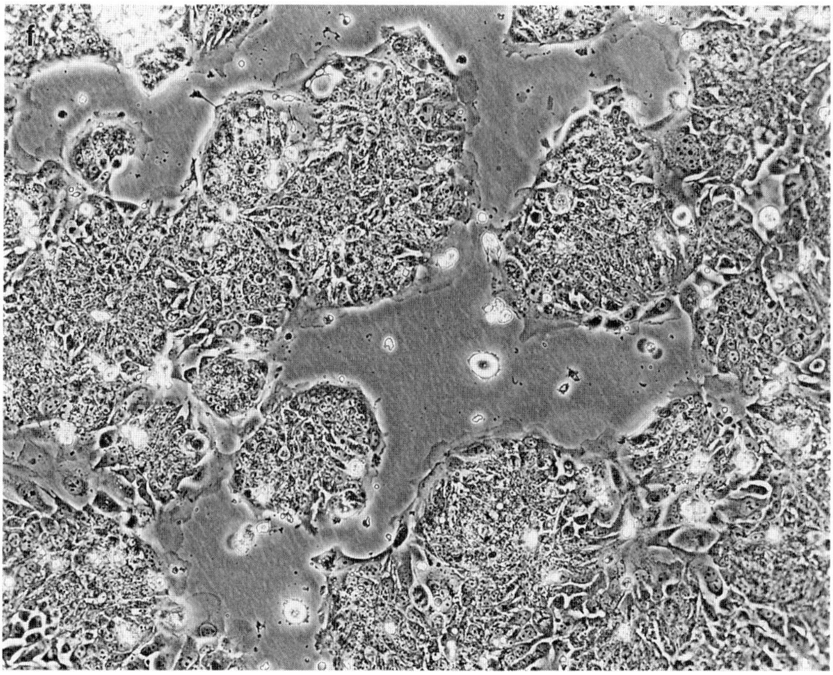

Fig. 1. *Continued* (f) T84 (CCL 248) is a transplantable line isolated from a metastatic lesion in the lung of a 72-year-old male. The cells proliferate as islands of tightly packed epithelia, with tight junctions and desmosomes, and eventually will form dense monolayers. Phase-contrast micrography. (See Hay *et al.*, 1992, for additional details.)

Although unmentioned, or only briefly referred to in published studies, the data clearly imply that seeding single cells that have been dissociated harshly by enzymes and chelating agents generally leads to poor viability and low success in culture initiation. Thus, almost all investigators who have cultured a large number of lines successfully indicated the use of "spill-out" or suspension cultures of multicellular aggregates with minimal disaggregation to single cells.

Rutzky and Moyer (1990) summarized key features that are relevant to successful cell culture: (1) nonenzymatic or minimal dissociation of tumor tissue, (2) seeding cultures as explants and at high cell densities, (3) removal of contaminating fibroblasts, usually after they have aided culture initiation, and (4) delaying passage until high cell densities have been achieved and plating cells at high density (both these procedures increase concentrations of conditioning or paracrine factors).

Our laboratory has extensive experience with serum-containing and fully defined media for the establishment of continuous cell lines derived from lung cancers. Our approach to the culture of colorectal tumors was based on our experience with lung cancer. ACL-4, the medium we formulated for bronchial adenocarcinomas, was devised after much trial and error and represented progressive improvements in formulation (Brower *et al.*, 1986; Gazdar and Oie, 1986). This medium is complex, consisting of 12 additives to basal medium. HITES, a medium for the selective growth of small-cell lung carcinomas (SCLC) was used earlier and included insulin, transferrin, hydrocortisone, estradiol, and selenium added to the basal medium RPMI 1640. Of the 5 HITES additives, 4 are present in ACL-4; this medium lacks only estradiol. Epidermal growth factor (EGF) is also present. Bovine serum albumin helps compensate for some of the high molecular weight proteins present in serum, and aids growth in semi-solid media. In addition, we added triiodothyronine (a selective mitogen), ethanolamine and phosphorylethanolamine (precursors of membrane lipids), HEPES buffer (to compensate for the loss of buffering properties of serum), glutamine, and sodium pyruvate (added empirically). Because ACL-4 lacks the high molecular weight attachment factors present in serum, cells cultured in this medium usually grow as floating aggregates. This characteristic is advantageous, since most nontransformed cells (other than lymphoid cells) cannot propagate without substrate attachment. However, the initial replication of some tumors also is aided by substrate attachment. Thus, we perform initial culture of human tumors in replicate flasks, half of which are precoated with Vitrogen, a commercial preparation of purified collagen. (Collagen precoating is not required after initial tumor growth has occurred.) Finally, tumors are heterogeneous in their basal medium requirements, so we routinely use an ACL-4 medium prepared with two basal media: RPMI 1640 or a 1 : 1 mixture of Ham's F-12 and Dulbecco's modified essential medium (MEM). All our established colorectal lines can be adapted readily from a routine serum-supplemented medium (SSM) to the ACL-4 medium.

III. Culture Characteristics

We selected 9 cell lines from our collection of more than 20 established cell lines for which to present detailed characteristics (Table II). Four (SNU-C1, SNU-C2A, SNU-C4, SNU-C5) of the lines were originated from Korean patients in Korea. Three lines (SNU-C2A, SNU-C4, SNU-C5) were cultured from xenografts in athymic nude mice. The remainder were cultured directly from tumor material. If tumor from the mucosal surface had to be used, the surface was cleansed with 70% ethanol- or iodine-containing solutions. Solid tumors were minced finely with scissors and dissociated into small aggregates by pipetting. Approximately $1-5 \times 10^6$ cells were seeded into 25-cm^2 flasks. Tumors were cultured in RPMI 1640 supplemented with 10% heat inactivated fetal bovine serum (FBS; designated medium R10) or in ACL-4 medium, and were maintained in a humidified incubator at 37°C in an atmosphere of 5% CO_2 and 95% air. The minimum number of viable tumor cells necessary for successful culture in either medium was estimated to be 5–10 $\times$ 10^6. In both media, growth occurred either immediately or after a dormant period lasting up to 8 wk. Initial cell passage was performed when heavy tumor cell growth was observed (usually 2–4 wk after initial culture). Once cell growth commenced, it tended to be progressive without periods of acute cell death (crises). Subsequent passages were performed weekly. Nonadherent cultures were passaged by transfer of floating multicellular aggregates. Adherent cultures were passaged at subconfluence after trypsinization. Once established, the lines could be cultured readily in SSM or ACL-4 media. For the sake of uniformity, however, the characterization was performed on sublines maintained in SSM, except for morphological characteristics, which could be assessed most accurately in the ACL-4 medium. All cultures expressed only human forms of several enzymes, and were negative for murine viruses as determined by the mouse antibody production (MAP) test. These cells were also free of mycoplasma contamination, as determined by multiple tests and methods. The lines had relatively long doubling times and low plating and cloning efficiencies. Population doubling times were determined by seeding 3 $\times$ 10^5 viable cells into replicate 25-cm^2 flasks and performing counts every 3 days for 4 wk or longer. Cultures were fed every 3 or 4 days and 24 hr before counting. For plating efficiencies, 10^3 cells were plated in 5 replicate 100-mm dishes and colonies consisting of more than 50 cells were enumerated 21–30 days later, after staining with 0.5% crystal violet. To determine colony-forming efficiency in semi-solid medium, 10^5 viable single cells were plated in 3 ml R10 medium containing 0.3% agarose over a base layer of R10 medium containing 0.5% agarose in 5 duplicate 60-mm dishes. Tumorigenicity was tested by inoculating 5–10 $\times$ 10^6 cells subcutaneously into each of 5 male athymic nude mice (BALB/c background) and observing twice weekly for progressive growth. Histological examination was performed on all tumors so obtained.

IV. Morphological Characteristics

Most lines demonstrate substrate adherence in SSM and lack adherence in ACL-4, unless the culture dishes are precoated with collagen.

In R10 medium, one culture, SNU-C1, displays both adherent and floating subpopulations. Cell line NCI-H498 grows as floating cell aggregates surrounded by a halo of easily visualized mucinous material. If the mucin coat is removed and the cells are dispersed by trypsinization, the line can be adapted to monolayer culture. Cell line NCI-H716 completely lacks substrate adherence and grows as amorphous floating cell aggregates in R10 medium.

In ACL-4, which lacks attachment factors, 7 of 9 cultures lack substrate adherence and grow as floating gland-like structures or amorphous aggre-

Table II

Characteristics of Cell Lines

	NCI-H548	NCI-H630	NCI-H508	SNU-C1	SNU-C2A	SNU-C4	SNU-C5	NCI-H716	NCI-H498
ATCC number	CCL 249	—	CCL 253	—	CCL 250.1	—	—	CCL 251	CCL 254
Date of initiation	1/83	9/83	10/82	3/84	9/84	12/84	11/84	4/84	9/82
Age	52	60	55	71	43	35	77	33	56
Sex	M	M	M	M	F	M	F	M	M
Race[a]	W	W	W	O	O	O	O	W	W
Blood type	A+	A+	A+	O+	A+	AB+	AB+	O+	A+
Prior therapy[b]	None	FAM & R	5-FU	None	None	None	None	5-FU	None
Primary tumor site	Sigmoid	Rectum	Cecum	Descending	Cecum	Transverse colon	Cecum	Cecum	Ileocecum
Cultured tumor site	Primary	Liver	Abdominal wall	Peritoneum	Primary, xenograft	Primary, xenograft	Primary, xenograft	Ascites	Peritoneum
Differentiation[c]	WD	WD	MD	MD	PD	PD	PD	PD	Mucinous
Original tumor differentiation	WD; carcinoid	MD	WD	MD	MD	MD	PD	PD	MD/WD, mucinous
Substrate adherence									
R10	+	+	+	+−	+	+	+	−	+
ACL-4	−	−	−	+−	−	+	+	−	−
Culture appearance[d]	G,D	G	G	G,U	U	U	U	U	G,D,ECM
Doubling time (hr)	73	46	53	45	82	34	67	67	94
Plating efficiency (%)[e]	0	3.5	5.6	F	0.5	6.5	3.3	F	0.2
Cloning efficiency (colonies/10^5 cells)	0	800	8	1300	5	310	380	2500	5
Tumorigenicity	+	+	+	+	+	+	+	+	+
CEA									
Cell extract	(−)	(+)	(−)	(+)	(+)	(+)	(+−)	(−)	(+)
Supernatant	(−)	(−)	(−)	(+)	(−)	(+)	(−)	(−)	(+)
CA19-9									
Cell extract	(−)	(−)	(−)	(+)	(−)	(+)	(+)	(−)	(+)
Supernatant	(−)	(−)	(−)	(+)	(−)	(+)	(+)	(−)	(+)
TAG-72									
Cell extract	(−)	(−)	(−)	(+)	(+)	(+)	(−)	(−)	(+)

...oS cell line[a]	32		24	133	5	10	6	64	99
DCG[g]	(−)	(−)	(−)	(−)	(−)	(−)	(−)	(+)	(−)
CGA[h] RNA expression	(−)	(−)	(−)	(−)	(−)	(−)	(−)	(+)	(−)
Chromosome numbers									
Modal	59	65	102	78	46	47	48	61	51
Range	55–63	63–69	71–131	76–83	44–48	46–47	46–53	55–64	47–54
Polyploid (%)	3	7	4	4	16	2	9	1	5
Metaphases with double minute chromosome (%)	62	100	94	100	4	0	4	100	0
c-myc	+	+	+	+	+	+	+	+	+
N-myc	+	0	0	0	0	0	0	0	0
L-myc	0	0	0	0	0	0	0	0	0
myb	+	+	+	++	0	+	tr	tr	+
EGF-R	+	tr	tr	tr	++	tr	++	+	+
K-ras	+	+++	+	+	+	+	+	+	+
Her-2	+	+	+	+	+	+	+	0	ND
sis	0	0	0	0	0	0	0	0	0
IGF-II	+	0	0	0	0	0	0	0	0
GRP	0	0	0	0	0	0	0	0	0
IGF-β	+	tr	+	0	tr	+	ND	++	+
MDR1(units[i]) RNA expression	8	10	5	2.7	0.7	18	1.2	20	3
ID$_{50}$ (μg/ml) 5-FU [Mean (SD)]	233(133)	43(30)	39(60)	0.60(0.56)	5.1(3.3)	0.93(0.06)	5.0(5.3)	6.6(3.9)	1.4(0.84)

[a] W, white; O, oriental.

[b] FAM, 5-fluorouracil + doxorubicin + mitomycin-C; R, radiotherapy; 5-Fu, 5-fluorouracil.

[c] WD, well differentiated; MD, moderately differentiated; PD, poorly differentiated.

[d] G, gland-like structures; D, dome formation; ECM, extracellular mucin secretion; U, undifferentiated.

[e] F, not done due to lack of substrate adherence.

[f] L-Dopa, decarboxylase activity.

[g] Dense core granule.

[h] Chromogranin A.

[i] Unit: A value of 1 arbitrary unit was assigned to the signal: given by 10 μg RNA of drug-sensitive KB-3-1 control cells. KB-8-5 cell line was used as positive control and a value of 30 U was assigned to the signal given by 10 μg RNA of drug-resistant KB-8-5 control cells.

gates. Most of the floating cultures can be grown as monolayers in ACL-4 medium if the culture dishes are precoated with collagen.

Evidence of glandular differentiation is present in 5 cultures. In SSM, 2 of 9 cultures demonstrate prominent dome formation (Kirkland, 1985). Culture morphology is determined most accurately in ACL-4 medium, although some features also can be discerned in R10 medium. Dome formation can be evaluated only in confluent monolayers grown in R10 medium. One gland- and dome-forming culture, NCI-H498, secretes large amounts of extracellular mucin. This feature is best appreciated in floating cultures, in which cellular aggregates are surrounded by a halo of extracellular mucin that can be seen readily by phase microscopy.

Based on culture, xenograft, and ultrastructural morphologies, the lines can be divided into four subtypes. Well- and moderately differentiated lines and the mucinous carcinoma line retain many morphological features of colorectal cells. Poorly differentiated lines retain epithelial cell features, but otherwise cannot be identified readily as being of colorectal origin. The division of cell lines into the moderately differentiated subtype is arbitrary, since these lines often have features that overlap with those characteristic of the other subtypes. With relatively minor differences, the cultures do reflect the morphologies of the tumors from which they were derived.

A. Well Differentiated

In ACL-4 medium, well-differentiated cells such as NCI-H548 and NCI-H630 grow as floating cell aggregates, and sometimes as acinar structures surrounding central lumini. A characteristic feature is the uniform radial orientation of tall columnar cells (Fig. 2a). In R10 medium, these lines consist of relatively large epithelioid cells. Prominent dome formation is present at confluence for two of the cell lines (Fig. 2b). In addition to domes, three-dimensional gland-like structures are present in confluent adherent cultures.

The ultrastructural appearances of gland-forming cultures demonstrate features of colonic mucosa, including uniform microvilli with prominent filamentous core rootlets, glycocalycal bodies, and well-formed junctional complexes that are present between cells that line glands.

Xenografts of well-differentiated cultures reflect the features of their respective cultures. Well-formed glands are lined by tall columnar cells with prominent apical borders.

B. Moderately Differentiated

SNU-C1 displays both adherent and floating subpopulations in R10 medium. In ACL-4 medium, the cells form spheroidal or irregular masses rather than gland-like structures (Fig. 3). In R10 medium, dome formation, if present, is inconspicuous. The cells tend to be cuboidal rather than columnar, and gland formation is less prominent.

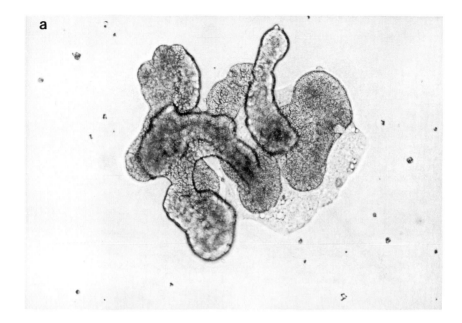

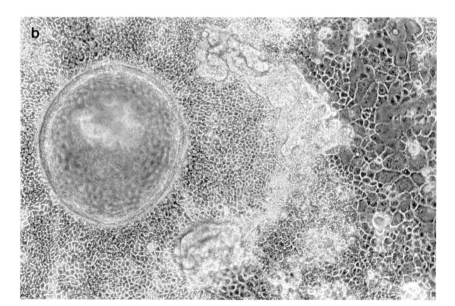

Fig. 2. Morphology of well-differentiated cultures. (a) Phase-contrast photomicrograph of cell line NCI-H548 in ACL-4 medium demonstrating partial substrate attachment (lighter areas). The floating component consists of tubuloglandular structures lined by tall columnar cells. (b) Cell line NCI-H548 in R10 medium, demonstrating attached epithelioid morphology and prominent dome formation.

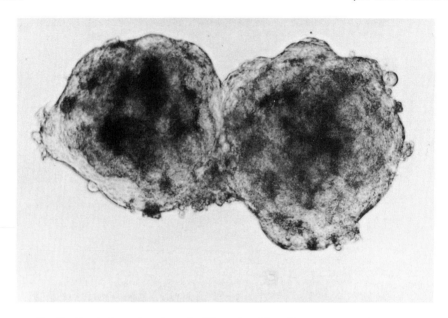

Fig. 3. Morphology of moderately differentiated line. Phase-contrast photomicrograph of cell line NCI-H508 in ACL-4 medium. The floating cell aggregates are attempting to form glandular structures.

Microvilli are less uniform in size, and filamentous core rootlets, glycocalycal bodies, and well-formed junctional complexes are present only occasionally.

Xenografts of moderately differentiated cultures produce glands that are less well organized than those of well-differentiated cultures; they are lined by cuboidal cells and vary considerably in size.

NCI-H508 is another moderately differentiated cell line.

C. Poorly Differentiated

In ACL-4, these cell lines (SNU-C2A, SNU-C4, SNU-C5, NCI-H716) grow as single cells or as small amorphous aggregates. In R10 medium, they display varying degrees of substrate adherence, but cell line NCI-H716 does not attach at all. No evidence of gland or dome formation is observed. These lines grow as sheets of attached epithelial cells lacking dome formation and other distinguishing characteristics (Fig. 4a) or as amorphous floating cell aggregates lacking central lumina (Fig. 4b). Abrupt shifting of medium to serum-free simple RPMI is possible for SNU-C5. This characteristic might be useful to study autocrine factors.

Ultrastructural studies confirm the epithelial nature of these lines (microvilli, desmosomes) and demonstrate occasional attempts to form inter- and intracellular glands. No characteristic ultrastructural features are seen, except in cell line NCI-H716. The undifferentiated cell line NCI-H716 demonstrates a unique ultrastructural feature: the presence of intracellular dense core granules (Fig. 4c). The granules, which are approximately 300 nm in size, are bound by a unit membrane and contain electron-dense cores surrounded by an electron-lucent halo. These structures are characteristic of neuroendocrine (NE) cells and tumors.

Xenografts usually demonstrate undifferentiated epithelial tumors. Occasional xenografts demonstrate intracellular mucin in a few cells, or make feeble attempts at gland formation. Xenografts of the NCI-H716 cell line have additional unusual features: subcutaneous tumors frequently invade and destroy surrounding muscle tissue, a feature not noted with other colorectal xenografts. In addition, when injected intraperitoneally, cell line NCI-H716 forms solid tumor masses with invasion and destruction of the pancreas and other organs.

D. Mucinous

In ACL-4 medium, cell line NCI-H498 grows as floating aggregates surrounded by a halo of varying diameter, easily visible by phase contrast microscopy (Fig. 5a). The halos stain strongly positive with alcian blue, mu-

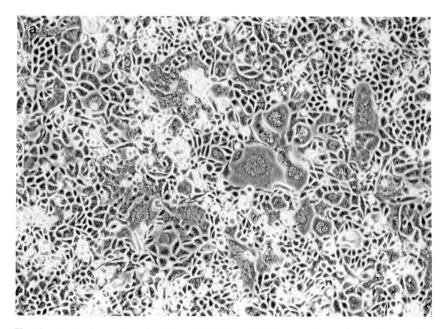

Fig. 4. Morphology of poorly differentiated lines. (a) Phase-contrast photomicrograph of cell line SNU-C2A in R10 medium, demonstrating solid sheets of attached epithelial cells. These cells often fuse together to form multinucleated giant cells.

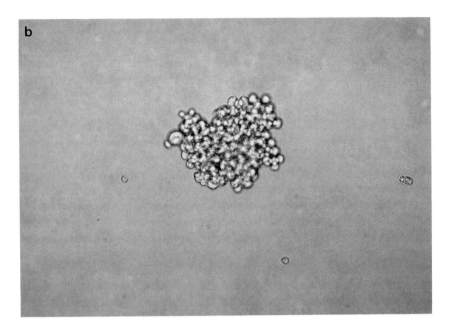

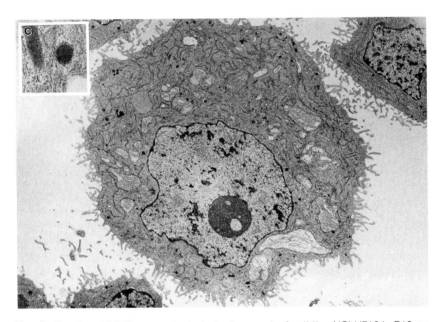

Fig. 4. *Continued* (b) Phase-contrast photomicrograph of cell line NCI-H716 in R10 medium, demonstrating growth in amorphous solid floating cell masses. (c) Electron micrograph of NCI-H716 cell with prominent nucleolus, small irregular microvilli, prominent mitochondria, endoplasmic reticulum, free ribosomes, and many dense core granules. (*Inset*) Higher magnification of a granule bound by a unit membrane and having an electron-dense core surrounded by a clear halo.

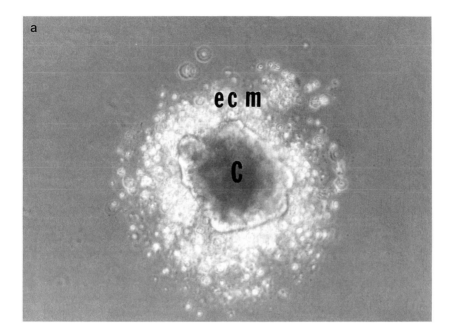

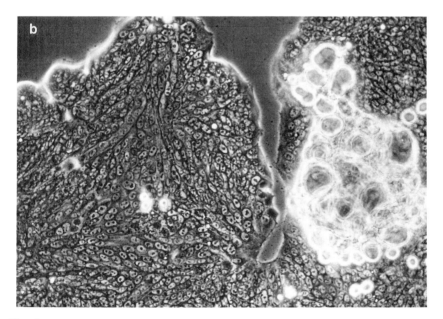

Fig. 5. Morphology of mucinous carcinoma line NCI-H498. (a) Culture morphology in R10 medium before trypsinization. A central cell aggregate is surrounded by a halo of extracellular mucin (ecm). (b) Culture morphology in R10 medium after trypsinization. The large adherent cells are filled with mucin vacuoles.

cicarmine, and periodic acid-Schiff (PAS), confirming the mucinous nature of the secretion. Mucin stains indicate varying degrees of intracellular mucin production by other cell lines, especially the well-differentiated lines, and their corresponding xenografts. However, prominent extracellular mucin secretion is a feature present only in NCI-H498 cells.

The aggregates do not attach to the substrate in R10 medium unless the mucinous coat is dispersed by trypsinization. Monolayer cells are large and often distended with mucin droplets (Fig. 5b). Gland and dome formation are occasionally present.

Xenograft histology and ultrastructural morphology indicate that the cell line has characteristics of a well-differentiated adenocarcinoma, and confirm the prominent mucin secretion.

Ultrastructural examination demonstrates abundant intra- and extracellular mucin. These cells can be adapted as an adherent line if the extracellular mucin coat is removed by trypsinization (and if attachment factors are present in the medium).

Inoculation of this line intraperitoneally into athymic nude mice results in a similar situation—glandular structures surrounded by a mucin coat. The appearances closely mimic pseudomyxoma peritonei, a clinical condition characteristic of mucinous carcinoma, showing extensive peritoneal involvement without invasion of abdominal organs. The tumor cells form floating cell aggregates with gland formation surrounded by mucinous coats, closely mimicking their cultural appearance in ACL-4 medium. Pleural metastases are noted frequently after intraperitoneal inoculation.

V. Other Characteristics

Our bank of colorectal cell lines has several unique features: (1) expression and secretion of high concentrations of three gastrointestinal cell-associated antigens; (2) possibly unique phenotypes (mucinous carcinoma and dense core granule-containing lines); (3) high frequency of expression of the NE cell marker dopa decarboxylase (DDC); and (4) high frequency of cytogenetic evidence for gene amplification.

The lines have relatively low cloning and plating efficiencies, perhaps because of our culture conditions. All lines are tumorigenic when inoculated subcutaneously into athymic nude mice; at least two lines grow intraperitoneally.

Carcinoembryonic antigen (CEA), CA19-9, and TAG-72 are expressed by 67%(6/9), 56%(5/9), and 44%(4/9) of the lines, respectively. Lines expressing CEA and CA19-9 actively secrete these antigens into the supernatant fluids, whereas TAG-72 antigen is not secreted. Our findings suggest that CEA and CA19-9 may function as more sensitive serum markers than TAG-72.

Many of our cell lines express varying concentrations of DDC, a key NE cell marker (Gazdar *et al.,* 1981). DDC is essential for formation of biogenic amines, including serotonin and catecholamines. The incidence and concentration of DDC increases after *in vitro* growth. These data agree with a large survey of human tumors that indicated that non-small-cell lung cancers (12%) and colorectal tumors (51%) were the only non-NE tumor types to express DDC frequently. The range of DDC values in the colorectal cell lines was similar to that in cell lines derived from small-cell carcinoma of the lung (Carney *et al.,* 1985), a typical NE tumor.

Cell line NCI-H716 may constitute another possibly unique subtype: an undifferentiated line, with cytoplasmic dense core granules that are characteristic of endocrine secretion present in every cell. Although osmiophilic granules have been described in COLO 320 and COLO 321 (Quinn *et al.,* 1979) derived from the same tumor, the illustrations available are of low power and cannot be evaluated critically. Because dense core granules characteristic of NE cells are not present in other DDC-positive lines, the latter only express part of the NE program.

Cytogenetic studies indicate that one cell line is pseudodiploid whereas the other 8 examined are aneuploid with modal chromosome numbers ranging from 47 to 102. An unexpected finding of great potential interest is the presence of double minute (DM) chromosomes in 7 of 9 lines. In one cell line, some of the DMs are larger than usual, as described by Bullerdiek *et al.* (1983), and are termed chromatic blocks. In addition, 3 lines, all from previously treated patients, contain HSRs (homogenously staining regions). DMs and HSRs are characteristic of cells with amplified gene sequences, especially proto-oncogenes and genes associated with drug resistance. Previous reports indicated that the finding of DMs and HSRs in colon carcinoma lines is infrequent (Chen *et al.,* 1982). However, COLO 320, a colon line reputed to have NE properties, has many DMs that are replaced later by an HSR (Quinn *et al.,* 1979).

The *myc* gene family contains many of the proto-oncogenes frequently associated with cytogenetic evidence of gene amplification. *c-myc* amplification is present in COLO 320 (Alitalo *et al.,* 1983). Although all our lines express *c-myc* mRNA, only one line, NCI-H716, has amplified gene sequences.

Our unpublished data (LaRocca *et al.,* 1990) provide the results of examination of our colorectal cell lines for proto-oncogene and growth factor mRNA expression. Analysis of proto-oncogene expression reveals detectable levels of *c-myc,* EGF-R, and K-*ras* message in all cell lines (Table II). The degree of *c-myc* expression is relatively uniform between the cell lines, regardless of the source (primary versus metastasis), the histological grade, or the location of the primary tumor. EGFR expression, however, varies in intensity. K-*ras* mRNA expression is uniform among cell lines, with the exception of NCI-H630, which demonstrates a strikingly increased level of K-*ras* expression.

Southern blotting analysis confirms the presence of 100-fold amplification of the K-*ras* gene. This line also is remarkable for the presence of HSRs. This cell line is the only one that originated from a site that had been irradiated previously and progressed subsequently. Carmichael *et al.* (1989) analyzed the radiosensitivity of four of these cell lines including NCI-H630, which turned out to be the most radioresistant of the four. Thus, the presence of increased K-*ras* mRNA expression may predict increased resistance of the tumor to radiotherapy.

Detectable levels of the 4.6-kb *Her*-2 message can be seen in 7 of the 9 colorectal cell lines, whereas detectable c-*sis* message (4.2 kb) is found in 2 of the 9 cell lines (SNU-C2 and SNU-C5, both of which are poorly differentiated cell lines and were derived from xenografts). c-*myb* expression is found in 8 of the 9 colorectal cell lines, whereas N-*myc* expression is limited to one of the cell lines (NCI-H548). No evidence of L-*myc* expression is detected in any of the 9 cell lines.

Significant variability among the colorectal cell lines is observed with respect to growth factor mRNA expression. Insulin-like growth factor II (IGF-II) transcripts are detected clearly in NCI-H548. Transforming growth factor β (TGFβ) mRNA is found in 7 of 8 cell lines tested, with marked variability in the degree of expression among those 8 cell lines. The highest level of expression is seen in NCI-H716, whereas only trace levels are seen in SNU-C2 and NCI-H630. No gastrin-releasing peptide (GRP) mRNA could be detected.

On the basis of their relative signal intensity and frequency of expression, proto-oncogenes and growth factor genes can be grouped into four categories: (1) genes whose level of expression is relatively uniform in all the colorectal cell lines (e.g., c-*myc*); (2) genes that are detectable in most, if not all, cell lines but vary in signal intensity (e.g., K-*ras*, *Her*-2, c-*myb*, EGFR, TGFβ); (3) genes whose level of expression is detected in less than 50% of the colorectal cell lines (e.g., N-*myc*, c-*sis*, IGF-II); and (4) genes whose expression is not detectable in any of the cell lines (e.g., L-*myc*, GRP). Although heterogeneity of proto-oncogene, particularly growth factor gene, expression is detected among the colorectal cell lines, no statistically significant associations can be made between the pattern of gene expression of a cell line and the known biological and clinical parameters. However, K-*ras* is amplified in a cell line derived from a radioresistant tumor (NCI-H630), whereas detectable N-*myc* expression is found in a cell line derived from a tumor with NE features (NCI-H548).

The NCI-H716 cell lines, derived from a poorly differentiated mucoid carcinoma whose primary tumor was originally in the cecum, does contain cytoplasmic dense core granules by electron microscopy, elevated levels of DDC activity, and detectable chromogranin A mRNA. This line lacks N-*myc* message but does demonstrate amplification of the c-*myc* gene and elevated levels of TGFβ message, and is the only line that lacks detectable *Her*-2 message. Subcutaneous xenografts of this cell line readily invade and destroy surrounding muscle tissue, unlike the xenografts from the other colorectal cell lines. Also, this cell line has the highest cloning efficiency of all the cell lines tested. The NCI-H716 cell line was derived from a patient with a history of ulcerative colitis. Additional lines clearly must be studied to establish whether this pattern of gene amplification and expression (c-*myc* amplification, elevated TGFβ mRNA, absence of *Her*-2 mRNA) reproducibly characterizes this biological behavior.

The pCEA1 probe was hybridized with 30 μg total mRNA from the 9 colorectal cell lines for 10 days and a signal was detectable in 3 lines (SNU-C1, NCI-H498, NCI-H508). When these levels are compared with the CEA antigen expression of each cell line, a correlation between detectable CEA message and a concentration of CEA antigen greater than 500 ng/10^6 cells is noted, implying that the quantity of CEA present in these cell lines is not regulated posttranscriptionally to a significant degree.

We used the semi-automated tetrazolium-based MTT colorimetric assay to test the chemosensitivity of the 9 cell lines (Park *et al.*, 1987b). Carmichael *et al.* (1987) reported that the MTT assay yields results very similar to those of the clonogenic tumor stem cell assay. ID_{50} value was defined as the concentration of drug that produced 50% reduction of absorbance at 540 nm. The range of drug concentrations that produced 50% inhibition of cell growth was greatest with 5-fluorouracil (5-FU; 388-fold vs 5- to 30-fold with the other six agents); the area under the curve (AUC) that produced 50% growth inhibition was within a clinically achievable range only for 5-FU. From assay results, we predicted 5-FU to be the sole active agent of the seven anticancer drugs tested.

We tested the effects of 5-FU and 5-fluoro-2'-deoxyuridine (FdUr) with and without leucovorin (LV) on our bank of 11 human colorectal carcinoma cell lines using the MTT test (Park *et al.*, 1988). The effect of LV on 5-FU and FdUr was quantitatively similar. A clinically achievable level of LV (20 μM) increased the cytotoxicity in all three replicate experiments in 10 of the 11 cell lines ($p < 0.05$, binomial test). LV alone at a concentration of 20 μM had no effect on cell survival. In 3 cell lines, 50% inhibition of growth occurred at a clinically achievable AUC for 5-FU alone. With the addition of LV, one additional cell line showed 50% growth inhibition at a clinically achievable level of 5-FU.

Intermediate or high levels of *mdr*1 mRNA were shown to be present in most of the colorectal carcinoma cell lines using a slot blot assay.

Fearon *et al.* (1990) reported that one of the postulated tumor suppressor genes, DCC (deleted in colorectal carcinomas) on 18q21, is expressed in most normal tissue, including colonic mucosa. Expression of this gene is reduced greatly or absent in most colorectal carcinomas tested. A somatic mutation, DNA insertion within the DCC gene, was found in the NCI-H630 cell line.

338

References

Alitalo, K., Schwab, M., Lin, C. C., Varmus, H. E., and Bishop, J. M. (1983). Homogeneously staining chromosomal regions contain amplified copies of an abundantly expressed cellular oncogen (c-myc) in malignant neuroendocrine cells from a human colon carcinoma. *Proc. Natl. Acad. Sci. U.S.A.* **80,** 1707–1711.

Brattain, M. G., Brattain, D. E., Fine, W. D., Khaled, F. M., Marks, M. E., Kimball, P. M., Arcolano, L. A., and Danbury, B. H. (1982). Initiation and characterization of cultures of human colonic carcinoma with different biological characteristics utilizing feeder layers of confluent fibroblasts. *Oncodev. Biol. Med.* **2,** 355–366.

Brattain, M. G., Marks, M. E., McComb, J., Finely, W., and Brattain, D. E. (1983). Characterization of human colon carcinoma cell lines isolated from a single primary tumor. *Br. J. Cancer* **47,** 373–381.

Bullerdiek, J., Bartnitzke, S., and Schloot, W. (1983). Cells with double minutes divided into two categories. *Cancer Genet. Cytogenet.* **10,** 301–304.

Brower, M., Carney, D. N., Oie, H. K., Gazdar, A. F., and Minna, J. D. (1986). Growth of cell lines and clinical specimens of human non-small cell lung cancer in a serum-free defined medium. *Cancer Res.* **46,** 798–806.

Carmichael, J., Park, J. G., DeGraff, W. G., Gamson, J., Gazdar, A. F., and Mitchell, J. B. (1989). Radiation sensitivity and study of glutathione and related enzymes in human colorectal cancer cell lines. *Eur. J. Cancer Clin. Oncol.* **24,** 1219–1224.

Carney, D. N., Gazdar, A. F., Bepler, G., Guccion, J. G., Mrangos, P. J., Moody, T. W., Zweig, M. H., and Minna, J. D. (1985). Establishment and identification of small cell lung cancer lines having classic and variant features. *Cancer Res.* **45,** 2913–2923.

Carrel, S., Sordat, B., and Merenda, C. (1976). Establishment of a cell line (Co-115) from a human colon carcinoma transplanted into nude mice. *Cancer Res.* **36,** 3978–3984.

Chen, T. R., Har, R. J., and Macy, M. L. (1982). Karyotype consistency in human colorectal carcinoma cell lines established in vitro. *Cancer Genet. Cytogenet.* **6,** 93–117.

Dexter, D. L., Barbosa, J. A., and Calabresi, P. N. (1979). N,N-Dimethylformamide-induced alteration of cell culture characteristics and loss of tumorigenicity in cultured human colon carcinoma cells. *Cancer Res.* **39,** 1020–1025.

Dexter, D. L., Spremulli, E. N., Fligiel, Z., Barbosa, J. A., Vogel, R., Van Voorhees, A., and Calabresi, P. (1981). Heterogenicity of cancer cells from a single human colon carcinoma. *Am. J. Med.* **71,** 949–956.

Drewinko, B., Romsdahl, M. M., Yang, L. Y., Ahern, M. J., and Trujillo, J. M. (1976). Establishment of a human carcinoembryonic antigen-producing colon adenocarcinoma cell line. *Cancer Res.* **36,** 467–475.

Durrant, L. G., Robins, R. A., Pimm, M. V., Perkins, A. C., Armitage, N. C., Hardcastle, J. D., and Baldwin, R. W. (1986). Antigenicity of newly established colorectal carcinoma cell lines. *Br. J. Cancer* **53,** 37–45.

Fearon, E. R., Cho, K. R., Nigro, J. M., Kern, S. E., Simons, J. W., Ruppert, J. M., Hamilton, S. R., Preisinger, A. C., Thomas, G., Kinzler, K. W., Vogelstein, B. (1990). Identification of a chromosome 18q gene that is altered in colorectal cancers. *Science* **247,** 49–56.

Fogh, J., and Trempe, G. (1975). New human tumor cell lines. *In* "Human Tumor Cells *in Vitro*" (J. Fogh, ed.), pp. 115–154. Plenum Press, New York.

Fogh, J., Fogh, J. M., and Orfeo, T. (1977a). One hundred and twenty-seven cultured human tumor cell lines producing tumors in nude mice. *J. Natl. Cancer Inst.* **59,** 221–225.

Fogh, J., Wright, W. C., and Loveless, J. D. (1977b). Absence of HeLa cell contamination in 169 cell lines derived from human tumors. *J. Natl. Cancer Inst.* **58,** 209–214.

Fogh, J. M., Orfeo, T., Tiso, J., Sharkey, F. E., and Daniels, W. P. (1980). Twenty-three new human tumor lines established in nude mice. *Exp. Cell Biol.* **48,** 229–239.

Gazdar, A. F., and Oie, H. K. (1986). Cell culture methods for human lung cancer. *Cancer Genet. Cytogenet.* **19,** 5–10.

Gazdar, A. F., Carney, D. N., Cuccion, J. G., and Baylin, S. B. (1981). Small cell carcinoma of the lung: Cellular origin and relationship to other pulmonary tumors. *In* "Small Cell Lung Cancer" (F. A. Greco, P. A. Bunn, and R. K. Oldham, eds.), pp. 145–175. Grune & Stratton, New York.

Goldenberg, D. M., Witte, S., and Elster, K. (1966). GW-39: A new human tumor serially transplantable in the golden hamster. *Transplantation* **4,** 760–763.

Hay, R. J., Caputo, J., Chen, T. R., Macy, M. L., McClintock, P., and Reid, Y. A. (1992). "Catalogue of Cell Lines and Hybridomas," 7th Ed. American Type Culture Collection, Rockville, Maryland.

Kirkland, S. C., and Bailey, I. G. (1986). Establishment and characterization of six human colorectal adenocarcinoma cell lines. *Br. J. Cancer* **53,** 779–785.

Leibovitz, A., Stinson, J. C., McCombs, W. B., III, McCoy, C. E., Mazur, K. C., and Mabry, N. O. (1976). Classification of human colorectal cell lines. *Cancer Res.* **36,** 4562–4569.

McBain, J. A., Weese, J. L., Meisner, L. F., Wolberg, W. H., and Wilson, J. K. V. (1984). Establishment and characterization of human colorectal cancer cell lines. *Cancer Res.* **44,** 5813–5821.

Murakami, H., and Masui, H. (1980). Hormonal control of human colon carcinoma cell growth in serum-free medium. *Proc. Natl. Acad. Sci. U.S.A.* **77,** 3464–3468.

Noguchi, P. D., Johnson, J. B., O'Donnell, R., and Petricciani, J. C. (1978). Chick embryonic skin as a rapid organ culture assay for cellular neoplasia. *Science* **199,** 980–983.

Park, J. G., Oie, H. K., Sugarbaker, P. H., Henslee, J. G., Chen, T. R., Johnson, B. E., and Gazdar, A. (1987a). Characteristics of cell lines established from human colorectal carcinoma. *Cancer Res.* **47,** 6710–6718.

Park, J. G., Kramer, B. S., Steiberg, S. M., Carmichael, J., Collins, J. M., Minna, J. D., and Gazdar, A. F. (1987b). Chemosensitivity testing of human colorectal carcinoma cell lines using a tetrazolium-based colorimetric assay. *Cancer Res.* **47,** 5875–5879.

Quinn, L. A., Moore, G. E., Morgan, R. T., and Woods, L. K. (1979). Cell lines from human colon carcinoma with unusual cell products, double minutes and homogeneously staining regions. *Cancer Res.* **39,** 4914–4924.

Reid, L. M., Holland, J., Jones, C., Wolf, B., Niwayama, G., Williams, R., Kaplan, N. O., and Sato, G. (1978). Some of the variables affecting the success of transplantation of human tumors into the athymic nude mouse. *In* "The Use of Athymic (Nude) Mice in Cancer Research" (D. P. Houchens and A. A. Ovejera, eds.), pp. 107–121. Fischer, Stuttgart.

Rutzky, L. P., and Moyer, M. P. (1990). Human cell lines in colon cancer research. *In* "Colon Cancer Cells" (M. P. Moyer, ed.), pp. 155–202. Academic Press, San Diego.

Rutzky, L. P., Giovanella, B. C., Tom, B. H., Kaye, C. I., Noguchi, P. D., and Kahan, B. D. (1983). Characterization of a new human colonic cell line, LS123. *In Vitro* **19,** 99–107.

Semple, T. V., Quinn, L. A., Woods, L. K., and Moore, G. E. (1978). Tumor and lymphoid cell lines from a patient with carcinoma of the colon for a cytotoxicity model. *Cancer Res.* **38,** 1345–1355.

Silverberg, E., Boring, C. C., and Squires, T. S. (1990). Cancer statistics, 1990. *CA Cancer Clin.* **40,** 9–26.

Tom, B. H., Rutzky, L. P., Jakstys, M. A., Oyasu, R., Kaye, C. I., and Kahan, B. D. (1976). Human colonic adenocarcinoma cells. I. Establishment and description of a new line. *In Vitro* **12,** 180–191.

Tompkins, W. A. F., Watrach, A. M., Schmale, J. D., Schultz, R. M., and Harris, J. A. (1974). Cultural and antigenic properties of newly established cell strains derived from adenocarcinomas of the colon and rectum. *J. Natl. Cancer Inst.* **52,** 1101–1110.

Watkins, J. F., and Sanger, C. (1977). Properties of a cell line from human adenocarcinoma of the rectum. *Br. J. Cancer* **35,** 785–794.

Whitehead, R. H., Macrae, F. A., St. John, D. J., and Ma, J. (1985). A colon cancer cell line (LIM 1215) derived from a patient with nonpolyposis colorectal cancer. *J. Natl. Cancer Inst.* **74,** 759–765.

Whitehead, R. H., Brown, A., and Bhathal, P. S. (1987). A method for the isolation and culture of human colonic crypts in collagen gels. *In Vitro Cell. Mol. Biol.* **23,** 436–442.

Wilson, J. K. V., Bittner, G. N., Oberley, T. D., Meisner, L. F., and Weese, J. L. (1987). Cell culture of human colon adenomas and carcinomas. *Cancer Res.* **47,** 2704–2713.

Yaniv, A., Altiboum, Z., Gazit, A., Bloch-Shtacher, N., and Eylan, E. (1978). Establishment and characterization of a cell line derived from human colon adenocarcinoma (HuCCL-14). *Exp. Cell Biol.* **46,** 220–230.

Cell Lines from Urinary Bladder Tumors

13

Sonny L. Johansson
Department of Pathology and Microbiology and
Eppley Institute for Research on Cancer and Allied Diseases
University of Nebraska, Medical Center, Omaha, Nebraska 68198-3135

Bertil Unsgaard
Department of General Oncology, Sahlgren Hospital
S-41345 Göteborg, Sweden

Carol M. O'Toole
Department of Medical Microbiology
London Hospital Medical College
London E1.2AD, England

I. Background 342

II. Patient History and History of Original Tumor 342
   A. RT4 (HTB2) 342
   B. T24 (HTB4) 342
   C. SCaBER (HTB3) 342
   D. TCC SUP (HTB5) 343
   E. J82 (HTB1) 343

III. Histological Examination of the Original Tumors and Developed Cell Cultures 343
   A. RT4 343
   B. T24 344
   C. SCaBER 345
   D. TCC SUP 345
   E. J82 346

IV. Establishment and Maintenance of Bladder Cancer Lines 349

V. Mycoplasma Testing 349

VI. Cryopreservation of Cell Lines 349

VII. Characterization of lines 350
   A. Ultrastructure 350
   B. HLA Phenotyping 350
   C. HLA Class II Antigen Expression 351
   D. Lymphocyte Cytotoxicity 351
   E. Human Antibodies 352
   F. Isoenzyme Pattern 354
   G. Chromosome Analysis 354

VIII. Summary 355

References 356

I. Background

Establishing permanent cell lines of bladder cancer has been relatively difficult (Burrows *et al.*, 1917). The first established line was reported in 1970 by Rigby and Franks. Subsequently, many additional cell lines have been described (Williams, 1980). However, few cell lines have been characterized fully. This chapter is a critical review of five established cell lines, including a review of the original biopsy material from which the cell lines were derived. These five cell lines are those present in the American Type Culture Collection (ATCC) cell bank. Clinical and morphological data pertaining to the origin of bladder cancer cell lines as well as the case histories are presented. The published data on HLA and isoenzyme profiles are summarized, as is information on lymphocyte cytotoxicity and karyotying antibody responses to these lines.

II. Patient History and History of Original Tumor

A. RT4 (HTB2)

The patient was a 63-year-old Caucasian male with a bladder tumor who was treated with open bladder excision and insertion of radioactive gold grains. In March, 1968, 10 months later, a local recurrence was treated by fulguration; 4 months later, the patient underwent total cystectomy. Several papillary tumors were present in the bladder. Specimens were taken for tissue culture and histology. The patient died after 2 yr from metastatic disease (Rigby and Franks, 1970).

B. T24 (HTB4)

The patient was a 73-year-old Caucasian female who had a partial bladder resection in 1962 for bladder cancer. Her follow-up included regular cystoscopic examinations with multiple fulgurations and/or transurethral resection (TUR) of bladder tumors. In March, 1970, after an episode of gross hematuria, cystoscopy revealed a broad-based 3- × 4-cm tumor located in the anterior wall of the bladder. TUR of the tumor was performed and material was submitted for tissue culture and histology. The patient died 7 months later from myocardial infarction. Clinically, no evidence of metastatic disease was found, but an autopsy was not performed (Bubenik *et al.*, 1973).

C. SCaBER (HTB3)

The patient was a 58-year-old Afro-American male who, in 1968, underwent treatment for obstruction of the lower urinary tract associated with urinary

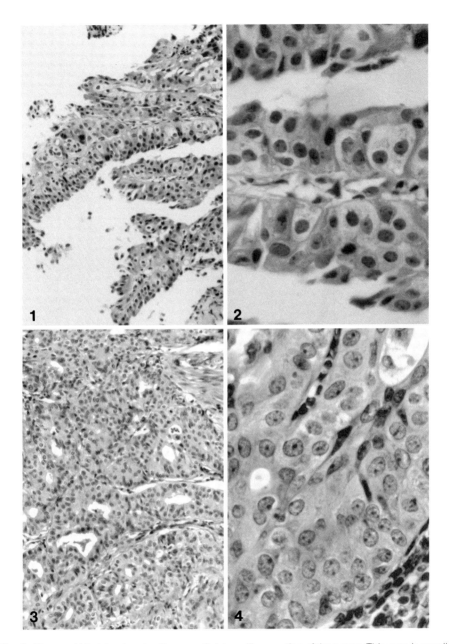

Fig. 1. Biopsy of bladder showing the superficial papillary portion of the tumor. This area is a well-differentiated transitional cell carcinoma Grade II/IV (H&E, 16×). **Fig. 2.** Higher magnification of Fig. 1 showing the well-differentiated tumor cells (H&E, 309×). **Fig. 3.** Deeper portion of the tumor showing a solid and pseudoglandular pattern. The tumor is still Grade II/IV. Anaplasia is not pronounced. The polarity is relatively well preserved and the nucleus:cytoplasm ratio is increased only moderately. Note smooth muscle involvement (right upper corner) (H&E,77×). **Fig. 4.** Higher magnification of Fig. 3 showing preserved urothelial differentiation and only moderate anaplasia (H&E, 309×).

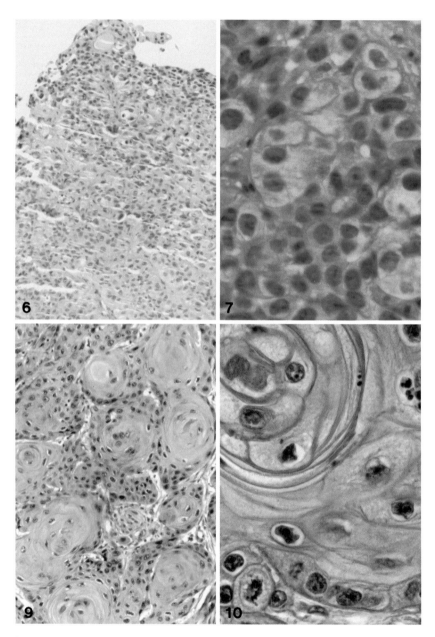

Fig. 6. Biopsy of bladder showing a solid invasive transitional cell carcinoma Grade III/IV (H&E, 77×). **Fig. 7.** Higher magnification of Fig. 6 showing pleomorphic hyperchromatic polygonal cells with relatively abundant cytoplasm (H&E, 309×). **Fig. 9.** Bladder tissue showing large-cell, keratinizing, moderately well-differentiated squamous cell carcinoma Grade II/IV (H&E, 77×). **Fig. 10.** Higher magnification of Fig. 9 showing keratin pearl formation and distinct cell borders (H&E, 309×).

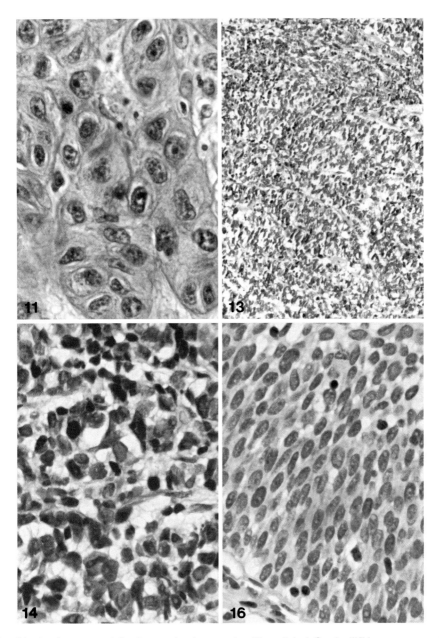

Fig. 11. Another area of the tumor showing poorly differentiated Grade III/IV squamous cell carcinoma lacking obvious keratinization (H&E, 309×). **Fig. 13.** Original biopsy of bladder showing undifferentiated carcinoma. This tumor has the features of a small cell carcinoma of intermediate cell type (H&E, 77×). **Fig. 14.** Higher magnification of Fig. 13 showing tumor cells with marked nuclear hyperchromasia, high nucleus:cytoplasm ratio, and relatively sparse cytoplasm (H&E, 309×) **Fig. 16.** The main bulk of this tumor showed features of a transitional cell carcinoma Grade II/IV. Note the well-preserved moderately increased nucleus:cytoplasm ratio and moderate hyperchromaticity (H&E, 309×).

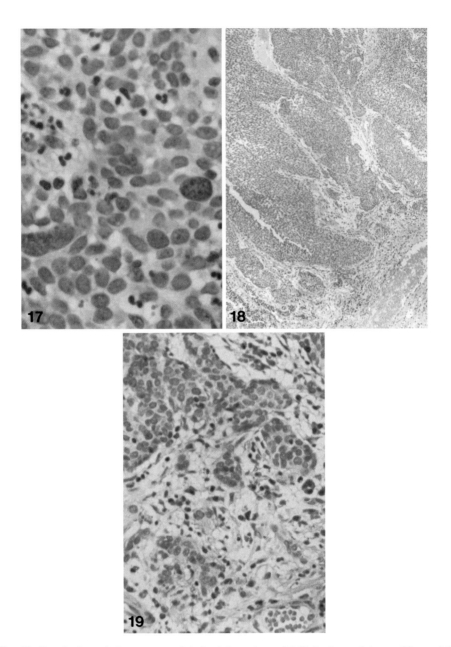

Fig. 17. Focally, the polarity was more disturbed. A greater variability in size and shape of the nuclei and increased hyperchromaticity corresponded to the Grade III/IV tumor (H&E, 309×). **Fig. 18.** Biopsy of bladder showing a papillary transitional cell carcinoma Grade II/IV with invasion into the lamina propria (H&E, 31×). **Fig. 19.** Higher magnification of Fig. 18. Note the early lamina propria invasion (H&E, 154×).

tract infection. Subsequently, he developed a urethral cutaneous fistula. The patient presented with gross hematuria in October of 1974. A biopsy 1 month later revealed squamous cell carcinoma. In December, 1974, the patient was treated with radical cystectomy. A 3- × 2-cm tumor was present in the posteriolateral wall extending to the dome and trigone. Material was taken for tissue culture, histology, and electron microscopy. The patient died 3 months later with extensive residual carcinoma in the pelvis that also encased segments of the ileum, including part of the ileal conduit bladder. Metastases were present in the periaortic lymph nodes and liver. The patient also had a confirmed diagnosis of multiple myeloma (O'Toole *et al.*, 1976).

D. TCC SUP (HTB5)

This patient was a 67-year-old Caucasian female who presented with a 4-month history of gross hematuria in December of 1974. She underwent cystoscopy and a TUR biopsy of a tumor localized in the bladder neck was performed. Material was taken for tissue culture and histology. The patient developed hemorrhagic diathesis and died 3 wk after admission. Metastatic disease involving the bone system was confirmed. Cerebral spread was suspected but not confirmed. No autopsy was performed (Nayak *et al.*, 1977).

E. J82 (HTB1)

This 58-year-old Caucasian male suffered from low back pain and intermittent gross hematuria for 1 yr. Cystoscopic examination revealed relatively extensive growth of papillary tumor with a base of approximately 3 cm. In November, 1972, the patient underwent TUR of the tumor and specimens were taken for tissue culture and histology. The patient received preoperative irradiation with rotational cobalt constituting a total of 4200 rads. The patient underwent total cystectomy and cutaneous uretero-entero anastomosis, according to Bricker, 3 wk postradiation. Status post-TUR was identified in the bladder; however, no residual tumor was present. The patient is still alive and well 20 years later (March, 1992) (O'Toole *et al.*, 1978).

III. Histological Examination of the Original Tumors and Developed Cell Cultures

A. RT4

RT4 is a papillary and solid transitional cell carcinoma (TCC) that, on the surface, shows fibrovascular cores lined with moderately pleomorphic urothelial cells (Figs. 1,2). The invasive part involving the lamina propria and the detrusor muscle is solid or shows pseudoglandular features (Fig. 3). Histological grade is a Grade II TCC (Figs. 3,4; Berqkvist *et al.*, 1965; Mostofi *et al.*, 1973). Figure 5 is a phase contrast photograph of a cell culture of RT4 with aggregates of epithelial tumor cells.

B. T24

The initial bladder resection specimen from 1962 shows an undifferentiated carcinoma with neuroendocrine features invading the detrusor muscle. The tumor has two cell populations, one showing a small distinct hyperchromatic nuclei and sparse cytoplasm. The other cell type is more oval and sometimes spindly. A tendency toward pseudoglandular formation is identified. Immunoperoxidase staining reveals keratin expression (MAK-6) in both cell types, predominantly in the more oat-cell-like areas. Immunoperoxidase staining against neuron-specific enolase is also positive in the majority of tumor cells. The two biopsy specimens from 1970 show predominantly solid, poorly differentiated TCCs of Grade III/IV (Fig. 6). Mitotic figures are identified easily. The tumor cells contain distinct nucleoli and frequently relatively abundant clear or somewhat foamy cytoplasm (Fig. 7). The biopsy is relatively superficial, but the tumor is clearly invasive into the detrusor muscle. Figure 8 is a phase contrast photograph of the T24 cell line, showing the relatively spindly nature of the tumor cells.

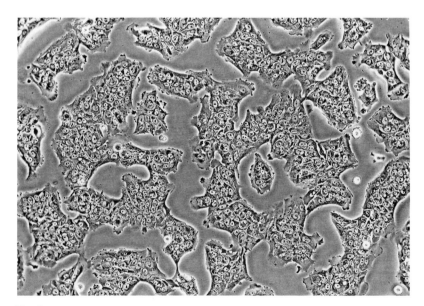

Fig. 5. Phase-contrast photograph of tissue cultures showing the aggregates of epithelial tumor cells (H&E, 100×).

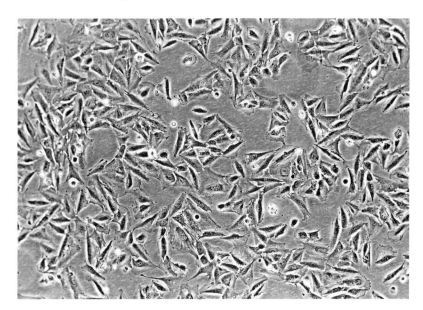

Fig. 8. Phase-contrast photograph of tissue culture showing the spindly character of the majority of the tumor cells (H&E, 100×).

C. SCaBER

The biopsy and the cystectomy specimen show similar features. Thus, the tumor is mostly a moderately well-differentiated squamous cell carcinoma (SCC) Grade II/IV with areas of prominent keratinization (Figs. 9,10). In the deeper areas, the tumor invades as clusters of cells or single cells and is more pleomorphic, and corresponds to a Grade III/IV SCC (Fig. 11). Cell borders are generally distinct and intercellular bridges are identified. The tumor invades diffusely into and through the bladder wall, involving the perivesical fat. The bladder mucosa shows extensive squamous metaplasia, frequently associated with dysplasia and severe chronic inflammation. Figure 12 is a phase contrast photograph of the SCaBER cell culture, displaying aggregates of polygonal epithelial cells.

D. TCC SUP

This tumor shows features of a small-cell carcinoma of intermediate cell type (Fig. 13). Thus, the tumor cells are markedly hyperchromatic with sparse cytoplasm (Fig. 14). Some of the tumor cells are oval or somewhat spindly. The tumor shows markedly invasive growth and the invasion involves both the lamina propria and the detrusor muscle. Focally, extensive necrosis and

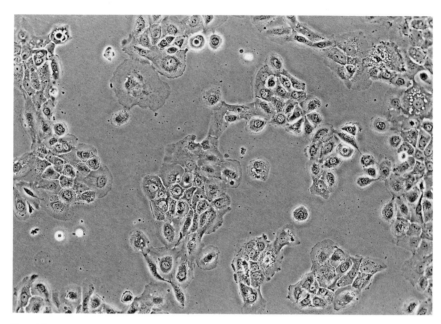

Fig. 12. Phase-contrast photograph of tissue culture specimen. Note the aggregates of the epithelial polygonal cells (H&E, 100×).

calcification is present, as is hemorrhage. Figure 15 is a phase-contrast photograph of the TCC SUP cell line, displaying relatively spindly tumor cells. Interestingly, small-cell carcinoma of the bladder was described first in 1981 (Cramer et al., 1981). This tumor generally is seen in individuals over 50 years of age; male patients predominate. The tumor is aggressive and most patients succumb to their disease within 2 yr. The TCC SUP donor died shortly after diagnosis. Rare long-term survivors have been reported (Podesta and True, 1989). The initial tumor in patient T24 was a small-cell carcinoma that was cured by partial bladder resection in 1962. The T24 cell line was established from a Grade III urothelial carcinoma 8 years later.

E. J82

The material for J82 is composed of abundant TUR fragments that show features of a predominantly papillary TCC that varies somewhat in different areas. Thus, focally, the tumor is very well differentiated, corresponding to a Grade I lesion. In other areas, the tumor is a Grade II tumor (Fig. 16), the predominant pattern, and focally the tumor is more pleomorphic, corresponding to a Grade III tumor (Fig. 17). The tumor invades superficially in the lamina propria focally, but no evidence of invasion into the detrusor muscle is seen (Figs. 18,19). Figure 20 is a phase-contrast photograph of the J82 cell

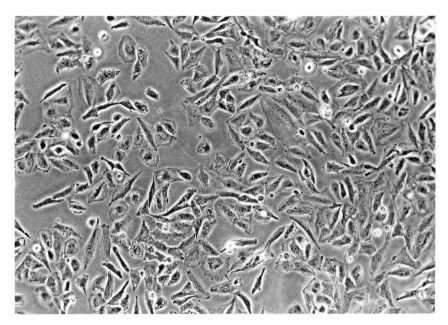

Fig. 15. Phase-contrast photograph of the tissue culture of the small cell carcinoma. Note the spindly configuration of the tumor cells (H&E, 100×).

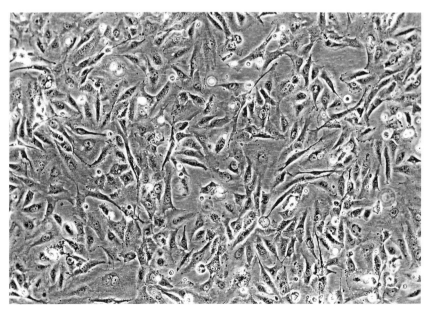

Fig. 20. Phase-contrast photograph of tissue culture showing tumor cells with spindly, round or polygonal shape (H&E, 100×).

Table I
Bladder Cancer Cell Lines

Cell line	Origin[a]	Tumor				Donor		
		Clinical stage	Pathological stage	Grade[b]	Blood group	Sex	Age (yr)	Karyotype
J82	TCC	T3	pT1	III/IV	A−	M	58	XYY + 5 markers
TCCSUP	SmCC	T4	pT3b	IV/IV	O+	F	67	XX + 9 markers
T24	TCC	T3	pT2-3	III/IV	O+	F	81	XX + 5 markers
RT4	TCC	T2	pT2	II/IV	O+	M	63	Y + 2 markers
SCaBER	SCC	T4	pT3b	III/IV	O+	M	58	XYY + 8 markers

[a] TCC, Transitional cell carcinoma; SCC, squamous cell carcinoma; SmCC, small cell carcinoma.
[b] According to Bergkvist et al. (1965), this grading system can be translated easily into the WHO classification. Grade IV, Undifferentiated carcinoma.

line composed of spindly, round, or polygonal cells. Table I summarizes the clinical data concerning the five estblished bladder cancer cell lines.

IV. Establishment and Maintenance of Bladder Cancer Lines

Each line was initiated from explants. Tissue culture medium was Medium 199 with Hanks' salts supplemented with 0.3 mg/ml glutamine, 100 IU penicillin, and 100 μg/ml streptomycin. Heat-inactivated (56°C/1 hr) fetal bovine serum (FBS) was used at 10 or 20% final volume. The tumor tissue was dissected into 1- to 2-mm cubes and placed on a medium-moistened surface in a 25-cm^2 culture flask. Flasks were incubated for 1 hr at 37°C in air and 5% CO_2. Culture medium (3 ml/flask) containing 20% FBS was added without detaching the explants. The medium was replaced every 2–5 days. After outgrowth from explants to monolayer, the cells were passaged by treatment at 37°C with a Ca^{2+}- and Mg^{2+}-free salt solution containing 0.05% trypsin and 0.02% ethylene diamine tetraacetic acid (EDTA). After washing in culture medium, the cells were seeded at $0.5 \times 10^6/25$ cm^2. After 5–10 passages, the concentration of FBS was reduced to 10%. On later passages, the cells were seeded routinely at densities as low as 5×10^3 cells/25 cm^2. Note that no bladder cancer cell line has been initiated *in vitro* to date from single-cell preparations directly from tumor tissue (Rigby and Franks, 1970; C.M. O'Toole, personal observation).

V. Mycoplasma Testing

Cultures were monitored routinely for mycoplasma contamination using the fluorescent DNA staining method of Russell *et al.* (1975) or by electron microscopy. Cells were grown on coverslips and incubated with 4,6-diamidino-2-phenylindole (DAPI) solution at 1 μg/ml in phosphate-buffered saline (PBS) for 30 min at 37°C. The coverslips were washed in PBS and examined under a fluorescence microscope using an incident excitation of 365 nm and emission at 450 nm. Plasma membrane fluorescence would reveal the presence of mycoplasma. Only the RT4 cell line was found to be mycoplasma infected. The agent was *Mycoplasma orale*. Ciprofloxicin (10 μg/ml) is now the most successful means of controlling mycoplasma contamination in cell cultures.

VI. Cryopreservation of Cell Lines

Cells from all *in vitro* passages were cryopreserved successfully using a freezing mixture of Medium 199 containing 10% FBS and 10% dimethyl sulfoxide (DMSO). Cells were trypsinized, washed in culture medium, and resuspended in precooled 4°C freezing solution at 0.5–2 × 10^6 cells/ml. A controlled freezing rate of 1°C/min was used from 4°C to −70°C. The cells then were transferred to the gas phase of liquid nitrogen. To recover cryopreserved cells, ampules were warmed rapidly to 37°C and then diluted immediately with a 10-fold volume of warm Medium 199 containing 25% FBS. After centrifuging at 300 *g* for 8–10 min, medium was removed and cells were placed in fresh medium in culture.

VII. Characterization of Lines

A. Ultrastructure

Electron microscopic examination of the tumor and tissue culture specimens was performed only on SCaBER; the two specimens are in good agreement. Abundant tonofilaments are oriented tangentially toward the cell surface; numerous desmosomes are present as well. The tumor cells show prominent nucleoli and lipid bodies are seen in the perinuclear area of the cytoplasm (O'Toole *et al.*, 1976).

Lines T24, J82, RT4, and TCC SUP, when examined by electron microscopy, lacked desmosomes but displayed pleomorphic nuclei, variable nucleus:cytoplasm ratios, numerous nucleoli, lipid bodies, and microfilaments.

B. HLA Phenotyping

HLA Class I antigen expression by these lines could not be detected with the standard National Institutes of Health (NIH) technique using complement-dependent cytotoxicity (CDC) (Festenstein *et al.*, 1973) because of complement toxicity and low titer of most alloantisera in CDC. However, HLA A, B, and C alloantigens were identified reproducibly using antibody-dependent cell-mediated cytotoxicity (ADCC) in a ^{51}Cr isotope release assay (O'Toole *et al.*, 1982,1983a,b). Cells were collected from monolayer culture by trypsinization, washed, and placed in culture at 0.25 × 10^6/25 cm^2 flask for 24 hr. The medium was replenished and 50 μCi $Na_2{}^{51}CrO_4$ (specific activity 350–600 mCi/mg Cr) was added to each flask. After an 18-hr incubation, the cells were harvested and used in cytotoxicity assays. For ADCC, lymphocytes from healthy donors were used as effector cells in a 50:1 effector:target cell ratio with a 4- to 8-hr assay. Fresh lymphocytes were available for HLA typing from the donors of J82, SCaBER, and TCC SUP lines. These lines were typed by CDC. Table II shows that full agreement on HLA A, B, and C antigens on lymphocytes and bladder cancer cells was found for these patients. Each of the five bladder cancer cell lines showed a distinct HLA Class I profile. The ADCC method of HLA antigen identification proved invaluable in establishing the cross-contamination of other putative bladder cancer cell lines by the fast growing T24 line (O'Toole *et al.*, 1983b).

C. HLA Class II Antigen Expression

Five lines—T24, RT4, J82, TCC SUP, and SCaBER—were compared before and after treatment with recombinant interferon-γ (r-IFNγ). The lines were seeded at a density of 10^6 cells/25 cm^2. After attachment, purified r-IFNγ (13×10^6 U/ml; specific activity 7×10^7 U/mg protein; a gift from Biogen, Geneva) was added. The cultures were incubated 3–4 days before assay.

To detect HLA Class II antigen, cell lines were harvested with trypsin/ EDTA, then washed in medium. Mouse monoclonal antibodies (MAB) against monomorphic epitopes on HLA DR, DP, and DQ were used in an indirect radiobinding assay in which 0.5×10^6 cells were reacted with MAB for 1 hr at 4°C. The cells were washed extensively in PBS; then 150,000 cpm ^{125}I-labeled F(ab')$_2$ fragment of sheep anti-mouse antibody was added for 1 hr at 4°C. After three washes, the residual bond radioactivity was measured. The results are seen in Table II and Figs. 21,22. An Epstein–Barr virus-transformed cell line was established from blood lymphocytes from the donor of HTB1. This line was used for comparison with the tumor cell line in HLA studies (O'Toole *et al.*, 1983a,b).

D. Lymphocyte Cytotoxicity Assays

Peripheral blood lymphoid cells were separated from defibrinated or heparinized blood. The sample was diluted 1/1 in Tris Hanks' solution (TH) and separated on Ficoll–Hypaque. The interface cells were collected and

Table II

Summary of HLA Class I and Class II Antigens Detected on Established Cell Lines of Urothelial Origin

Line	Class I[a]			Class II	
	HLA-A	HLA-B	HLA-Cw	Untreated	Interferon-γ treated
RT4	2,3	12	5	DP, DQ, DR	DP, DQ, DR
J82[b]	2,w32	5,12	5	Negative	DR
TCCSuP[b]	2,3	7,12	5	Negative	DR
SCaBER[b]	2	5,17	?	Negative	DR
T24	1,3	18	5	Negative	Negative

[a] For details on detection of HLA Class I antigens on cell lines, see O'Toole *et al.* (1982; 1983ab).
[b] HLA Class I antigen typing confirmed on patient's lymphocytes by conventional HLA typing.

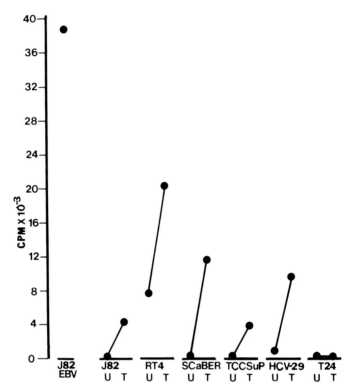

Fig. 21. Indirect radiobinding assay using HLA Class II anti-HLA-DR monoclonal antibody (Q5/13). U, untreated cell lines incubated with medium; T, interferon-γ (IFNγ) treated cells. Cell lines were incubated for 4 days with r-IFNγ. The RT4 line received 500 units r-IFNγ/10^6 cells; all other lines received 2500 units r-IFNγ/10^6 cells.

washed twice in TH. Adherent cells were removed by incubation on nylon fiber or tissue culture grade plastic. After thorough washing, lymphocytes were cultured for 12–18 hr and added to target cells in ratios of 100 : 1, 50 : 1, and 25 : 1. Targets were prepared from bladder cancer cell lines by trypsinization. The cells were labeled with ^{51}Cr and placed in microplates at 5 × 10^3/well. Cytolysis was measured after an 18-hr incubation. Lymphocytes from patients with TCC were compared with those from healthy donors and other nontumor control groups (Table III) (O'Toole, 1977).

The results show that T24 and RT4 cell lines were lysed selectively by lymphocytes from patients with localized TCC. To date, the surface structures on T24 and RT4 that are recognized in this reaction have not been identified.

E. Human Antibodies

Sera from patients with advanced or metastatic SCC contain IgM antibodies that react with the SCaBER cell line and another squamous cell line

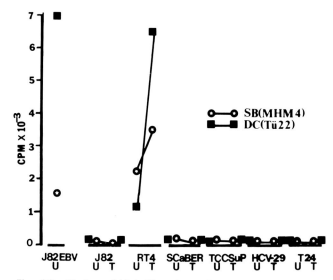

Fig. 22. Binding of HLA Class II anti-DP (MHM4) and anti-DQ (Tü22) monoclonal antibodies to cell lines before and after exposure to r-IFNγ. Other conditions as in Fig. 21.

(COLO 16), but not with J82, RT4, TCC SUP, or T24. IgM antibody is detected in a quantitative indirect immunofluorescence assay using fluorescence-labeled goat anti-human-IgM antibodies. Serum from the donor of the SCaBER cell line, obtained 2 months after cystectomy when the patient had a metastatic tumor, showed specific IgM binding to the SCaBER line in this assay (Sofen and O'Toole, 1978).

Antibodies in sera from patients with TCC that recognized TCC cell lines have not been shown convincingly, with the exception of the report by

Table III

Comparison of Cytotoxicity of Lymphocytes from Patients with Transitional Cell Carcinoma and Those from Normal Healthy Donors on Bladder Cancer Cell Lines[a]

		Cell Line		
Effector cells[b]	E/T	T24	RT4	J82
Transitional cell carcinoma	100/1	19 ± 15^{c}	36 ± 19^{d}	8 ± 6
	50/1	15 ± 11^{c}	26 ± 19^{d}	
Normal	100/1	5 ± 4	22 ± 13	3 ± 2
	50/1	4 ± 3	14 ± 7	1 ± 1

[a] Percentage cytotoxicity given as mean $\pm$ SD.
[b] ≥ 12 samples per group. Differences between groups compared by Student's t test.
[c] $p < 0.001$.
[d] $p < 0.05$.

Table IV
Isoenzyme Patterns of Human Bladder Cancer Cell Lines[a]

Cell line	Me-2	PGM3	PGM1	ESD	AK1	GLO	G6PD
J82 (0.0041)	1–2	2	1	1	1	2	B
RT4 (0.0050)	1	1–2	1–2	1–2	1	1–2	B
ScaBER (0.0008)	2	1–2	1–2	1	1	1–2	A
T24 (0.0216)	1–2	1	1	1	1	1	B
TCCSUP (0.0005)	1	1	2	1	1–2	1–2	B

[a] Phenotype frequency indicated in parentheses. Abbreviations: Me-2, mahc enzyme; PGM3,1, phosphoglucomutase bars 3 and 1; ESD, esterase D; AK1, adenylate kinase; GLO, glyoxylase; G6PD, glucose 6-phosphate dehydrogenase.

Bubenik *et al.* (1973) in which serum from the donor of the T24 line showed weak complement-dependent lysis of autologous targets from the T24 cell line.

F. Isoenzyme Pattern

These enzyme patterns are presented in Table IV (O'Toole *et al.*, 1983b).

G. Chromosome Analysis

A summary of the results of the karyotyping is given in Table V. No consistent pattern of chromosomal aberrations has been identified. The karyotypes derived from cell line J82 are illustrated in Fig. 23. Note the presence of an extra Y chromosome. The patient's blood lymphocytes showed a normal diploid karyotype (O'Toole *et al.*, 1978).

Table V
Karyotypes of Bladder Tumor Cell Lines[a]

Cell line	Chromosomal aberrations	Reference
TCC SUP	No mode number; hypotetraploid; XX; ≥9 markers	Nayak *et al.* (1977)
SCaBer	No mode number; hypotetraploid; XY and in some cells (XYY) after 25 passages; ≥8 markers	O'Toole *et al.* (1976)
J82	Mode 72; hypotetraploid; XYY; ≥5 markers	O'Toole *et al.* (1978)
T24	Stemline 86; XX; ≥5 markers	Bubenik *et al.* (1973); O'Toole (unpublished observations)
RT4	Mode 48; 1 Y identified; ≥2 markers	Rigby and Franks (1970)

[a] XYY also found by Grace Cannon in ATCC HTB9 cell line derived from a primary epidermoid carcinoma Grade II.

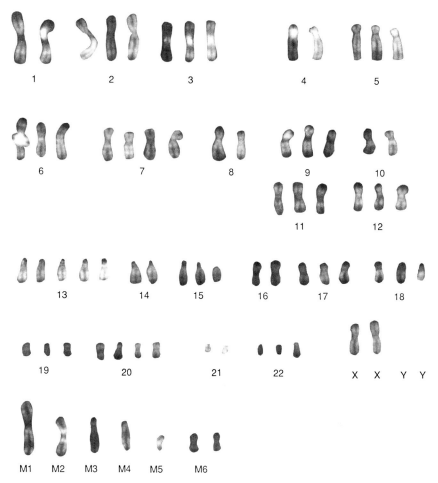

Fig. 23. The karyotype derived from J82 at *in vitro* passage 35. Note the presence of an extra Y chromosome.

VIII. Summary

Cell lines derived from five human bladder tumors are presented here. Each line has a distinct HLA profile and isoenzyme pattern. The lines differ in susceptibility to cytotoxicity of lymphoid cells isolated from patients with bladder cancer. IgM antibodies specific to the SCC cell line HTB3/SCaBER are found in patients with advanced squamous cell cancers.

The correspondence in morphology when comparing the light microscopic appearance of the original tumors with the appearance of the cells in the cultures is excellent in RT4 (HTB2) and SCaBER (HTB3). The epithelial nature of the cell cultures is striking and is very similar to the histology. (T24

(HTB4) has large polygonal cells histopathologically, but more elongated spindly cells in the culture, as is true for J82 (HTB1), although cells with epithelial features are present. In TCC SUP (HTB5), the cell culture shows a predominance of spindle-shaped cells with relatively few of the round to oval cells that predominate histopathologically.

Acknowledgments

We acknowledge the financial support from the FRF and PKF foundations and from the Foundation of the Clinical Oncological Laboratory at the Department of General Oncology, Sahlgren Hospital, Göteborg, Sweden.

References

Bergkvist, A., Ljungqvist, A., and Moberger, G. (1965). Classification of bladder tumors based on the cellular pattern. *Acta Chir. Scand.* **130,** 371–378.

Bubenik, J., Baresova, M., Viklicky, V., Jakoubkova, J., Sainerova, H., and Donner, J. (1973). Established cell line of urinary bladder carcinoma (T24) containing tumor specific antigen. *Int. J. Cancer* **11,** 765–773.

Burrows, M. T., Burns, J. E., and Suzuki, Y. (1917). Cultivation of bladder and prostatic tumors outside the body. *Johns Hopkins Hosp. Bull.* **28,** 178.

Cramer, S. F., Aikawa, M., and Cebelin, M. (1981). Neurosecretory granules in small cell invasive carcinoma of the urinary bladder. *Cancer* **47,** 724–730.

Festenstein, H., Adams, E., Burke, J., Oliver, R. T. D., Sachs, J. A., and Wolfe, E. (1973). The distribution of HLA-A antigens in expatriates from East Bengal living in London. *In* "Histocompatibility Testing 1972" (J. Dansset and J. Colombani, eds.) pp. 175–180. Munksgaard, Copenhagen.

Mostofi, F. K., Torloni, H., and Sobin, L. H. (1973). Histological typing of bladder tumours. *In* "International Classification of Tumours." World Health Organization, Geneva.

Nayak, S. K., O'Toole, C., and Price, Z. H. (1977). Characteristics of a cell line (TCCSUP) derived from an anaplastic transitional cell carcinoma of human urinary bladder. *Br. J. Cancer* **35,** 142–151.

O'loole, C. (1977). A [51]chromium isotope release assay for detecting cytotoxicity to human bladder carcinoma. *Int. J. Cancer* **19,** 324–331.

O'Toole, C. M. (1986). Human bladder cancer cell lines: HLA class I and class II antigen expression and susceptibility to cytostatic and cytolytic effects *in vitro*. *In* "*In Vitro* Models for Cancer Research" (M. M. Webber and L. I. Sekely, eds.), Vol. IV, 103–125. CRC Press, Boca Raton, Florida.

O'Toole, C., Nayak, S., Price, Z., Gilbert, W. H., and Waisman, J. (1976). A cell line (SCaBER) derived from squamous cell carcinoma of the human urinary bladder. *Int. J. Cancer* **17,** 707–714.

O'Toole, C., Price, Z. H., Ohnuki, Y., and Unsgaard, B. (1978). Ultrastructure, karyology and immunology of a cell line originated from a human transitional cell carcinoma. *Br. J. Cancer* **38,** 64–76.

O'Toole, C. M., Tiptaft, R. C., and Stevens, A. (1982). HLA antigen expression on urothelial cells: Detection by antibody-dependent cell-mediated cytotoxicity. *Int. J. Cancer* **29,** 391–395.

O'Toole, C. M., Lewis, C. M., and Wolf, E. (1983a). Detection of HLA antigens on lymphoblastoid and epithelial cell lines and cross-reactivity of HLA-Cw5 and HLA-Cw8. *Hum. Immunol.* **6,** 119–131.

O'Toole, C. M., Povey, S., Hepburn, P., and Franks, L. M. (1983b). Identity of some human bladder cancer cell lines. *Nature* (*London*) **301,** 429–430.

Podesta, A. H., and True, L. D. (1989). Small cell carcinoma of the bladder. Report of five cases with immunohistochemistry and review of the literature with evaluation of prognosis according to stage. *Cancer* **64,** 710–714.

Ribgy, C. C., and Franks, L. M. (1970). A human tissue culture cell line from a transitional cell tumour of the urinary bladder: Growth, chromosome pattern and ultrastructure. *Br. J. Cancer* **24,** 746–754.

Russell, W. C., Newman, C., and Williamson, D. H. (1975). A simple cytochemical technique for demonstration of DNA in cells infected with mycoplasma and virus. *Nature* (*London*) **253,** 461–466.

Sofen, H., and O'Toole, C. (1978). Anti-squamous tumor antibodies in patients with squamous cell carcinoma. *Cancer Res.* **38,** 199–203.

Williams, R. D. (1980). Human urologic cancer cell lines. *Invest. Urol.* **17,** 359–370.

The Female Reproductive System: Cell Lines from Tumors of the Human Ovary and Uterus

14

Y.-C. Hung and P. G. Satyaswaroop

Department of Obstetrics and Gynecology
Milton S. Hershey Medical Center
Pennsylvania State University, Hershey, Pennsylvania 17033

S. Tabibzadeh

Department of Pathology
Moffitt Cancer Center and
University of South Florida Health Science Center
Tampa, Florida 33682

I. Introduction 359

II. Methods of Establishment and Maintenance of Cell Lines 360
  A. Ovarian Cell Lines 360
  B. Endometrial Carcinoma Cell Lines 362

III. Morphological Aspects 363

  A. Ovarian Cancer Cell Lines 364
  B. Uterine Cancer Cell Lines 371

IV. Growth and Other Characteristics 377

V. Future Prospects 384

References 384

I. Introduction

The ovary and uterus are the major sites of malignancies of the female genital tract in the United States. Ovarian carcinoma is currently the fourth most fatal form of cancer in American women. Indeed, more women die of ovarian cancer than of all other gynecological malignances. About 12,000 women die of this disease every year (American Cancer Society, 1991). The median survival is about 1.2 yr and the 5-yr survival rate for patients with advanced disease is about 20%. Ovarian tumors are difficult to diagnose. At the time of diagnosis, ~70% of patients present with advanced stage disease (Stage III or IV). Effective treatment strategies are currently nonexistent for this disease, primarily because of the poor understanding of the biological behavior of this tumor.

Endometrial carcinoma is the most common gynecological malignancy; about 39,000 new cases are diagnosed every year (Silverberg, 1982). The

359

majority of the neoplasms are diagnosed in postmenopausal women, and surgery alone results in a cure rate of 85–90% of patients. About 3300 deaths occur in women with endometrial carcinoma, mostly because of advanced, recurrent, or metastatic disease. Since the endometrium is a steroid hormone responsive tissue, and about 25% of patients with endometrial carcinoma respond to progestin therapy, many studies have focused on the steroid hormone receptor determinations and steroid sensitivity of these tumors. One of the goals of these studies is the development of a predictive test for steroid sensitivity of endometrial tumors, to enable the clinician to select patients for hormonal or cytotoxic chemotherapy.

As for cancers of various other organs, establishment of permanent human ovarian and uterine tumor cell lines in continuous culture and the study of their growth regulation *in vitro* has proven to be one of the most useful techniques toward understanding these cells. The established tumor cell lines (Table I) serve as a valuable resource for a multitude of investigations. They have been used extensively in the identification of growth factors, growth factor receptors, and oncogene expression; in the study of the mechanisms of action of steroids, antisteroids, and other drugs; in the development of predictive tests of steroid and drug sensitivity; in the characterization of tumor markers; and in the generation of specific monoclonal antibodies with potential applications in tumor diagnosis, monitoring disease progression, and development of novel treatment strategies. In addition to being a ready source material for initial screening and for use as a positive standard in a variety of biochemical analysis, these cell lines have been valuable in the development of experimental *in vivo* model systems for these diseases. In this chapter, we review the various established ovarian and uterine cell lines, their growth and morphological characteristics, and their utility in the development of different tests with potential application in tumor biology.

II. Methods of Establishment and Maintenance of Cell Lines

A. *Ovarian Cell Lines*

Since 1970, several permanent ovarian cancer cell lines have been established in culture. Tumors of human ovarian surface epithelium constitute more than 80% of all ovarian tumors. Most patients with such afflictions seek medical attention at advanced stages of the disease, primarily because of the lack of diagnostic tests and the absence of any overt symptoms. At this stage, the tumors are large and have considerable malignant ascites. Culturing and establishing ovarian carcinoma cell lines from this material has proven relatively easy.

In general, the solid ovarian tumors were minced into 1- to 2-mm^3 pieces and dissociated enzymatically into small clusters of cells using collagenase

Table I
Established Cell Lines

Tumor	Origin	Reference
Human ovarian carcinoma cell lines		
SK-OV-3	Ovarian adenocarcinoma from malignant ascites	Fogh *et al.* (1977a)
SW 626	Ovarian adenocarcinoma from poorly differentiated cystadenocarcinoma	Fogh *et al.* (1977b)
CaOV-3	Ovarian adenocarcinoma	Hay *et al.* (1992)
CaOV-4	Ovarian adenocarcinoma from a fallopian tube metastasis	Hay *et al.* (1992)
PA-1	Ovarian teratocarcinoma from malignant ascites	Giovanella *et al.* (1974)
NIH:OVCAR 2	Ovarian adenocarcinoma	Hamilton *et al.* (1984)
NIH:OVCAR 3	Ovarian adenocarcinoma from malignant ascites	Hamilton *et al.* (1983)
NIH:OVCAR 4	Ovarian adenocarcinoma	Hamilton *et al.* (1984)
NIH:OVCAR 5	Ovarian adenocarcinoma	Hamilton *et al.* (1984)
NIH:OVCAR 7	Ovarian adenocarcinoma	Hamilton *et al.* (1984)
NIH:OVCAR 8	Ovarian adenocarcinoma	Hamilton *et al.* (1984)
NIH:OVCAR 9	Ovarian adenocarcinoma	Hamilton *et al.* (1984)
OVCA 420	Ovarian carcinoma	Bast *et al.* (1981)
OVCA 429	Ovarian carcinoma	Bast *et al.* (1981)
OVCA 432	Ovarian carcinoma	Bast *et al.* (1981)
OVCA 433	Ovarian carcinoma	Bast *et al.* (1981)
HEY	HY-62, a nude mouse xenograft of a moderately differentiated papillary cystadenocarcinoma of the ovary	Buick *et al.* (1985)
HEY A8	HEY after intraperitoneal growth in nude mice in the presence of ascitic fluid of ovarian cancer patients	Mills *et al.* (1990)
Human endometrial carcinoma cell lines		
HEC-1A	Endometrial adenocarcinoma from primary tumor	Kuramoto *et al.* (1972)
HEC-1B	Substrain of HEC-1A	Kuramoto *et al.* (1972)
SCRC	Moderately differentiated endometrial adenocarcinoma	Gorodecki *et al.* (1979)
AN3CA	Endometrial carcinoma from lymph node metastasis	Daine *et al.* (1964)
RL 95-2	Endometrial adenosquamous carcinoma from primary tumor	Way *et al.* (1983)
KLE	Endometrial adenocarcinoma from primary tumor	Richardson *et al.* (1984)
Ishikawa	Endometrial adenocarcinoma from primary tumor	Nishida *et al.* (1985)
ECC-1	Endometrial adenocarcinoma from primary tumor	Satyaswaroop *et al.* (1988)
Human uterine tumor cell lines		
SK-UT-1	Uterine mixed mesodermal tumor	Fogh and Trempe (1975)
SK-UT-1B	Substrain of SK-UT-1	Fogh and Trempe (1975)

with or without DNAse at 37°C for 1–4 hr. When malignant ascites were used as the source material, the cells were pelleted, extensively washed, and cultured in medium with 2–20% serum. In some cases, the clumps of tumor cells were filtered through nylon or metal screens (35–50 μm pore size). The tumor clumps are retained on the filters while the single blood and immune system cells pass through the pores. The tumor clumps were backwashed and cultured in nutrient medium. The tumor cells adhered to the plastic dish and began to outgrow nonmalignant cells during continuous culture. After several passages, the culture dishes contained essentially malignant ovarian cells, which then were used for morphological and growth characterization. In some cases, the monolayer cultures were transplanted subcutaneously into athymic nude mice. The solid tumors were excised from the animals and monolayer cultures were reestablished. Human ovarian tumor cell lines also have been established from human tumor xenografts in athymic mice. In this case, the ovarian tumors (solid or ascitic tumors) were grown after direct transplantation. Monolayer cultures were established after serial passages of the growing tumors.

B. Endometrial Carcinoma Cell Lines

In contrast to the relative ease of establishing ovarian carcinoma cell lines, human endometrial carcinoma cell lines have proven surprisingly difficult to establish. Relatively few permanent endometrial carcinoma cell lines are available for investigation. One of the first endometrial carcinoma cell lines was established in continuous monolayer cultures using the plasma clot culture method (Kuramoto, 1972). In general, the excised tumors from the primary or metastatic sites were minced into 2- to 3-mm^3 pieces and cultured as explants in nutrient medium containing 10–20% fetal bovine serum (FBS) until confluent monolayers were formed. The monolayer cells were trypsinized and passaged repeatedly until homogeneous epithelial cultures, free of fibroblasts, were established. The tumor explants appear to yield more active cultures than do enzymatic procedures, which yield isolated endometrial glands or epithelial cells (Satyaswaroop et al., 1979).

Permanent carcinoma cell lines also were established after direct subcutaneous transplantation of human endometrial tumors into athymic mice. Solid tumors, excised from athymic mice, were cultured in vitro in nutrient medium to derive cell lines for continuous culture (Merenda et al., 1975; Satyaswaroop et al., 1988).

Note that, although endometrium is exquisitely sensitive to steroid hormones in vivo, only two established endometrial carcinoma cell lines, Ishikawa and ECC-1, exhibit steroid responsiveness in vitro (Holinka et al., 1986; Tabibzadeh et al., 1990). This behavior may be attributed to the lability of steroid receptors and the loss of functional receptors under culture conditions (Satyaswaroop and Mortel, 1982). The absence of cell lines with their complement of receptors and responses to steroid hormones has limited significantly our understanding of the hormonal regulation of growth of endometrial carcinomas.

We have derived a series of cell lines, including one estrogen receptor (ER)-positive, estrogen-sensitive ECC-1 cell line (Satyaswaroop *et al.*, 1988), from endometrial tumors maintained by serial transplantation in the nude mouse model that was developed in our laboratory (Tabibzadeh *et al.*, 1988). The selection procedure used to derive the estrogen-sensitive cell line is summarized here:

EnCa-101 (Transplant 31, ER, progesterone receptor (PR) positive)
  $\downarrow$ collagenase digestion
Glandular units
  $\downarrow$ cultured in 10% FBS medium containing insulin, cholera toxin, transferrin, and
    estradiol (E_2)
Epithelial monolayer (101 AE7)
  $\downarrow$ cytokerin positive, Ki-67 positive, ER positive, PR negative, and E_2 insensitive for 25
    passages for 1 yr
101 AE7 (Passage 26)
  $\downarrow$ 5% cells intensely PR positive by immunocytochemistry; cloned twice and selected for
    PR positivity
ECC-1 cell line in continuous culture (ER-positive cells), respond to E_2 with stimulation of growth and PR levels

Although EnCa-101, the tumor from which this cell line was derived, consistently exhibited E_2 sensitivity in nude mice during serial transplantation for the past 8 yr, systematic monitoring of steroid receptors and their E_2 sensitivity during the various steps of the culture procedure, in addition to the selection and cloning of PR-positive cells, was vital to the establishment of this steroid-responsive cell line.

III. Morphological Aspects

After establishing tumor cell lines in continuous cultures, their morphological characterization at the light microscopic and ultrastructural levels are determined routinely. The epithelial, cobblestone-like morphology of the monolayer cultures usually is recognized as evidence of the epithelial origin of these cells. The immunocytochemical localization of cytokeratin with specific monoclonal antibodies has become mandatory in demonstrating the epithelial lineage of the established cells. Characterization, at the electron microscopic level, of the formation of desmosomes and microvilli provides additional confirmation of the epithelial characteristics of these cell lines. Karyotyping these cells establishes the female origin of these cells, in addition to generating information on the chromosomal patterns.

A. Ovarian Cancer Cell Lines

Monolayer cultures of various established ovarian carcinoma cell lines—OVCA 420, OVCA 429, OVCA 432, OVCA 433 (Bast *et al.*, 1981), NIH:OVCAR 3 (Hamilton *et al.*, 1983), NIH:OVCAR 5 (Hamilton *et al.*, 1984a), SK-OV-3 (Fogh *et al.*, 1977a), CaOV-3, CaOV-4 (Hay *et al.*, 1992), SW 626 (Fogh *et al.*, 1977b), A 2780 (Eva *et al.*, 1982), HEY (Buick *et al.*, 1985), and HEY A8 (Mills *et al.*, 1990)—exhibit a polygonal or cobblestone-like growth pattern (Figs. 1–13). Monolayer cultures of PA-1, a cell line derived from the ascites fluid of a patient with ovarian teratocarcinoma (Giovanella *et al.*, 1974), also exhibit epithelial morphology (Fig. 14).

Essentially all these ovarian cell lines form tumors when transplanted subcutaneously in athymic mice. Hamilton *et al.* (1984b) developed an intraperitoneal model of ovarian carcinoma that resembles the disease in women. After an intraperitoneal injection of 10–40 million NIH:OVCAR 3 cells into female athymic mice, the animals developed malignant ascites and intraabdominal tumor growth within 2–4 wk (Fig. 15). Death occurred in these animals between 40 and 80 days after injection. The malignant cells within

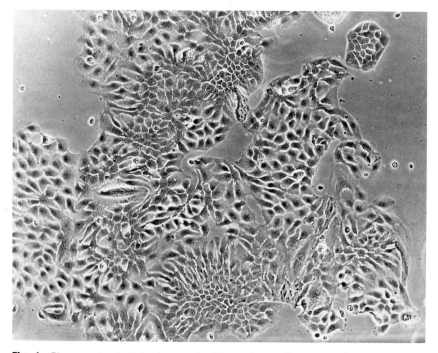

Fig. 1. Phase-contrast photomicrograph of live cultures of human ovarian carcinoma cell line OVCA 420. The cells growing as monolayers show a polygonal morphology (100×).

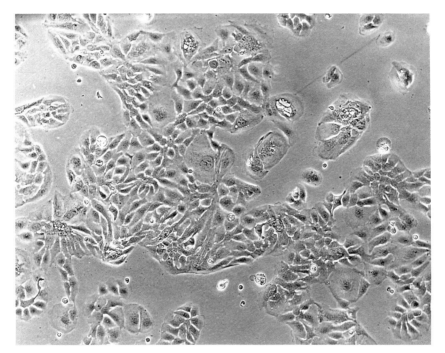

Fig. 2. Phase-contrast photomicrograph of live cultures of human ovarian carcinoma cell line OVCA 429. The cells growing as monolayers show a polygonal morphology (100×).

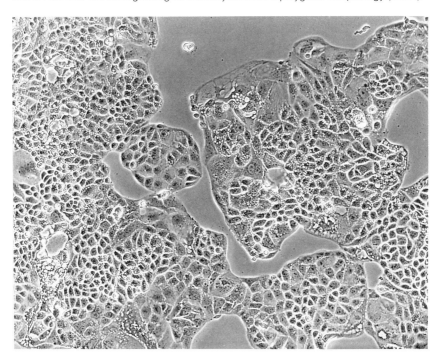

Fig. 3. Phase-contrast photomicrograph of live cultures of human ovarian carcinoma cell line OVCA 432. The cells growing as monolayers show a polygonal morphology (100×).

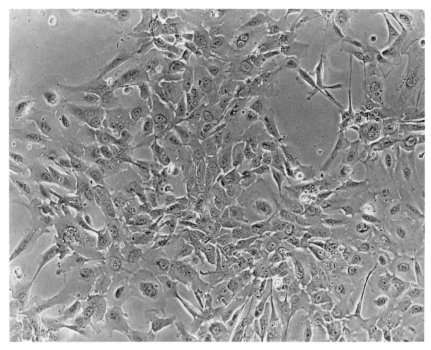

Fig. 4. Phase-contrast photomicrograph of live cultures of human ovarian carcinoma cell line OVCA 433. The cells growing as monolayers show a polygonal morphology (100×).

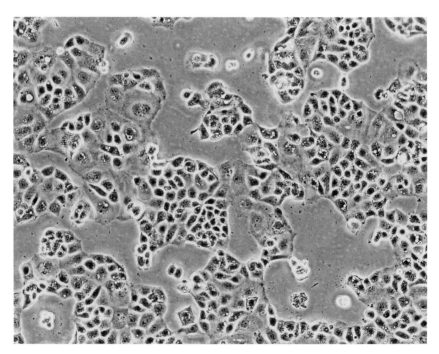

Fig. 5. Phase-contrast photomicrograph of live cultures of human ovarian carcinoma cell line OVCAR 3. The cells growing as monolayers show a polygonal morphology (125×).

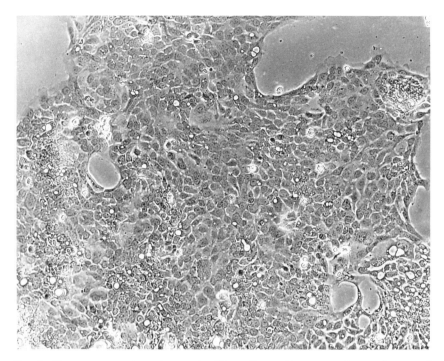

Fig. 6. Phase-contrast photomicrograph of live cultures of human ovarian carcinoma cell line OVCAR 5. The cells growing as monolayers show a polygonal morphology (100×).

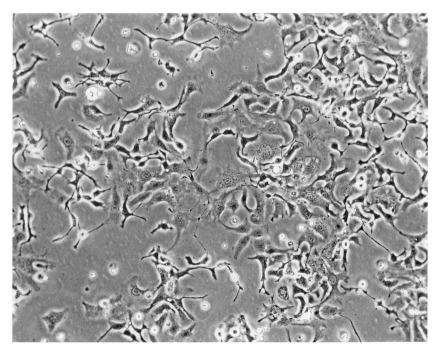

Fig. 7. Phase-contrast photomicrograph of live cultures of human ovarian carcinoma cell line SK-OV-3. The cells growing as monolayers show a polygonal morphology (125×).

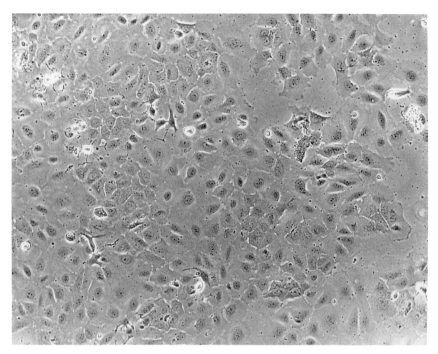

Fig. 8. Phase-contrast photomicrograph of live cultures of human ovarian carcinoma cell line CaOV-3. The cells growing as monolayers show a polygonal morphology (100×).

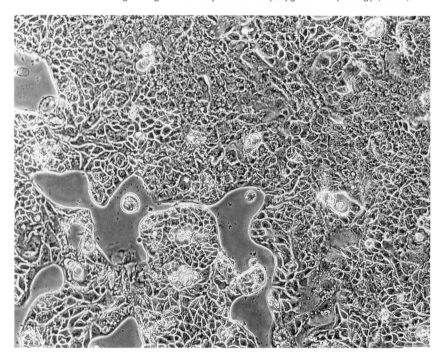

Fig. 9. Phase-contrast photomicrograph of live cultures of human ovarian carcinoma cell line CaOV-4. The cells growing as monolayers show a polygonal morphology (100×).

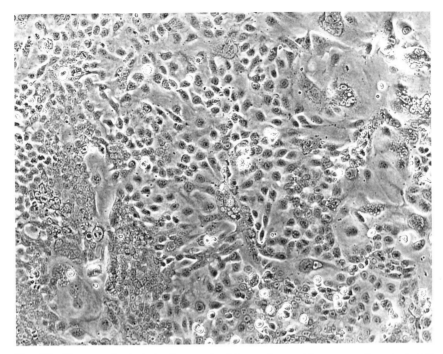

Fig. 10. Phase-contrast photomicrograph of live cultures of human ovarian carcinoma cell line SW 626. The cells growing as monolayers show a polygonal morphology (100×).

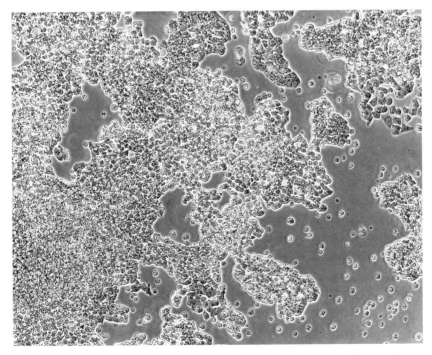

Fig. 11. Phase-contrast photomicrograph of live cultures of human ovarian carcinoma cell line A 2780. The cells growing as monolayers show a polygonal morphology (100×).

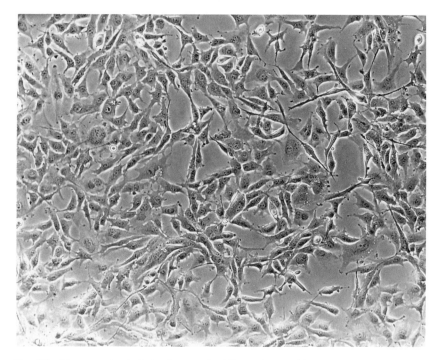

Fig. 12. Phase-contrast photomicrograph of live cultures of human ovarian carcinoma cell line HEY. The cells growing as monolayers show a polygonal morphology (100×).

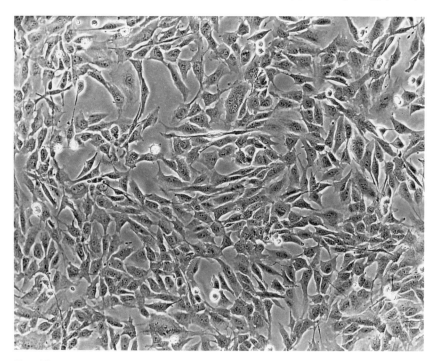

Fig. 13. Phase-contrast photomicrograph of live cultures of human ovarian carcinoma cell line HEY A8. The cells growing as monolayers show a polygonal morphology (100×).

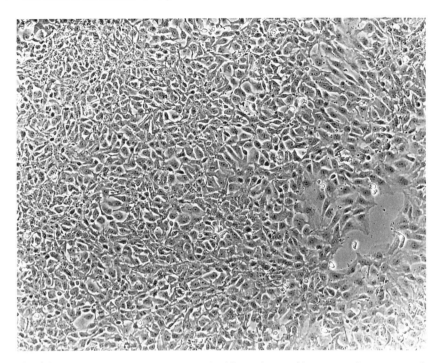

Fig. 14. Phase-contrast photomicrograph of live cultures of human ovarian teratocarcinoma cell line PA-1. The cells growing as monolayers show a polygonal morphology (100×).

the ascites fluid are transplantable. The intraperitoneal model is widely used in studying the cytotoxic effects of various drugs and in preclinical testing of various treatment strategies (Pirker *et al.*, 1985; FitzGerald *et al.*, 1986). A requirement for growth factor (presumably in the ascites fluid of ovarian cancer patients) for development of ascites in nude mice was reported using the HEY cell line (Mills *et al.*, 1988).

The morphological and growth characteristics of ovarian carcinoma cell lines also were examined in three-dimensional spheroid cultures *in vitro* and in subcutaneous tumors in athymic mice *in vivo*. Electron microscopic evaluation of the spheroid cultures and of the nude mouse-grown solid tumors of NIH : OVCAR 5 cells reveals the interaction of these cells to form tubular structures with a central lumen (Figs. 16,17).

B. Uterine Cancer Cell Lines

The SK-UT-1 cell line, derived from human uterine mixed mesodermal tumor (Fogh and Trempe, 1975), exhibits a fibroblast-like growth in monolayer cultures and a characteristic spindle-shaped appearance (Fig. 18). The American Type Culture Collection (ATCC), however, indicates the morphology as epithelial-like for the stocks transferred initially from the Sloan–

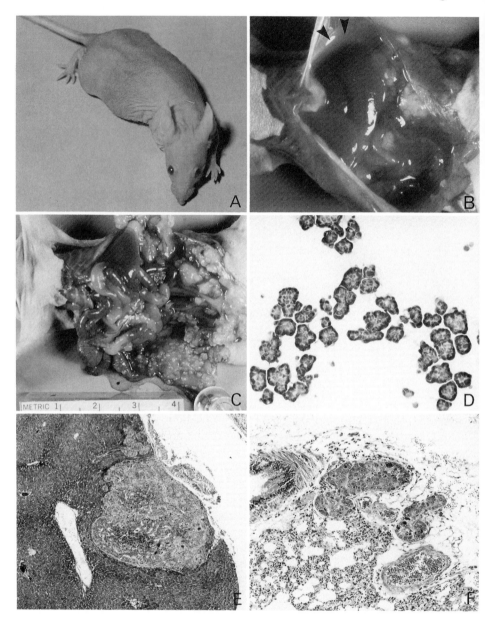

Fig. 15. Intraabdominal tumor growth and malignant ascites 45 days following intraperitoneal injection of 10 million NIH:OVCAR 3 cells. (A) Female mouse with abdominal distension. (B) Presence of ascites shown in A. (C) Presence of ovarian tumors throughout the peritoneal cavity. (D) Malignant cells in the ascites (Papanicolaou, 220×). (E) Surface invasion of the liver (H&E, 54×). (F) Metastatic lesion in lung of animal shown in A (H&E, 130×). (Figure kindly provided by Dr. T. C. Hamilton, Fox Chase Cancer Center, PA.)

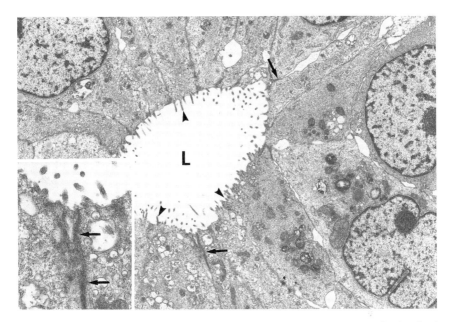

Fig. 16. Electron micrograph of a three-dimensional spheroid cultures (7 days) of NIH : OVCAR 5. The neoplastic cells within spheroid are arranged around a central lumen (L). Microvilli (arrowheads) protrude into the lumen. Epithelial cells exhibit junctions adjacent to the lumen (arrows) (7800×). (*Inset*) Presence of junctions including desmosomes (arrows) (23,000×).

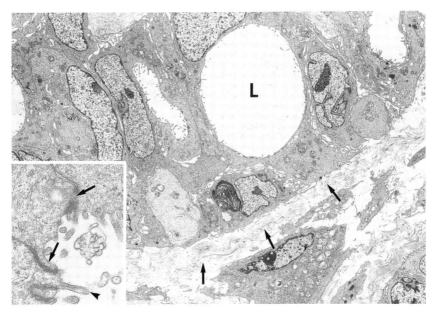

Fig. 17. Electron micrograph of NIH : OVCAR 5 cells grown as subcutaneous tumor in athymic mouse. The neoplastic cells exhibit lumen formation (L). Arrows point to the collagen bundles within the host stroma (9300×). (*Inset*) Neoplastic cells exhibit junctions (arrows). Microvilli (small arrowhead) protrude into the lumen (23,000×).

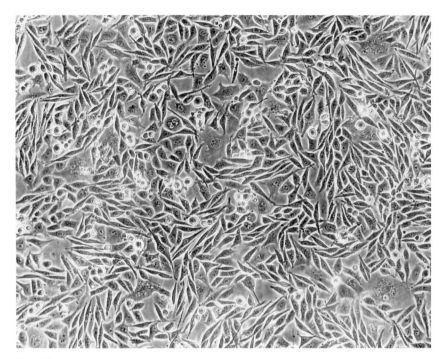

Fig. 18. Phase-contrast photomicrograph of live cultures of human uterine mixed meso-dermal tumor cell line SK-UT-1. The cells growing as monolayers show spindle-shaped fibroblast-like morphology (100×).

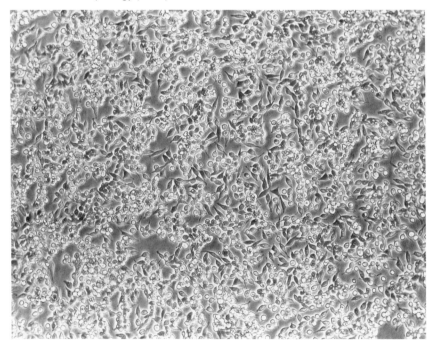

Fig. 19. Phase-contrast photomicrograph of live cultures of SK-UT-1B, a substrain of SK-UT-I cells. The monolayer cultures are composed of both spindle-shaped cells and cells with bead-like morphology (100×).

Kettering originators. These cells are reported to form a spindle cell carcinoma in nude mice. SK-UT-1B, a substrain of SK-UT-1, shows a less pronounced spindle-shaped appearance when grown as monolayers (Fig. 19). However, these cells differ from the parent line in their karyology and are reported to form well-differentiated adenocarcinomas in nude mice.

The established human endometrial carcinoma cell lines HEC 1A and 1B (Kuramoto *et al.*, 1972), SCRC (Gorodecki *et al.*, 1979), KLE (Richardson *et al.*, 1984), RL 95.2 (Way *et al.*, 1983), AN3Ca (Daine *et al.*, 1964), Ishikawa (Nishida *et al.*, 1985), and ECC-1 (Satyaswaroop *et al.*, 1988) were derived from primary as well as metastatic sites of different histological grades. These cells exhibit the typical cobblestone pattern of growth in monolayer cultures, a distinguishing characteristic of epithelial cells (Figs. 20–27). The epithelial nature of ECC-1 cells, one of the steroid receptor-positive E_2-responsive cell lines, is evident from the homogeneous immunoreactivity to cytokeratin (Fig. 28).

The morphological features of endometrial carcinoma cells also were examined in spheroid cultures and on growth of these cells as solid tumors in nude mice. When Ishikawa cells are allowed to organize themselves as three-dimensional spheroids by plating them over a layer of agarose, these endometrial adenocarcinoma cells form sheet-like structures and fail to form glandular structures (Fig. 29). Identical sheet-like structures also were ob-

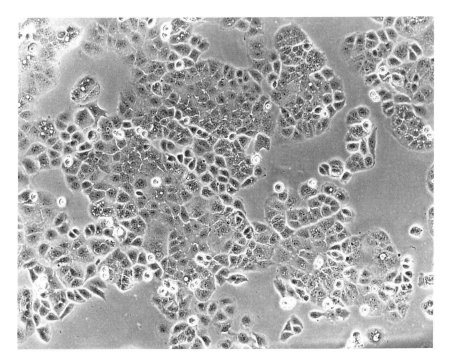

Fig. 20. Phase-contrast photomicrograph of live cultures of human endometrial carcinoma cell line HEC-1A. The cells growing as monolayers show a polygonal morphology (100×).

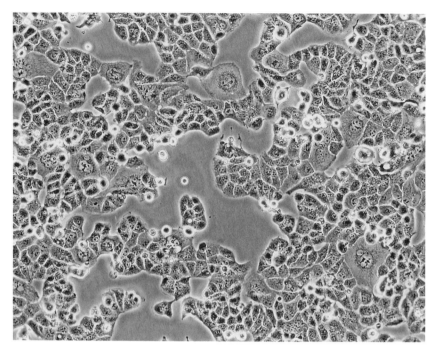

Fig. 21. Phase-contrast photomicrograph of live cultures of human endometrial carcinoma cell line HEC-1B. The cells growing as monolayers show a polygonal morphology (125×).

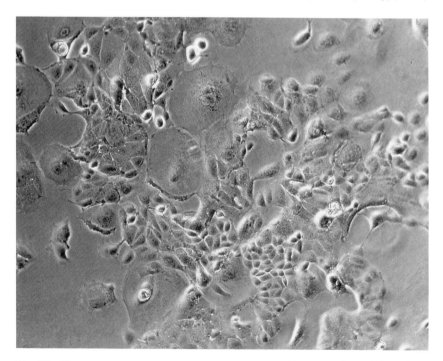

Fig. 22. Phase-contrast photomicrograph of live cultures of human endometrial carcinoma cell line SCRC. The cells growing as monolayers show a polygonal morphology (100×).

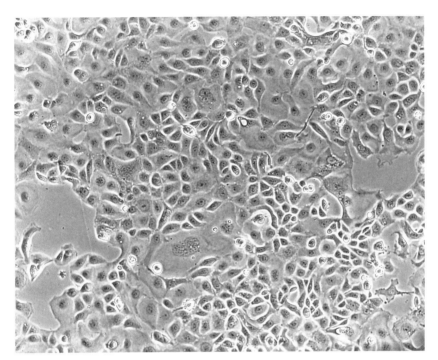

Fig. 23. Phase-contrast photomicrograph of live cultures of human endometrial carcinoma cell line KLE. The cells growing as monolayers show a polygonal morphology (100×).

served on growth of Ishikawa cells as subcutaneous tumors in athymic mice (Fig. 30). In contrast, the glandular organization of ECC-1 cells is observed when they are grown in the extracellular matrix Matrigel *in vitro* or as subcutaneous tumors in athymic mice (Satyaswaroop and Tabibzadeh, 1991).

IV. Growth and Other Characteristics

Growth regulation of ovarian and uterine cancers is the major focus of all studies that use these established cell lines. The direct or indirect effects of various growth factors, their synthetic agonists and antagonists, specific antibodies, steroids, antisteroids, and other drugs; the role of oncogenes and anti-oncogenes; and the autocrine, paracrine, and endocrine mechanisms regulating the growth of these tumor cells are investigated widely with the permanent cell lines. The growth rates of these cells are monitored routinely by cell counting using a hemocytometer or electronic cell counters over a period of several days. Several studies use [^{3}H]thymidine incorporation into DNA at specific time intervals in the control and treated cells to monitor the effects of various agents on growth. Nonradioisotopic methods, such as bro-

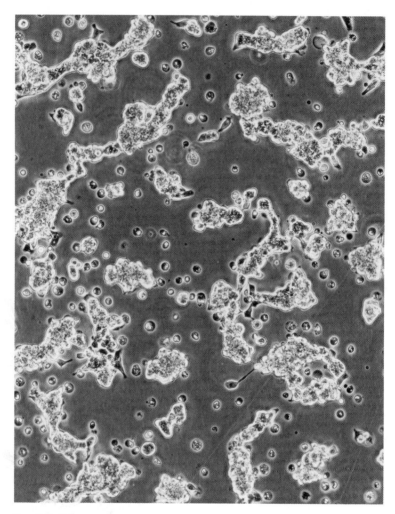

Fig. 24. Phase-contrast photomicrograph of live cultures of human endometrial carcinoma cell line RL 95.2. The cells growing as monolayers show a polygonal morphology (125×).

modeoxyuridine (BrdU) incorporation into DNA of cells in S phase followed
by immunolocalization of BrdU with specific monoclonal antibodies to determine the percentage of immunoreactive cells in each treatment group, are
also widely employed. Immunocytochemical localization of the proliferation
marker Ki-67 (Gerdes *et al.*, 1983), which is expressed by cells in the late
G_1, S, and G_2/M phases of the cell cycle, with specific monoclonal antibody
has gained popularity in studies addressing the growth effects of drugs and
other agents on cultured cancer cells.

Studies of the steroid regulation of growth of the human endometrial carcinoma cell line ECC-1 show the dose-dependent increase in growth in re-

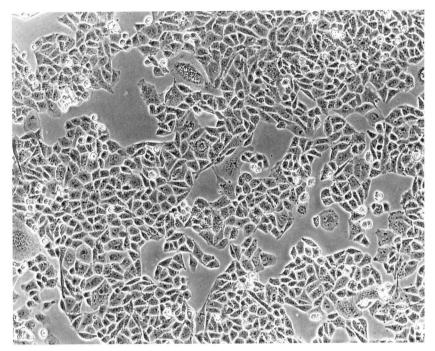

Fig. 25. Phase-contrast photomicrograph of live cultures of human endometrial carcinoma cell line AN3Ca. The cells growing as monolayers show a polygonal morphology (100×).

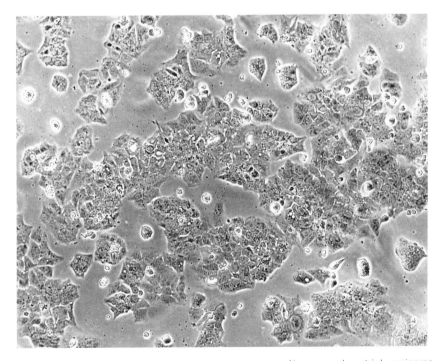

Fig. 26. Phase-contrast photomicrograph of live cultures of human endometrial carcinoma cell line Ishikawa. The cells growing as monolayers show a polygonal morphology (100×).

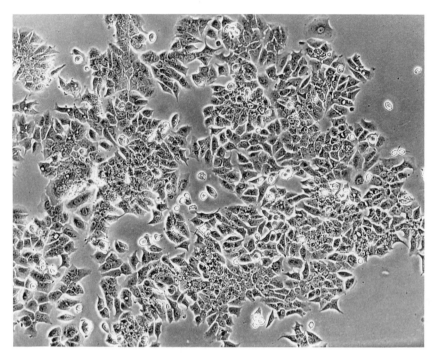

Fig. 27. Phase-contrast photomicrograph of live cultures of human endometrial carcinoma cell line ECC-101. The cells growing as monolayers show a polygonal morphology (100×).

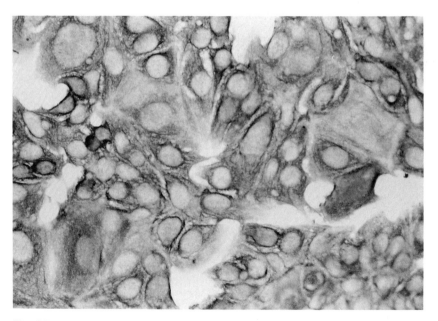

Fig. 28. The E_2-responsive ECC-1 cell line exhibits low molecular weight cytokeratins in their cytoplasm (300×).

Fig. 29. Electron micrograph of three-dimensional spheroid cultures (7 days) of the human endometrial adenocarcinoma cell line Ishikawa. The neoplastic cells within spheroids have formed sheet-like structures. Gland formation is lacking. The morphological appearance is similar to that observed when these cells are grown as subcutaneous tumors in athymic mice (1800×).

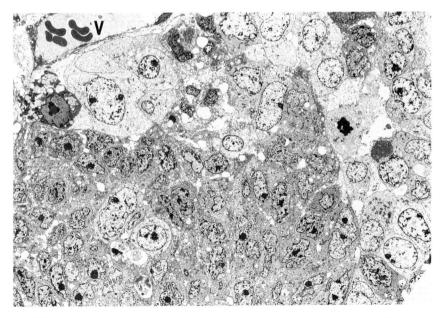

Fig. 30. Electron micrograph of a human endometrial adenocarcinoma cell line, Ishikawa, grown as subcutaneous tumors in athymic mice. The neoplastic cells of this adenocarcinoma form sheet-like structures and do not exhibit gland formation (1800×). V, Vessel.

sponse to exogenous E_2 (Fig. 31). Although the original cell line 101 AE7, established in monolayer culture from the nude mouse-grown tumors, failed to respond to exogenous E_2, cultures of the cloned cell line ECC-1, selected for its ability to express PR, demonstrated the growth response to increasing concentrations of the steroid. Parallel increases in the concentrations of PR also are observed consistently. Addition of E_2 to ECC-1 cultures results in enhanced nuclear immunostaining for PR, an estrogen-induced protein. Western blot analysis of PR using monoclonal antibodies generated against human PR (Clarke *et al.*, 1987) further confirmed the increased PR levels in ECC-1 cells after E_2 exposure. Enhanced immunoreactivity of protein bands 116 and 81 kDa in size, a characteristic feature of PR positivity, can be visualized in E_2-treated cells compared with controls (Fig. 32).

The growth of ECC-1 cells is regulated negatively by the cytokine interferon γ (IFNγ). Using the immunostaining procedure with Ki-67 and BrdU monoclonal antibodies, in addition to cell counting, the dose-dependent inhibition of ECC-1 growth by IFNγ was demonstrated *in vitro* (Tabizadeh *et al.*, 1990). The Ki-67-positive ECC-1 cells show characteristic punctate nuclear immunostaining with monoclonal antibody (Fig. 31).

The identification of an ovarian tumor marker, CA 125, and the generation of monoclonal antibodies against this protein were accomplished with estab-

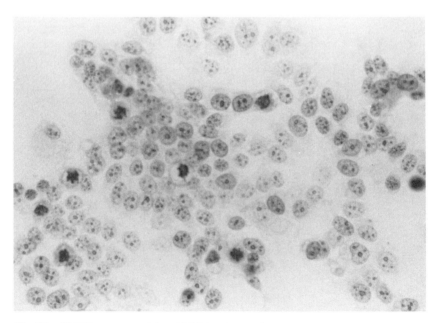

Fig. 31. Ki-67 immunoreactivity of ECC-1 cells proliferating in the presence of 200 pg/ml estradiol. The proliferative cells show characteristic punctate nuclear immunoreactivity (300×).

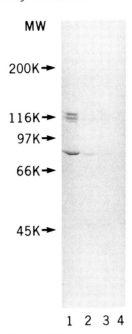

Fig. 32. Induction of PR protein by estradiol in ECC-1 cells. Cytosolic proteins of cell homogenates were separated by SDS-PAGE on 7.5% acrylamide gels (100 μg protein per lane). Gels were blotted onto nitrocellulose and immunoreactive proteins were revealed after incubation with $hPRa_1$, a monoclonal antibody raised against human PR. Lane 1, ECC-1 cells grown in the presence of 200 pg/ml estradiol; lane 2, ECC-1 grown in estradiol-free medium; lane 3, parent cell line 101 AE grown in the presence of estradiol; lane 4, 101 AE grown in the absence of estradiol.

lished ovarian cancer cell lines. Use of this monoclonal antibody to monitor disease progression in ovarian cancer patients is a significant advance. Bast *et al.* (1981) demonstrated the presence of a high molecular weight mucin-like protein in a series of cell lines they established from ovarian cancer patients. A murine monoclonal antibody against this high molecular weight protein, OC 125, was generated using the ovarian carcinoma cell line OVCA 433 (Bast *et al.*, 1981). A radioimmunoassay using this monoclonal antibody was developed later to monitor the disease course in patients with epithelial ovarian carcinomas (Bast *et al.*, 1983). The antibody kit has been commercialized and is used routinely to determine CA 125 levels in patients undergoing treatment for epithelial ovarian cancer. ^{125}I-labeled OC 125 also was used to radioimage small subcutaneously transplanted human ovarian tumors in nude mice. The applicability to this protein visualizing tumor distribution for maximal surgical excision in ovarian cancer patients has been suggested (Manetta *et al.*, 1987).

V. Future Prospects

One of the major problems confronting clinicians and researchers in the field of ovarian cancers is the absence of a noninvasive test for early detection of this malignancy. As indicated earlier, the low survival figures are primarily the result of the identification of this malignancy at very late stages. Survival of patients with early stage disease is significantly higher. Therefore, early detection of ovarian carcinoma will impact the high mortality figures in this disease significantly. In addition to OC 125, several clinically useful ovarian tumor-related monoclonal antibodies have been generated. Anticipating that more highly specific and sensitive ovarian tumor-specific antibodies will be developed is reasonable. The tumor markers, either measured singly or in a group with specific antibodies, will permit the early detection of the disease. Established human ovarian tumor cell lines should play a major role in hastening the development of such tests and furthering our understanding of the biology of this "silent slayer." Further, the newly developed in vivo model systems using the permanent ovarian carcinoma cell lines may be expected to facilitate the preclinical screening of novel treatment strategies.

The quest to establish endometrial carcinoma cell lines that are responsive to sex steroids has been thwarted by the lability of the steroid receptors in these cells under current culture conditions. Investigators are acutely aware of the need for repeated monitoring of the receptor status, not only of the original tumor tissue but also during the various steps involved in establishing permanent cell lines. The availability of highly specific monoclonal antibodies will simplify this process, and we can anticipate the establishment of more endometrial cell lines with their complement of receptors and steroid responses. Such novel cell lines, in conjunction with those currently available, should increase our understanding of the steroid regulation of human endometrial carcinoma growth and aid in the development of improved treatment modalities in the management of this disease.

Acknowledgments

The work presented here was supported by the Gustavus and Louise Pfeiffer Research Foundation. We thank Jeri Miller and Shane Miller for their technical assistance and Barbara S. Hynum in the preparation of the manuscript.

References

American Cancer Society (1991). Primary Care Newsletter. Volume 2, issue 1.
Bast, R. C., Jr., Feeney, M., Lazarus, H., Nadler, L. M., Colvin, R. B., and Knapp, R. C. (1981). Reactivity of a monoclonal antibody with human ovarian carcinoma. J. Clin. Invest. **68**, 1331–1337.

Bast, R. C., Klug, T. L., St. John, E., Jenison, E., Niloff, J. M., Lazarus, H., Berkowitz, R. S., Leavitt, T., Griffiths, C. T., Parker, L., Zurawski, V. R., and Knapp, R. C. (1983). A radioimmunoassay using a monoclonal antibody to monitor the course of epithelial ovarian cancer. *N. Engl. J. Med.* **309,** 883–887.

Buick, R. N., Pullano, R., and Trent, J. M. (1985). Comparative properties of five human ovarian carcinoma cell lines. *Cancer Res.* **45,** 3668–3676.

Clarke, C. L., Zaino, R. J., Feil, P. D., Miller, J. V., Steck, M. E., Ohlsson-Wilhelm, B., and Satyaswaroop, P. G. (1987). Monoclonal antibodies to human progesterone receptor: Characterization by biochemical and immunohistochemical techniques. *Endocrinology* **121,** 1123.

Daine, C. J., Banfield, W. G., Morgan, W. D., Slatik, M. S., and Curth, H. O. (1964). Growth in continuous culture and in hamsters of cells from a neoplasm associated with acanthosis nigricans. *J. Natl. Cancer Inst.* **33,** 441–456.

Eva, A., Robbins, K. C., Andersen, P. R., Srinivasan, A., Tronick, S. R., Reddy, E. P., Ellmore, N. W., Galen, A. T., Lautenberger, J. A., Papas, T. S., Westin, E. H., Wong-Staal, F., Gallo, R. C., and Aaronson, S. A. (1982). Cellular genes analagous to retroviral onc genes are transcribed in human tumor cells. *Nature (London)* **295,** 116–119.

FitzGerald, D. J., Willingham, M. C., and Pastan, I. (1986). Antitumor effects of an immunotoxin made with *Pseudomonas* exotoxin in a nude mouse model of human ovarian cancer. *Proc. Natl. Acad. Sci. U.S.A.* **83,** 6627–6630.

Fogh, J., and Trempe, G. (1975). New human tumor cell lines. *In* "Human Tumor Cells *In Vitro*" (J. Fogh, ed.), pp. 115–159. Plenum Press, New York.

Fogh, J., Wright, W. C., and Loveless, J. D. (1977a). Absence of HeLa cell contamination in 169 cell lines derived from human tumors. *J. Natl. Cancer Inst.* **58,** 209–214.

Fogh, J., Fogh, J. M., and Orfeo, T. (1977b). One hundred and twenty-seven cultured human tumor cell lines producing tumors in nude mice. *J. Natl. Cancer Inst.* **59,** 221–226.

Gerdes, J., Schwab, U., and Lemke, H. (1983). Production of a mouse monoclonal antibody reactive with a human nuclear antigen associated with cell proliferation. *Int. J. Cancer* **31,** 13–20.

Giovanella, B. C., Stehlin, J. S., and Williams, L. J., Jr. (1974). Heterotransplantation of human malignant tumors in "nude" thymusless mice. II. Malignant tumors induced by injection of cell cultures derived from human solid tumors. *J. Natl. Cancer Inst.* **52,** 921–930.

Gorodecki, J., Mortel, R., Ward, S. P., Geder, L., and Rapp, F. (1979). Establishment and characterization of a new endometrial carcinoma cell line (SCRC-1). *Am. J. Obstet. Gynecol.* **135,** 671–679.

Hamilton, T. C., Young, R. C., McCoy, W. M., Grotzinger, K. R., Green, J. A., Chu, E. W., Whang-Peng, J., Rogan, A. M., Green, W. R., and Ozols, R. F. (1983). Characterization of a human ovarian carcinoma cell line (NIH : OVCAR-3) with androgen and estrogen receptors. *Cancer Res.* **43,** 5379–5389.

Hamilton, T. C., Young, R. C., and Ozols, R. F. (1984a). Novel in vivo and in vitro models for the study of human ovarian cancers. *In* "Human Tumor Cloning" (E. E. Salmon and J. M. Trent, eds.), pp. 311–337. Grune & Stratton, New York.

Hamilton, T., Young, R. C., Louie, K. G., Behrens, B. C., McCoy, L. M., Grotzinger, K. R., and Ozols, R. F. (1984b). Characterization of a Xenograft model of human ovarian carcinoma which produces ascites and intra-abdominal carcinomatosis in mice. *Cancer Res.* **44,** 5286–5290.

Hay, R. J., Caputo, J., Chen, T. R., Macy, M. L., McClintock, P., and Reid, Y. A. (1992). "Catalogue of Cell Lines and Hybridomas," 7th Ed. American Type Culture Collection, Rockville, Maryland.

Holinka, C. F., Hata, H., Kuramoto, H., and Gurpide, E. (1986). Responses to estradiol in a human endometrial adenocarcinoma cell line (Ishikawa). *J. Steroid Biochem.* **24,** 85–89.

Kuramoto, H. (1972). Studies on the growth and cytogenetic properties of human endometrial adenocarcinoma in culture and its development into an established line. *Acta Obstet. Gynaec. Jap.* **19,** 47–58.

Kuramoto, H., Tamura, S., and Notake, Y. (1972). Establishment of a cell line of human endometrial adenocarcinoma *in vitro. Am. J. Obstet. Gynecol.* **114,** 1012–1019.

Manetta, A., Satyaswaroop, P. G., Hamilton, T., Ozols, R., and Mortel, R. (1987). Radioimaging of human ovarian carcinoma. *Gynecol. Oncol.* **28,** 292.

Merenda, C., Sordat, B., Mach, J. P., and Carrel, S. (1975). Human endometrial carcinoma serially transplanted in nude mice and established in continued cell lines. *Int. J. Cancer* **16,** 559–570.

Mills, G. B., May, C., McGill, M., Roifman, C. M., and Mellors, A. (1988). A putative new growth factor in ascitic fluid from ovarian cancer patients: Identification, characterization, and mechanism of action. *Cancer Res.* **48,** 1066–1071.

Mills, G. B., May, C., Hill, M., Campbell, S., Shan, P., and Marks, A. (1990). Ascitic fluid from human ovarian cancer patients contains growth factors necessary for intraperitoneal growth of human ovarian adenocarcinoma. *J. Clin. Invest.* **86,** 851–855.

Nishida, M., Kasahara, K., Kaneko, M., and Iwasaki, H. (1985). Establishment of a new human endometrial adenocarcinoma cell line, Ishikawa cells, containing estrogen and progesterone receptors. *Acta Obstet. Gynaec. Jap.* **37,** 1103–1111.

Pirker, R., Fitzgerald, D. J. P., Hamilton, T., Ozols, R. F., Laird, W., Frankel, A. E., Willingham, M. C., and Pastan, I. (1985). Characterization of immunotoxins against ovarian cancer cell lines. *J. Clin. Invest.* **76,** 1261–1267.

Richardson, G. S., Dickersin, G. R., Atkins, L., MacLaughlin, D. T., Raam, S., Merk, L. P., and Bradley, F. M. (1984). KLE: A cell line with defective estrogen receptor derived from undifferentiated endometrial cancer. *Gynecol. Oncol.* **17,** 213–230.

Satyaswaroop, P. G., Bressler, R. S., de la Pena, M. M., and Gurpide, E. (1979). Isolation and culture of human endometrial glands. *J. Clin. Endocrinol. Metab.* **48,** 639.

Satyaswaroop, P. G., and Mortel, R. (1982). Failure of progestins to induce E_2DH in endometrial carcinoma *in vitro. Cancer Res.* **42,** 1322.

Satyaswaroop, P. G., and Tabibzadeh, S. S. (1991). Extracellular matrix and the patterns of differentiation of human endometrial carcinomas *in vitro* and *in vivo. Cancer Res.* **51,** 5661–5666.

Satyaswaroop, P. G., Zaino, R., and Mortel, R. (1983). Human endometrial adenocarcinoma transplanted into nude mice: Growth regulation by estradiol. *Science* **219,** 58.

Satyaswaroop, P. G., Sivarajah, A., Zaino, R. J., and Mortel, R. (1988). Hormonal control of growth of human endometrial carcinoma in the nude mouse model. *Prog. Cancer Res. Ther.* **35,** 430–435.

Silverberg, E. (1982). Cancer Statistics. *Cancer* **32,** 18.

Tabibzadeh, S. S., Satyaswaroop, P. G., and Rao, P. N. (1988). Antiproliferative effect of interferon-gamma in human endometrial epithelial cells *in vitro:* Potential local growth modulatory role in endometrium. *J. Clin. Endocrinol. Metab.* **67,** 131.

Tabibzadeh, S., Kaffka, K. L., Kilian, P. L., and Satyaswaroop, P. G. (1990). Human endometrial epithelial cell lines for studying steroid and cytokine action. *In Vitro Cell. Dev. Biol.* **26,** 1173–1179.

Way, D. L., Grosso, D. S., Davis, J. R., Surwit, E. A., and Christian, C. D. (1983). Characterization of a new human endometrial carcinoma (RL95.2) established in tissue culture *In Vitro* **19,** 147.

The Male Reproductive System: Prostatic Cell Lines

15

Donna M. Peehl
Department of Urology
Stanford University School of Medicine
Stanford, California 94305–5118

I. Introduction 387

II. Methods of Establishment and Maintenance 388
  A. Immortal Cell Lines 388
  B. Short-Term Cell Strains 390

III. Morphology 391
  A. Monolayer Cultures 391
  B. Three-Dimensional Cultures 396

IV. Other Characteristics 396
  A. Antigenic Determinants 396
  B. Karyotype 400

C. Oncogenes and Suppressor Genes 401
D. Invasive Properties *in Vitro* 402
E. Expression of and Response to Growth Factors 402
F. Response to Inhibitory Factors 402
G. Hormonal Response 404
H. Tumorigenicity in Host Animals 405

V. Discussion and Future Prospects 405

References 407

I. Introduction

Prostate cancer is the second leading cause of death from cancer for males in the United States (Carter and Coffey, 1990). The establishment of human prostatic epithelial cells in culture is an essential research tool for understanding the nature of this disease and for developing effective diagnostic, prognostic, and therapeutic strategies. After many years of frustration and difficulty, the contributions of numerous investigators have led to the availability of several immortal prostate cancer cell lines (Table I), as well as to the ability to establish primary cultures of cells directly from normal, benign hyperplastic, or malignant prostatic tissues. Although improvements still remain to be made, *in vitro* models for prostate cancer are now accessible and provide exciting experimental opportunities.

Table I
Reported Immortal Prostate Cancer Cell Lines

Line	Reference	Source
DU 145	Mickey et al. (1977)	ATCC
PC-3	Kaighn et al. (1979)	ATCC
LNCaP	Horoszewicz et al. (1980)	ATCC
TSU-Pr1	Iizumi et al. (1987)	Iizumi
LRVA-4	Fan (1988)	Fan
PPC-1 (PC-3)	Brothman et al. (1989)	Brothman
JCA-1	Muraki et al. (1990)	Muraki
DuPro-1	Gingrich et al. (1991)	Gingrich
PC 82/PC EG/PC EW	Hoehn et al. (1980,1984)	Hoehn
HONDA	Ito and Nakazato (1984)	Ito
SV40 neonatal	Kaighn et al. (1989)	Kaighn
SV40 adult	Cussenot et al. (1991)	Cussenot

II. Methods of Establishment and Maintenance

A. Immortal Cell Lines

Despite improvements in culture techniques, the development of an immortal prostate cancer cell line still remains a rare event. With the exception of experimentally induced immortal cell lines derived from transfection of normal cells with oncogenes, most of the spontaneously arising immortal prostate cancer cell lines have been derived from metastases, not from primary tumors. Rather than reflecting random chance, this phenomenon actually may have a biological basis. Although the exact nature of the process leading to an immortal phenotype *in vitro* is understood incompletely, several findings point toward mutation or loss of specific genes as the basis for immortality (Goldstein, 1990). The chance of a cell developing these specific genetic changes probably correlates with genetic instability, and the degree of genetic instability is likely to relate to the stage of malignant progression. An experiment by Smith et al. (1987) illustrates this concept. Tumor biopsies from a patient with breast cancer were placed into culture consecutively over a period of time as the cancer progressed. Only in the advanced stages of disease progression did the biopsies give rise to immortal cell lines. The same principle may apply to prostate cancer, and perhaps explains why immortal cell lines are obtained so rarely from primary tumors.

1. DU 145

DU 145 was one of the first immortal cell lines to be established from a human prostate cancer (Stone et al., 1978). Tissue derived from a brain metastasis of a 69-year-old Caucasian male was placed into culture in Medium 199 with 20% bovine bull serum. The metastatic lesion was classified as a moderately differentiated adenocarcinoma with poorly differentiated focal areas. An immortal population of cells grew out from tissue explants and from "spill" created by mincing the tissue. This line is available from the American Type Culture Collection (ATCC), and is cultured routinely in Eagles' minimal essential medium (MEM) with 10% fetal bovine serum (FBS).

2. PC-3

In 1979, Kaighn and colleagues reported the establishment of an immortal cell line from a prostatic metastasis to the bone of a 62-year-old Caucasian male. The metastatic lesion was described as undifferentiated with characteristics of Gleason Grade 4 (Gleason, 1977). The tissue was minced and dissociated with a solution of collagenase/trypsin/chicken serum, followed by culture in medium PFMR-1 supplemented with 20% FBS. The immortal cell line that developed is obtainable from the ATCC; recommended culture medium is F-12K with 7% FBS.

3. LNCaP

LNCaP is the only available human androgen-responsive prostate cancer cell line that can be grown in vitro. Horoszewicz et al. (1980) described the isolation of this line from a metastasis to the lymph node. The line was initiated in medium RPMI 1640-GA supplemented with 5% heat-inactivated FBS. Clone FGC of LNCaP (Berns et al., 1986) is available from the ATCC; RPMI 1640 with 10% FBS is recommended for cultivation. LNCaP-FGC was isolated from a fast-growing colony; a doubling time of approximately 73 hr has been reported.

4. Other Immortal Cell Lines

The isolation of several other spontaneously immortal prostate cancer cell lines has been reported, but few of these lines have been distributed widely. The line TSU-Pn1 was derived from a lymph node metastasis in a 73-year-old patient (Iizumi et al., 1987). The tissue was minced in 0.25% trypsin, then inoculated into DM160 with 15% FBS. TSU-Pr1 cells have a doubling time of ~36 hr. Fan (1988) described the establishment of line LRVA-4 from a primary tumor with a Gleason score of 8 from a 70-year-old male. Explants of the autopsy tissue were established in a mixture of Ham's F-12/Eagle's MEM with 25% FBS. Another line, PPC-1, was derived from a poorly differentiated primary tumor of a 67-year-old black male with stage D2 disease by Brothman et al. (1989). Tissue was obtained after a transurethral resection of the prostate and explants were started in RPMI 1640 with 20% FBS. One report (Chen, 1993) indicates, however, that PPC-1 was derived through contamination by PC-3. Line JCA-1 was reported by Muraki et al. (1990). Tissue from a primary tumor with a Gleason score of 8 was obtained following radical prostatectomy of a 67-year-old male. After mincing, the tissue was inoculated into RPMI 1640 plus 10% FBS. The doubling time of this line is ~28 hr.

A novel approach has been used to establish a cell line called DuPro-1 from a human prostatic xenograft (Gingrich et al., 1991). The xenograft originally was established in athymic mice from a poorly differentiated metastasis to the lymph node. After the 15th serial passage, the xenograft was minced and digested with Dispase. The disaggregate then was dispersed in Matrigel®, and the cells were grown in strands of Matrigel® until they could be passaged as monolayers in RPMI 1640 with 10% bovine serum. DuPro-1 grows rapidly with a doubling time of 22–24 hr.

Other serially transplantable xenograft lines that have not yet been passaged successfully in vitro include PC 82, PC EG, and PC EW (Hoehn et al., 1980, 1984). Of these, the properties of PC 82 have been described in most detail. Fragments of a moderately differentiated tumor of the cribriform type were obtained from a 58-year-old patient and transplanted into male athymic mice. Although slow-growing (doubling time 18 days), the resultant xenograft has been passaged serially and maintains cribriform histology. Another serially transplantable xenograft is the line HONDA, established in male mice from a prostatic metastasis to the testicle of a 46-year-old Japanese male (Ito and Nakazato, 1984).

Immortal cell lines also have been created experimentally from normal prostate cells by the introduction of oncogenes. The transfection of SV40 viral DNA into neonatal prostatic epithelial cells resulted in the development of clonal immortal cell lines that remained nontumorigenic in athymic mice (Kaighn et al., 1989). Normal adult prostatic epithelial cells also have been transformed to immortality with SV40 viral DNA that was introduced into the cells by lipofection (Cussenot et al., 1991). Like the transformed neonatal cells, the transformed adult cells also remain nontumorigenic.

B. Short-Term Cell Strains

Despite the rarity of immortal cell lines arising from primary prostatic tumors, cell strains can be initiated readily from these tumors as well as from normal and benign prostatic hyperplasia (BPH) tissues. These cultures can be passaged serially until the cells become senescent, usually after 15–25 population doublings. Tissues generally are obtained from radical prostatectomy specimens or needle biopsies and are subjected to collagenase digestion before inoculation into primary culture (Peehl et al., 1991b; Peehl, 1992). Either of two serum-free culture systems is commonly used: one uses WAJC 404 as the basal medium (Chaproniere and McKeehan, 1986) whereas the other uses PFMR-4A (Peehl, 1992). Both media are supplemented with a variety of factors including cholera toxin, epidermal growth factor (EGF), bovine pituitary extract (BPE), hydrocortisone, and insulin.

III. Morphology

A. Monolayer Cultures

The cell lines DU 145, PC-3 and LNCaP manifest morphological traits that are characteristic of immortal cancer cell cultures (Fig. 1). Of the three lines, DU 145 displays perhaps the most "normal" morphology, consisting of small, flat polygonal cells that form discrete islands when subconfluent and a cobblestone pattern when confluent (Stone et al., 1978). PC-3 cells tend to be more rounded and loosely attached to the substratum than DU 145 cells (Kaighn et al., 1979). LNCaP cells possess the most disorderly appearance, consisting of somewhat elongated cells that pile on each other and easily detach from the substratum (Horoszewicz et al., 1980). None of the cell lines is contact inhibited, and all can pile up on surrounding cells to form typical "transformed foci." Note that morphology is dependent on culture conditions, as discussed elsewhere in this volume, and that the appearance of cells may differ considerably depending on the amounts of calcium, serum, or other factors present.

Morphometry provides a more sophisticated approach to characterization of cell morphology. Image analysis of whole cell parameters could not distinguish the PC-3 line from DU 145 because of large variability within each line (Carruba et al., 1989). However, nuclear parameters, particularly the nuclear roundness factor (the degree to which a nucleus approximates a perfect circle), proved useful in describing features that could discriminate between the cell lines.

The other cell lines that have been reported apparently have no unique morphological traits. TSU-Pr1 has been described as forming a tightly packed epithelial monolayer (Iizumi et al., 1987) whereas DuPro-1, like DU 145, forms isolated islands when subconfluent (Gingrich et al., 1991). Fan (1988) described the morphology of LRVA-4 as heterogeneous, with "normal" cells and transformed foci. SV40-transformed neonatal prostatic epithelial cells also display a "transformed" morphology. Although each clonal population is subtly different, all generally consist of cuboidal, somewhat loosely attached cells.

In contrast to immortal cell lines, cultures of prostatic epithelial cells derived directly from tumors do not have a "transformed" morphology. In fact, cancer-derived cultures are indistinguishable in appearance from cultures derived from normal or BPH tissues. The cells attach tightly to the substratum in an orderly array; only mitotic cells are rounded (Figs. 2, 3). Figure 4 pro-

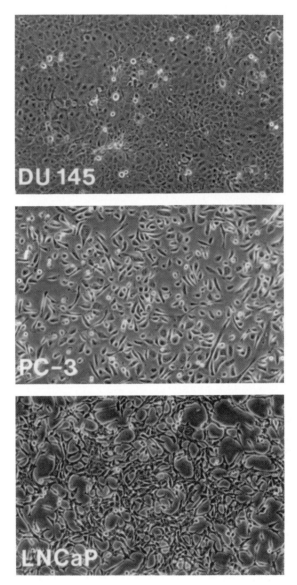

Fig. 1. Morphology of cell lines. DU 145 cells were grown to confluency in Dulbecco's modified Eagle's medium with 2% fetal bovine serum (FBS). PC-3 cells were grown in PFMR-4A with 5% FBS. LNCaP cells were cultured in MCDB 105 with 10% FBS (800×).

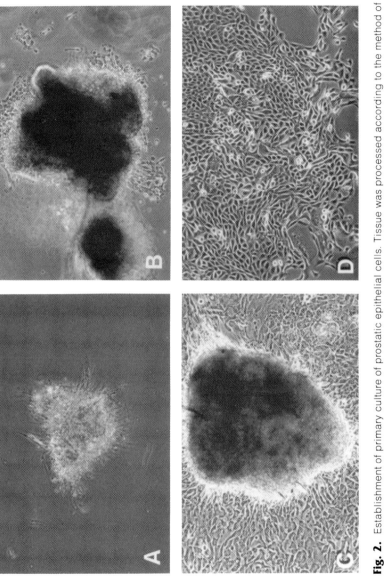

Fig. 2. Establishment of primary culture of prostatic epithelial cells. Tissue was processed according to the method of Peehl (1992) and cultures were established in serum-free medium (Peehl, 1992). (A) Typical clump of digested tissue after overnight incubation with collagenase. (B) Attached clump and cell outgrowth after 1 day in culture. (C) Clump and portion of outgrowth after 1 week in culture. (D) Edge of outgrowth shown in C (200×).

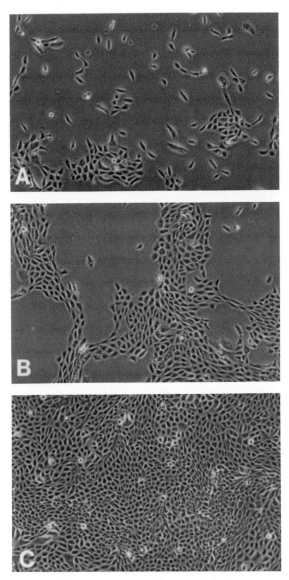

Fig. 3. Morphology of cell strain growing in culture. (A) Sparse culture, characterized by many single migratory cells. (B) Culture of intermediate density, with extensive intercellular adhesions beginning to form. (C) Confluent culture (200×).

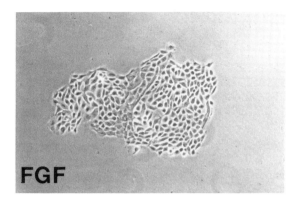

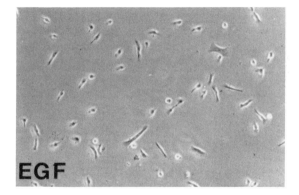

Fig. 4. Morphological alterations induced by culture conditions. Normal prostatic epithelial cells were cultured in
serum-free medium containing either epidermal growth factor (EGF) or fibroblast growth factor (FGF). In medium with
EGF (*bottom*), colonies consisted of migratory cells forming
few intercellular bonds. In FGF-containing medium (*top*),
colonies were very tight with concise edges. Approximately
equal numbers of cells were present in each colony, regardless of which factor was present (400×).

vides a dramatic illustration of morphological changes that can occur with
changes in medium composition. If EGF is provided as a mitogen, the cells
are very migratory and become widely spread. If EGF is replaced with fibroblast growth factor (FGF), growth is equivalent but the cells remain closely
adherent to each other and form patches. Other changes, such as the deletion of all growth factors or glucose or the addition of transforming growth
factor β (TGFβ) or suramin, also markedly alter cell interactions and morphology (Peehl *et al.*, 1991a).

B. Three-Dimensional Cultures

Anchorage-independent growth long ago was established as a common characteristic of the transformed phenotype. Although the ability to grow unattached to a solid substrate does not correlate directly with the ability to form tumors (Stanbridge and Wilkinson, 1980), and even normal cells can be induced to undergo anchorage-independent growth if the appropriate factors are provided (Moses et al., 1987), the ability to form colonies in soft agar still remains a hallmark of most immortal cell lines. The lines DU 145, PC-3, and LNCaP are all capable of growth in soft agar, as are the lines DuPro-1 and TSU-Pr1. Analysis of the capacity for anchorage-independent growth of SV40-transformed neonatal cells has not been reported, but SV40-transformed adult prostatic epithelial cells apparently are not capable of growth in agar (Cussenot et al., 1991). Cell strains derived from normal, BPH, or malignant tissues do not form colonies in soft agar (D. M. Peehl, unpublished observations).

Semi-solid substrates other than agar also are used to provide conditions that more closely mimic the environment in vivo and provide an opportunity for three-dimensional cellular organization. Suspended in a complex extracellular matrix called Matrigel, normal or BPH cells do not proliferate but form rounded structures and secrete increased amounts of prostate-specific antigen (Albini et al., 1987; Fong et al., 1991). DU 145 cells, on the other hand, grow to form tight colonies in Matrigel, whereas PC-3 cells form branching invasive colonies (Albini et al., 1987). Pigskin is another substrate that has been used to create three-dimensional cell outgrowths, particularly of prostate tumor tissue (Freeman and Hoffman, 1986; Perrapato et al., 1990).

IV. Other Characteristics

A. Antigenic Determinants

Because morphology is altered so readily by variations in culture conditions, morphology alone is a rather poor criterion for cellular identification. Verification of antigenic determinants can be a valuable adjunct to morphological description. Prostate-associated proteins that can be useful markers include keratins, prostate-specific antigen, and prostatic acid phosphatase. Other antibodies that show some specificity to or cross-reactivity with prostate cells are being developed, but in many cases little is known about the corresponding antigen. The labeling patterns of prostate cell lines and strains with various antibodies, which will be described in the subsequent sections, are summarized in Table II.

Table II

Antigenic Determinants

Cell line or strain	D83.21	Keratin 5	Keratins 8/18	PSA	TURP 27	PR92
LNCaP	−	−	+	+	−	NA
PC-3	+	−	+	−	−	−
DU 145	+	−	+	−	−	+
Normal/BPH	−	+	+	+	−	NA
Cancer	−	+	+	+	−	NA
PC 82	NA[a]	NA	NA	+	+	NA

[a] NA, not available.

1. Keratin

Of the 20 or more different keratins that have been recognized (Moll *et al.*, 1982), a subset of three keratins is perhaps of most utility with respect to prostate cells. In tissue, basal cells of normal or BPH glands express keratin 5 whereas luminal cells express keratins 8 and 18; cancer tisues also express keratins 8 and 18 (Brawer *et al.*, 1985).

The cell lines DU 145, PC-3, and LNCaP do not express keratin 5, but produce keratins 8 and 18 (Nagle *et al.*, 1987; Sherwood *et al.*, 1990). Figure 5 illustrates labeling of the cell lines by antibody against keratin 8. TSU-Pr1 (Djakiew *et al.*, 1991) also expresses keratin, but no details are available. SV40-transformed neonatal prostate cells express keratins 8 and 18, but not keratin 5, as do the other cell lines (Kaighn *et al.*, 1989).

Cell strains derived from normal, BPH, or malignant tissues all express keratin 5 as well as keratins 8 and 18; these keratins all are expressed simultaneously by individual cells (Sherwood *et al.*, 1989; D. M. Peehl, unpublished observations). This pattern of keratin expression by cell strains resembles that described for regenerating prostatic epithelia (Verhagen *et al.*, 1988). An extensive compilation of other keratins that are expressed by cultured prostate cells is available in a paper by Sherwood *et al.* (1989).

2. Prostate-Specific Antigen

Prostate-specific antigen (PSA) is a serine protease (Watt *et al.*, 1986), the expression of which is confined solely to the luminal cells of the prostate (Sinha *et al.*, 1986). PSA also is expressed commonly by prostate tumors. Of the common cell lines, only LNCaP expresses PSA, as well as the closely related human glandular kallikrein-1 (Riegman *et al.*, 1991; Fig. 6). Of the serially transplantable prostate cancer xenografts, PC 82 and HONDA both produce PSA (Ito and Nakazato, 1984; Riegman *et al.*, 1991). Although the xenograft from which the cultured line DuPro-1 was recovered generates

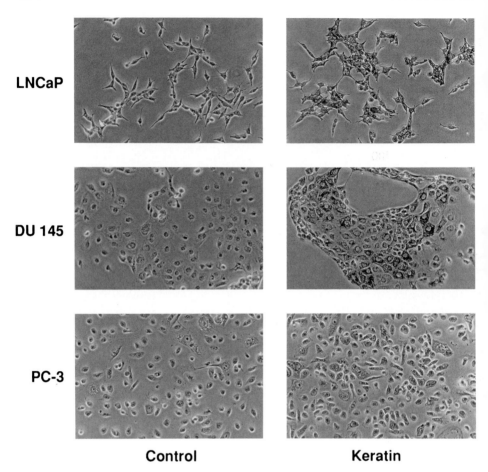

LNCaP

DU 145

PC-3

Control   **Keratin**

Fig. 5. Labeling of cell lines with anti-keratin. Immunoperoxidase techniques were used to label LNCaP, PC-3, and DU 145 cells with antibodies specific for keratin 8. Negative controls were prepared by omitting the primary antibody (800×).

PSA, DuPro-1 apparently has lost this ability (Gingrich et al., 1991). Cell strains also express PSA (Peehl and Stamey, 1984; Brothman et al., 1990).

3. Prostatic Acid Phosphatase

The term "prostatic acid phosphatase" (PAP), as commonly used in the past, is probably a misnomer. Widely used antibodies against PAP probably cross-react with lysosomal acid phosphatases, and enzymatic assays for PAP also may not be specific. Although most prostate cell lines and strains have been reported to express "PAP," when hybridized with a probe specific for PAP only LNCaP and not PC-3 cells were positive (Solin et al., 1990).

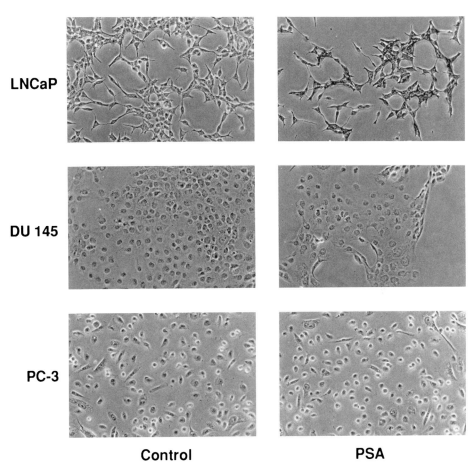

Control   PSA

Fig. 6. Labeling of cell lines with anti-PSA. LNCaP, PC-3, and DU 145 cells were labeled with antibody specific for PSA using immunoperoxidase techniques. Negative controls were prepared by omitting the primary antibody. Only LNCaP cells were labeled by anti-PSA (800×).

4. Other Antibodies

Other antibodies with various specificities for prostate tissues and cell cultures have been described. The TURP 27 antibody does not label any cultured prostate cancer cell lines or strains, but does label the xenografts PC 82 and PC EG (Wright *et al.*, 1990). D83.21 labels DU 145 and PC-3 cells, but not LNCaP cells or cell strains (Campbell *et al.*, 1985). Antibody PR92, described by Kim *et al.* (1988), labels DU 145 but not PC-3; no information is available regarding the reactivity of prostate cell strains with PR92.

B. Karyotype

With the advent of DNA fingerprinting (Thein et al., 1987), the origin and identity of cell lines and strains can be determined precisely. However, karyotyping has been the most common method of cytogenetic characterization in the past, and chromosomal content has been determined for most of the prostate cell lines, as well as for many cell strains (Table III). LNCaP appears to have the least complex karyotype of the cell lines (Gibas et al., 1984). With a modal number of chromosomes near 46, LNCaP cell populations also have a subset with a near tetraploid content. Seven marker chromosomes have been noted. The PC-3 line reportedly has a modal chromosomal number of 55; 11 marker chromosomes have been described, and the Y chromosome is absent (Ohnuki et al., 1980). Conflicting information about the karyotype of DU 145 has been reported. In their publication describing the isolation and properties of DU 145, Mickey et al. (1977) reported that the cell line had a modal number of 64 chromosomes and 4 marker chromosomes. Hartley-Asp et al. (1989), using a more sensitive C-banding method, reported that the DU 145 cell line had a modal number of 59 chromosomes, and that the Y chromosome was present in approximately 50% of the cells. Nelson-Rees and Flandermeyer (1978) found that a higher multiplicity of marker chromosomes was present in DU 145 cells than previously described, and a rearranged Y chromosome was noted.

Detailed karyotypic analyses have been reported for other lines. LRVA-4 has a modal number of 46 chromosomes, and structural abnormalities are

Table III

Karyotype

Cell line or strain	Modal number of chromosomes	Y chromosome	Marker chromosomes
LNCaP	46	+	7
PC-3	55	−	11
DU 145	59–64	+/−	>4
LRVA-4	46	NA[a]	NA
DuPro-1	82–89	−	10–12
TSU-Pr1	80	+	10
SV40 neonatal	Aneuploid	NA	NA
SV40 adult	Aneuploid	NA	NA
Normal/BPH strains	46	+	−
Cancer strains	46[b]	+/−	−

[a] NA, not available.

[b] The majority of cancer strains that have been karyotyped appear diploid. However, approximately one-third had clonal aberrations, including heteroploidy, translocations, and double minute chromosomes.

apparent (Fan, 1988). DuPro-1 has 82–89 chromosomes, with 3–4 copies of each autosome. No Y chromosome is detectable, but 10–12 marker chromosomes are present (Gingrich et al., 1991). Iizumi et al. (1987) described the karyotype of TSU-Pr1 as aneuploid, with a modal number of 80 chromosomes, 10 marker chromosomes, and a Y chromosome. SV40-transformed prostate cells are also aneuploid (Kaighn et al., 1989; Cussenot et al., 1991), as is characteristic of SV40-transformed cells.

The transplantable cell line PC 82 is diploid (Hoehn et al., 1980). The majority of cell strains cultured from primary tumors also have been diploid, although clonal aberrations have been noted for about one-third of the cell strains that have been karyotyped (Brothman et al., 1990,1991). No consistent cytogenetic changes have been observed to date.

C. Oncogenes and Suppressor Genes

Genetic changes are believed to be the basis for the initiation and progression of malignancy. Although the activation of specific oncogenes or the loss of particular tumor suppressor genes has been correlated with the development of certain cancers (Bishop, 1991), no specific changes have been recognized for prostate cancer. Nevertheless, several of the prostatic cell lines have altered oncogenes or suppressor genes (Table IV). With respect to oncogenes, LNCaP, DU 145, and PC-3 cell lines all express sis (Smith and Nag, 1987; Sitaras et al., 1988), and LNCaP cells have a rearranged c-myc gene with elevated expression (Nag and Smith, 1989). DU 145 cells have alterations in two tumor suppressor genes, RB1 and p53. The cell lines PC-3 and TSU-Pr1 also have mutant p53 genes (Bookstein et al., 1990; Isaacs et al., 1991).

Table IV
Oncogenes and Suppressor Genes

	Oncogene		Suppressor gene	
Line	sis	myc	RB1	p53
LNCaP	+[a]	+	wt[b]	wt
PC-3	+	NA	wt	mutant
DU 145	+	NA	mutant	mutant
TSU-Pr1	NA[c]	NA	NA	mutant

[a] +, inappropriate or excessive expression.
[b] wt, wild type.
[c] NA, not available.

D. Invasive Properties in Vitro

Various invasion assays have been developed in an effort to find an *in vitro* correlate of metastatic potential. Using a basement membrane assay, Albini *et al.* (1987) reported that cultured normal or BPH cells were incapable of invasion. In the same assay, DU 145 cells showed a baseline level of invasion, whereas PC-3 cells were highly invasive. Gaylis *et al.* (1989) also found PC-3 cells to be invasive in a chemoinvasion assay, and noted that PC-3 cells secreted plasminogen activator, believed to be a mediator of invasion.

E. Expression of and Response to Growth Factors

Endocrine, paracrine, and autocrine factors all may have a role in the malignant process. Therefore, delineating mitogenic factors for prostate cells and identifying the sources of these factors is essential. The development of serum-free media has facilitated such studies with normal, BPH, or malignant cell strains greatly (Peehl and Stamey, 1986). EGF, FGF, and insulin-like growth factor (IGF) have been identified as mitogenic for these cells (Peehl *et al.,* 1989; Cohen *et al.,* 1991; Rubin *et al.,* submitted for publication).

Serum-free media permitting clonal growth have not been optimized for most of the prostatic cell lines, complicating the interpretation of mitogenic assays. For example, PC-3 cells were thought not to respond to EGF (Lechner and Kaighn, 1979), but now it is known that PC-3 as well as DU 145 and LNCaP cells synthesize EGF (or TGFα, a homologous factor) and that it is likely to be an autocrine growth factor (Schuurmans *et al.,* 1988b; Connolly and Rose, 1990).

Extensive information regarding the expression of or response to other growth factors by prostatic cell lines is not available. Perkel *et al.* (1990) reported that PC-3 cells synthesize IGF I and II; the mitogenic effect of these factors on PC-3 cells has not been described, although researchers have suggested that LNCaP cells do not respond to IGFs (Schuurmans *et al.,* 1988a). PC-3 and DU 145 cells, but not LNCaP cells, synthesize basic FGF; LNCaP and DU 145, but not PC-3 cells, respond to basic FGF (Nakamoto *et al.,* 1992).

A summary of the response to and expression of growth factors by cell lines and strains is provided in Table V.

F. Response to Inhibitory Factors

Identification of factors inhibitory to the growth of prostate cells may facilitate the development of new strategies for cancer therapy. TGFβ inhibits the growth of most epithelial cells, including normal, BPH, and malignant prostatic cell strains (Peehl *et al.,* 1989). PC-3 and DU 145 cell lines produce and respond to TGFβ, whereas LNCaP cells may not (Wilding *et al.,* 1989b).

Table V

Growth Factors

Line or strain	EGF Synthesize	EGF Respond	FGF Synthesize	FGF Respond	IGF Synthesize	IGF Respond
LNCaP	+	+	−	+	NA	−
PC-3	+	+	+	−	+	NA
DU 145	+	+	+	+	NA	NA
Normal/BPH	NA[a]	+	NA	+	−	+
Cancer	NA	+	NA	+	−	+

[a] NA, not available.

Schuurmans *et al.* (1988a) and Gleave *et al.* (1991), however, reported that the growth of LNCaP cells was inhibited by TGFβ. Tumor necrosis factor (TNF) is toxic to all three cell lines (Sherwood *et al.*, 1988) as well as to normal, BPH, and cancer-derived cell strains (Fig. 7). An interesting observa-

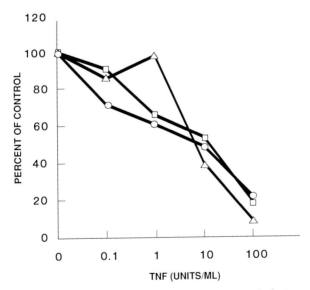

Fig. 7. Response of cell strains to tumor necrosis factor (TNF). Normal (○), BPH (△), or malignant (□) cells were inoculated at 200 cells per 60-mm collagen-coated dish containing serum-free medium and concentrations of TNF as indicated. After 10 days of incubation, the cells were fixed and stained, and growth was measured with an Artek image analyzer. Growth of each cell strain without TNF was standardized as 100%. Each point represents the average of duplicate dishes.

tion has been that DU 145, PC-3, and LNCaP cells have receptors for interleukin 6 (IL-6) and also secrete IL-6 (Siegall *et al.,* 1990). The potential significance of this observation is not clear, but Pai and colleagues (L. H. Pai, D. M. Peehl, E. Lovelace, and I. Pastan, unpublished observations) showed that cultured prostate cancer cell strains were very sensitive to IL-6 chimeric toxins. An intriguing finding from that study was that normal prostate cells were unaffected by IL-6 chimeric toxins, indicating a possible absence of IL-6 receptors. Vitamin A is another potentially important biological modulator of prostate growth and is inhibitory to normal, BPH, and malignant cell strains (Peehl *et al.,* 1989). However, PC-3 and LNCaP cells have been reported to be unresponsive to the inhibitory effects of vitamin A (Hasenson *et al.,* 1985; Halgunset *et al.,* 1987). Table VI summarizes the responses of cell lines and strains to inhibitory factors.

G. Hormonal Response

Despite the fact that the prostate is known to be an androgen-dependent organ, demonstrating androgen sensitivity *in vitro* has been difficult. The only cultured prostate cells to show strong androgen responsiveness are LNCaP cells (Horoszewicz *et al.,* 1983). Androgen receptors are not detectable in PC-3 or DU 145 cells, although the gene appears to be intact (Tilley *et al.,* 1990). LNCaP cells have a mutated androgen receptor (Veldscholte *et al.,* 1990), which may explain the stimulatory effect of antiandrogens as well as androgens on these cells (Wilding *et al.,* 1989a; Olea *et al.,* 1990). Perhaps a mutant androgen receptor also explains the responsiveness of LNCaP cells to progestins and estrogen, since most investigators have been unable to demonstrate either the progesterone or the estrogen receptor on these cells (Schuurmans *et al.,* 1988b; de Launoit *et al.,* 1991). Androgen-unresponsive LNCaP cell derivatives also have been reported (Hasenson *et al.,* 1985).

Table VI
Inhibitory Factors

Line or strain	TGF-β	TNF	IL-6	Vitamin A
LNCaP	+[a]	+	R[b]	—
PC-3	+	+	R	—
DU 145	+	+	R	NA[c]
Normal/BPH	+	+	—	+
Cancer	+	+	R	+

[a] +, factor inhibits growth.
[b] R, cells have receptors for factor. Effect of factor on cell growth is unknown.
[c] NA, not available.

H. Tumorigenicity in Host Animals

The development of animal models for human prostate cancer has been problematic. When prostate cancer tissue is implanted into athymic mice, xenograft tumor formation is very rare (Reid et al., 1980). Cancer-derived cell strains also fail to grow in athymic mice. Pretlow et al. (1991) reported growth of prostate cancer tissue in athymic mice when the tissue was first mixed with Matrigel before injection. If validated by other investigators, this method finally may provide in vivo models for the study of human prostate cancer.

The cell lines DU 145 and PC-3 consistently form tumors when injected subcutaneously into male or female athymic mice (Mickey et al., 1977; Kaighn et al., 1979). LNCaP cells form tumors only in male mice; some researchers have reported the inability of LNCaP cells to form tumors at all after standard subcutaneous injection. In these cases, investigators have found it necessary to co-inject LNCaP cells with prostatic or bone stromal cells (Gleave et al., 1991) or to suspend cells in Matrigel® before injection (Pretlow et al., 1991). The histologies of tumors formed by LNCaP, DU 145, or PC-3 cells after injection in Matrigel are shown in Fig. 8. Tumors formed by DU 145 or PC-3 cells are undifferentiated adenocarcinomas; LNCaP tumors are better differentiated and express PSA.

DuPro-1 (Gingrich et al., 1991) and TSU-Pr1 (Iizumi et al., 1987) form undifferentiated adenocarcinomas in male and female mice. Of the xenografts, HONDA (Ito and Nakazato, 1984) and PC 82 (van Steenbrugge et al., 1985) are hormone dependent and only grow in male mice.

The metastatic properties of prostate cancer cells in athymic mice have been reviewed by Ware (1987). Of the cultured lines and xenografts, only PC-3 appears to have metastatic potential. This property has enabled investigators to isolate a number of sublines from PC-3 with metastatic propensity for disparate sites including the lymph node, liver, and lungs.

V. Discussion and Future Prospects

Compared with the accomplishments in some other organ systems, the development of in vitro cell culture models for the human prostate has been slow and tedious. This lack of success may be explained in part by a relative absence of close interaction between basic scientists and urologists or pathologists, but is also the result of the biological nature of the prostate itself. Extensive heterogeneity within the prostate does not facilitate simple dissection of easily characterized tissue. The mechanism of androgen action in the prostate is not understood, making it difficult to establish hormone-responsive cell cultures. Prostate cancer develops very slowly, which is perhaps counterindicative to the chance of deriving immortal cell lines in vitro, and metastatic tissues are not readily obtainable for culture.

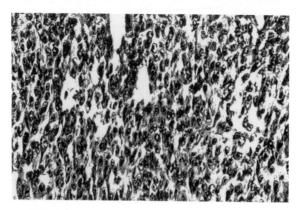

LNCaP

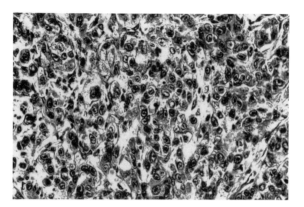

PC-3

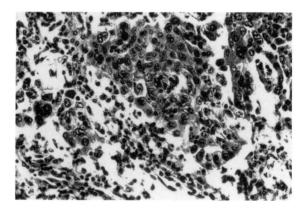

DU 145

Fig. 8. Tumor formation by cell lines. LNCaP, PC-3, and DU 145 cells were suspended in Matrigel and injected subcutaneously into male nude mice at 5×10^6 cells per site. The resulting tumors were retrieved 1 month later and fixed for histological analysis of H & E-stained sections (200×).

Nevertheless, steady progress is being made. Studies of the available cell lines and strains are revealing a great deal of information about how the proliferation and differentiation of prostate cells are regulated. Developments in the optimization of three-dimensional cell cultures and animal models of human prostate cancer promise to offer avenues to extend *in vitro* studies to clinically relevant methodologies.

References

Albini, A., Iwamoto, Y., Kleinman, H. K., Martin, G. R., Aaronson, S. A., Kozlowski, J. M., and McEwan, R. N. (1987). A rapid in vitro assay for quantitating the invasive potential of tumor cells. *Cancer Res.* **47,** 3239–3245.

Berns, E. M. J. J., de Boer, W., and Mulder, E. (1986). Androgen-dependent growth regulation of and release of specific protein(s) by the androgen receptor containing human prostate tumor cell line LNCaP. *Prostate* **9,** 247–259.

Bishop, J. M. (1991). Molecular themes in oncogenesis. *Cell* **64,** 235–248.

Bookstein, R., Shew, J., Chen, P., Scully, P., and Lee, W. (1990). Suppression of tumorigenicity of human prostate carcinoma cells by replacing a mutated RB gene. *Science* **247,** 712–715.

Brawer, M. K., Peehl, D. M., Stamey, T. A., and Bostwick, D. G. (1985). Keratin immunoreactivity in the benign and neoplastic human prostate. *Cancer Res.* **45,** 3663–3667.

Brothman, A. R., Lesho, L. J., Somers, K. D., Wright, G. L., Jr., and Merchant, D. J. (1989). Phenotypic and cytogenetic characterization of a cell line derived from primary prostatic carcinoma. *Int. J. Cancer* **44,** 898–903.

Brothman, A. R., Peehl, D. M., Patel, A. M., and McNeal, J. E. (1990). Frequency and pattern of karyotypic abnormalities in human prostate cancer. *Cancer Res.* **50,** 3795–3803.

Brothman, A. R., Peehl, D. M., Patel, A. M., MacDonald, G. R., McNeal, J. E., Ladaga, L. E., and Schellhammer, P. F. (1991). Cytogenetic evaluation of 20 primary prostatic tumors. *Cancer Genet. Cytogenet.* **55,** 79–84.

Campbell, A. E., Beckett, M. L., Starling, J. J., Sieg, S. M., and Wright, G. L., Jr. (1985). Antiprostate carcinoma monoclonal antibody (D 83.21) cross reacts with a membrane antigen expressed on cytomegalovirus-transformed human fibroblasts. *Prostate* **6,** 205–215.

Carruba, G., Pavone, C., Pavone-Macaluso, M., Mesiti, M., d'Aquino, A., Vita, G., Sica, G., and Castagnetta, L. (1989). Morphometry of in vitro systems. An image analysis of two human prostate cancer cell lines (PC-3 and DU 145). *Pathol. Res. Pract.* **185,** 704–708.

Carter, H. B., and Coffey, D. S. (1990). The prostate: an increasing medical problem. *Prostate* **16,** 39–48.

Chaproniere, D. M., and McKeehan, W. L. (1986). Serial culture of single adult human prostatic epithelial cells in serum-free medium containing low calcium and a new growth factor from bovine brain. *Cancer Res.* **46,** 819–824.

Chen, T. R. (1993). Chromosome identity of human prostate cancer cell lines, PC-3 and PPC-1. *Cytogenet. Cell Genet.* **62,** 183–184.

Cohen, P., Peehl, D. M., Lamson, G., and Rosenfeld, R. G. (1991). Insulin-like growth factors (IGFs), IGF receptors and IGF binding proteins in primary cultures of prostate epithelial cells. *J. Clin. Endocrinol. Metab.* **73,** 401–407.

Connolly, J. M., and Rose, D. P. (1990). Production of epidermal growth factor and transforming growth factor-alpha by the androgen-responsive LNCaP human prostate cancer cell line. *Prostate* **16,** 209–218.

Cussenot, O., Berthon, P., Berger, R., Mowszowicz, I., Faille, A., Hojman, F., Teillac, P., Le Duc, A., and Calvo, F. (1991). Immortalization of human adult normal prostatic epithelial cells by liposomes containing large T-SV40 gene. *J. Urol.* **143,** 881–886.

de Launoit, Y., Veilleux, R., Dufour, M., Simard, J., and Labrie, F. (1991). Characteristics of the biphasic action of androgens and of the potent antiproliferative effects of the new pure anti-estrogen EM-139 on cell cycle kinetic parameters in LNCaP human prostate cancer cells. *Cancer Res.* **51**, 5165–5170.

Djakiew, D., Delsite, R., Pflug, B., Wrathall, J., Lynch, J. H., and Onoda, M. (1991). Regulation of growth by a nerve growth factor-like protein which modulates paracrine interactions between a neoplastic epithelial cell line and stromal cells of the human prostate. *Cancer Res.* **51**, 3304–3310.

Fan, K. (1988). Heterogeneous subpopulations of human prostatic adenocarcinoma cells: Potential usefulness of p21 protein as a predictor for bone metastasis. *J. Urol.* **139**, 318–322.

Fong, C.-J., Sherwood, E. R., Sutkowski, D. M., Abu-Jawdeh, G. M., Yokoo, H., Bauer, K. D., Kozlowski, J. M., and Lee, C. (1991). Reconstituted basement membrane promotes morphological and functional differentiation of primary human prostatic epithelial cells. *Prostate* **19**, 221–235.

Freeman, A. E., and Hoffman, R. M. (1986). In vivo-like growth of human tumors *in vitro*. *Proc. Natl. Acad. Sci. U.S.A.* **83**, 2694–2698.

Gaylis, F. D., Keer, H. N., Wilson, M. J., Kwaan, H. C., Sinha, A. A., and Kozlowski, J. M. (1989). Plasminogen activators in human prostate cancer cell lines and tumors: Correlation with the aggressive phenotype. *J. Urol.* **142**, 193–198.

Gibas, Z., Becher, R., Kawinski, E., Horoszewicz, J. and Sandberg, A. A. (1984). A high-resolution study of chromosome changes in a human prostatic carcinoma cell line (LNCaP). *Cancer Genet. Cytogenet.* **11**, 399–404.

Gingrich, J. R., Tucker, J. A., Walther, P. J., Day, J. W., Poulton, S. H. M., and Webb, K. S. (1991). Establishment and characterization of a new human prostatic carcinoma cell line (DuPro-1). *J. Urol.* **146**, 915–919.

Gleason, D. F. (1977). Histologic grading and clinical staging of prostatic carcinoma. *In* "Urologic Pathology: The Prostate" (M. Tannenbaum, ed.), pp. 171–197. Lea & Febiger, Philadelphia.

Gleave, M., Hsieh, J.-T., Gao, C., von Eschenbach, A. C., and Chung, L. W. K. (1991). Acceleration of human prostate cancer growth in vivo by factors produced by prostate and bone fibroblasts. *Cancer Res.* **51**, 3753–3761.

Goldstein, S. (1990). Replicative senescence: The human fibroblast comes of age. *Science* **249**, 1129–1133.

Halgunset, J., Sunde, A., and Lundmo, P. I. (1987). Retinoic acid (RA): An inhibitor of 5-alpha-reductase in human prostatic cancer cells. *J. Steroid Biochem.* **28**, 731–736.

Hartley-Asp, B., Billstrom, A., and Kruse, E. (1989). Identification by C-banding of two human prostate tumour cell lines, 1013L and DU 145. *Int. J. Cancer* **44**, 161–164.

Hasenson, M., Hartley-Asp, B., Kihlfors, C., Lundin, A., Gustafsson, J.-A. and Pousette, A. (1985). Effect of hormones on growth and ATP content of a human prostatic carcinoma cell line, LNCaP-r. *Prostate* **7**, 183–194.

Hoehn, W., Schroeder, F. H., Riemann, J. F., Joebsis, A. C., and Hermanek, P. (1980). Human prostatic adenocarcinoma: Some characteristics of a serially transplantable line in nude mice (PC 82). *Prostate* **1**, 95–104.

Hoehn, W., Wagner, M., Riemann, J. F., Hermanek, P., Williams, E., Walther, R., and Schrueffer, R. (1984). Prostatic adenocarcinoma PC EW, a new human tumor line transplantable in nude mice. *Prostate* **5**, 445–452.

Horoszewicz, J. S., Leong, S. S., Chu, T. M., Wajsman, Z. L., Friedman, M., Papsidero, L., Kim, J., Chai, L. S., Kakati, S., Arya, S. K., and Sandberg, A. A. (1980). The LNCaP cell line—A new model for studies on human prostatic carcinoma. *In* "Models for Prostate Cancer" (G. P. Murphy, ed.), pp. 115–132. Liss, New York.

Horoszewicz, J. S., Leong, S. S., Kawinski, E., Karr, J. P., Rosenthal, H., Chu, T. M., Mirand, E. A., and Murphy, G. P. (1983). LNCaP model of human prostatic carcinoma. *Cancer Res.* **43**, 1809–1818.

Iizumi, T., Yazaki, T., Kanoh, S., Kondo, I., and Koiso, K. (1987). Establishment of a new prostatic carcinoma cell line (TSU-PR1). *J. Urol.* **137,** 1304–1306.

Isaacs, W. B., Carter, B. S., and Ewing, C. M. (1991). Wild-type p53 suppresses growth of human prostate cancer cells containing mutant p53 alleles. *Cancer Res.* **51,** 4716–4720.

Ito, Y. Z., and Nakazato, Y. (1984). A new serially transplantable human prostatic cancer (HONDA) in nude mice. *J. Urol.* **132,** 384–387.

Kaighn, M. E., Narayan, K. S., Ohnuki, Y., Lechner, J. F., and Jones, L. W. (1979). Establishment and characterization of a human prostatic carcinoma cell line (PC-3). *Invest. Urol.* **17,** 16–23.

Kaighn, M. E., Reddel, R. R., Lechner, J. F., Peehl, D. M., Camalier, R. F., Brash, D. E., Saffioti, U., and Harris, C. C. (1989). Transformation of human neonatal prostate epithelial cells by strontium phosphate transfection with a plasmid containing SV40 early region genes. *Cancer Res.* **49,** 3050–3056.

Kim, Y. D., Robinson, D. Y., and Tomita, J. T. (1988). Monoclonal antibody PR92 with restricted specificity for tumor-associated antigen of prostate and breast carcinoma. *Cancer Res.* **48,** 4543–4548.

Lechner, J. F., and Kaighn, M. E. (1979). Application of the principles of enzyme kinetics to clonal growth rate assays: An approach to delineating interactions among growth promoting agents. *J. Cell Physiol.* **100,** 519–530.

Mickey, D. D., Stone, K. R., Wunderli, H., Mickey, G. H., Vollmer, R. T., and Paulson, D. F. (1977). Heterotransplantation of a human prostatic adenocarcinoma cell line in nude mice. *Cancer Res.* **37,** 4049–4058.

Moll, R., Franke, W. W., Schiller, D. L., Geiger, B., and Krepler, R. (1982). The catalog of human cytokeratins: patterns of expression in normal epithelia, tumors and cultured cells. *Cell* **31,** 11–24.

Moses, H. L., Coffey, R. J., Jr., Leof, E. B., Lyons, R. M., and Keski-Oja, J. (1987). Transforming growth factor beta regulation of cell proliferation. *J. Cell Physiol. Suppl.* **5,** 1–8.

Muraki, J., Addonizio, J. C., Choudhury, M. S., Fischer, J., Eshghi, M., Davidian, M. M., Shapiro, L. R., Wilmot, P. L., Nagamatsu, G. R., and Chiao, J. W. (1990). Establishment of new human prostatic cancer cell line (JCA-1). *Urology* **36,** 79–84.

Nag, A., and Smith, R. G. (1989). Amplification, rearrangement, and elevated expression of c-*myc* in the human prostatic carcinoma cell line LNCaP. *Prostate* **15,** 115–122.

Nagle, R. B., Ahmann, F. R., McDaniel, K. M., Paquin, M. L., Clark, V. A., and Celniker, A. (1987). Cytokeratin characterization of human prostatic carcinoma and its derived cell lines. *Cancer Res.* **47,** 281–286.

Nakamoto, T., Chang, C., Li, A., and Chodak, G. W. (1992). Basic fibroblast growth factor in human prostate cancer cells. *Cancer Res.* **52,** 571–577.

Nelson-Rees, W. A., and Flandermeyer, R. R. (1978). Letter to the Editor. *Int. J. Cancer* **21,** 796–797.

Ohnuki, Y., Marnell, M. M., Babcock, M. S., Lechner, J. F., and Kaighn, M. E. (1980). Chromosomal analysis of human prostatic adenocarcinoma cell lines. *Cancer Res.* **40,** 524–534.

Olea, N., Sakabe, K., Soto, A. M., and Sonnenschein, C. (1990). The proliferative effect of "antiandrogens" on the androgen-sensitive human prostate tumor cell line LNCaP. *Endocrinology* **126,** 1457–1463.

Peehl, D. M. (1992). Culture of human prostatic epithelial cells. In "Culture of Epithelial Cells" (R. I. Freshney, ed.), pp. 159–180. Wiley-Liss, New York.

Peehl, D. M., and Stamey, T. A. (1984). Serial propagation of adult human prostatic epithelial cells with cholera toxin. *In Vitro* **20,** 981–986.

Peehl, D. M., and Stamey, T. A. (1986). Serum-free growth of adult human prostatic epithelial cells. *In Vitro* **22,** 82–90.

Peehl, D. M., Wong, S. T., Bazinet, M., and Stamey, T. A. (1989). In vitro studies of human prostatic epithelial cells: Attempts to identify distinguishing features of malignant cells. *Growth Factors* **1,** 237–250.

Peehl, D. M., Wong, S. T., and Stamey, T. A. (1991a). Cytostatic effects of suramin on prostate cancer cells cultured from primary tumors. *J. Urol.* **145,** 624–630.

Peehl, D. M., Wong, S. T., Terris, M. K., and Stamey, T. A. (1991b). Culture of prostatic epithelial cells from ultrasound-guided needle biopsies. *Prostate* **19,** 141–147.

Perkel, V. S., Mohan, S., Herring, S. J., Baylink, D. J., and Linkhart, T. A. (1990). Human prostatic cancer cells, PC-3, elaborate mitogenic activity which selectively stimulates human bone cells. *Cancer Res.* **50,** 6902–6907.

Perrapato, S. D., Slocum, H. K., Huben, R. P., Ghosh, R., and Rustum, Y. (1990). Assessment of human genitourinary tumors and chemosensitivity testing in 3-dimensional collagen gel culture. *J. Urol.* **143,** 1041–1045.

Pretlow, T. G., Delmoro, C. M., Dilley, G. G., Spadafora, C. G., and Pretlow, T. P. (1991). Transplantation of human prostatic carcinoma into nude mice in Matrigel. *Cancer Res.* **51,** 3814–3817.

Reid, L. C. M., Minato, N., and Rojkind, M. (1980). Human prostatic cells in culture and in conditioned animals. *In* "Male Accessory Sex Glands" (E. Spring-Mills and E. S. E. Hafez, eds.), Vol. 4, pp. 617–640. Elsevier/North Holland Biomedical Press, New York.

Riegman, P. H. J., Vlietstra, R. J., van der Korput, H. A. G. M., Romijn, J. C., and Trapman, J. (1991). Identification and androgen-regulated expression of two major human glandular kallikrein-1 (hGK-1) mRNA species. *Mol. Cell. Endocrinol.* **76,** 181–190.

Schuurmans, A. L. G., Bolt, J., and Mulder, E. (1988a). Androgens and transforming growth factor beta modulate the growth response to epidermal growth factor in human prostatic tumor cells (LNCaP). *Mol. Cell. Endocrinol.* **60,** 101–104.

Schuurmans, A. L. G., Bolt, J., Voorhorst, M. M., Blankenstein, R. A., and Mulder, E. (1988b). Regulation of growth and epidermal growth factor receptor levels of LNCaP prostate tumor cells by different steroids. *Int. J. Cancer* **42,** 917–922.

Sherwood, E. R., Fike, W., Kozlowski, J. M., and Lee, C. (1988). The cytotoxic/cytostatic effect of recombinant human tumor necrosis factor alpha on experimental human prostate cancer. *Proc. Am. Urol. Assoc.* **139,** 175A.

Sherwood, E. R., Berg, L. A., McEwan, R. N., Pasciak, R. M., Kozlowski, J. M., and Lee, C. (1989). Two-dimensional protein profiles of cultured stromal and epithelial cells from hyperplastic human prostate. *J. Cell Biochem.* **40,** 201–214.

Sherwood, E. R., Berg, L. A., Mitchell, N. J., McNeal, J. E., Kozlowski, J. M., and Lee, C. (1990). Differential cytokeratin expression in normal, hyperplastic and malignant epithelial cells from human prostate. *J. Urol.* **143,** 167–171.

Siegall, C. B., Schwab, G., Nordan, R. P., Fitzgerald, D. J., and Pastan, I. (1990). Expression of the interleukin 6 receptor and interleukin 6 in prostate carcinoma cells. *Cancer Res.* **50,** 7786–7788.

Sinha, A. A., Hagen, K. A., Sibley, R. K., Wilson, M. J., Limas, C., Reddy, P. K., Blackard, C. E., and Gleason, D. F. (1986). Analysis of fixation effects on immunohistochemical localization of prostatic specific antigen in human prostate. *J. Urol.* **136,** 722–727.

Sitaras, N. M., Sariban, E., Bravo, M., Pantazis, P., and Antoniades, H. N. (1988). Constitutive production of platelet-derived growth factor-like proteins by human prostate carcinoma cell lines. *Cancer Res.* **48,** 1930–1935.

Smith, H. S., Wolman, S. R., Dairkee, S., Hancock, M. C., Lippman, M., Leff, A., and Hackett, A. J. (1987). Immortalization in culture: Occurrence at a late stage in the progression of breast cancer. *J. Natl. Cancer Inst.* **78,** 611–615.

Smith, R. G., and Nag, A. (1987). Regulation of c-*sis* expression in tumors of the male reproductive tract. *Prog. Clin. Biol. Res.* **239,** 113–122.

Solin, T., Kontturi, M., Pohlmann, R., and Vihko, P. (1990). Gene expression and prostate specificity of human prostatic acid phosphatase (PAP): Evaluation by RNA blot analyses. *Biochim. Biophys. Acta* **1048,** 72–77.

Stanbridge, E. J., and Wilkinson, J. (1980). Dissociation of anchorage independence from tumorigenicity in human cell hybrids. *Int. J. Cancer* **26**, 1–8.

Stone, K. R., Mickey, D. H., Wunderli, H., Mickey, G. H., and Paulson, D. F. (1978). Isolation of a human prostate carcinoma cell line (DU 145). *Int. J. Cancer* **21**, 274–281.

Thein, S. L., Jeffreys, A. J., Gooi, H. C., Cotter, F., Flint, J., O'Connor, N. T. J., Weatherall, D. J., and Wainscoat, J. C. (1987). Detection of somatic changes in human cancer DNA by DNA fingerprint analysis. *Br. J. Cancer* **55**, 353–356.

Tilley, W. D., Wilson, C. M., Marcelli, M., and McPhaul, M. J. (1990). Androgen receptor gene expression in human prostate carcinoma cell lines. *Cancer Res.* **50**, 5382–5386.

van Steenbrugge, G. J., Groen, M., Bolt-de Vries, J., Romijn, J. C., and Schroeder, F. H. (1985). Human prostate cancer (PC 82) in nude mice: A model to study androgen regulated tumor growth. *Prog. Clin. Biol. Res.* **185A**, 23–50.

Veldscholte, J., Ris-Stalpers, C., Kuiper, G. G. J. M., Jenster, G., Berrevoets, C., Claassen, E., van Rooij, H. C. J., Trapman, J., Brinkman, A. O., and Mulder, E. (1990). A mutation in the ligand binding domain of the androgen receptor of human LNCaP cells affects steroid binding characteristics and response to anti-androgens. *Biochem. Biophys. Res. Commun.* **173**, 534–540.

Verhagen, A. P. M., Aalders, T. W., Ramaekers, F. C. S., Debruyne, F. M. J., and Schalken, J. A. (1988). Differential expression of keratins in the basal and luminal compartments of rat prostatic epithelium during degeneration and regeneration. *Prostate* **13**, 25–38.

Ware, J. L. (1987). Prostate tumor progression and metastasis. *Biochim. Biophys. Acta* **907**, 279–298.

Watt, K. W. K., Lee, P.-J., M'Timkulu, T., Chan, W.-P., and Loor, R. (1986). Human prostate-specific antigen: Structural and functional similarity with serine proteases. *Proc. Natl. Acad. Sci. U.S.A.* **83**, 3166–3170.

Wilding, G., Chen, M., and Gelmann, E. P. (1989a). Aberrant response in vitro of hormone-responsive prostate cancer cells to antiandrogens. *Prostate* **14**, 103–115.

Wilding, G., Zugmeier, G., Knabbe, C., Flanders, K., and Gelmann, E. (1989b). Differential effects of transforming growth factor beta on human prostate cancer cells in vitro. *Mol. Cell. Endocrinol.* **62**, 79–87.

Wright, G. L., Jr., Haley, C. L., Csapo, Z., and van Steenbrugge, G. J. (1990). Immunohistochemical evaluation of the expression of prostate tumor-association markers in the nude mouse human prostate carcinoma heterotransplant lines PC 82, PC EW, and PC EG. *Prostate* **17**, 301–316.

Melanocyte and Melanoma Cell Lines

16

Tibor Györfi and Meenhard Herlyn
Wistar Institute of Anatomy and Biology
Philadelphia, Pennsylvania 19104-4268

I. Introduction 413
II. Establishment and Maintenance 414
 A. Melanocyte Cultures 414
 B. Nevus and Melanoma Cultures 414
III. Morphology of Cultured Melanocytes and Melanoma Cells 416
IV. Growth Characteristics 416
V. Growth Factor Production by Melanoma Cells 419

VI. Chromosomal Abnormalities in Melanoma Cells 420
VII. Antigen Expression by Melanocytic Cells 421
VIII. Invasion and Metastasis of Melanoma Cells 422
IX. Conclusions 424
X. Origin of Cell Lines 425
 References 426

I. Introduction

Melanomas often evolve and progress through sequential steps that have been defined clinically and histopathologically (Clark *et al.*, 1984,1989). Lesions representing each step include (1) common acquired melanocytic nevus; (2) dysplastic nevus; (3) radial growth phase (RGP) primary melanoma; (4) vertical growth phase (VGP) primary melanoma; and (5) metastatic melanoma. Only cells from the VGP of primary melanomas are competent to metastasize.

Cell lines derived from primary melanomas and early metastatic lesions are from patients without prior exposure to radiation therapy or drugs that could induce DNA damage. On the other hand, patients with advanced metastatic melanomas may have undergone chemotherapy or radiation therapy prior to surgery. Therefore, metastatic cells derived from these patients may have been exposed to DNA-damaging agents.

Atlas of Human Tumor Cell Lines

413

II. Establishment and Maintenance

A. Melanocyte Cultures

Foreskins from newborns are immersed in 70% ethanol for 15 sec and washed four times with sterile phosphate-buffered saline (PBS) (Herlyn et al., 1987). After removal of subcutaneous tissues, the skin is cut into 2 × 4-mm fragments and incubated overnight at 4°C in Hanks' balanced salt solution (BSS) without Ca^{2+} and Mg^{2+} and with 30 mM 4-(2-hydroxethyl)-1-piperazineethanesulfonic acid (HEPES; modified HBSS, pH 7.4) containing 0.25% trypsin (Herlyn et al., 1987). The epidermis is then separated mechanically from the dermis, immersed in PBS, and incubated while slowly rotating for 3 hr at 37°C in a 15-ml plastic tube containing 5 ml modified HBSS with 0.033% versene, 1.25 U/ml Dispase (neutral protease, grade II; Boehringer Mannheim Biochemials, Indianapolis, Indiana), and 0.01% hyaluronidase type 1S from bovine testis (Sigma, St. Louis, Missouri). After incubation, the epidermal tissue is vortexed vigorously, and the resulting single cells are washed twice in modified HBSS, seeded in culture medium at 50,000 cells/ cm^2, and maintained in a 5% CO_2/95% air atmosphere at 37°C. The culture medium for melanocytes consists of 4 parts MCDB 153 medium supplemented with 2 mM Ca^{2+} and 1 part L15 medium (W489 medium), 10 ng/ml 12-O-tetradecanoyl phorbol-13-acetate (TPA), 5 μg/ml insulin, 5 ng/ml epidermal growth factor (EGF), 40 μg/ml bovine pituitary extract, and 2% fetal bovine serum (FBS). Medium is changed twice each week. W489 medium with or without supplements is relatively unstable and should be prepared fresh at least once each week. Subconfluent cultures are subcultured, usually after 2–3 wk, using 0.025% trypsin and 0.01% versene in modified HBSS. Trypsin is inactivated with 0.01% soybean trypsin inhibitor or L15 medium containing 10% FBS. If after 3–5 wk dermal fibroblasts begin to contaminate cultures, cells are grown for 48–72 hr in the presence of 200–300 μg/ml G418 (Halaban and Alfano, 1987). Whereas fibroblasts are killed at this concentration of G418, melanocytes survive. The success rate for culture of foreskin melanocytes is approximately 80%; for melanocytes from adult skin the yield is approximately 50%, although cell yield is lower for adult skin.

B. Nevus and Melanoma Cultures

Premalignant nevi and RGP primary melanomas are isolated from skin in a procedure similar to that described for melanocytes, except the cells from the epidermal/dermal junction and from the dermis are cultured also by scraping the junctional area and by dissociating the dermis with collagenase, hyaluronidase, and Dispase in modified HBSS for 5 hr at 37°C (Mancianti et al., 1988,1989). Whereas the success rate for cultures of congenital nevi is high (80%), that for dysplastic nevi and RGP primary melanomas is very low; cell lines have been established only rarely (Herlyn, 1990). The medium for nevi is the same as for melanocytes, and RGP melanomas are cultured in the presence as well as the absence of TPA.

Culture techniques for VGP primary melanomas depend on the thickness of the tumor and the size of the lesion. Thin (<1 mm tumor thickness) and small (2 × 2 mm) lesions are treated for dissociation as described for nevi, whereas thick (>2 mm) and large lesions are not dissociated enzymatically but are, after removal of excess skin, minced finely in glass petri dishes with crossed scalpels. Single cells and small fragments are seeded into 2- to 4-cm^2 cells freshly coated with 1% gelatin in 0.85% NaCl. Metastatic lesions generally also do not require enzymes for disaggregation. Trypsin treatment of lesions may increase fibroblast contamination in cultures. Cells of VGP primary and metastatic lesion are cultured initially in the same medium as melanocytes and nevus cells, except that TPA is omitted. The cultures are fed twice each week and are split only when fully confluent. After several weeks, pituitary extract and EGF can be omitted.

The success rate for culture of VGP primary melanomas is largely dependent on the size of the lesions. Thin lesions with low mitotic rates and brisk lymphocytic infiltrates are rarely established in culture, whereas thick lesions are cultured with a high success rate (>60%). Our laboratory has established cell lines from more than 40 VGP primary melanomas.

Melanomas from metastases frequently can be established in culture. Based on the success rate for culture of several hundred melanoma metastases and the establishment of over 250 permanent cell lines, we have divided cultures into five subgroups, each representing 10–30% of all metastatic melanomas (Table I). From 75 to 80% of all melanoma specimens cell lines can be established in culture. However, cells of subgroups 2 and 3 often are

Table I

Subgrouping of Cultured Metastatic Melanoma Cells Based on Success Rate for Cell Culture

Subgroup	Initial growth	Time for establishment of cell lines	Percentage of total
1	Rapid	3–6 weeks	25–30
2	Rapid, but fibroblast contamination	2–4 months	25–30
3	Crisis after 3 weeks, slow emergence of proliferating clones	5–10 months	20
4	Spontaneous differentiation	—	10–15
5	No growth (hyperpigmentation, no mitosis, too few cells, etc.)	—	10–15

lost because of fibroblast contamination or growth that is too slow during the initial months in culture. Therefore, studies in most laboratories are done using cells from subgroup 1.

III. Morphology of Cultured Melanocytes and Melanoma Cells

Melanocytes cultured in TPA-containing medium have a bi- to tripolar morphology (Fig. 1a). Addition of dibutyryl cyclic AMP renders cells multidendritic (Herlyn et al., 1988). If TPA is omitted from the culture medium, the cells lose their dendrites and become flat and elongated. TPA-stimulated melanocytes are pigmented and have elevated levels of the pigment cell-specific enzyme tyrosinase.

Melanoma cells are pigmented less frequently than melanocytes (20–30% of all melanoma cultures) and heavily pigmented melanoma cells may not proliferate (Table I). Over time in culture, cells may lose their pigmentation further and become pigmented only when not proliferating. Pigmentation of melanoma cells also may depend on the culture conditions. Overall, pigmentation is not a reliable marker for melanoma cells.

Houghton et al. (1982) classified melanoma cell lines in three groups based on differences in morphology and antigenic expression: group 1— cells expressing early melanocyte markers, an epithelioid morphology, and no pigmentation (subgroup 1 in Table I belongs to this group); group 2— cells expressing intermediate melanocyte markers, a bipolar spindle-shaped morphology, and little or no pigmentation; and group 3—cells expressing late melanocyte markers, with long dendritic processes and heavy pigmentation.

Figure 1b–f shows examples of epithelioid, spindle, polygonal, and dendritic melanoma cells. No significant morphological differences exist between primary and metastatic melanoma cells (Shih and Herlyn, 1992). The morphology of cultured melanoma cells may vary depending on growth conditions. Over time in culture (>10 passages), melanoma cell lines may change their morphology due to the emergence of subpopulations. Such changes may occur in original mass cultures or in clones derived by single-cell cloning, indicating the instability of the morphological phenotype in melanoma.

IV. Growth Characteristics

Normal melanocytes have a finite life-span, do not grow in soft agar, and are nontumorigenic (Herlyn, 1990). Melanoma cells in culture proliferate indefinitely, grow at clonal cell densities, proliferate anchorage independently

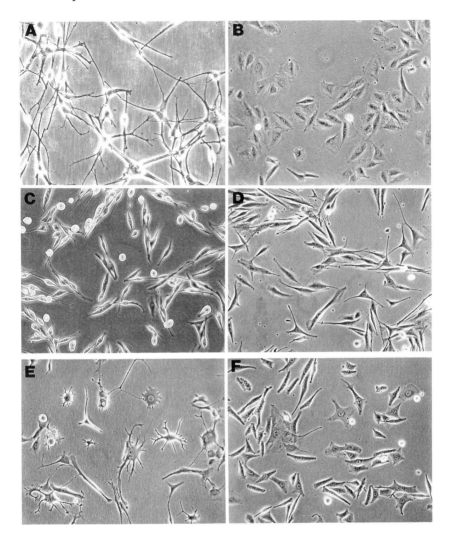

Fig. 1. Morphology of melanocytes and melanoma cells. (A) FM723 newborn foreskin melanocytes are bi- or tripolar and have long dendrites. (B) WM1361A melanoma cells have epithelioid morphology. The cells are round to polygonal, with occasional tripolar cells. (C) SKMEL23 melanoma cells are spindle shaped and uniformly bipolar with dendritic extensions. (D) WM266-4 melanoma cells form a heterogeneous cell population with bipolar, tripolar, and polygonal cells. (E) WM75 melanoma cells have dendritic morphology. Cell bodies are round or polygonal, and the cells have numerous short or long dendrites. (F) WM46 cells are heavily pigmented and granulation can be seen in the cytoplasm. The cells have spindle or polygonal morphology.

in soft agar, and are tumorigenic in athymic nude mice or other immunodeficient mice and rats (Herlyn et al., 1985a,b). Primary melanoma cells show less vigorous growth than cells from metastases (Table II). Differences in growth characteristics are most obvious in VGP primary melanoma cells from early and intermediate groups, whereas those from the advanced group do not show significant differences from metastases (Kath et al., 1990).

The minimal growth requirement in vitro in serum-free W489 medium reflects the stage of tumor progression better than any other available marker. Table III summarizes the growth factor and mitogen requirements for melanocytic cells. Normal melanocytes require at least four mitogens: insulin [or insulin-like growth factor (IGF)-I], TPA, α-melanocyte-stimulating hormone (α-MSH) or another cyclic AMP enhancer such as cholera toxin, and basic fibroblast growth factor (bFGF) (Herlyn et al., 1988). Nevus cells are less dependent on bFGF because they synthesize this growth factor (Mancianti et al., 1993). These cells also survive for several passages in the absence of TPA (Mancianti et al., 1988). Primary melanoma cells from early and intermediate groups require at least insulin or IGF-I in medium, whereas metastatic

Table II

Growth Characteristics of Primary and Metastatic Melanoma Cell Lines[a, b]

Group	Melanoma lesion	Cell line	Population doublings per day[c]	CFE[d] at clonal densities (%)	CFE in soft agar (%)	Tumor growth in nude mice[e]
I (Early)	Primary	WM793	0.29[f]	1.0[f]	7.2	s
	Primary	WM902B	0.31	1.8	6.5	s
II (Intermediate)	Primary	WM75	0.36	1.5	10.8	s
	Metastasis	WM373[g]	0.30	9.6	29.8	r
	Primary	WM115	0.22	1.0	6.0	s
	Metastasis	WM239A[g]	0.38	3.6	9.0	r
III (Advanced)	Primary	WM983A	0.51	21.0	75.0	rr
	Metastasis	WM983B[g]	0.65	24.0	85.0	rr
	Metastasis	WM793C[g]	0.65	23.0	72.0	rr
	Primary	WM1361A	0.35	16.0	0.8	s
	Metastasis	WM1361C[g]	0.29	14.0	1.0	s

[a] Reproduced from Kath et al. (1990) with permission; see Table X for origin of cells.
[b] Melanoma cells from primary lesions representing three different groups and metastatic lesions of four patients were assayed for growth at optimal and clonal densities, anchorage-independent growth in soft agar, and tumorigenicity in athymic nude mice.
[c] After seeding at high density ($10^4/cm^2$) and determined between days 1 and 8.
[d] CFE, colony forming efficiency.
[e] Growth after 10 weeks: s, <0.5 g; r, 0.5–1.0 g; rr, >1.0 g.
[f] Mean of triplicates with less than 10% standard deviation.
[g] From same patient as primary cell line.

Table III
Minimal Requirements for Growth Factors and Other Mitogens by Cultured Melanocytic Cells[a]

Melanocytes	Nevus cells	Primary melanoma cells (VGP)	Metastatic melanoma cells
Insulin (or IGF-I)[b]	Insulin (or IGF-I)[b]	Insulin (or IGF-I)[b]	—
TPA[c]	(TPA)[d]	—[e]	—[e]
α-MSH[f] or other cAMP enhancer	α-MSH	—	—
bFGF[g]	(bFGF)[h]	—	—

[a] Fetal bovine serum (FBS) is mitogenic for cells from all stages of tumor progression. Epidermal growth factor (EGF) is mitogenic for cells from all stages of tumor progression during the first 2–3 weeks only.

[b] Insulin and IGF-I both bind to the IGF-I receptor.

[c] TPA, 12-O-tetradecanoyl phorbol-13-acetate.

[d] Cells survive longer than melanocytes in absence of TPA.

[e] TPA is inhibitory for primary and metastatic melanoma cells.

[f] α-MSH, α-melanocyte-simulating hormone.

[g] Bovine pituitary basic fibroblast growth factor (bFGF) is more mitogenic than recombinant bFGF.

[h] Response is heteorgeneous among cultures; some cultures are independent.

cells can be grown continuously (>6 months) in the absence of any exogenous mitogen (Rodeck et al., 1987).

Primary melanoma cells can be adapted gradually to proliferate in medium without growth factors, even at clonal cell densities (Kath et al., 1991). Once adapted, these cells have population doubling times of 24–72 hr, similar to those of cells in serum-containing medium. Despite the growth factor independence of metastatic cells or selected primary cells, melanoma cells remain sensitive to stimulation by exogenous growth factors (Table IV). FBS is the strongest mitogen; insulin, EGF, and transferrin are also mitogenic, either individually or in combination.

V. Growth Factor Production by Melanoma Cells

Aberrant production of autostimulatory growth factors might explain the decreased growth factor requirements and growth autonomy of melanoma cells (Sporn and Roberts, 1985). These factors could replace those normally provided by the host and thus allow tumor cells to escape normal growth control.

Melanoma cells in culture secrete various growth factors that are potentially self-stimulatory, including transforming growth factor (TGF)-α (Marquard and Todaro, 1982), TGF-β (DeLarco et al., 1985), platelet-derived growth factor (PDGF)-A and PDGF-B (Westermark et al., 1986), interleukin

Table IV

Stimulation of Primary and Metastatic Melanoma Cell Lines by Exogenous Mitogens[a]

		Increase in cell number (%)[b, c]					
Source	Cell line	Base medium	Insulin (5 g/ml)	EGF (5 ng/ml)	Transferrin (10 μg/ml)	Combination insulin, EGF, transferrin	2% FBS
Primary	WM115	0	35	0	0	30	300
tumor	WM75	0	0	0	0	0	590
	WM39	0	70	0	0	75	415
	WM902B	0	50	40	0	60	460
	WM793	0	110	0	0	100	420
Metastatic	WM239A	110	290	220	260	370	650
melanoma	WM266-4	140	370	210	170	400	510
	WM373	120	460	190	200	400	430
	WM164	360	790	760	620	660	720
	WM852	240	820	270	310	770	880
	WM9	0	0	0	0	0	1740

[a] Reproduced from Rodeck et al. (1987) with permission.
[b] Cells were seeded at $1-2 \times 10^4/cm^2$ on gelatin-coated plastic.
[c] EGF, Epidermal growth factor; FBS, fetal bovine serum.

(IL) 1 (Bennicelli et al., 1989; Köck et al., 1989), bFGF (Halaban et al., 1988), and melanoma growth stimulatory activity (MGSA) (Richmond et al., 1988). Individual cell lines constitutively, that is, without stimulation by serum or other mitogens, express several growth factors (Table V). Most frequently, melanoma cells express bFGF mRNA and synthesize the protein. In addition, TGFα, TGFβ, PDGF-A, and IL-1β are produced (Rodeck et al., 1991). The regulatory role of each growth factor awaits further study, but bFGF and MGSA appear to be synthesized for autocrine stimulation, whereas TGFβ, TGFα, and PDGF are apparently important for paracrine stimulation of cells in the microenvironment (Rodeck and Herlyn, 1991).

VI. Chromosomal Abnormalities in Melanoma Cells

Cytogenetic studies reveal nonrandom chromosomal abnormalities involving chromosomes 1, 6, and 7, and possibly 9 in human melanomas (Table VI). The abnormalities are deletions, translocations, or amplifications (Balaban, 1984,1986). Comparison of primary and metastatic cells from the same patient often reveals identical abnormalities, but metastatic cells frequently have additional random abnormalities. The similarities in chromosomal abnormalities of primary and metastatic cells provide strong evidence for a clonal evolution of metastases in melanoma.

Table V

Constitutive Expression of Growth Factor Genes by Melanoma Cells[a]

Growth factor[b]	Cell line in growth factor-free medium[c]				
	WM983B	WM164	WM239A	WM852	WM35
PDGF-A	+	−	+	+	+
PDGF-B	−	−	−	+	−
TGF-β	+	−	+	+	−
TGF-α	−	+	+	+	+
bFGF	+	+	+	+	+
MGSA	+	−	−	−	−
IL-1α	−	−	−	−	+
IL--1β	+	−	−	+	+
IL-3	−	−	−	−	−
Total	6/9	2/9	4/9	6/9	5/9

[a] Reproduced from Rodeck and Herlyn (1991) with permission.
[b] PDGF, platelet derived growth factor; TGF, transforming growth factor; bFGF, basic fibroblast growth factor; MGSA, melanocyte growth stimulatory activity; IL, interleukin.
[c] +, expression of mRNA and/or protein; −, no expression.

VII. Antigen Expression by Melanocytic Cells

The antigenic profile of human melanocytic cells has been studied extensively using monoclonal antibodies (MAb) derived from spleen cells of mice immunized with melanoma cells. Approximately 60 different antigenic systems have been identified on melanoma cells with MAb (Herlyn and Koprowski, 1988; Kath and Herlyn, 1989). HLA Class II antigens are highly expressed on melanoma cells (Table VII). HLA-DR is involved in the immunological recognition of tumor cells by the host immune system. Of the growth factor receptors, expression of the nerve growth factor (NGF) receptor has been studied most extensively, although its biological role, if any, in melanoma development is not known. The EGF receptor may be used for autocrine and paracrine stimulation. Other growth factor receptors on melanoma cells are the IGF-I receptor, bFGF receptor, and α-MSH receptor. Cation binding and transport proteins on melanomas include p97 melanotransferrin (iron), S100 (calcium), and ceruloplasmin (copper).

Melanoma cells express a variety of antigens associated with adhesion, including gangliosides (GD$_2$, GD$_3$, 9-O-acetyl GD$_3$) and integrins such as the $\alpha_v\beta_3$ vitronectin receptor. The high molecular weight chondroitin sulfate proteoglycan also might be involved in adhesion. Melanoma cells secrete extracellular matrix proteins such as fibronectin and tenascin (Herlyn *et al.*,

Table VI

Similarities of Chromosomal Abnormalities in Primary and Metastatic Melanoma Cells from the Same Patient[a]

Patient	Melanoma/cell line	1	6	7
			Abnormality of chromosome	
1	Primary			
	WM115	t(1p;9q)	6p−	+7
	Metastasis			
	WM165-1	t(1p;9q)	6p−	+7
	5 others	t(1p;9q)	6p−	+7
2	Primary			
	WM75	0	0	7q−
	Metastasis			
	WM373	1q−	0	7q−
3	Primary			
	WM740	0	6q−	0
	Metastasis			
	WM858	1p−	6q−	+7
	3 others	1p−	6q−	+7
4	Primary			
	WM983A	1p+,1q−	6q+,6q−	+7q−
	Metastasis			
	WM983B	1p+,1q−	6q+,6q−	+7q−
	WM983C	1p+,1q−	6q+,6q−	+7q−
5	Primary			
	WM1361A	t(1;14)	6q−	+7
	Metastasis			
	WM1361C	t(1;14)	6q−	+7

[a] Reproduced from Herlyn et al. (1990) with permission.

1991), which appear to have an antagonistic function for melanoma cell adhesion and separation.

VIII. Invasion and Metastasis of Melanoma Cells

Primary and metastatic melanoma cells have similar adhesive and invasive properties in vitro (Table VIII). When injected subcutaneously into nude mice, most cell lines grow locally and do not metastasize. However, exceptional cell lines metastasize without prior selection of variants, whereas others must be selected by continuous passage in animals (Table IX). Selected cell lines show an increased independence from exogenous growth factors (Herlyn et al., 1990), and secrete higher concentrations of the proteolytic

Table VII

Expression of Cell Surface Antigens on Cultured Primary Melanoma Cells

		Positive cells (%)[a, b]						
Group	Cell line	HLA-DR[c]	p97	Proteo-glycan	Vitronectin receptor	NGF receptor	GD$_2$[d]	9-O-acetyl GD$_3$[d]
I (Early)	WM35	30	90	100	100	90	60	100
	WM793	70	50	90	100	80	20	70
	WM902B	40	80	80	100	80	30	70
II (Intermediate)	WM75	60	80	50	60	100	30	40
	WM115	100	90	50	30	100	70	40
	WM278	100	30	100	60	90	20	80
	WM98-1	80	70	30	100	100	60	20
	WM853-2	100	90	100	100	100	80	70
III (Advanced)	WM983A	0	100	0	100	100	0	80
	WM1361A	0	0	60	100	80	20	0

[a] Average of five experiments.
[b] NGF, nerve growth factor; GD, ganglioside.
[c] Expression may vary depending on culture conditions.
[d] Values from cells in suspension; values from adherent cells were lower.

Table VIII

Attachment to and Invasion of Reconstructed Basement Membrane Matrigel by Primary and Metastatic Melanoma Cells[a, b]

			Attachment (%)		
Group	Melanoma lesion	Cell line	60 min	90 min	Invasion[c]
I (Early)	Primary	WM793	52.3	58.4	19
		WM902B	72.9	69.6	16
		WM35	42.2	51.1	9
II (Intermediate)	Primary	WM75	42.8	46.7	11
	Metastasis	WM373	40.4	45.7	8
	Primary	WM115	13.7	18.7	14
	Metastasis	WM239A	11.6	12.2	10
III (Advanced)	Primary	WM983A	44.5	46.6	14
	Metastasis	WM983B	56.5	65.4	15
	Metastasis	WM983C	65.2	68.7	16
	Primary	WM1361A	20.4	34.3	4
	Metastasis	WM1361C	7.6	10.8	9

[a] Reproduced from Kath et al. (1990) with permission.
[b] Cell lines were tested for attachment after 60 and 90 min and for chemoinvasion after 5 hr.
[c] Mean number of migrated cells/field ($\times$10) in triplicate tests with less than 10% standard deviation.

Table IX

Cell Lines Spontaneously Metastatic in Nude Mice

Cell line	Route of injection	Metastases in organ	Time after injection	Reference
A375M	Subcutaneous	Lung	4 weeks	Kozlowski et al. (1984)
MeWo	Subcutaneous	Lung, lymph nodes	3 months	Cornil et al. (1989)
451Lu	Subcutaneous	Lung, lymph nodes	8 weeks	Iliopoulos et al. (1989)
HT168-M1	Intrasplenic	Liver	20 days	Ladanyi et al. (1990)
MV3	Subcutaneous	Lung	7 weeks	VanMuijen et al. (1991)
C81 61	Subcutaneous	Lymph nodes, other	42–69 days	Welch et al. (1991)
BL3	Subcutaneous	Lung	3–4 months	Quax et al. (1991)
Lox	Subcutaneous	Lung	20 days	Shoemaker et al. (1991)
M21	Subcutaneous	Lung	16 days	Mueller et al. (1991)
UCT-Mel3	Subcutaneous	Lung, kidney	160 days	Wilson et al. (1988)

enzymes plasminogen activator and collagenase type IV. Cell lines selected for growth factor independence or invasion of reconstructed basement membrane are more metastatic *in vivo* than the parental cells (Kath et al., 1991).

IX. Conclusions

In recent years, lesions of melanocytic cells have provided excellent models for the study of human tumor progression. Detailed histopathological and clinical investigations of familial and sporadic melanocytic lesions led to the delineation of various stages of neoplastic development, extending from benign to malignant lesions with increasingly aberrant features. However, little is known about the molecular events leading to melanoma development and progression. Genetic aberrations, although present, have not been defined clearly. Therefore, experimental studies require the combined use of multiple markers.

Cells derived from advanced melanoma lesions are propagated relatively easily *in vitro*. On the other hand, dysplastic nevus cells and RGP primary melanoma cells are still very difficult to grow *in vitro*. Such cell lines also have a limited life-span. Because of the ease of culturing one subgroup of melanomas (see Table I), most laboratories work only with cells representing one phenotype. Additional improvement in culture techniques and the use of the same well-selected cell lines in different laboratories would facilitate comparison of results obtained in different laboratories.

Chromosomal analyses of melanoma cells at the time of isolation, after 1–2 passages in culture and after prolonged (>6 months) culture, indicate that melanoma cells in culture maintain their properties in a stable manner, although changes over time in culture may occur (Herlyn *et al.*, 1985a). Freezing of large cell batches at an early time point improves the reproducibility of results. Single-cell cloning may enhance homogeneity of cell lines, but diversity still occurs after cloning.

Melanoma cells can contaminate other cultures, and karyotypic analyses may not reveal the cross-contamination. Therefore, we have tested the primary and metastatic cell lines discussed here (and listed in Table X) by DNA fingerprinting to insure that they are derived from the same patient.

X. Origin of Cell Lines

Approximately 1000 cell lines derived from human malignant melanomas have been reported. Major cell banks were established not only by our laboratory but also by Old and colleagues at Sloan–Kettering Cancer Center, New York, Morton and colleagues at the University of Southern California, Los

Table X

Primary and Metastatic Melanoma Cell Lines Isolated from Lesions of Patients with Long-Term Clinical Follow-Up[a]

Primary melanoma		Metastatic melanoma[b]		
Group	Cell line[c]	Recurrence (months)[d]	Site	Cell line
I (Early)	WM793	None (76)	—	None
	WM902B	None (68)	—	None
	WM35	None (126)	—	None
II (Intermediate)	WM75	33	Cutis	WM373
	WM115	9	Cutis	WM165-1, WM1652
		16	Lymph node, cutis	WM239A, WM239B
		18	Cutis	WM266-1, WM266-2
			—	WM266-3, WM266-4
	WM278	84	Cutis	WM1617
	WM98-1, WM98-2	60	Cutis	None
	WM853-2	36	Cutis	None
	WM740V[e]	9	Cutis	WM858
		11	Cutis	WM873-1, WM873-2, WM873-3
III (Advanced)	WM983A	0	Lymph nodes	WM983B, WM983C
	WM1361A	0	Cutis	WM1361B, WM1361C

[a] Reproduced from Kath *et al.* (1990) with permission.
[b] From the same patient.
[c] VGP or complex (RGP and VGP) primary melanoma lesions, except WM35 (RGP).
[d] Number of months of clinical follow-up.
[e] Establishment of permanent cell line unsuccessful or material unavailable.

Angeles, and Leibovitz, in Temple, Texas. Investigators studying T cell immunity in melanoma have begun to require autologous melanoma cultures. Laboratories with interests in T cell immunity with increasingly large cell banks are headed by Parmiani in Milano, Italy; by Rosenberg at the NIH, Bethesda, Maryland; by Lotze in Pittsburgh, Pennsylvania; and by Hersey in Newcastle, Australia.

Table X summarizes the origin of primary and metastatic melanoma cell lines established in our laboratory. All cell lines have been karyotyped. The properties of several of these cell lines are described in this chapter.

Acknowledgement

Some of the studies described here were supported by Grant CA-25874 from the National Institutes of Health.

References

Balaban, G., Herlyn, M., Guerry, D., IV, Bartolo, R., Koprowski, H., Clark, W. H., and Nowell, P. C. (1984). Cytogenetics of human malignant melanoma and premalignant lesions. *Cancer Genet. Cytogenet.* **11,** 429–439.

Balaban, G. B., Herlyn, M., Clark, W. H., Jr., and Nowell, P. C. (1986). Karyotypic evolution in human malignant melanoma. *Cancer Genet. Cytogenet.* **19,** 113–122.

Bennicelli, J. L., Elias, J., Kern, J., and Guerry, D., IV (1989). Production of interleukin 1 activity by cultured human melanoma cells. *Cancer Res.* **49,** 930–935.

Clark, W. H., Elder, D. E., Guerry, D., Epstein, M., Greene, M. H., and Van Horn, M. (1984). A study of tumor progression: The precursor lesions of superficial spreading and nodular melanoma. *Hum. Pathol.* **15,** 1147–1165.

Clark, W. H. Jr., Elder, D. E., Guerry, D., Braitman, L. E., Trock, B. J., Schultz, D., Synnestvedt, M., and Halpern, A. C. (1989). Model predicting survival in stage I melanoma based on tumor progression. *J. Natl. Cancer Inst.* **81,** 1893–1904.

Cornil, I., Man, S., Fernandez, B., and Kerbel, R. S. (1989). Enhanced tumorigenicity, melanogenesis, and metastases of a human malignant melanoma after subdermal implantation in nude mice. *J. Natl. Cancer Inst.* **81,** 938–944.

DeLarco, J. E., Pigott, D. A., and Lazarus, J. A. (1985). Ectopic peptides released by a human melanoma cell line that modulate the transformed phenotype. *Proc. Natl. Acad. Sci. U.S.A.* **82,** 5015–5019.

Halaban, R., and Alfano, E. D. (1984). Selective elimination of fibroblasts from cultures of normal human melanocytes. *In Vitro* **20(5),** 447–450.

Halaban, R., Kwon, B. S., Ghosh, S., Delli Bovi, P., and Baird, A. (1988). bFGF as an autocrine growth factor for human melanomas. *Oncogene Res.* **3,** 177–186.

Herlyn, M. (1990). Human melanoma: Development and progression. *Cancer Metab. Rev.* **9,** 101–112.

Herlyn, M., and Koprowski, H. (1988). Melanoma antigens: Immunological and biological characterization and clinical significance. *Ann. Rev. Immunol.* **6,** 283–308.

Herlyn, M., Balaban, G., Bennicelli, J., Guerry, D., IV, Halaban, R., Herlyn, D., Elder, D. E., Maul, G. G., Steplewski, Z., Nowell, P. C., Clark, W. H., and Koprowski, H. (1985a). Primary melanoma cells of the vertical growth phase: Similarities to metastatic cells. *J. Natl. Cancer Inst.* **74,** 283–289.

Herlyn, M., Thurin, J., Balaban, G., Bennicelli, J. L., Herlyn, D., Elder, D. E., Bondi, E., Guerry, D.,

Nowell, P., Clark, W. H., and Koprowski, H. (1985b). Characteristics of cultured human melanocytes isolated from different stages of tumor progression. *Cancer Res.* **45**, 5670–5676.

Herlyn, M., Rodeck, U., Mancianti, M. L., Cardillo, F. M., Lang, A., Ross, A. H., Jambrosic, J., and Koprowski, H. (1987). Expression of melanoma-associated antigens in rapidly dividing human melanocytes in culture. *Cancer Res.* **47**, 3057–3061.

Herlyn, M., Mancianti, M. L., Jambrosic, J., Bolen, J. B., and Koprowski, H. (1988). Regulatory factors that determine growth and phenotype of normal human melanocytes. *Exp. Cell Res.* **179**, 322–331.

Herlyn, M., Kath, R., Williams, N., Valyi-Nagy, I., and Rodeck, U. (1990). Growth regulatory factors for normal, premalignant, and malignant human cells. *Adv. Cancer Res.* **54**, 213–234.

Herlyn, M., Graeven, U., Speicher, D., Sela, B.-A., Bennicelli, J. B., Kath, R., and Guerry, D., IV. (1991). Characterization of tenascin secreted by melanoma cells. *Cancer Res.* **51**, 4853–4858.

Houghton, A. N., Eisinger, M., Albino, A. P., Cairncross, J. G., and Old, L. J. (1982). Surface antigens of melanocytes and melanomas. *J. Exp. Med.* **156**, 1755–1766.

Iliopoulos, D., Ernst, C., Steplewski, Z., Jambrosic, J. A., Rodeck, U., Herlyn, M., Clark, W. H., Jr., Koprowski, H., and Herlyn, D. (1989). Inhibition of metastases of a human melanoma xenograft by monoclonal antibody to the GD_2/GD_3 gangliosides. *J. Natl. Cancer Inst.* **81**, 440–444.

Kath, R., and Herlyn, M. (1989). Molecular biology of tumor antigens. *Curr. Opin. Immunol.* **1**, 863–866.

Kath, R., Rodeck, U., Parmiter, A., Jambrosic, J., and Herlyn, M. (1990). Growth factor independence in vitro of primary melanoma cells from advanced but not early or intermediate lesions. *Cancer Ther. Control* **1**, 179–191.

Kath, R., Jambrosic, J., Holland, L., Rodeck, U., and Herlyn, M. (1991). Development of invasive and growth factor independent cell variants from primary melanomas. *Cancer Res.* **51**, 2205–2211.

Köck, A., Schwarz, T., Urbanski, A., Peng, Z., Vetterlein, M., Miksche, M., Ansel, C., Kung, H. F., and Luger, T. A. (1989). Expression and release of interleukin-1 by different melanoma cell lines. *J. Natl. Cancer Inst.* **81**, 36–42.

Kozlowski, J. M., Fidler, I. J., Campbell, D., Xu, Z., Kaighn, M. E., and Hart, I. R. (1984). Metastatic behavior of human tumor cell lines grown in the nude mouse. *Cancer Res.* **44**, 3522–3529.

Ladanyi, A., Timar, J., Paku, S., Molnar, G., and Lapis, K. (1990). Selection and charcterization of human melanoma lines with different liver-colonizing capacity. *Int. J. Cancer* **46**, 456–461.

Mancianti, M. L., Herlyn, M., Weil, D., Jambrosic, J., Rodeck, U., Becker, D., Diamond, L., Clark, W. H., and Koprowski, H. (1988). Growth and phenotypic characteristics of human nevus cells in culture. *J. Invest. Dermatol.* **90**, 134–141.

Mancianti, M. L., Clark, W. H., Hayes, F. A., and Herlyn, M. (1990). Malignant melanoma simulants arising in congenital melanocytic nevi do not show experimental evidence for a malignant phenotype. *Am. J. Pathol.* **136**, 817–829.

Mancianti, M. L., Györfi, T., Shih, I.-M., Levengood, G., Valyi-Nagy, I., Menssen, H.-D., Halpern, A. C., Elder, D. E., and Herlyn, M. (1993). Growth regulation of cultured human nevus cells. *J. Invest. Dermatol.* **100**, 281$_s$–287$_s$.

Marquard, H., and Todaro, D. (1982). Human transforming growth factor production by a melanoma cell line, purification and initial characterization. *J. Biol. Chem.* **257**, 5220–5227.

Mueller, B. M., Romerdahl, C. A., Trent, J. M., and Reisfeld, R. A. (1991). Suppression of spontaneous melanoma metastasis in SCID mice with an antibody to the epidermal growth factor receptor. *Cancer Res.* **51**, 2193–2198.

Quax, F. H., Van Muijen, G. N., Weening-Verboeff, E. J., Lund, L. R., Dano, K., Ruiter, D. J., and Verheijen, J. H. (1991). Metastatic behavior of human melanoma cell lines in nude mice correlates with urokinase-type plasminogen activator, its type-1 inhibitor, and urokinase-mediated matrix degradation. *J. Cell Biol.* **115**, 191–199.

Richmond, A., Balentien, E., Thomas, H. G., Flaggs, G., Barton, D. E., Speiss, J., Bordoni, R., Francke, U., and Derynck, R. (1988). Molecular characterization and chromosomal mapping of melanoma growth stimulatory activity, a growth factor structurally related to beta-thromboglobulin. *EMBO J.* **7,** 2025–2033.

Rodeck, U., and Herlyn, M. (1991). Growth factors in melanoma. *Cancer Met. Rev.* **10,** 89–101.

Rodeck, U., Herlyn, M., Menssen, H.-D., Furlanetto, R. W., and Koprowski, H. (1987). Metastatic but not primary melanoma cell lines grow in vitro independently of exogenous growth factors. *Int. J. Cancer* **40,** 687–690.

Rodeck, U., Melber, K., Kath, R., Menssen, H.-D., Varello, M., Atkinson, B., and Herlyn, M. (1991). Constitutive expression of multiple growth factor genes by melanoma cells but not normal melanocytes. *J. Invest. Dermatol.* **97,** 20–26.

Shih, I. M., and Herlyn, M. (1992). The role of growth factors and their receptors in the development and progression of melanoma. *J. Invest. Dermatol.* **100,**196$_s$–203$_s$.

Shoemaker, R. H., Dykes, D. J., Plowma, J., Harrison, S. D., Jr., Griswold, D. P., Jr., Abbott, B. J., Mayo, J. G., Fodstad, D., and Boyd, M. R. (1991). Practical spontaneous metastasis model for in vivo therapeutic studies using a human melanoma. *Cancer Res.* **51,** 2837–2841.

Sporn, M. B., and Roberts, A. B. (1985). Autocrine growth factors and cancer. *Nature (London)* **313,** 745–747.

Van Muijen, G. N., Jansen, K. F., Cornelissen, I. M., Smeets, D. F., Beck, J. L., and Ruiter, D. J. (1991). Establishment and characterization of a human melanoma cell line (MV3) which is highly metastatic in nude mice. *Int. J. Cancer* **48,** 85–91.

Welch, D. R., Bisi, J. E., Miller, B. E., Conaway, D., Seftor, E. A., Yohem, K. H., Gilmore, L. B., Seftor, R. E., Nakajima, M., and Hendrix, M. J. (1991). Characterization of a highly invasive and spontaneously metastatic human malignant melanoma cell line. *Int. J. Cancer* **47,** 227–237.

Westermark, B., Johnsson, A., Paulsson, Y., Betsholtz, C., Heldin, C.-H., Herlyn, M., Rodeck, U., and Koprowski, H. (1986). Human melanoma cell lines of primary and metastatic origin express the genes encoding the chains of platelet-derived growth factor (PDGF) and produce a PDGF-like growth factor. *Proc. Natl. Acad. Sci. U.S.A.* **83,** 7197–7200.

Wilson, E. L., Gartner, M. F., Campbell, J. A., and Dowdle, E. B. (1988). Metastasis of a human melanoma cell line in the nude mouse. *Int. J. Cancer* **41,** 83–86.

Exocrine Pancreatic Tumor Cell Lines

Richard S. Metzgar
Department of Immunology, Duke University Medical Center
Durham, North Carolina 27710

I. Introduction 429

II. Methods of Establishment and
Maintenance 432

III. Morphology 432
A. Cell Lines 432
B. Clones 435

IV. Other Characteristics 436

V. Induction of
Differentiation 436

VI. Discussion 439

VII. Future Prospects 440

References 440

I. Introduction

Most (>80%) human exocrine pancreatic tumors are ductular adenocarinomas that are situated in the head of the gland. Histologically, the tumors often show a ductal architecture in certain sections, but some degree of anaplasia is also a common feature of the primary neoplasm. Because the tumor is often small and fibrotic, and in many cases is surrounded by pancreatitis, establishing tissue culture cell lines from primary pancreatic adenocarcinomas was difficult. Many of the pancreatic tumor cell lines currently maintained and distributed by the American Type Culture Collection (ATCC) are derived from metastatic cells (Table I). The stages of differentiation cited for the primary, metastatic, or xenografted pancreatic tumors from which the tissue culture cell lines were derived varied from poorly differentiated to moderately differentiated to well differentiated (Table I). However, in most cases, considerable intratumor heterogeneity existed in the tumor grades of the primary, metastatic, and xenografted exocrine pancreatic tumors.

The focus of this chapter is the morphological and ultrastructural properties of a pleomorphic human pancreatic adenocarcinoma cell line, HPAF, established at Duke University Medical Center (Metzgar *et al.*, 1982) and of clones of this cell line that had characteristics of different grades of differentiation (Kim *et al.*, 1989). The PANC-1 cell line (Lieber *et al.*, 1975) is used as

Table I
Human Pancreatic Tumor Cell Lines[a]

Cell line designation	Source of human tumor cells for culture	Description of original tumor	Grading of xenografts of tumor cell line	Reference
AsPC-1	Ascites cells grown as xenograft in athymic mice	Adenocarcinoma of head of pancreas	Well-differentiated to poorly differentiated adenocarcinoma	Tan et al. (1981); Chen et al. (1982)
BxPC-3	Biopsy of primary tumor and re-established from xenograft grown in athymic mice	Adenocarcinoma of body of pancreas	Moderately well- to poorly differentiated adenocarcinoma	Tan et al. (1986)
Capan-1	Pancreatic adenocarcinoma metastatic to liver	Information not readily available	Not graded; in vitro cytopathology described as well differentiated	Levrat et al. (1988)
Capan-2	Pancreatic adenocarcinoma	Information not readily available	Well-differentiated adenocarcinoma	
CFPAC-1	Pancreatic adenocarcinoma metastatic to liver, from cystic fibrosis patient	Well-differentiated ductal adenocarcinoma	Cells grown as tumor xenograft in athymic mice, but not graded	Schoumaker et al. (1990)
Hs700T	Pancreatic or intestinal adenocarcinoma metastatic to pelvis	Mucin-producing adenocarcinoma; primary unknown	Well-differentiated adenocarcinoma in immunosuppressed mice	Owens et al. (1976)

Hs766T	Pancreatic carcinoma metastatic to lymph node	Pancreatic carcinoma	Information not readily available	Owens et al. (1976)
MIA PaCa-2	Pancreatic carcinoma	Undifferentiated carcinoma	Information not readily available	Yunis et al. (1977)
PANC-1	Pancreatic carcinoma	Undifferentiated carcinoma	Anaplastic carcinoma	Lieber et al. (1975)
SU.86.86	Pancreatic ductal carcinoma metastatic to liver	Ductal carcinoma	Moderately well-differentiated adenocarcinoma	Drucker et al. (1988)
HPAF	Ascites from a patient with pancreatic adenocarcinoma	Adenocarcinoma of pancreas	Moderately well- to well-differentiated adenocarcinoma in athymic mice	Metzgar et al. (1982)
HPAF Clone CD11	Clone of HPAF tumor line	Adenocarcinoma of pancreas	Well-differentiated adenocarcinoma in athymic mice	Kim et al. (1989)
HPAF Clone CD18	Clone of HPAF tumor line	Adenocarcinoma of pancreas	Poorly differentiated adenocarcinoma in athymic mice	Kim et al. (1989)

[a] The cell lines in this Table, except for HPAF and the HPAF clones CD11 and CD18, are listed and described in the 1991 edition of the ATCC Catalog of Cell Lines and Hybridomas. The parental HPAF line, but not the CD11 or CD18 clones, is now available from ATCC.

the prototype of an undifferentiated or poorly differentiated exocrine pancreatic tumor line.

II. Methods of Establishment and Maintenance

Several different techniques were used to establish the pancreatic tumor lines listed in Table I, including initially growing tumor fragments as explants and then dispersing the explanted cells and growing them as monolayer cultures. Other approaches used ascites as a source of tumor cells. Ascites samples were centrifuged and the cells seeded in suspension directly onto plastic or glass surfaces. In some cases, primary or metastatic tumor fragments were transplanted directly as tumor xenografts, or early-passage tissue cultured explant cells from these fragments were transplanted as xenografts in immunodeficient or athymic mice. Depending on the degree of fibrosis of the starting tumor sample, which is a hallmark of this tumor, fibroblast contamination and growth was often a critical problem. Various manual approaches to remove the fibroblasts or to isolate the epithelial tumor cells were used to resolve this problem.

Once monolayer cultures of the pancreatic adenocarcinomas were established and passaged a few times, continuous cultures often were derived, many of which have been passaged continuously for 10 or more years. Many of the cultures were established initially and grown on RPMI 1640 or a modified Eagle's medium (MEM) with 10–20% fetal bovine serum (FBS). Our experience at Duke University with many of the cell lines listed in Table I is that the growth requirements, once the cells were established as continuous cultures, are not demanding. For example, the HPAF cell line now grows well on 3–5% FBS and on 5% human serum, and has been adapted to growth on serum-free medium.

III. Morphology

A. Cell Lines

The morphologies of the established exocrine pancreatic tumor cell lines and clones are commensurate with the differentiation grade of the tumor of the patient from which they were derived. In tissue culture monolayers, the cells were often pleomorphic, especially when established from tumors that were moderately well to well differentiated. For example, the HPAF monolayer cultures (Metzgar et al., 1982; Fig. 1A) contained cells that ranged from small mononuclear cells to large multinucleated ones (Fig. 1B), some of which stained prominently for mucin (Fig. 1C). The HPAF cells, as well as other cell lines derived from moderately to well-differentiated pancreatic

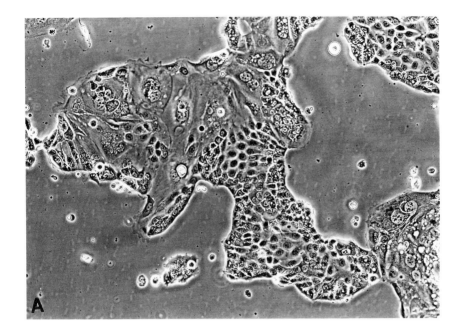

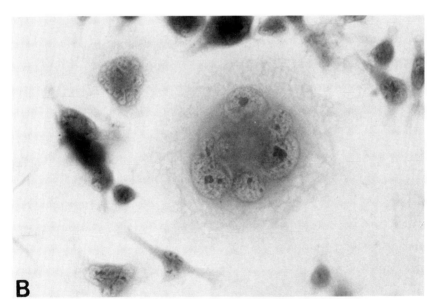

Fig. 1. Monolayer cultures of human pancreatic tumor cell lines. (A) HPAF culture growing as a monolayer on plastic (phase-contrast photomicrograph, 125×). (B) HPAF cells growing on plastic coverslips (H & E, 250×).

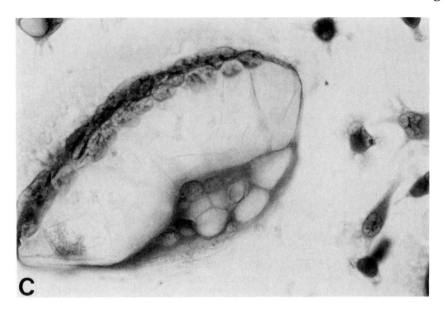

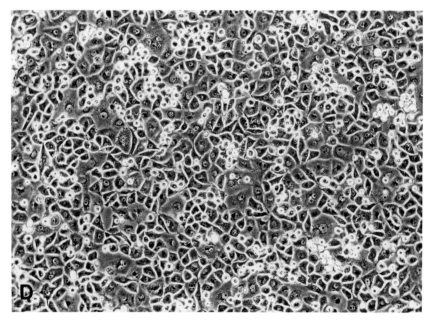

Fig. 1. *Continued* (C) HPAF cells stained for mucin production (250×), (D) Confluent monolayer of PANC–1 cells (100 ×).

ductal carcinomas, were characterized by their pleomorphism and their syncytial growth pattern (Fig. 1A).

In contrast the PANC-1 cell line, which was derived from an undifferentiated pancreatic carcinoma, had a more uniform epithelial appearance and organized growth pattern in monolayer culture, but also contained some multinucleate giant cells (Fig. 1D). More recent studies on PANC-1 cells by Madden and Sarras (1988) indicated that this cell line has some characteristics attributable to normal pancreatic ductal epithelial cells.

B. Clones

Since the pleomorphism noted for the HPAF cell line established from moderately to well-differentiated pancreatic tumors persisted in culture after years of passage, this line was cloned by a limiting dilution technique (Kim *et al.*, 1989). The cloning studies were anticipated to provide some evidence that would help determine whether the polymorphism was the result of spontaneous differentiation of a clonal population or whether multiple clones representing different differentiation stages were present in the HPAF cell line. Several clones were isolated that had different phenotypic properties, but identical isozyme patterns. Two of these clones (CD11 and CD18) were selected for study in more detail because nude mouse xenografts of these clones had morphologies indicating distinct differentiation stages. Light microscopic morphologies of the monolayer cultures of the two clones are shown in Fig. 2. The CD18 cells (Fig. 2A) appeared somewhat smaller than the CD11 cells (Fig. 2B) and grew at a higher density, having a more uniform growth pattern. The CD11 cells, in contrast, were more fusiform and demonstrated less organized monolayer growth *in vitro*. Nude mouse xenografts of these tissue culture clones showed striking morphological differences in their differentiation grade. The tumors resulting from the injection of the CD11 cells were well-differentiated adenocarcinomas (Fig. 3B), in contrast to the poorly differentiated appearance of the tumors resulting from cells of clone CD18 (Fig. 3A).

The differences in differentiation grade noted at the light microscopic level were also distinguishable when ultrastructural electron microscopic studies were done (Mullins *et al.*, 1991). The monolayer cultures of clone CD11 contained some rough endoplasmic reticulum that was associated with a well-developed Golgi complex; intercellular and intracytoplasmic lumens with microvilli were noted (Fig. 4A). Clone CD18 cells (Fig. 4B), in contrast, were characterized by few profiles of rough endoplasmic reticulum, poorly developed Golgi, and a paucity of lumenal structures, indicative of a low level of secretory activity. The ultrastructural differences of the xenografted tumors produced by these clones was also striking. The CD11 tumors had secretory properties of well-differentiated adenocarcinomas (Fig. 4C), whereas CD18 tumors (Fig. 4D) were much less differentiated.

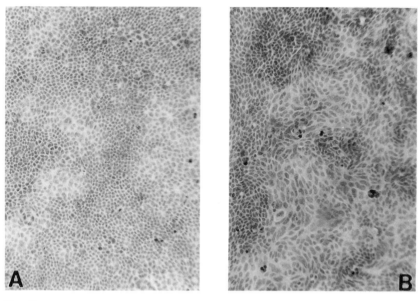

Fig. 2. Monolayer cultures of two different clones of HPAF pancreatic tumor lines. (A) CD18 clone (75×). (B) CD11 clone (75×).

IV. Other Characteristics

Biochemical differences also can be used to distinguish some of the pancreatic adenocarcinoma lines, although often these differences are quantitative rather than qualitative. In some cases, the biochemical differences can be associated loosely with the differentiation grades of the cell lines, or can be considered markers of pancreatic ductal cells that distinguish them from acinar or islet cells. Mucin expression and secretion, alkaline phosphatase, carbonic anhydrase, and Na^+/K^+-ATPase activity are the most studied of the biochemical markers (Githens, 1988).

V. Induction of Differentiation

Two pancreatic adenocarcinoma cell lines, CAPAN-1 (Bloom *et al.*, 1989) and PANC-1 (Mullins *et al.*, 1991), as well as the HPAF clones (Mullins *et al.*, 1991) were treated with sodium butyrate, a compound reported to induce differentiation *in vitro*. Treatment of CAPAN-1 cells with this compound caused an increase in *in vitro* growth of these cells and quantitative changes in their cell surface carbohydrates (Bloom *et al.*, 1989). Similar studies with

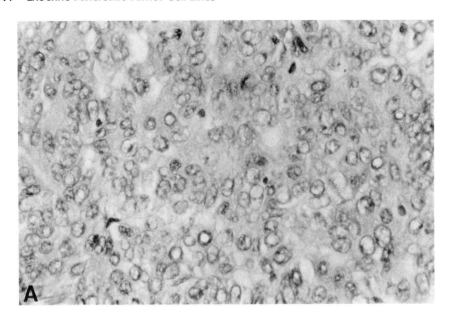

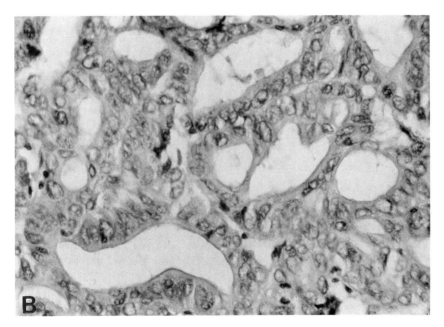

Fig. 3. Nude mouse tumor xenografts of HPAF clones CD18 (A) and CD11 (B) (400×).

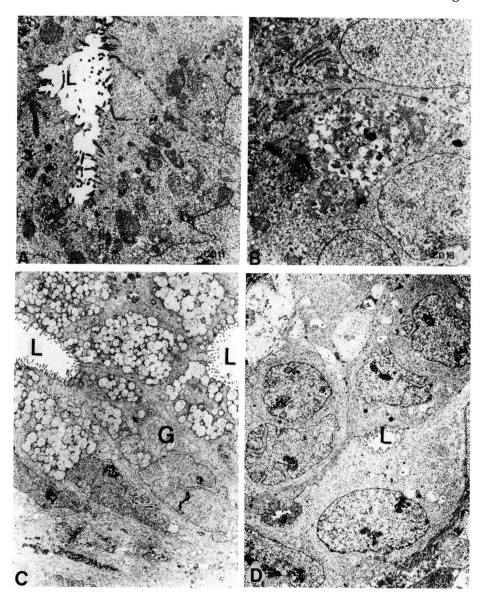

Fig. 4. Ultrastructure of HPAF clone CD11 and CD18 cells as monolayer cultures and nude mouse tumor xenografts. Clone CD11 monolayer culture (A; 10,000×) and tumor xenograft (C; 4,200×). Clone CD18 monolayer culture (B; 10,000×) and tumor xenograft (D; 4200×). L, lumen; G, Golgi.

clones of HPAF cells demonstrated that the CD18 clone that was poorly differentiated showed morphological alterations that were consistent with increased secretory activity, in which intercellular and intracytoplasmic lumen formation that was characteristic of the more differentiated CD11 clone (Mullins *et al.*, 1991) occurred. In contrast, PANC-1 cells derived from an undifferentiated or very poorly differentiated pancreatic tumor, when treated with sodium butyrate, showed a slight increase in cytoplasmic secretory elements. However, these cells showed no morphological evidence of the lumen formation or other ultrastructural changes noted for the sodium butyrate-treated poorly differentiated HPAF clone CD18 (Mullins *et al.*, 1991).

VI. Discussion

Several cell lines from exocrine pancreatic tumors have been established. Since most of the exocrine pancreatic tumors are adenocarcinomas of presumed ductal cell origin with some differentiated properties, this type of tumor cell line is the most common. Xenografts of most of these cell lines in athymic mice are characterized as moderately to well-differentiated adenocarcinomas. Studies of clones of one of these cell lines, HPAF, suggest that some spontaneous differentiation of the tumor cells occurs *in vitro* (Kim *et al.*, 1989). Some poorly differentiated clones, as well as well-differentiated clones, could be isolated that maintained these stages of differentiation for several generations. The PANC-1 cell line may represent the most poorly differentiated or undifferentiated stage of the available pancreatic tumor cell lines. PANC-1 cells, when treated with sodium butyrate, failed to show the same ultrastructural changes associated with differentiation that were reported for the poorly differentiated HPAF clone CD18, despite the fact that, initially, the untreated poorly differentiated HPAF clone had a morphology similar to that of PANC-1 cells. However, we have noted that some of the HPAF clone CD18 cells, after 50 or more passages *in vitro,* are apparently undergoing spontaneous differentiation (R. S. Metzgar, unpublished observations). PANC-1 cells are also difficult to grow as tumor xenografts in athymic mice (R. S. Metzgar, unpublished observation), which may reflect their low differentiation or undifferentiated grade or their inability to become established in the murine dermal stromal elements. In contast, the pancreatic tumor lines established from moderately to well-differentiated tumors readily form xenografts in athymic mice. These xenografted tumors have morphological characteristics that are often indistinguishable from the human tumors from which the cell lines were derived.

The pancreatic tumor lines also express many of the biochemical properties of normal pancreatic ductal cells, including the expression and secretion of mucins (Metzgar *et al.*, 1993). The similarities of the morphological, ultrastructural, phenotypic, and biochemical properties of the exocrine pancreatic tumor lines to the *in situ* pancreatic ductal adenocarcinomas make these tumor lines extremely useful in a variety of basic and clinical research areas that are related to study of normal and malignant pancreatic ductal cells.

VII. Future Prospects

A considerable need still exists to define better those factors directly and indirectly influencing growth and differentiation of normal and malignant pancreatic ductal cells. The absence of stable long-term cultures of normal pancreatic ductal cells indicates our knowledge and shortcomings with respect to normal pancreatic duct cell growth and regulation, and suggests that the cell lines from pancreatic ductal cell tumors probably reflect stable changes in normal growth requirements of these cell types. Progress in establishing short-term primary cultures of pancreatic ductal cells, however, indicates that these cultures may be useful in defining some of the potential autocrine growth factors provided by pancreatic adenocarcinoma cell lines.

References

Bloom, E. J., Siddiqui, B., Hicks, J. W., and Kim, Y. S. (1989). Effect of sodium butyrate, a differentiating agent, on cell surface glycoconjugates of a human pancreatic cell line. *Pancreas* **4,** 59–64.

Chen, W. H., Horoszewicz, J. S., Leong, S. S., Shimano, T., Penetrante, R., Sanders, W. H., Berjian, R., Douglass, H. O., Martin, E. W., and Chu, M. T. (1982). Human pancreatic adenocarcinoma: In vitro and in vivo morphology of a new tumor line established from ascites. *In Vitro* **18,** 24–34.

Drucker, B. J., Marincola, F. M., Siao, D. Y., Donlon, T. A., Bangs, C. D., and Holder, W. D. (1988). A new pancreatic adenocarcinoma cell line developed for adaptive immunotherapy studies with lymphokine-activated killer cells in nude mice. *In Vitro Cell. Dev. Biol.* **24,** 1179–1187.

Githens, S. (1988). The pancreatic duct cell: Proliferative capabilities, specific characteristics, metaplasia, isolation and culture. *J. Pediatr. Gastroenterol. Nutr.* **7,** 486–506.

Kim, Y. W., Kern, H. F., Mullins, T. D., Koriwchak, M. J., and Metzgar, R. S. (1989). Characterization of clones of a human pancreatic adenocarcinoma cell line representing different stages of differentiation. *Pancreas* **4,** 353–362.

Levrat, J. H., Palevody, C., Daumas, M., Ratovo, G., and Hollande, E. (1988). Differentiation of the human pancreatic adenocarcinoma cell line (CAPAN-1) in culture and co-culture with fibroblasts dome formation. *Int. J. Cancer* **42,** 615–621.

Lieber, M., Mazzetta, J., Nelson-Rees, W., Kaplan, M., and Todaro, G. (1975). Establishment of a continuous tumor-cell line (PANC-1) from a human carcinoma of the exocrine pancreas. *Int. J. Cancer* **15,** 741–747.

Madden, M. E., and Sarras, M. P. (1988). Morphological and biochemical characterization of a human pancreatic ductal cell line (PANC-1). *Pancreas* **3,** 512–528.

Metzgar, R. S., Gaillard, M. T., Levine, S. J., Tuck, F. L., Bossen, E. H., and Borowitz, M. J. (1982). Antigens of human pancreatic adenocarcinoma defined by murine monoclonal antibodies. *Cancer Res.* **42,** 601–608.

Metzgar, R. S., Hollingsworth, M. A., and Kaufman, B. (1993). Pancreatic mucins. *In* "The Pancreas: Biology, Pathobiology, and Disease" (V. L. Go *et al.*, eds.), 2nd ed., pp. 351–367. Raven Press, New York.

Mullins, T. D., Kern, H. F., and Metzgar, R. S. (1991). Ultrastructural differentiation of sodium butyrate-treated human pancreatic adenocarcinoma cell lines. *Pancreas* **6,** 578–587.

Owens, R. B., Smith, H. S., Nelson-Rees, W. A., and Springer, E. L. (1976). Epithelial cell cultures from normal and cancerous human tissues. *J. Natl. Cancer Inst.* **56,** 843–849.

Schoumaker, R. A., Ram, J., Iannuzzi, M. C., Bradbury, N. A., Wallace, R. W., Hon, C. T., Kelly, D. R., Schmid, S. M., Gelder, F. B., Rado, T. A., and Frizzell, R. A. (1990). A cystic fibrosis pancreatic adenocarcinoma cell line. *Proc. Natl. Acad. Sci. U.S.A.* **87,** 4012–4016.

Tan, M. H., Shimano, T., and Chu, M. T. (1981). Differential localization of human pancreas cancer-associated antigen and carcinoembryonic antigen in homologous pancreatic tumoral xenograft. *J. Natl. Cancer Inst.* **67,** 563–569.

Tan, M. H., Nowak, N. J., Loor, R., Ochi, H., Sandberg, A. A., Lopez, C., Pickren, J. W., Berjian, R., Douglass, H. A., and Chu, M. T. (1986). Characterization of a new primary human pancreatic tumor line. *Cancer Invest.* **4,** 15–23.

Yunis, A. A., Arimura, G. K., and Russin, D. J. (1977). Human pancreatic carcinoma (MIA PaCa-2) in continuous culture: Sensitivity to asparaginase. *Int. J. Cancer* **19,** 128–135.

Cell Lines from Human Germ-Cell Tumors

18

Peter W. Andrews[1]
Wistar Institute of Anatomy and Biology
Philadelphia, Pennsylvania 19104

Ivan Damjanov
Department of Pathology
Thomas Jefferson University Medical School
Philadelphia, Pennsylvania 19107

I. Introduction 443

II. Establishment and Maintenance of Cell Lines 445
   A. Embryonal Carcinomas 445
   B. Cell Lines Corresponding to Other Cell Types 448
   C. Growth Conditions 449

III. Morphology 451
   A. Light Microscopy 451
   B. Ultrastructure 454
   C. Histology of Xenograft Tumors 457

IV. Other Characteristics 458
   A. Cell Surface Antigens 458
   B. Secreted Products 462
   C. Intracellular Antigens 462

V. Discussion and Future Prospects 463

VI. Reported Cell Lines Derived from Human Germ-Cell Tumors 466

References 470

I. Introduction

Although relatively rare, teratomas long have fascinated medical scientists because of their curious manifestations, in which more or less well-formed embryonic tissues and body parts may be found. Many early pathologists recognized the relationship of these tumors to the developing embryo and produced various schemes to account for their origins (for reviews, see Dam-

---

[1] Present address: Department of Biomedical Science, University of Sheffield, Western Bank, Sheffield S10 2TN, Great Britain.

janov and Solter, 1974; Wheeler, 1983). However, experimental testing of these ideas became possible only after Stevens discovered that spontaneous testicular teratomas occur in the 129 strain of laboratory mice (reviewed by Stevens, 1967). These studies demonstrated a germ-cell origin for testicular teratomas. Subsequent discovery of ovarian teratomas in LT mice (Stevens and Varnum, 1974) proved that ovarian teratomas are also of germ-cell origin. Other studies by Kleinsmith and Pierce (1964) confirmed the earlier developed notion that tumor cells classified histologically as "embryonal carcinoma" (EC) cells are the pluripotent stem cells that give rise to the multitude of differentiated tissues found in teratomas (Friedman and Moore, 1946; Dixon and Moore, 1953). EC cells also were recognized as the most common malignant cell type found in teratomas of laboratory mice; in their absence, teratomas are typically benign tumors. The term "teratocarcinoma" therefore was applied to malignant teratomas in which EC cells persisted, whereas the term "teratoma," in a strict sense, was reserved for benign tumors devoid of EC cells. EC cells are malignant but can differentiate into nonproliferating benign tissues.

During the 1970s, murine teratocarcinomas became well-established models for early mouse development when researchers found that these tumors also could be derived from ectopically transplanted embryos and that EC cells closely resemble cells of the primitive ectoderm (Solter and Damjanov, 1979; Martin, 1980). This conclusion was demonstrated dramatically by the observation that EC cells differentiate in a controlled way after injection into a blastocyst, and that their derivatives can colonize the host embryo to give rise to a chimeric, but otherwise normal, mouse (Brinster, 1974; Papaioannou et al., 1975). Later, techniques for isolating embryonic stem cell lines directly from explanted embryos were perfected (Evans and Kaufman, 1981). These cells, which form teratomas when injected into adult mice, are very efficient at chimerizing all embryonic tissues, including the germ line, after injection into blastocysts. EC cells, presumably because of selection during tumor growth, very often have lost part of their developmental repertoire; indeed, "nullipotent" EC cells that can no longer differentiate are common.

The experimental study of human teratomas has lagged behind that of their murine counterparts. However, as in the mouse, pluripotent human EC cell lines can be anticipated to offer a useful model for the generally inaccessible early human embryo. Further, although rare in the population as a whole, testicular germ-cell tumors are the most common malignancy in young adult men. Thus, studies of cell lines derived from these human tumors may provide useful insights into clinical significance, as well as into human developmental biology.

Germ-cell tumors in humans differ in a number of respects from the murine teratomas and teratocarcinomas, suggesting that the mouse tumors are not a complete replica of the human disease. Thus, testicular germ-cell tumors in humans typically present after puberty and are usually highly malignant, whereas the testicular teratomas of 129 mice originate in fetal gonads soon after the genital ridge is populated with germ cells. In contrast to human testicular tumors, which are mostly malignant, the mouse tumors are more often benign teratomas rather than teratocarcinomas. Also, the most common type of human testicular germ-cell tumor, seminoma, which resembles primitive premeiotic germ cells, does not occur spontaneously in mice and cannot be induced experimentally in any animals (Damjanov, 1986). Further, choriocarcioma, which consists of malignant cells resembling trophoblast, frequently is found in human but not murine germ-cell tumors whereas many of the differentiated cell types of human teratomas often resemble "immature" embryonic tissues rather than the "mature" tissues seen in murine teratomas, and exhibit a malignant phenotype themselves.

In humans, most teratomas arise in the gonads. Several studies point to a germ-cell origin for these tumors in humans, as in the laboratory mouse (Skakkebaek *et al.,* 1987). Nevertheless, extragonadal tumors that closely resemble the gonadal germ-cell tumors in their pathology also occur. The origin of these tumors has been debated widely; some researchers have argued strongly that some, but not all, tumors are derived from germ cells misplaced during embryogenesis (Oosterhuis *et al.,* 1985).

Numerous cell lines derived from human germ-cell tumors have been reported to date. In this chapter, we summarize the general features that characterize these cells, the techniques used to study them, and their relationship to the complex histology of the tumors from which they have been derived. A list of the reported cell lines is provided at the end of the chapter.

II. Establishment and Maintenance of Cell Lines

Several classification schemes for germ-cell tumors have been devised, for example, by Dixon and Moore (1952), by the British testicular tumor panel (Pugh, 1976), and by the World Health Organization (Mostofi and Sobin, 1976; Table I). We adhere to the World Health Organization's scheme. A detailed review of germ-cell tumor pathology is not possible here. The interested reader is referred to several other well illustrated monographs and articles (Mostofi and Price, 1973; Jacobsen and Talerman, 1989; Damjanov, 1991).

A. Embryonal Carcinomas

The key to understanding germ-cell tumors is the EC cell which, as discussed earlier, is the developmentally pluripotent stem cell that differentiates to form the various somatic tissues found in teratomas (Damjanov, 1991). Nevertheless, human germ-cell tumors often contain cells that morphologically resemble EC cells but have lost their capacity for differentiation. These

Table I

Comparison of Various Classifications of Teratomas and Related Tumors

Dixon and Moore (1952)	British testicular tumor panel (Pugh, 1976)	WHO, 1975 (Mostofi and Sobin, 1976)	Mostofi (1980)
Group III teratoma, pure	Teratoma, differentiated (TD)	Teratoma, mature, immature	Teratoma, mature, immature
Group IV teratoma with carcinoma or sarcoma	Malignant teratoma, intermediate (MTI)	Teratoma with malignant transformation	Teratoma with malignant areas other than seminoma, embryonal carcinoma, or choriocarcinoma
Group IV teratoma with embryonal carcinoma or choriocarcinoma	Malignant teratoma, intermediate (MTI)	Teratocarcinoma (embryonal carcinoma and teratoma)	Embryonal carcinoma and teratoma (teratocarcinoma)
Group II embryonal carcinoma	Malignant teratoma, undifferentiated (MTU)	Embryonal carcinoma	Embryonal carcinoma adult-type
Group V choriocarcinoma	Malignant teratoma, trophoblastic (MTT)	Choriocarcinoma	Choriocarcinoma, pure
Not listed	Yolk sac tumor; orchioblastoma	Yolk sac tumor (endodermal sinus tumor)	Infantile embryonal carcinoma
Not listed	Not listed	Polyembryoma	Polyembryoma

cells correspond to "nullipotent" mouse EC cells. Many cell lines established *in vitro* from germ-cell tumors probably correspond to such EC cells. Evidence for this view came initially from their ability to grow in immunodeficient athymic (nu/nu) mice and to form xenograft tumors that are histologically consistent with the EC cells recognized by pathologists in clinical specimens of germ-cell tumors (Bronson *et al.*, 1980; Andrews *et al.*, 1980,1982; Cotte *et al.*, 1981,1982; McIlhinney *et al.*, 1983).

These putative human EC cells resemble murine EC cells morphologically and in their expression of high levels of alkaline phosphatase (e.g., Benham *et al.*, 1981; Cotte *et al.*, 1981,1982). However, these cells differ with respect to markers often considered objective criteria for defining EC cells in the mouse. For example, stage specific embryonic antigen 1 (SSEA-1) is expressed characteristically by murine EC cells (Solter and Knowles, 1978), which lack another embryonic antigen, SSEA-3 (Shevinsky *et al.*, 1982). In contrast, Andrews *et al.* (1980,1982) reported that human EC cells in culture appear to be SSEA-3$^+$ and SSEA-1$^-$. Subsequently, EC cells in clinical specimens of human germ-cell tumors similarly were found to express SSEA-3 but not SSEA-1 (Damjanov *et al.*, 1982). Further, other human EC-like cell lines were derived and expressed similar phenotypes. A more detailed summary of human EC cell markers is provided in the following discussion.

Many of the EC cell lines that were characterized initially show only limited capacities for differentiation. For example, 2102Ep cells retain an EC phenotype when cultured at high cell densities ($>5 \times 10^6$ cells per 75-cm^2 flask), but some cells undergo morphological differentiation with the disappearance of SSEA-3 and the appearance of SSEA-1 when cultured at low density (10^5 cells per 75-cm^2 flask). Under such conditions, fibronectin synthesis is induced (Andrews, 1982) and some trophoblastic giant cells appear (Damjanov and Andrews, 1983). This requirement for a high cell density to maintain an EC phenotype seems to be a common feature of other human EC cell lines (e.g., see Bronson *et al.*, 1984b; Casper *et al.*, 1987). However, some human EC cell lines show a tendency to form domes and floating vesicles when cultured at high density. In one case, these domes were reported to resemble blastocysts with trophectodermal cells surrounding an inner core of undifferentiated EC cells (Izhar *et al.*, 1986). However, in other cases, we have seen no compelling evidence for such differentiation in these structures (Andrews *et al.*, 1980,1984b).

A differentiation similar to that seen at low cell densities has been observed with other human EC cell lines treated with certain drugs. For example, LICR-LON HT39/3 EC cells differentiated in response to phorbol esters (McIlhinney *et al.*, 1983) and TERA-1 differentiated in response to α-difluoromethylornithine (Uhl *et al.*, 1986). However, many human EC cell lines do not respond to retinoic acid (RA) (Matthaei *et al.*, 1983), which is widely used as an inducer of murine EC cell differentiation and may function as a morphogen during embryogenesis.

Nevertheless, several human EC cell lines that can be induced to undergo more extensive differentiation have been isolated. Of these, the most widely studied are sublines of the teratocarcinoma cell line TERA-2 (Fogh and Trempe, 1975; Andrews et al., 1984b; Thompson et al., 1984). These cells do respond to RA, which induces differentiation into neurons and other cell types (Andrews, 1984; Lee and Andrews, 1986; Rendt et al., 1989). These cells also can be induced to differentiate along a quite distinct lineage by exposure to another agent, hexamethylene bisacetamide (Andrews et al., 1990).

Although most of the cells arising during NTERA-2 differentiation remain to be identified, none appears to correspond to the extraembryonic cell types of the trophoblast and yolk sac. On the other hand, lines of EC cells that undergo this type of differentiation, in addition to somatic differentiation, include 1777N (Bronson et al., 1983a), NCCIT (Teshima et al., 1988), and GCT27 (Pera et al., 1987,1989). In 1777N and GCT27, maintenance of an EC phenotype requires continual growth on feeder layers of embryonic fibroblasts; removal from the feeder layers causes the EC cells to differentiate. However, NCCIT EC cells can be maintained without feeder cells and can be induced to differentiate with RA.

B. Cell Lines Corresponding to Other Cell Types

Many cell lines that have been derived from human germ-cell tumors appear to correspond to EC cells, but some are more likely to represent other cell types found in these tumors. For example, several lines of yolk sac carcinoma cells have been isolated, notably 1411H (Vogelzang et al., 1985), GCT44, GCT46, and GCT72 (Pera et al., 1987). Pera et al. (1987) argued for the definition of two types of yolk sac carcinoma-derived cell lines. One, of which GCT72 was the prototype, forms solid yolk sac tumors and is strongly positive for α-fetoprotein when grown as a xenograft in athymic mice. These cells are similar to rodent visceral endoderm and express patterns of markers that are characteristic of EC cells (e.g., SSEA-3 and SSEA-4). The other type of yolk sac carcinoma-derived cell line, typified by GCT44 and GCT46, more closely resembles rodent parietal endoderm, does not express EC-like markers, and forms xenograft tumors with the appearance of endodermal sinus tumor.

On the other hand, only one pure choriocarcinoma cell line appears to have been derived from a (ovarian) germ-cell tumor (Sekiya et al., 1983), although several other lines do produce human chorionic gonadotropin, a marker of the trophoblastic cells of which choriocarcinomas are a malignant analog. In most germ-cell tumor cell lines in which human chorionic gonadotropin has been detected, the primary cell is likely to be an EC cell that differentiates spontaneously at a significant frequency into trophoblastic cell types (e.g., 1777N and 1075L; Bronson et al., 1983a,b,1984b). However, several cell lines from gestational choriocarcinomas, although not strictly germ-cell tumors, have been isolated (e.g., Pattillo and Gey, 1968; Patillo et al., 1971) and provide a useful tool for study of this tumor cell type.

A number of cell lines that have been derived probably correspond to somatic cell types found in teratomas, although their exact identity is difficult to establish. One example is the set of 577 lines derived by Bronson and his colleagues (Andrews et al., 1980); another is the HAZ lines derived from an extragonadal germ-cell tumor by Oosterhuis et al. (1985). A particularly intriguing example is the PA1 line derived from an ovarian germ-cell tumor (Giovanella et al., 1974). After extensive study, Zeuthen et al. (1980) described this line as a pluripotent EC cell with a propensity for neuroectodermal differentiation. However, these cells do not closely resemble other human EC cells in culture, either morphologically (Fig. 1) or in their expression of characteristic surface antigen markers. They also form xenograft tumors that appear to be neuroectodermal in nature. A review of slides of the original tumor from which the line was derived led to the conclusion that the line may be better characterized as an example of the immature neuroectodermal teratomas commonly found in the ovaries of young women (I. Damjanov, personal observation). Thus, the PA1 cells correspond more to a committed neuroectodermal stem cell from a later stage of embryogenesis than to the early embryonic cells considered to be the normal equivalents of EC cells. In any case, the PA1 cells do seem to undergo differentiation under certain conditions, including in response to RA (Tainsky et al., 1988).

C. Growth Conditions

Various media have been used to derive and maintain cell lines from human germ-cell tumors, including McCoy's 5A, Dulbecco's modified Eagle's medium (DME) in its high glucose formulation, RPMI 1640, and the Alpha modification of Eagle's medium. In each case, the medium has been supplemented with 10–15% fetal bovine serum (FBS) and sometimes with 10% tryptose phosphate broth. Primarily we use DME with 10% FBS and have found that most lines grow satisfactorily under these conditions. However, occasionally we have observed better success with other media, such as RPMI for the NCCIT line.

As can other cells, these lines usually can be passaged after harvesting using 0.25% trypsin and 2 mM ethylenediamine tetraacetic acid (EDTA). However, a common observation we have made is that human EC cells best retain an EC phenotype if passaged at high cell densities ($>5 \times 10^6$ cells per 75-cm^2 flask), as described for 2102Ep (Andrews et al., 1982). Further, in some cases the cells are maintained better if they are kept in clumps by scraping rather than trypsinization at harvest. We have found it convenient to use glass beads to detach the cells—about twenty 3-mm acid-washed glass beads are rolled gently over the cell monolayer in the presence of some

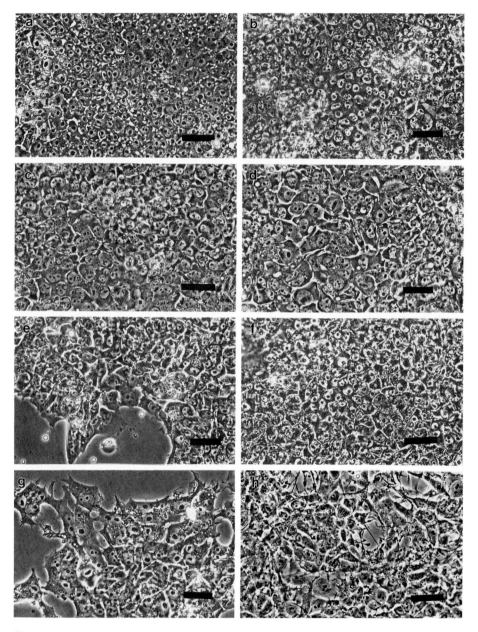

Fig. 1. (a–g) Various human teratocarcinoma-derived cell lines exhibiting typical embryonal carcinoma (EC) cell morphology. (a) 1156QE. (b) TERA-1. (c) SuSa. (d) 833KE. (e) 1777NRPmet. (f) 1618K. (g) NCCIT. All these cultures were passaged using trypsin, and replated at high cell densities (>5 × 10^6 per 75-cm^2 flask), except for SuSa, which was passaged by scraping. (h) PA1 cells. These cells have been described as EC cells (Zeuthen *et al.*, 1980), but they exhibit a somewhat different morphology and growth pattern from the other lines and may correspond to a different stem cell type (see text for discussion). Bar, 50 μm.

medium. Under these conditions, the cells are detached efficiently in large clumps with good viability.

Feeder layers of human lung fibroblasts or transformed mouse cells (e.g., STO) also have been advocated by some investigators for maintaining undifferentiated EC cells (Bronson *et al.*, 1983a,1984b; Pera *et al.*, 1987). Although this requirement may not be universal for all human EC cells, it should be met in the case of cells that have been maintained in that manner.

III. Morphology

A. *Light Microscopy*

Photomicrographs of several human germ-cell tumor cell lines, illustrating the morphology and growth patterns commonly seen for these cells, are shown in Figs. 1–4.

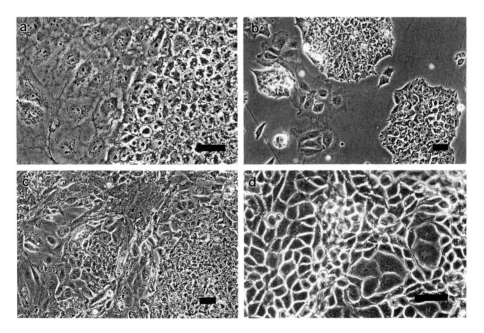

Fig. 2. Morphological differentiation of human embryonal carcinoma (EC) cell lines induced by altered growth conditions. (a) 2102Ep cl.4D3 EC cells in cultures plated at low density (10^5 cells per 75-cm^2 flask). Note the large flat cells surrounding a colony of typical EC cells. (b) 1156QE cells. Note the few large flat cells that have arisen spontaneously in this culture containing predominantly EC cells that form typical tightly packed colonies. (c) Cells in a culture of SuSa passaged by trypsinization rather than scraping (cf. Fig. 1c). Note the morphological heterogeneity of this culture. (d) 1777NRPdiff cells (cf. Fig. 1e). This stable differentiated cell line was derived by repeatedly passaging 1777NRPmet cells at low densities (Bronson *et al.*, 1983a). Bar, 50 μm.

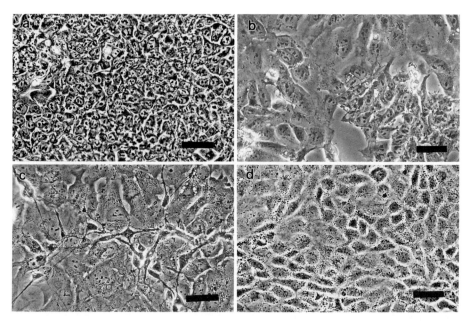

Fig. 3. TERA-2—a pluripotent human embryonal carcinoma (EC) cell line. (a) NTERA-2 cl.D1 EC cells growing at high cell density. These cells are a cloned subline of TERA-2 (Andrews *et al.*, 1984b). Note their typical EC morphology. (b) A culture of uncloned TERA-2 cells, as available from the ATCC. Note the colony of EC cells, but the predominance of other morphologically distinct cell types. (c) Differentiated NTERA-2 cl.D1 cells derived by exposure to 10^{-5} *M* all-*trans*-retinoic acid for 3 weeks, followed by replating and growth in the absence of inducer (Andrews, 1984). Note the contrast with the EC cells and the presence of neuron-like cells. (d) Differentiated NTERA-2 cl.D1 cells derived by exposure to 3 m*M* hexamethylene bisacetamide for 3 weeks. Note the contrast with the EC cells and retinoic acid-induced cells. Bar, 50 μm.

Human EC cells in culture resemble their murine counterparts, typically forming tightly packed clusters in which individual cells may be difficult to discern. Depending on the particular cell line, these clusters may grow to form dense layers of cells covering the culture dish or they may tend to remain isolated, growing upward rather than outward. In either case, clusters of cells may bud off and form floating hollow vesicles, as discussed earlier. EC cells generally exhibit a high nucleus : cytoplasm ratio. Under phase contrast microscopy, the nuclei are pale and well differentiated from a darker cytoplasm. In many cases, the nuclei contain only one or two quite prominent nucleoli. These features are illustrated by the seven EC cell lines shown in Fig. 1. Note the different growth patterns and morphology of PA-1 cells (Fig. 1h), which tend to form dendritic processes and be less tightly packed.

Morphological differentiation of many human EC cells occurs when growth conditions are altered, for example, when cells are grown at low density. Typically, the cells flatten out and appear to enlarge, as shown in Fig. 2a,

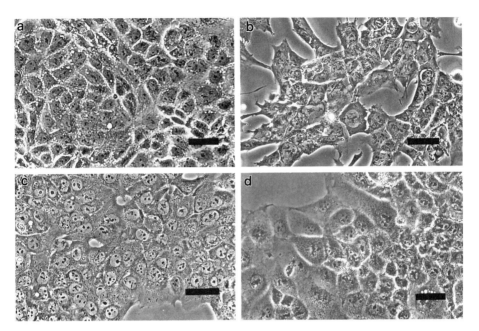

Fig. 4. Nonembryonal carcinoma (EC) cell lines. 1411H, a yolk sac carcinoma cell line, and 577MF, a line derived from a teratoma. Both are clearly distinct from the EC cells shown in Fig. 1. (c,d) Bewo and Jar gestational choriocarcinoma cell lines, respectively (Pattillo and Gey, 1968; Pattillo *et al.*, 1971). Bar, 50 μm.

which illustrates 2102Ep EC cells after plating at low density. Note the colony of typical EC cells adjacent to an area of distinct large flat cells. Often, similar cells may be seen around the edges of clusters of otherwise typical EC cells growing at high density, as in 1156QE (Fig. 2b). Dispersing cells by trypsinizing rather than scraping also may lead to the appearance of a variety of morphologies and growth patterns in some cases. For example, SuSa cells in a trypsinized culture as shown in Fig. 2c. In this culture, many of the cells exhibit diverse non-EC morphologies. In our experience, SuSa is best maintained in its EC-like form by harvesting cells for passaging by scraping rather than by trypsinizing. Another example is provided by the 1777NRPdiff cells (Fig. 2d). This stable line of non-EC cells was derived from 1777NRP EC cells (Fig. 1) by repeated passaging at low density (Bronson *et al.*, 1983a).

The TERA-2 line provides an unusual example of pluripotent human EC cells. The NTERA-2 cl.D1 cells (Fig. 3a) were isolated clonally after passage of TERA-2 through a xenograft tumor in a nude mouse. Maintained at high cell density, these cells exhibit a typical EC morphology (Andrews *et al.*, 1984b). Similar cells can be found with non-EC-like cells in the parent TERA-2 cell line from the American Type Culture Collection (ATCC; Fig. 3b). NTERA-2 cl.D1 cells are sensitive to altered growth conditions which, if suboptimal, can lead to the accumulation of morphologically heterogeneous populations of cells, many of which have lost their typical EC-like appearance. Such changes are avoided best by passaging at high density after scraping to harvest the cells, by selection in FBS, and by adequate feeding of the cultures. Unlike many other human EC cells, NTERA-2 cells respond to chemical agents such as RA or hexamethylene bisacetamide (Fig. 3c,d). Neurons are especially evident after exposure to RA (Fig. 3c).

The morphologies of non-EC cell lines derived from germ-cell tumors are mixed. Describing common specific features is difficult, except to point out differences from typical EC morphology. Examples of some such lines are shown in Fig. 4.

B. Ultrastructure

Typical EC cells measure 12–15 μm in diameter and have a high nucleus : cytoplasm ratio. The nuclei vary in shape and may be round, oval, irregularly shaped, and deeply indented (Fig. 5). The nucleoli are prominent and

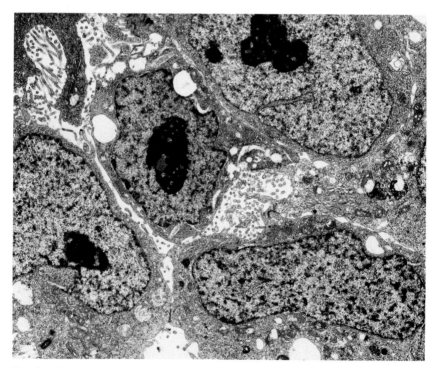

Fig. 5. Electron micrograph of 2102Ep embryonal carcinoma cells. The cells have irregularly shaped nuclei with prominent nucleoli. Their cytoplasm is relatively scant and they form microvilli and intercellular junctions (3200×).

show a skein-like appearance or a complex nucleolonema, with intertwining of granular and fibrillar components. Euchromatin usually predominates over heterochromatin.

The cytoplasm is developed to a variable degree; in most cells it contains only a few organelles and is filled mostly with free ribosomes. The mitochondria are scarce and usually are located in the perinuclear area. A few short profiles of rough endoplasmic reticulum and a few lysosomal granules may be seen in the vicinity of the Golgi apparatus. Several human EC cell lines (e.g., TERA-2, 833KE, and SuSa) contain large amounts of glycogen in the cytoplasm (Fig. 6).

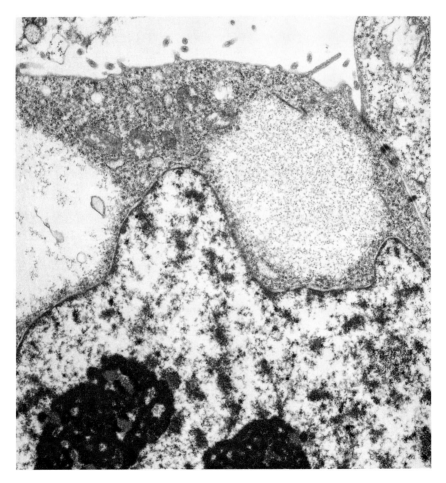

Fig. 6. Electron micrograph of 833KE embryonal carcinoma cells. The cytoplasm contains aggregates of glycogen. Also note the complex nucleoli, the intercellular junctions, and the short microvilli (18,000×).

In the cytoplasm of many EC cells are prominent intermediate filaments, frequently assembled into bundles of tonofilaments or inserted into well-formed desmosomes (Fig. 7). Other forms of intercellular junctions, such as tight junctions, are seen also. The cell surface is usually smooth, although in many cell lines the surface extends into short microvilli.

EC cells grow in small aggregates that often become dome-like and even detach, forming floating bodies (Bronson *et al.*, 1983b). These bodies have been called embryoids, analogous to the murine embryoid bodies (Zeuthen *et al.*, 1980). Ultrastructurally these bodies consist of typical EC cells forming the inner cell mass and elongated EC cells wrapping around them and forming an outer layer (Fig. 8). We have called these structures morules (Damjanov and Andrews, 1983), since we could not find any definitive evidence that the outer cells have differentiated into yolk sac cells, as one would expect on the basis of the analogy with mouse embryoid bodies. Thus, although the morules formed in cultures of human EC cells represent a peculiar form of morphogenesis *in vitro*, equating these structures with early embryos would be unwise.

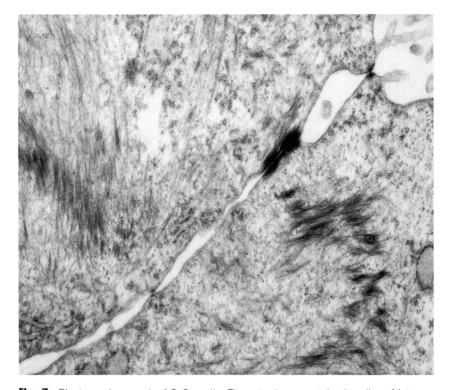

Fig. 7. Electron micrograph of SuSa cells. The cytoplasm contains bundles of intermediate filaments (keratin). A desmosome and a tight junction interconnect the adjacent cells (25,000×).

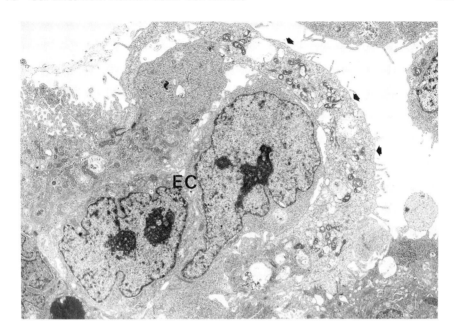

Fig. 8. Electron micrograph of a morule in a culture of 2102Ep cells. The embryonal carcinoma cells of typical morphology are enveloped with another cell that has elongated cytoplasm and apparently contains more organelles (3400×). (Reproduced from Damjanov and Andrews, 1983, with permission.)

Cultures of EC cells treated with RA, as well as untreated cell cultures, may contain cells that show signs of differentiation. Such cells are larger and their cytoplasm contains abundant organelles such as mitochondria and rough and smooth endoplasmic reticulum (Fig. 9), as in mouse EC. Yolk sac differentiation is marked by the appearance of prominent profiles of rough endoplasmic reticulum and absorptive cytoplasmic vacuoles (Damjanov et al., 1990; Fig. 10).

C. Histology of Xenograft Tumors

Many human cell lines will grow and form tumors in immunosuppressed animals, most commonly athymic (nu/nu) mice. Histological examination of these xenograft tumors provides an important link with the standard histopathology of the tumors from which the cell line in question was derived, and provides one method of defining the nature of particular cell lines. We and others have used this approach to determine whether germ-cell tumor-derived cell lines correspond to EC cells or to other cell types found in these tumors (e.g., Andrews et al., 1982). The reader is referred elsewhere for detailed summaries of germ-cell tumor pathology (Mostofi and Price, 1973; Pugh, 1976; Talerman and Roth, 1986; Jacobsen and Talerman, 1989).

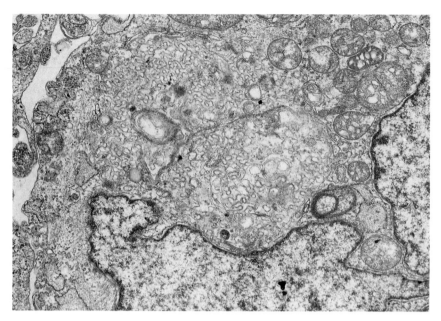

Fig. 9. Electron micrograph of a giant cell formed in a low density (differentiated) culture of 2102Ep. The well-developed cytoplasm contains prominent profiles of endoplasmic reticulum and more mitochondria than the typical stem cells (14,000×). (Reproduced from Damjanov and Andrews, 1983, with permission.)

IV. Other Characteristics

Morphology, although a useful quick way of assessing the nature of cell lines, is a subjective and unreliable approach to defining cells in culture. More objective criteria are afforded by assays of products formed by the cells. The useful markers for such studies fall into three groups, namely, cell surface antigens, secreted products, and internal cell molecules, especially those that form the cytoskeleton. Few if any markers are expressed uniquely by a given cell type and, in most cases, a single marker is not diagnostic. Instead, the pattern of expression of several markers must be considered.

A. Cell Surface Antigens

Immunoassays for the expression of certain cell surface antigens defined by monoclonal antibodies provide one of the most useful techniques for characterizing cell lines. Such assays can be carried out on small numbers of cells, and can be coupled with immunofluorescence and fluorescence-activated cell sorting to assess the heterogeneity of cultures and to isolate subsets of live cells for further growth and study in isolation.

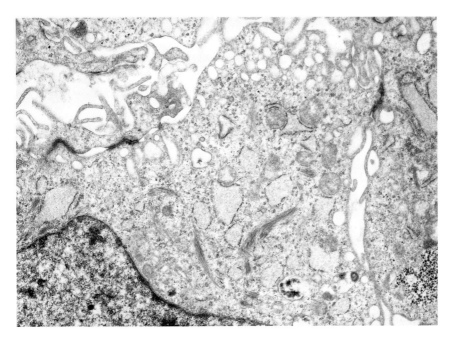

Fig. 10. Electron micrograph of differentiated 1777N cells, showing prominent dilated profiles of rough endoplasmic reticulum filled with finely granular material, bundles of tonofilaments, absorptive vacuoles, and long microvilli, consistent with yolk sac differentiation (18,000×).

Several cell surface antigens have been described as characteristically expressed by human EC cells (Table II), including glycolipid antigens of the globoseries (Table III) as well as several high molecular weight glycoprotein antigens (Andrews *et al.*, 1984a; Rettig *et al.*, 1985; Pera *et al.*, 1988). For the latter group, the epitopes of several, if not all, appear to be composed of oligosaccharides (Fukuda *et al.*, 1986; Andrews *et al.*, 1991). High molecular weight lactosaminoglycans appear to be common to mouse and human EC cells (Muramatsu *et al.*, 1982); an antigen recognized by monoclonal antibody GCTM2 on EC cells was shown to be keratan sulfate (Pera *et al.*, 1988). Although some antigens are expressed commonly by human EC cells in culture and in tumors, they are not expressed invariably. For example, several sublines of EC cells isolated from the TERA-2 line do not express cell surface SSEA-3 or SSEA-4 (Thompson *et al.*, 1984; Andrews *et al.*, 1985). Nevertheless, globoseries glycolipids carrying the appropriate epitopes can be isolated from these cells. The reasons for lack of surface reactivity are unclear (Fenderson *et al.*, 1987).

Alkaline phosphatases provide other useful markers of EC cells. These enzymes (which are expressed at a high level by EC cells) are cell surface molecules and can be detected as surface antigens as well as by enzymatic

Table II

Expression of Cell Surface Antigens Useful in Characterizing Cell Lines Derived from Germ-Cell Tumors

Antibody/antigen	Cell type[a]			Reference
	EC	YSC	Choriocarcinoma	
MC631/SSEA-3	+	−	−	Shevinsky et al. (1982); Andrews et al. (1983)
MC813-70/SSEA-4	+	−	−	Kannagi et al. (1983a)
TRA-1-60	+	−	−	Andrews et al. (1984a)
TRA-1-81	+	−	−	Andrews et al. (1984a)
8-7D	+	−	−	Andrews et al. (1984a)
K21	+	−	−	Rettig et al. (1985)
K4	+	−	−	Rettig et al. (1985)
GCTM2	+	(+)[b]	−	Pera et al. (1988)
MC480/SSEA-1	−	±	+	Solter and Knowles (1978); Andrews et al. (1983)

[a] EC, embryonal carcinoma; YSC, yolk sac carcinoma.
[b] GCTM2 reacts with cell line GCT72, which resembles rodent visceral yolk sac carcinomas, but not with cell lines GCT44 and GCT46, which resemble parietal yolk sac carcinoma and form endodermal sinus tumors in athymic mice (Pera et al., 1988).

activity. In human EC cells, most of the activity is attributable to the liver/bone/kidney isoenzyme (Benham et al., 1981), which can be detected by monoclonal antibodies such as TRA-2-49 and TRA-2-54 (Andrews et al., 1984c). A form of alkaline phosphatase resembling the placental isoenzyme is also often detectable in EC cell lines, but this isoenzyme usually only accounts for a small fraction of the total enzymatic activity (Benham et al., 1981; Cotte et al., 1981,1982).

Major histocompatibility complex (MHC) Class I antigens usually are expressed only at low levels by human EC cells, although these antigens (unlike in murine EC cells) are often detectable (Andrews et al., 1980,1982,1984b; Pera et al., 1987). In several cases, these antigens are also inducible by interferon γ (IFNγ) (Andrews et al., 1987).

Few antigens characteristic of other components of germ-cell tumors have been described, but the tumors often can be defined in part by their lack of EC antigens. However, choriocarcinoma cells often express high levels of SSEA-1, which is not expressed by human EC cells (Andrews et al., 1980,1982). Also, an antigen characteristic of yolk sac carcinomas has been described (Tanaka et al., 1989).

Table III

Cell Surface Glycolipid Antigens Useful in Following Differentiation of NTERA-2 EC Cells[a]

Antigen/glycolipid	Glycolipid structure	Antibody	Reference
Globoseries			
P^k, GL3	Galα1 → 4Galβ1 → 4Glcβ1 → Cer		Naiki and Marcus (1974)
P, globoside	GalNAcβ1 → 3Galα1 → 4Galβ1 → 4Glcβ1 → Cer		Naiki and Marcus (1974)
SSEA-3	Galβ1 → 3GalNAcβ1 → 3Galα1 → 4Galβ1 → 4Glcβ1 → Cer		Shevinsky et al. (1982)
SSEA-4 and SSEA-3	NeuAcα2 → 3Galβ1 → 3GalNAcβ1 → 3Galα1 → 4Galβ1 → 4Glcβ1 → Cer	MC631, MC813-70	Kannagi et al. (1983a,b)
Globo-H	Fucα1 → 2Galβ1 → 3GalNAcβ1 → 3Galα1 → 4Galβ1 → 4Glcβ1 → Cer	MBr1	Bremer et al. (1984)
Globo-A	Fucα1 → 2Galβ1 → 3GalNAcβ1 → 3Galα1 → 4Galβ1 → 4Glcβ1 → Cer, with GalNAcα1 at 3	HH5	Clausen et al. (1986)
Lactoseries			
SSEA-1,Le^x	Galβ1 → 4GlcNAcβ → 3Galβ1 → 4Glcβ1 → Cer, with Fucα1 at 3	MC480	Solter and Knowles (1978); Kannagi et al. (1982)
Le^y	Galβ1 → 4GlcNAcβ1 → 3Galβ1 → 4Glcβ1 → Cer, with Fucα1 at 2 and Fucα1 at 3	AH6	Abe et al. (1983)
Ganglioseries			
GD3	NeuNAcα2 → 8NeuNAcα2 → 3Galβ1 → 4Glcβ1 → Cer	VIN-IS-56	Andrews et al. (1990)
9-O-acetyl-GD3	9-O-acetyl NeuNAcα2 → 8NeuNAcα2 → 3Galβ1 → 4Glcβ1 → Cer	ME311	Thurin et al. (1985)
GT3	NeuNAcα2 → 8NeuNAcα2 → 8NeuNAcα2 → 3Galβ1 → 4Glcβ1 → Cer	A2B5	Eisenbarth et al. (1979); Fenderson et al. (1987)
GD2	NeuNAcα2 → 8NeuNAcα2 → 3Galβ1 → 4Glcβ1 → Cer, with GalNAcβ1 at 4	VIN-2PB-22	Andrews et al. (1990)

[a] See Fenderson et al. (1987) and Andrews et al. (1990) for details.

In studying the differentiation in culture of pluripotent EC cells, especially the NTERA-2 EC cells, we have identified a number of other antigens that are induced on differentiation. Several of these are glycolipid antigens (Table III). We concluded that one characteristic of embryonic cell differentiation is a switch in oligosaccharide synthesis from globoseries to lacto- and ganglio-series core structures (Fenderson et al., 1987; Chen et al., 1989).

B. Secreted Products

Two proteins, human chorionic gonadotropin and α-fetoprotein, often are produced by germ-cell tumors and are widely used as markers in clinical practice. These proteins are generally products of trophoblastic giant cells, typically found in choriocarcinoma and yolk sac carcinoma (endodermal sinus tumor), a pattern reflecting their sites of production in the conceptus. That these proteins are products of EC cells seems unlikely and their production in cultures of EC cells provides evidence of differentiation into trophoblastic or endodermal cells.

Extracellular matrix proteins, notably fibronectin, laminin, and the collagens, typically are synthesized by cells in yolk sac carcinomas, presumably reflecting their resemblance to the extraembryonic endoderm. EC cells do not seem to synthesize these proteins, except after differentiation (e.g., Andrews, 1982; Andrews et al., 1983). High levels of synthesis, especially of laminin, may indicate the presence of cells corresponding to yolk sac carcinoma (Cooper and Pera, 1988). However, conclusions should not be drawn on this basis alone, since a number of cell types may synthesize these proteins.

C. Intracellular Antigens

The cytoskeletal structures offer a plethora of epitopes the expression of which is restricted to limited numbers of cell types, and these antigens provide particularly useful tools for dissecting the heterogeneous groups of cells found in teratoma-derived cell lines and for relating these cells to those found in clinical specimens of germ-cell tumors..

Human EC cells have a well-developed cytoskeleton that consists of actin- and myosin-rich microfilaments, tubulin-rich microtubules, and intermediate filaments. In contrast to mouse EC cells, which typically do not contain keratin intermediate filaments, the intermediate filaments of most human EC lines are composed of keratins (Damjanov et al., 1984), an observation consistent with the fact that EC cells in human solid tumors react with antibodies against keratins (Miettinen et al., 1985; Lifschitz-Mercer et al., 1992). However, not all human EC cells are keratin positive (Table IV) and, even in those cell lines that typically express keratin polypeptides, not all cells express this cytoskeletal polypeptide to the same extent (Fig. 11). In some cell lines, the expression of keratin depends on the culture conditions and, in

Table IV

Expression of Keratin Polypeptides in Human Germ-Cell
Tumor Cell Lines[a]

Cell line	Percentage of cells showing immunofluorescence	
	Strong	Weak
SuSa	100	—
833KE	20	80
1218E	30	70
2866B	—	100
TERA-1	10	90
NTERA-2 cl.D1	10	90
2102Ep	10	90
PA1	0	0
NCCIT	0	0
UM-TC-1	5	95
1156QE	5	95
1411H	100	—
T84-5130	20	80
577MF	100	—

[a] The cells were grown on coverslips and stained with
antikeratin antibody AE-1 and AE-3 (Sun et al., 1983).

general, the conditions favoring differentiation tend to promote the expression of keratins. Cells showing mesenchymal differentiation express vimentin, those showing muscle differentiation express desmin, and those exhibiting neural or glial differentiation express neurofilaments or glial acidic fibrillary protein (Trojanowski and Hickey, 1984; Lee and Andrews, 1986).

Among antibodies raised against cell lines derived from human germ-cell tumors, Pera et al. (1988) described two with restricted tissue distribution. Antibody GCTM3 recognizes a 57-kDa cytoskeletal protein found in EC cells and in yolk sac carcinoma cells; another antibody, GCTM4, recognizes a 69-kDa protein associated with the lysosomal compartment and found only in EC cells.

V. Discussion and Future Prospects

A large number of cell lines has now been derived from human germ-cell tumors. Although only a few of these lines have been studied in any great detail, various characteristics of these cells (especially the EC cells) are

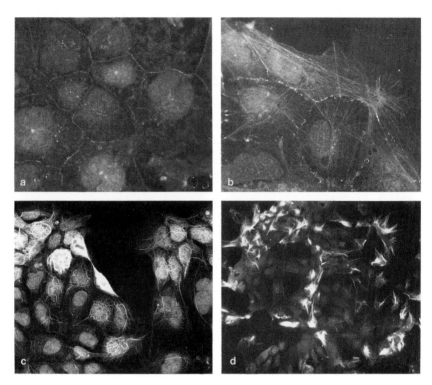

Fig. 11. Immunofluorescence microscopy of high density (a,b) and low density (c,d) cultures of 2102Ep cells stained with fluoresceinated antibodies against keratin (a,b,c) and vimentin (d). The high density cultures are composed mainly of embryonal carcinoma cells that contain only keratin filaments, which are either limited to intercellular junctions (a) or intercellular junctions and tension filaments in the cytoplasm (b). In low density cultures, spontaneous differentiation occurs and many cells contain abundant, dense keratin filaments (c) whereas some cells become vimentin positive (d) (320× for a,b; 220× for c,d). (Reproduced from Damjanov and Andrews, 1983, with permission.)

becoming clear, as we have discussed. In reviewing this work, we have focused on those studies that bear on the practical issues involved in using these cell lines. Studies that can be addressed only briefly here include reports of retroviruses in many of these cell lines (Harzmann *et al.*, 1982; Boller *et al.*, 1983; Bronson *et al.*, 1984a), karyotype analyses indicating the common occurrence of i(12)p (e.g., Geurts van Kessel *et al.*, 1991), investigations of various developmentally regulated genes expressed by these cells (Miller *et al.*, 1990; Simeone *et al.*, 1990), and the effects of growth factors on the growth and differentiation of EC cells (Engström *et al.*, 1991; Miller and Dmitrovsky, 1991; Mummery and Weima, 1991).

Much interest in teratomas is generated by their potential relationship to early embryonic cells and by the possibility that they can provide convenient experimental tools for studying the processes of cell differentiation in the early embryo. This possibility has proved true of murine teratocarcinoma cells, although perhaps not to the extent envisioned when the lines first became available for experimental study. In this respect, note that, despite their resemblance to embryonic stem cells, EC cells are tumor cells that have been subject to all the selection pressures that are to be expected when cells adapt to tumor growth. For example, since differentiation of EC cells usually results in a loss of malignant phenotype, strong selection pressure for EC cells that have lost their ability to differentiate will always exist. Interestingly, somatic cell hybrids of murine EC cells may exhibt an EC phenotype that possesses a greater capacity for differentiation than the parental EC cell (Andrews and Goodfellow, 1980; Atsumi *et al.*, 1982; Rousset *et al.*, 1983). Such a result could be explained by complementation of a recessive mutation that restricts differentiation in the parental EC cell. Thus, EC cells are unlikely to be precise models of early embryonic stem cells; their differentiation is more likely to be a "caricature of embryogenesis" (Pierce, 1975). Nevertheless, when properly interpreted, experiments with such cells can provide important information about the molecules and processes that control cell differentiation during development.

Such reservations expressed for murine EC cells apply more strongly to human EC cells. In these cases, the tumors probably appear after rather long periods of subclinical growth and, hence, selection. This fact is, perhaps, reflected in the greater proportion of "nullipotent" EC cells in human germ-cell tumors and by the high degree of aneuploidy in these cells. Nevertheless, human embryonic development is much less accessible to experimental study than that of the mouse, so teratoma cells can provide an important insight into the processes of cell differentiation in the human embryo. Several results now suggest differences between mouse and human teratomas and embryos; generalizing from one species to the other without specific experimental study would be unwise. Moreover, human cells, with a different spectrum of available markers and reagents, sometimes may provide practical advantages over their counterparts from other species. Also, some studies, such as investigations with human cytomegalovirus (e.g., Gönczöl *et al.*, 1984; Andrews *et al.*, 1989) which affects human embryos but does not replicate in cells from other species, require the use of human cells.

Apart from their appeal to developmental biologists, cell lines from germ-cell tumors can provide important tools for investigating clinically relevant issues. For example, we found that an antigen, TRA-1-60, characteristically expressed by EC cells can be used as a serum marker for the follow-up of treated germ-cell tumor patients (Marrink *et al.*, 1991). The etiology of germ-cell tumors also poses a large unanswered question that might be addressed in part by studies with appropriate cell lines. In particular, seminomas have been proposed to represent one stage of the progression from malignant transformation of germ cells to the appearance of pluripotent EC cells. Unfortunately, no seminoma cell lines with which to test this hypothesis have been successfully derived to date. Cells from these tumors are likely to require culture conditions that remain unidentified, presenting an important challenge for future studies.

VI. Reported Cell Lines Derived from Human Germ-Cell Tumors

Table V lists all the reported human germ-cell tumor-derived cell lines of which we are aware, indicating the origin, site of metastasis (if pertinent), histology of the explanted tumor (when known), and the reported histology of xenograft tumors grown from the established cell lines. We also provide key references, especially to the paper(s) in which the lines were first described. However, giving a full bibliography of each cell line or providing details of the various sublines that have been derived from them is not possible here; additional references are presented in the text. We apologize for any omissions. Cell lines established from the same patient are bracketed.

Table V

Reported Cell Lines from Human Germ-Cell Tumors

Cell line	Biopsy[a]		Xenograft histology	Reference
	Site	Histology		
Testis tumors				
1075L hep	Liver	EC, C		Bronson et al. (1983b)
1075L lung	Lung	EC, C		Bronson et al. (1983b)
1156QE	Primary	EC, C, S		Andrews et al. (1980); Wang et al. (1980,1981); Bronson et al. (1984a)
1218E	Primary	EC, S		Andrews et al. (1980); Wang et al. (1980,1981); Bronson et al. (1984a)
1242B		EC		Wang et al. (1980, 1981); Bronson et al. (1984a)
12550				Wang et al. (1980,1981); Bronson et al. (1984a)
1411Hp	Primary	EC, T, Y	EC, Y	Vogelzang et al. (1985)
1411HRQmet	RPLN	EC, Y	EC, Y	Vogelzang et al. (1985)

(continues)

Table V

Continued

Cell line	Biopsy[a]		Xenograft histology	Reference
	Site	Histology		
1428A		EC, Y		Wang et al. (1981); Bronson et al. (1984a)
1446S				Bronson et al. (1984a)
1685M	Metastasis	EC		Bronson et al. (1984b)
1777N-Pr	Primary	EC, T		Bronson et al. (1983a, 1984a)
1777N-RP	RPLN	EC		Bronson et al. (1983a, 1984a)
1777NRP-diff			Sarcoma	Bronson et al. (1983a); Hiraoka et al. (1988)
2044L		EC		Wang et al. (1980, 1981); Bronson et al. (1984a)
2061H		EC, T		Wang et al. (1980, 1981); Bronson et al. (1984a)
2102Ep	Primary	EC, T, Y	EC	Andrews et al. (1980, 1982); Wang et al. (1980, 1981); Andrews (1982); Bronson et al. (1984a)
2102ERP	RPLN	EC, T		Wang et al. (1980, 1981); Bronson et al. (1984a)
2806B		T		Hiraoka et al. (1988)
577MF	Forehead	EC, T	Undifferentiated carcinoma	Andrews et al. (1980); Wang et al. (1980, 1981)
577ML	Lung	EC, T		Wang et al. (1980, 1981)
577MR	RPLN	EC, T		Wang et al. (1980, 1981)
833KE	Primary	EC, T, C, S	EC	Bronson et al. (1978, 1980, 1984a); Andrews et al. (1980); Wang et al. (1980, 1981)
ER	Primary	EC, T		Harzmann et al. (1982); Boller et al. (1983)
GCT27	Primary	EC, T	EC, T, C, Y	Pera et al. (1987, 1988, 1989)
GCT35	Primary	EC, T	EC, C, Y	Pera et al. (1987, 1988)
GCT44	PALN	EC, Y	Y[b]	Pera et al. (1987, 1988, 1989); Cooper and Pera (1988)

(continues)

Table V

Continued

Cell line	Biopsy[a]		Xenograft histology	Reference
	Site	Histology		
GCT46	Lung	EC, Y	Y[b]	Pera et al. (1987,1988)
GCT48	Primary	EC	EC	Pera et al. (1987,1988)
GCT72	Primary	EC	Y[b]	Pera et al. (1987,1988)
GH	Primary	EC, T		Harzmann et al. (1982); Boller et al. (1983)
H12.1	Primary	EC, T, C, S	EC, T, C	Casper et al. (1987); Tesch et al. (1990)
H12.5	Primary	EC, T, C, S	EC	Casper et al. (1987); Tesch et al. (1990)
H12.7	Primary	EC, T, C, S	EC, T, C, Y	Casper et al. (1987); Tesch et al. (1990)
H23.1	Primary	EC	EC, Y	Tesch et al. (1990)
HL	Primary	EC, T		Harzmann et al. (1982); Boller et al. (1983)
ITO		T, Y, S	EC	Sekiya et al. (1985)
LICR-LON-HT1		EC, T, C		Cotte et al. (1981)
LICR-LON-HT3		EC, Y		Cotte et al. (1981)
LICR-LON-HT39/7		EC	EC	Cotte et al. (1981,1982); McIlhinney et al. (1983)
LICR-LON-HT5		EC, T, Y, S		Cotte et al. (1981)
LICR-LON-HT7		EC, Y		Cotte et al. (1981)
NCC-EC-1	Primary	EC, T	EC	Teshima et al. (1988)
NCC-EC-2	Primary	EC, S	EC	Teshima et al. (1988)
NCC-EC-3	Primary	EC, C, S	EC, C	Teshima et al. (1988)
NCR-G2		EC, S, T, Y	EC	Hata et al. (1993)
NCR-G3		EC, C, S, T, Y	EC, C, Y	Hata et al. (1993)
NEC14		EC, Y	EC	Sekiya et al. (1985,1990); Hasegawa et al. (1990)
NEC15		EC, C	EC	Sekiya et al. (1985)
NEC8		EC	EC	Yamamoto et al. (1979); Sekiya et al. (1985)
SuSa		EC, T		Hogan et al. (1977)
T84-5130		EC, S		Fenderson et al. (1987); J. W. Oosterhuis (unpublished observations)
TERA-1	Lung	EC,T		Fogh and Trempe (1975); Wang et al. (1981); Bronson et al. (1984a)
TERA-2	Lung	EC, T	EC, T	Fogh and Trempe (1975); Wang et al.

(continues)

Table V

Continued

Cell line	Biopsy[a]		Xenograft histology	Reference
	Site	Histology		
				(1981); Andrews (1984); Andrews et al. (1984b,1990); Gönczöl et al. (1984); Thompson et al. (1984); Fenderson et al. (1987); Simeone et al. (1990); Engström et al. (1991); Miller and Dmitrovsky (1991)
UM-TC-1	RPLN	EC, T, Y, S	EC	Grossman and Wedemeyer (1986)
Ovary tumors				
HUOT		IT	Anaplastic carcinoma	Ishiwata et al. (1985)
IMa		C	C	Sekiya et al. (1983)
PA1	Ascites	IT, T	IT, T	Giovanella et al. (1974); Zeuthen et al. (1980); Tainsky et al. (1984,1988)
YK		IT	EC	Kikuchi et al. (1984)
Extragonadal tumors				
NCC-IT	Primary	EC, T, Y	EC, T, C, Y	Teshima et al. (1988)
1618K		EC		Vogelzang et al. (1983); Hiraoka et al. (1988)
Haz1		T		Oosterhuis et al. (1985)
Haz2		IT		Oosterhuis et al. (1985)
Haz3		IT		Oosterhuis et al. (1985)
HOGT		T	T	Ishiwata et al. (1985)
U1161		IT, T, S		Sundstrom et al. (1980)

[a] Abbreviations: C, choriocarcinoma; EC, embryonal carcinoma; S, seminoma; T, teratoma (i.e., various somatic tissues present); IT, immature teratoma; Y, yolk sac carcinoma; RPLN, retroperitoneal lymph node; PALN, para-aortic lymph node. For a more complete description, the relevant references should be consulted.

[b] Pera et al. (1987) distinguish two forms of yolk sac carcinoma, namely, a solid form (cell line GCT72) and the more typical pattern of endodermal sinus tumor (cell lines GCT44 and GCT46).

Acknowledgments

Research presented in this chapter was supported by U.S. Public Health Services (USPHS) Grants CA29894, AI24943, and CA10815 (PWA) and by Grant HD-21355 (ID) from the National Institutes of Health.

References

Abe, K., McKibbin, J. M., and Hakomori, S. (1983). The monoclonal antibody directed to difuco-sylated type II chain (Fucα1→2Galβ1→4[Fucα1→3] GlcNAc; Y determinant). *J. Biol. Chem.* **258,** 11793–11797.

Andrews, P. W. (1982). Human embryonal carcinoma cells in culture do not synthesize fibronec-tin until they differentiate. *Int. J. Cancer* **30,** 567–571.

Andrews, P. W. (1984). Retinoic acid induces neuronal differentiation of a cloned human em-bryonal carcinoma cell line *in vitro. Dev. Biol.* **103,** 285–293.

Andrews, P. W., and Goodfellow, P. N. (1980). Antigen expression by somatic cell hybrids of a murine embryonal carcinoma cell with thymocytes and L cells. *Somat. Cell Genet.* **6,** 271–284.

Andrews, P. W., Bronson, D. L., Benham, F., Strickland, S., and Knowles, B. B. (1980). A compar-ative study of eight cell lines derived from human testicular teratocarcinoma. *Int. J. Cancer* **26,** 269–280.

Andrews, P. W., Goodfellow, P. N., Shevinsky, L., Bronson, D. L., and Knowles, B. B. (1982). Cell surface antigens of a clonal human embryonal carcinoma cell line: Morphological and anti-genic differentiation in culture. *Int. J. Cancer* **29,** 523–531.

Andrews, P. W., Goodfellow, P. N., and Bronson, D. L. (1983). Cell surface characteristics and other markers of differentiation of human teratocarcinomas in culture. *In* "Teratocarcinoma Stem Cells" (L. M. Silver, G. R. Martin, and S. Strickland, eds.), Cold Spring Harbor Confer-ences on Cell Proliferation, Vol. 10, pp. 579–590. Cold Spring Harbor Laboratory Press, Cold Spring Harbor, New York.

Andrews, P. W., Banting, G. S., Damjanov, I., Arnaud, D., and Avner, P. (1984a). Three monoclo-nal antibodies defining distinct differentiation antigens associated with different high molecu-lar weight polypeptides on the surface of human embryonal carcinoma cells. *Hybridoma* **3,** 347–361.

Andrews, P. W., Damjanov, I., Simon, D., Banting, G., Carlin, C., Dracopoli, N. C., and Fogh, J. (1984b). Pluripotent embryonal carcinoma clones derived from the human teratocarcinoma cell line Tera-2: Differentiation *in vivo* and *in vitro. Lab. Invest.* **50,** 147–162.

Andrews, P. W., Meyer, L. J., Bednarz, K. L., and Harris, H. (1984c). Two monoclonal antibodies recognizing determinants on human embryonal carcinoma cells react specifically with the liver isozyme of human alkaline phosphatase. *Hybridoma* **3,** 33–39.

Andrews, P. W., Damjanov, I., Simon, D., and Dignazio, M. (1985). A pluripotent human stem-cell clone isolated from the TERA-2 teratocarcinoma line lacks antigens SSEA-3 and SSEA-4 *in vitro,* but expresses these antigens when grown as a xenograft tumor. *Differentiation* **29,** 127–135.

Andrews, P. W., Trinchieri, G., Perussia, B., and Baglioni, C. (1987). Induction of class I major histocompatibility complex antigens in human teratocarcinoma cells by interferon without induction of differentiation, growth inhibition or resistance to viral infection. *Cancer Res.* **47,** 740–746.

Andrews, P. W., Gönczöl, E., Fenderson, B., Holmes, E. H., O'Malley, G., Hakomori, S.-I., and Plotkin, S. A. (1989). Human cytomegalovirus induces stage-specific embryonic antigen-1 in differentiating human teratocarcinoma cells and fibroblasts. *J. Exp. Med.* **169,** 1347–1359.

Andrews, P. W., Nudelman, E., Hakomori, S.-I., and Fenderson, B. A. (1990). Different patterns of glycolipid antigens are expressed following differentiation of TERA-2 human embryonal carci-noma cells induced by retinoic acid, hexamethylene bisacetamide (HMBA) or bromodeoxyu-ridine (BUdR). *Differentiation* **43,** 131–138.

Andrews, P. W., Marrink, J., Hirka, G., von Keitz, A., Sleijfer, D., and Gönczöl, E. (1991). The surface antigen phenotype of human embryonal carcinoma cells: modulation upon differentia-tion and viral infection. *Recent Res. Cancer Res.* **123,** 63–83.

Atsumi, T., Shirayoshi, Y., Takeichi, M., and Okada, T. S. (1982). Multipotent teratocarcinoma cells acquire the pluripotency for differentiation by fusion with somatic cells. *Differentiation* **23,** 83–86.

Benham, F. J., Andrews, P. W., Bronson, D. L., Knowles, B. B., and Harris, H. (1981). Alkaline phosphatase isozymes as possible markers of differentiation in human teratocarcinoma cell lines. *Dev. Biol.* **88,** 279–287.

Boller, K., Frank, H., Löwer, J., Lower, R., and Kurth, R. (1983). Structural organization of unique retrovirus-like particles budding from human teratocarcinoma cell lines. *J. Gen. Virol.* **64,** 2549–2559.

Bremer, E. G., Levery, S. B., Sonnino, S., Ghidoni, R., Canevari, S., Kannagi, R., and Hakomori, S. (1984). Characterization of a glycosphingolipid antigen defined by the monoclonal antibody MBr1 expressed in normal and neoplastic epithelial cells of human mammary gland. *J. Biol. Chem.* **259,** 14773–14777.

Brinster, R. L. (1974). The effect of cells transferred into the mouse blastocyst on subsequent development. *J. Exp. Med.* **140,** 1049–1056.

Bronson, D. L., Ritzi, D. M., Fraley, E. E., and Dalton, A. J. (1978). Morphologic evidence for retrovirus production by epithelial cells derived from a human testicular tumor metastasis. *J. Natl. Cancer Inst.* **60,** 1305–1308.

Bronson, D. L., Andrews, P. W., Solter, D., Cervenka, J., Lange, P. H., and Fraley, E. E. (1980). A cell line derived from a metastasis of a human testicular germ-cell tumor. *Cancer Res.* **40,** 2500–2506.

Bronson, D. L., Andrews, P. W., Vessella, R. L., and Fraley, E. E. (1983a). *In vitro* differentiation of human embryonal carcinoma cells. In "Teratocarcinoma Stem Cells" (L. M. Silver, G. R. Martin, and S. Strickland, eds.), Cold Spring Harbor Conferences on Cell Proliferation, Vol. 10, pp. 597–605. Cold Spring Harbor Laboratory Press, Cold Spring Harbor, New York.

Bronson, D. L., Clayman, R. V., and Fraley, E. E. (1983b). Human testicular germ cell tumors *in vitro. In* "The Human Teratomas. Experimental and Clinical Biology" (I. Damjanov, B. B. Knowles, and D. Solter, eds.), pp. 267–284. Humana Press, Englewood Cliffs, New Jersey.

Bronson, D. L., Saxinger, W. C., Ritzi, D. M., and Fraley, E. E. (1984a). Production of virions with retrovirus morphology by human embryonal carcinoma cells *in vitro. J. Gen. Virol.* **65,** 1043–1051.

Bronson, D. L., Vessella, R. L., and Fraley, E. E. (1984b). Differentiation potential of human embryonal carcinoma cell lines. *Cell Diff.* **15,** 129–132.

Casper, J., Schmoll, H-J., Schnaidt, U., and Fonatsch, C. (1987). Cell lines of human germinal cancer. *Int. J. Androl.* **10,** 105–113.

Chen, C., Fenderson, B. A., Andrews, P. W., and Hakomori, S. (1989). Glycolipid-glycosyltransferases in human embryonal carcinoma cells during retinoic acid-induced differentiation. *Biochemistry* **28,** 2229–2238.

Clausen, H., Levery, S. B., Nudelman, E., Baldwin, M., and Hakomori, S. (1986). Further characterization of type 2 and type 3 chain blood group A glycosphingolipids from human erythrocyte membranes. *Biochemistry* **25,** 7075–7085.

Cooper, S., and Pera, M. F. (1988). Vitronectin production by human yolk sac carcinoma cells resembling parietal endoderm. *Development* **104,** 565–574.

Cotte, C. A., Easty, G. C., and Munro-Neville, A. (1981). Establishment and properties of human germ cell tumors in tissue culture. *Cancer Res.* **41,** 1422–1427.

Cotte, C., Raghavan, D., McIlhinney, R. A. J., and Monaghan, P. (1982). Characterization of a new human cell line derived from a xenografted embryonal carcinoma. *In Vitro* **18,** 739–749.

Damjanov, I. (1986). Spontaneous and experimental testicular tumors in animals. *In* "Pathology of the Testis and Its Adnexa" (A. Talerman and L. M. Roth, eds.), pp. 193–206. Churchill Livingstone, New York.

Damjanov, I. (1991). Pathobiology of human germ cell neoplasia. *Recent Res. Cancer Res.* **123,** 1–19.

Damjanov, I., and Andrews, P. W. (1983). Ultrastructural differentiation of a clonal human embryonal carcinoma cell line in vitro. Cancer Res. **43**, 2190–2198.

Damjanov, I., and Solter, D. (1974). Experimental teratoma. Curr. Top. Pathol. **59**, 69–129.

Damjanov, I., Fox, N., Knowles, B. B., Solter, D., Lange, P. H., and Fraley, E. E. (1982). Immunohistochemical localization of murine stage-specific embryonic antigen in human testicular germ cell tumors. Am. J. Pathol. **108**, 225–230.

Damjanov, I., Clark, R. K., and Andrews, P. W. (1984). Cytoskeleton of human embryonal carcinoma cells. Cell Diff. **15**, 133–139.

Damjanov, A., Wewer, U. M., Tuma, B., and Damjanov, I. (1990). Basement membrane components secreted by mouse yolk sac carcinoma cell lines. Differentiation **45**, 84–95.

Dixon, F. J., Jr., and Moore, R. A. (1952). "Tumors of the Male Sex Organs." Armed Forces Institute of Pathology, Washington, D.C.

Dixon, F. J., Jr., and Moore, R. A. (1953). Testicular tumors: A clinico-pathologic study. Cancer **6**, 427–454.

Eisenbarth, G. S., Walsh, F. S., and Nirenberg, M. (1979). Monoclonal antibody to a plasma membrane antigen of neurons. Proc. Natl. Acad. Sci. U.S.A. **76**, 4913–4917.

Engström, W., Tally, M., Granerus, M., Hedley, E. P., and Schofield, P. (1991). Growth factors and the control of human teratoma cell proliferation. Recent Res. Cancer Res. **123**, 145–153.

Evans, M. J., and Kaufman, M. H. (1981). Establishment in culture of pluripotential cells from mouse embryos. Nature (London) **292**, 154–156.

Fenderson, B. A., Andrews, P. W., Nudelman, E., Clausen, H., and Hakomori, S.-I. (1987). Glycolipid core structure switching from globo- to lacto- and ganglio-series during retinoic acid-induced differentiation of TERA-2-derived human embryonal carcinoma cells. Dev. Biol. **122**, 21–34.

Fogh, J., and Trempe, G. (1975). New human tumor cell lines. In "Human Tumor Cells in Vitro" (J. Fogh, ed.), pp. 115–159. Plenum Press, New York.

Friedman, N. B., and Moore, R. A. (1946). Tumors of the testis. A report of 922 cases. Milit. Surg. **99**, 573–593.

Fukuda, M. N., Brothner, B., Lloyd, K. O., Rettig, N. J., Tiller, P. R., and Dell, A. (1986). Structures of glycosphingolipids isolated from human embryonal carcinoma cells. The presence of mono- and disialosyl glycolipids with blood group type 1 sequence. J. Biol. Chem. **261**, 5145–5153.

Geurts van Kessel, A., Suijkerbuijk, R., de Jong, B., and Oosterhuis, J. W. (1991). Molecular analysis of isochromosome 12p in testicular germ cell tumors. Recent Res. Cancer Res **123**, 113–118.

Giovanella, B. C., Stehlin, J. S., and Williams, L. J. (1974). Heterotransplantation of human malignant tumors in "nude" thymusless mice. II. Malignant tumors induced by injection of cell cultures derived from human solid tumors. J. Natl. Cancer Inst. **52**, 921–930.

Gönczöl, E., Andrews, P. W., and Plotkin, S. A. (1984). Cytomegalovirus replicates in differentiated but not undifferentiated human embryonal carcinoma cells. Science **224**, 159–161.

Grossman, H. B., and Wedemeyer, G. (1986). UM-TC-1, a new human testicular carcinoma cell line. Cancer J. **1**, 22–24.

Harzmann, R., Löwer, J., Löwer, R., Bichler, K., and Kurth, R. (1982). Synthesis of retrovirus-like particles in testicular teratocarcinomas. J. Urol. **128**, 1055–1059.

Hasegawa, T., Nakada, S., Nakajima, T., Oda, K., Kawata, M., Kimura, H., and Sekiya, S. (1990). Expression of various viral and cellular enhancer-promoters during differentiation of human embryonal carcinoma cells. Differentiation **42**, 191–198.

Hata, J., Fujimoto, J., Ishii, E., Umezawa, A., Kokai, Y., Matsubayashi, Y., Abe, S., Kusakari, S., Kikuchi, H., Yamada, T., and Maruyama, T. (1993). Differentiation of human germ cell tumor cells in vivo and in vitro. Acta Histochem. Cytochem. **25**, 563–576.

Hiraoka, A., Vogelzang, N. J., Rosner, M. C., and Golomb, H. M. (1988). Ultrastructure of four human germ cell tumor-derived cell lines: Effect of 12-O-tetradecanoylphorbol-13-acetate. *Cancer Invest.* **6,** 393–402.

Hogan, B., Fellous, M., Avner, P., and Jacob, F. (1977). Isolation of a human teratoma cell line which expresses F9 antigen. *Nature (London)* **270,** 515–518.

Ishiwata, I., Ishiwata, C., Soma, M., Tomita, K., Nozawa, S., and Ishikawa, H. (1985). Establishment and characterization of HUOT, a human ovarian malignant teratoma cell line producing α-fetoprotein. *J. Natl. Cancer Inst.* **75,** 411–422.

Izhar, M., Siebert, P. D., Oshima, R. G., DeWolf, W. C., and Fukuda, M. N. (1986). Trophoblastic differentiation of human teratocarcinoma cell line HT-H. *Dev. Biol.* **116,** 510–518.

Jacobsen, G. K., and Talerman, A. (1989). "Atlas of Germ Cell Tumours." Munksgaard, Copenhagen.

Kannagi, R., Nudelman, E., Levery, S. B., and Hakomori, S. (1982). A series of human erythrocyte glycosphingolipids reacting to the monoclonal antibody directed to a developmentally regulated antigen, SSEA-1. *J. Biol. Chem.* **257,** 14865–14874.

Kannagi, R., Cochran, N. A., Ishigami, F., Hakomori, S.-I., Andrews, P. W., Knowles, B. B., and Solter, D. (1983a). Stage-specific embryonic antigens (SSEA-3 and -4) are epitopes of a unique globoseries ganglioside isolated from human teratocarcinoma cells. *EMBO J.* **2,** 2355–2361.

Kannagi, R., Levery, S. B., Ishigami, F., Hakomori, S., Shevinsky, L. H., Knowles, B. B., and Solter, D. (1983b). New globoseries glycosphingolipids in human teratocarcinoma reactive with the monoclonal antibody directed to a developmentally regulated antigen, stage-specific embryonic antigen 3. *J. Biol. Chem.* **258,** 8934–8942.

Kikuchi, Y., Manose, E., Kizawa, I., Ishida, M., Sunaga, H., Mukai, K., Seki, K., and Kato, K. (1984). Characterization of an established cell line from human immature teratoma of the ovary and effects of retinoic acid on cell proliferation. *Cancer Res.* **44,** 2952–2958.

Kleinsmith, L. J., and Pierce, G. B. (1964). Multipotentiality of single embryonal carcinoma cells. *Cancer Res.* **24,** 1544–1551.

Lee, V. M.-Y., and Andrews, P. W. (1986). Differentiation of NTERA-2 clonal human embryonal carcinoma cells into neurons involves the induction of all three neurofilament proteins. *J. Neurosci.* **6,** 514–521.

Lifschitz-Mercer, G., Fogel, M., Moll, R., Jacob, N., Kushnir, I., Livoff, A., Waldherr, R., Franke, W. W., and Czernobilsky, B. (1991). Intermediate filament protein profiles of human testicular non-seminomatous germ cell tumors: Correlation of cytokeratin synthesis to cell differentiation. *Differentiation* **48,** 191–198.

McIlhinney, R. A. J., Patel, S., and Monaghan, P. (1983). Effects of 12-O-tetradecanoylphorbol-13-acetate (TPA) on a clonal human teratoma-derived embryonal carcinoma cell line. *Exp. Cell Res.* **144,** 297–311.

Marrink, J., Andrews, P. W., van Brummen, P. J., de Jong, H. J., Sleijfer, D., Schraffordt-Koops, H., and Oosterhuis, J. W. (1991). TRA-1-60: A new serum marker in patients with germ-cell tumors. *Int. J. Cancer* **49,** 368–372.

Martin, G. (1980). Teratocarcinomas and mammalian embryogenesis. *Science* **209,** 768–775.

Matthaei, K., Andrews, P. W., and Bronson, D. L. (1983). Retinoic acid fails to induce differentiation in human teratocarcinoma cell lines that express high levels of cellular receptor protein. *Exp. Cell Res.* **143,** 471–474.

Miettinen, M., Virtanen, I., and Talerman, A. (1985). Intermediate filaments in human testis and testicular germ cell tumors. *Am. J. Pathol.* **120,** 402–410.

Miller, W. H., and Dmitrovsky, E. (1991). Growth factors in human germ cell cancer. *Recent Res. Cancer Res.* **123,** 183–189.

Miller, W. H., Moy, D., Li, A., Grippo, J. F., and Dmitrovsky, E. (1990). Retinoic acid induces down-regulation of several growth factors and proto-oncogenes in a human embryonal cancer cell line. *Oncogene* **5,** 511–517.

Mostofi, F. K. (1980). Pathology of germ cell tumors. A progress report. *Cancer* **45,** 1735–1754.

Mostofi, F. K., and Price, E. B. (1973). "Tumors of the Male Genital System." Armed Forces Institute of Pathology, Washington, D.C.

Mostofi, F., and Sobin, L. H. (1976). Histological typing of testicular tumors. *In* "International Histologic Classification of Tumors." World Health Organization, Geneva.

Mummery, C. L., and Weima, S. M. (1991). Growth factors and receptors during differentiation: A comparison of human and murine embryonal carcinoma cells. *Recent Res. Cancer Res.* **123,** 165–182.

Muramatsu, H., Muramatsu, T., and Avner, P. (1982). Biochemical properties of the high molecular weight glycopeptides released from the cell surface of human teratocarcinoma cells. *Cancer Res.* **42,** 1749–1752.

Naiki, M., and Marcus, D. M. (1974). Human erythrocyte P and P^k blood group antigens: Identification as glycosphingolipids. *Biochem. Biophys. Res. Commun.* **60,** 1105–1111.

Oosterhuis, J. W., de Jong, B., van Dalen, I., van der Meer, I., Visser, M., de Leij, L., Mesander, G., Collard, J. G., Schrafford Koops, H., and Sleijfer, D. M. (1985). Identical chromosome translocations involving the region of the c-*myb* oncogene in four metastases of a mediastinal teratocarcinoma. *Cancer Genet. Cytogenet.* **15,** 99–107.

Papaioannou, V. E., McBurney, M. W., Gardner, R. L., and Evans, M. J. (1975). Fate of teratocarcinoma cells injected into early mouse embryos. *Nature (London)* **258,** 70–73.

Pattillo, R. A., and Gey, G. O. (1968). The establishment of a cell line of human hormone synthesizing trophoblastic cells *in vitro*. *Cancer Res.* **28,** 1231–1236.

Pattillo, R. A., Ruckert, A., Hussa, R., Bernstein, R., and Delfs, E. (1971). The Jar cell line—Continuous human multihormone production and control. *In Vitro* **6,** 398–399.

Pera, M. F., Blasco-Lafita, M. J., and Mills, J. (1987). Cultured stem cells from human testicular teratomas: The nature of human embryonal carcinoma, and its comparison with two types of yolk sac carcinoma. *Int. J. Cancer* **40,** 334–343.

Pera, M. F., Blasco-Lafita, M. J., Cooper, S., Mason, M., Mills, J., and Monoghan, P. (1988). Analysis of cell-differentiation lineage in human teratomas using new monoclonal antibodies to cytostructural antigens of embryonal carcinoma cells. *Differentiation* **39,** 139–149.

Pera, M. F., Cooper, S., Mills, J., and Parrington, J. M. (1989). Isolation and characterization of a multipotent clone of human embryonal carcinoma cells. *Differentiation* **42,** 10–23.

Pierce, G. B. (1975). Teratocarcinoma: Introduction and perspectives. *In* "Teratomas and Differentiation" (M. I. Sherman and D. Solter, eds.), pp. 3–12. Academic Press, New York.

Pugh, R. C. B. (ed). (1976). "Pathology of the Testis." Blackwell, Oxford.

Rendt, J., Erulkar, S., and Andrews, P. W. (1989). Presumptive neurons derived by differentiation of a human embryonal carcinoma cell line exhibit tetrodotoxin-sensitive sodium currents and the capacity for regenerative responses. *Exp. Cell Res.* **180,** 580–584.

Rettig, W. J., Cordon-Cardo, C., Ng, J. S. C., Oettgen, H. F., Old, L. J., and Lloyd, K. O. (1985). High molecular weight glycoproteins of human teratocarcinoma defined by monoclonal antibodies to carbohydrate determinants. *Cancer Res.* **45,** 815–821.

Rousset, J. P., Bucchini, D., and Jami, J. (1983). Hybrids between F9 nullipotent teratocarcinoma and thymus cells produce multidifferentiated tumors in mice. *Dev. Biol.* **96,** 331–336.

Sekiya, S., Kaiho, T., Shirotake, S., Iwasawa, H., Inaba, N., Kawata, M., Higaki, K., Ishige, H., Takamizawa, H., Minamihisamatsu, M., and Kuwata, T. (1983). Establishment and properties of a human choriocarcinoma cell line of ovarian origin. *In Vitro* **19,** 489–494.

Sekiya, S., Kawata, M., Iwasawa, H., Inaba, H., Sugita, M., Suzuki, N., Motoyama, T., Yamamoto, T., and Takamizawa, H. (1985). Characterization of human embryonal carcinoma cell lines derived from testicular germ-cell tumors. *Differentiation* **29,** 259–267.

Sekiya, S., Kimura, H., Yamazawa, K., Kera, K., Kawata, M., Takamizawa, H., and Oda, K. (1990). Induction of human embryonal carcinoma cell differentiation using N,N'-hexamethylene bisacetamide *in vitro*. *Gynecol. Oncol.* **36,** 69–78.

Shevinsky, L. H., Knowles, B. B., Damjanov, I., and Solter, D. (1982). Monoclonal antibody to murine embryos defines a stage-specific embryonic antigen expressed on mouse embryos and human teratocarcinoma cells. *Cell* **30,** 697–705.

Simeone, A., Acampora, D., Arcioni, L., Andrews, P. W., Boncinelli, E., and Mavilio, F. (1990). Sequential activation of human HOX2 homeobox genes by retinoic acid in human embryonal carcinoma cells. *Nature (London)* **346,** 763–766.

Skakkebaek, N. E., Berthelsen, J. G., Giwercman, A., and Müller, J. (1987). Carcinoma *in situ* of the testis: possible origin from gonocytes and precursor of all types of germ cell tumours except spermatocytoma. *Int. J. Androl.* **10,** 19–28.

Solter, D., and Damjanov, I. (1979). Teratocarcinoma and the expression of oncodevelopmental genes. *Meth. Cancer Res.* **18,** 277–332.

Solter, D., and Knowles, B. B. (1978). Monoclonal antibody defining a stage-specific mouse embryonic antigen (SSEA-1). *Proc. Natl. Acad. Sci. U.S.A.* **75,** 5565–5569.

Stevens, L. C. (1967). The biology of teratomas. *Adv. Morphog.* **6,** 1–31.

Stevens, L. C., and Varnum, D. S. (1974). The development of teratomas from pathenogenetically activated ovarian mouse eggs. *Dev. Biol.* **37,** 369–380.

Sun, T.-T., Eichner, R., Nelson, W. G., Tseng, S. C. G., Weiss, R. A., Jarvinen, M., and Woodcock-Mitchell, J. (1983). Keratin classes: Molecular markers for different types of epithelial differentiation. *J. Invest. Dermatol.* **81,** 1095–1155.

Sundstrom, C., Mark, J., and Westermark, B. (1980). An established human cell line derived from a malignant mediastinal teratoma. *Acta Pathol. Microbiol. Scand. Sect. A.* **88,** 1899–1904.

Tainsky, M. A., Cooper, C. S., Giovanella, B. C., and Van de Woude, G. F. (1984). An activated ras^N gene: Detected in late but not early passage human PA-1 teratocarcinoma cells. *Science* **225,** 643–645.

Tainsky, M. A., Krizman, D. B., Chiao, P. J., Yin, S. O., and Giovanella, B. C. (1988). PA-1, a human cell model for multistage carcinogenesis: Oncogenes and other factors. *Anticancer Res.* **8,** 899–914.

Talerman, A., and Roth, L. M. (eds). (1986). "Pathology of the Testis and Its Adnexa." Churchill Livingstone, New York.

Tanaka, S., Fujimoto, J., Ishii, E., and Hata, J. (1989). Human yolk sac tumor antigen 2G10: Biochemical characterization and significance as a serum antigen. *Int. J. Cancer* **44,** 788–794.

Tesch, H., Fürbas, R., Casper, J., Lyons, J., Bartram, C. R., Schmoll, H. J., and Bronson, D. L. (1990). Cellular oncogenes in human teratocarcinoma cell lines. *Int. J. Androl.* **13,** 377–388.

Teshima, S., Shimosato, Y., Hirohashi, S., Tome, Y., Hayashi, I., Kanazawa, H., and Kakizoe, T. (1988). Four new human germ cell tumor cell lines. *Lab. Invest.* **59,** 328–336.

Thompson, S., Stern, P. L., Webb, M., Walsh, F. S., Engström, W., Evans, E. P., Shi, W. K., Hopkins, B., and Graham, C. F. (1984). Cloned human teratoma cells differentiate into neuron-like cells and other cell types in retinoic acid. *J. Cell Sci.* **72,** 37–64.

Thurin, J., Herlyn, M., Hindsgaul, O., Stromberg, N., Karlsson, K., Elder, D., Steplewski, Z., and Koprowski, H. (1985). Proton NMR and fast atom bombardment mass spectrometry analysis of the melanoma-associated ganglioside 9-O-acetyl-GD3. *J. Biol. Chem.* **260,** 14556–14563.

Trojanowski, J. Q., and Hickey, W. F. (1984). Human teratomas express differentiated neural antigens: An immunochemical study with anti-neurofilament, anti-glial filament and anti-amyelin basic protein monoclonal antibodies. *Am. J. Pathol.* **115,** 383–389.

Uhl, L., Kelley, M., and Schindler, J. (1986). α-Difluoromethylornithine induces differentiation of a human embryonal carcinoma cell line *in vitro. Biochem. Biophys. Res. Commun.* **140,** 66–73.

Vogelzang, N., Andrews, P. W., and Bronson, D. (1983). An extragonadal human embryonal carcinoma cell line 1618K. *Proc. Am. Assoc. Cancer Res.* **24,** 3.

Vogelzang, N. J., Bronson, D. L., Savino, D., Vessella, R. L., and Fraley, E. E. (1985). A human embryonal-yolk sac carcinoma model system in athymic mice. *Cancer* **55,** 2584–2593.

Wang, N., Trend, B., Bronson, D. L., and Fraley, E. E. (1980). Nonrandom abnormalities in chromosome 1 in human testicular cancers. *Cancer Res.* **40,** 796–802.

Wang, N., Perkins, K. L., Bronson, D. L., and Fraley, E. E. (1981). Cytogenetic evidence for pre-meiotic transformation of human testicular cancers. *Cancer Res.* **41,** 2135–2140.

Wheeler, J. E. (1983). History of teratomas. *In* "The Human Teratomas: Experimental and Clinical Biology" (I. Damjanov, B. B. Knowles, and D. Solter, eds.), pp. 1–22. Humana Press, Englewood Cliffs, New Jersey.

Yamamoto, T., Komatsubara, S., Suzuki, T., and Oboshi, S. (1979). *In vitro* cultivation of human testicular embryonal carcinoma and establishment of a new cell line. *Gann* **70,** 677–680.

Zeuthen, J., Nørgaard, J. O. R., Avner, P., Fellous, M., Wartiovaara, J., Vaheri, A., Rosen, A., and Giovanella, B. C. (1980). Characterization of a human ovarian teratocarcinoma-derived cell line. *Int. J. Cancer* **25,** 19–32.

Index

A

Anaplastic astrocytoma-derived cell
lines
chromosomal abnormalities, 23–26
growth factors, 29–31
oncogenes and products, 27–29
phenotypic properties, 18–23
spheroid cultures, 31
tumor suppressor genes, 26–27
Antibodies
labeling of urinary bladder tumor cell
lines, 352–354
monoclonal, OC 125, 382–383
prostate tissue-specific, 399
Antigenic determinants
melanoma cell lines, 421–422
prostatic cancer cell lines, 396–399
Antigens
blood group, expression on SCC
cells, 113
cell surface, germ-cell tumor lines,
458–462
HLA, see Human leukocyte antigen
MHC, see Major histocompatiblity
complex antigen
phenotype expression by SCCs,
108–113
prostate-specific, 397–398

B

Bacterial contamination, cell culture
systems, 4
Bladder tumor cell lines, see Urinary
bladder tumor cell lines

Bovine viral diarrhea virus, transmission
during cell culture, 6–7
Breast-derived cell lines
establishment and maintenance, 162
media for, 181
from metastatic sites, 180
from primary breast carcinoma, 180
lists
breast tissue cell lines, 163
carcinomas metastatic to ascitic
fluid, 180
carcinomas metastatic to pleural
fluid, 175
carcinomas metastatic to solid
tissues, 173
markers, 181
morphology
breast tissue lines, 162–173
metastatic carcinomas to ascitic
fluid, 176–180
metastatic carcinomas to pleural
fluid, 175–176
metastatic carcinomas to solid
tissue, 173–175
prospects, 181
Bronchial carcinoid cell lines, 134–136

C

Carcinomas
choriocarcinoma cell lines, 448–449
embryonal, see Embryonal carcinoma
cell lines
endometrial, see Endometrial
carcinoma cell lines

mammary tissue
 breast tissue lines, 162–173
 metastatic to ascitic fluid, 176–180
 metastatic to pleural fluid, 175–176
 metastatic to solid tissue, 173–175
 non-small-cell lung, see Lung
  carcinoma cell lines,
  non-small-cell
 ovarian, see Ovarian carcinoma cell
  lines
 small-cell lung, see Lung carcinoma
  cell lines, small-cell
 uterine, see Uterine cancer cell lines
 yolk sac
  establishment and maintenance,
   448
  secreted products, 462
CD, see Clusters of differentiation
Cell lines
 bacterial contamination, 4
 cellular cross-contamination, 8–13
 DNA fingerprinting, 10–11
 fungal contamination, 4
 mycoplasma contamination, 4–5
 source tissue verification, 13–14
 viral contamination, 5–7
Choriocarcinoma cell lines,
  establishment and maintenance,
  448–449
Chorionic gonadotropin, production by
  germ-cell tumors, 462
Chromosome analysis, see also
  Karyotype analysis
 abnormalities in glioma cell lines,
  23–26
 esophageal tumor cell lines, 277–278
 melanoma cell lines, 420–421
 urinary bladder tumor cell lines,
  354–355
Chromosomes, double-minute
 in malignant glioma, 27
 in medulloblastoma cell lines, 32
Clusters of differentiation
 characterization, 225
 subclassification of leukemia cell
  lines, 236
Colony-stimulating factors, effects on
 hematopoiesis, 239

Colorectal cancer cell lines
 antigen expression, 334
 characteristics, 326–327
 culture characteristics, 324–327
 DOPA decarboxylase expression, 335
 establishment, 320–324
 gene amplification, 335
 growth factor expression, 335–337
 list, 318–319
 morphology, 325–328
  moderately differentiated, 328–329
  mucinous, 331–334
  poorly differentiated, 330–331
  well differentiated, 328
 proto-oncogene expression, 335–337
 pseudodiploid cell type, 335
Contamination
 cellular cross-contamination
  intraspecies, 9–13
  species verification, 9
  summary, 8
 esophageal tumor cell lines, 281
 microbial, cell culture systems, 3–7
 mycoplasma
  bladder tumor cell lines, 249
  cell culture systems, 4–5
Cross-contamination, intraspecies, 9–13
Cryopreservation, bladder tumor cell
  lines, 349–350
Cytochemistry
 hematopoietic cell lines, 225
 sarcoma cells, 255
Cytotoxicity, lymphocyte, bladder tumor
  cell lines, 351–352

D

DNA fingerprinting, human cell lines,
  10–11

E

EBV, see Epstein–Barr virus
Embryonal carcinoma cell lines
 antigens, intracellular, 462–463
 establishment and maintenance,
  445–448
 secreted products, 462

Endometrial carcinoma cell lines
  establishment and maintenance,
    362–363
  growth, 377–382
  list, 361
Epidermal growth factor, synthesis and
    response by prostatic cell lines,
    402–403
Epidermal growth factor receptor
  expression
    in gastric carcinomas, 311
    in glioma cell lines, 27–29
  role in esophageal cancer, 280
Epstein–Barr virus
  transformed B-lymphoblastoid cells,
    226
  transmission during cell culture, 6–7
Esophageal tumor cell lines
  chromosome analysis, 277–278
  contamination, 281
  epidermal growth factor receptors, 280
  establishment and maintenance
    major papers, 273
    TE series, 270
  growth and differentiation, 279
  hormone effects, 281
  immunocytochemistry, 277
  list, 271
  markers, 280–281
  models, 279–280
  morphology, 273–274
  oncogenes, 279–280
  prospects, 282
  therapeutic response, 281
  tumorigenicity, 278–279
  tumor suppressor genes, 279–280
  ultrastructure, 277
Extragonadal germ-cell tumors, 449, 469
Extrapulmonary small-cell carcinomas, 134

F

α-Fetoprotein, production by germ-cell
    tumors, 462
Fibroblast growth factor
  basic, autocrine effects in glioma cell
    lines, 29
  in prostatic cell lines, 402–403

Fibronectin, in glioma cell lines, 18–21
Fibrosarcoma HT-1080 cell line, 264
Fungal contamination, cell culture
    systems, 4

G

Galactocerebroside, expression on HOG
    and TC 620 cells, 31
Gastric tumor cell lines
  collagen production, 311
  cultures
    need for, 312
    serum-free media for, 312–313
  establishment and maintenance,
    288–292
  growth patterns, 293
  inhibitory effects of prostaglandins, 312
  list, 289–291
  MHC Class II antigen expression, 311
  morphology
    general features, 292
    KATO-III, 299–305
    MKN-1, 307–309
    MKN-7, 295–296
    MKN-28, 293–295
    MKN-45, 296–299
    MKN-74, 293–295
  oncogenes, 313
  tumorigenicity, 309
  ultrastructure, 293
    KATO-III, 305–307
    MKN-1, 309
    MKN-28, 296
    MKN-45, 299
    MKN-74, 296
    TSG-6, 305–307
Gene amplification, in glioma cell lines,
    27–29
Genes, see Oncogenes;
    Proto-oncogenes; Tumor suppressor
    genes
Genotyping, hematopoietic cell lines,
    236–237
Germ-cell tumor cell lines
  antigens
    cell surface, 458–462
    intracellular, 462–463

classification schemes, 446
establishment and maintenance
  choriocarcinomas, 448–449
  embryonal carcinomas, 445–448
  extragonadal germ-cell tumors, 449
  yolk sac carcinomas, 448
growth, 449–451
list, 466–469
morphology, 451–454
nomenclature, 444
prospects, 463–466
ultrastructure, 454–457
xenograft histology, 457
GFAP, see Glial fibrillary acidic protein
Glial fibrillary acidic protein, in glioma
  cell lines, 18–21
Glioblastoma multiforme-derived cell
  lines
  chromosomal abnormalities, 23–26
  growth factors, 29–31
  oncogenes and products, 27–29
  phenotypic properties, 18–23
  spheroid cultures, 31
  tumor suppressor genes, 26–27
Glioma cell lines
  chromosomal abnormalities, 23–26
  GFAP/fibronectin expression, 20–21
  growth factors, 29–31
  medulloblastoma-derived, 32–33
  oligodendroglioma-derived, 31
  oncogenes and products, 27–29
  phenotypic properties, 18–23
  primitive neuroectodermal
    tumor-derived, 32–33
  spheroid cultures, 31
  tumor suppressor genes, 26–27
Glycosylation, esophageal cancer cell
  products, 280
Growth
  esophageal tumor cell lines, 279
  gastric tumor cell lines, 293
Growth factors, see also specific growth
  factors
  autocrine effects in glioma cell lines,
    29–31
  in colorectal tumor cell lines, 335–337
  dependent leukemia cell lines,
    242–243
  hematopoietic, 239

production by melanoma cells,
  419–421
role in esophageal cancer, 280

H

Head and neck cancer
  characterization, 79–80
  etiology, 80–81
Head and neck tumor cell lines
  antigen phenotype, 108–113
  culture success rate, 93–95
  establishment and maintenance
    history, 81–86
    methodology, 86–95
    explant cultures, 86–91
    extracellular matrices, 92
    feeder layer cultures, 91–92
    growth factor effects, 92–93
  list, 82–85
  malignant types, 108
  morphology
    ATTC-SCC cell lines, 105
    monolayer nonstratifying types,
      97–103
    salivary gland cell lines, 106, 118
    squamous carcinomas: UM-SCC
      series, 96–105
    stratifying colonies, 103–105
    thyroid tumor cell lines, 116
  nomenclature, 96
  prospects, 115
Hematopoietic cells
  cell lineages, 215
  culture, 214–217
  hierarchy, 214
  leukemia cells, see Leukemia cell
    lines
HLA, see Human leukocyte antigen
HOG cell line, characterization, 31
Hogkin's disease-derived cell lines,
  243–247
Human immunodeficiency virus,
  transmission during cell culture, 6
Human leukocyte antigen
  alloantigens, bladder tumor cell lines,
    350–351
  Class II, bladder tumor cell lines, 351

I

Immunocytochemistry
  esophageal tumor cell lines, 277
  sarcoma cells, 255
Immunophenotyping, hematopoietic cell
    lines, 225–236
Insulin/insulin-like growth factor,
    autocrine effects in glioma cell
    lines, 30
Insulin-like growth factor, synthesis and
    response by prostatic cell lines,
    402–403
$\alpha6\beta4$ Integrin, expression by SCC cell
    lines, 115
Interleukin 6, prostatic cell response,
    404
Isoenzymes, bladder cancer cell lines,
    354

K

Karyotype analysis, *see also*
    Chromosome analysis
  for cell line verification, 12–13
  hematopoietic cell lines, 237–239
  prostatic cancer cell lines, 400–401
  sarcoma cells, 255
Keratin
  in embryonal carcinoma cells, 462
  in prostatic cancer cell lines, 397

L

Leukemia cell lines
  analytical characterization, 219–220
  collections, 247
  culture *in vitro*, 221–222
  cytochemistry, 225
  cytogenetics, 237–241
  cytokines, 239
  data bank, 244–246
  functional features, 239–240
  genotyping, 236–237
  growth factor-dependent, 242–243
  Hodgkin's disease-derived lines,
      243–247
  identification and description, 221
  immunophenotyping, 225, 236

  karyotype analysis, 237–239
  lists, 218, 240–241, 243, 249
  milestones, 218
  morphology, 222–225
  novel, establishment, 247–248
  origins, 220–221
Lung cancers
  cellular origins, 121–123
  major forms, 123–124
  tumor culture
    advances in, 126–128
    media for, 124–125
Lung carcinoma cell lines
  non-small-cell
    adenocarcinomas, 138–144
    large-cell carcinomas, 145
    NSCLC–NE tumors, 134, 137
    squamous cell carcinomas,
        137–139
  small-cell
    classic subtype, 132–133
    classic type, 130–132
    establishment and maintenance,
        125–132
    variant form, 133–134
Lymphocytes, cytotoxicity to bladder
    tumor cell lines, 351–352

M

Major histocompatiblity complex antigen
    Class I
      embryonal carcinoma cell lines, 460
      neuroblastoma cell lines, 55
    Class II, gastric tumor cell lines, 311
Mammary tumor-derived cell lines, *see*
    Breast-derived cell lines
Medulloblastoma-derived cell lines,
    characterization, 32–33
Melanocyte cell lines
  establishment and maintenance, 414
  growth, 416–419
  morphology, 416–419
Melanoma cell lines
  antigen expression, 421–422
  chromosome analysis, 420–421
  establishment and maintenance,
      414–416
  growth, 416–419

growth factor production, 419–420
invasive properties *in vitro*, 422–424
list, 425
metastasis, 422–424
morphology, 416–419
Mesothelioma cell lines, 146
Micrographs
  breast-related cell lines
    DU4475, 174
    HMT-3909, 171
    HMT-3909S8, 166
    Hs578T, 169
    MDA-MB-134, 179
    MDA-MB-453, 178
    RW-972, 170
    UAC 245, 177
    UACC 265, 172
    UACC 812, 164
    UACC 893, 167
    UACC 2116, 165
  central nervous system-derived cell
      lines
    U-343 MG, 19
    U-343 MGa, 24–25
    U-348 MG, 19
  colorectal cancer cell lines
    Caco-2, 320
    COLO 205, 321
    HT-29, 321
    LoVo, 322
    NCI-H498, 333
    NCI-H508, 330
    NCI-H548, 329
    NCI-H716, 332
    SNU-C2A, 331
    SW-480, 322
    T84, 323
  endometrial carcinoma cell lines
    AN3Ca, 379
    ECC-1, 380, 382–383
    HEC-1A, 375
    HEC-1B, 376
    Ishikawa, 379, 381
    KLE, 377
    RL 95.2, 378
    SCRC, 376
  esophageal tumor cell lines
    KYSE-30, 275
    KYSE-50, 275

KYSE-200, 275–276
KYSE-520, 276
TE-1, 274
TE-2, 274
TE-3, 274
TE-4, 274
TE-5, 274
  gastric tumor cell lines
    KATO-III, 304–308
    MKN-1, 309–310
    MKN-7, 296–297
    MKN-28, 294–295, 297–298
    MKN-45, 300–303
    MKN-74, 299
    TSG-6, 306
  germ-cell tumor lines
    177NRPmet, 450
    833KE, 450, 455
    1156QE, 450–451
    1411H, 453
    1618K, 450
    1777N, 459
    1777NRPdiff, 451
    2102Ep, 457–458
    2102Ep cl.4D3, 451
    2120Ep, 454
    NCCIT, 450
    NTERA-2 cl.D1, 452
    SuSa, 450–451, 456
    TERA-1, 450
    TERA-2, 452
  head and neck tumor cell lines
    Detroit 526, 114
    FaDu, 114
    RPMI 2650, 114
    SCC-15, 114
    SW 579, 116
    TT, 116
    UM-SCC-5, 98
    UM-SCC-6, 89
    UM-SCC-9, 107
    UM-SCC-11A, 102
    UM-SCC-11B, 102
    UM-SCC-16, 107, 109
    UM-SCC-17A, 112
    UM-SCC-17B, 112
    UM-SCC-18, 91, 94
    UM-SCC-20, 104
    UM-SCC-22A, 99

UM-SCC-22B, 99
UM-SCC-25, 100
UM-SCC-29, 89
UM-SCC-30, 101
UM-SCC-31, 110
UM-SCC-32, 110
UM-SCC-37, 111
UM-SCC-38, 101
UM-SCC-55, 101
hematopoietic cell lines
   380, 229
   697, 228
   CML-T1, 227
   EBV-transformed B-lymphoblastoid
      cells, 226, 229
   EM-2, 223, 232
   HDLM-2, 245–246
   HL-60, 224, 231
   JOSK-I, 232
   JVM-2, 230
   K-562, 233
   KM-H2, 234
   M-07e, 233
   MEG-01, 223
   MKB-1, 226
   MOLT-3, 228
   P12/Ichikawa, 227
   RC-2A, 231
   U-266, 229
lung-related cell lines
   A549, 140
   adenocarcinomas, 140–144
   Darmouth group SCLC cells, 131
   NCI-H82, 133
   NCI-H441, 142–143
   NCI-H727, 135
   NCI-H810, 137
   NCI-H920, 133
   NCI-H1404, 141
   NCI series, classic type, 130
   SCC-5, 145
   SCC-15, 139
melanocytes and melanoma cells
   FM723, 417
   SKMEL23, 417
   WM266-4, 417
ovarian carcinoma cell lines
   A 2780, 369
   CaOV-3, 368

CaOV-4, 368
   HEY, 370
   HEY A8, 370
   NIH:OVCAR 3, 372
   NIH:OVCAR 5, 373
   OVCA 420, 364
   OVCA 429, 365
   OVCA 432, 365
   OVCA 433, 366
   OVCAR 3, 366
   OVCAR 5, 367
   PA-1, 371
   SK-OV-3, 367
   SW 626, 369
pancreatic tumor cell lines
   HPAF, 433–434
   HPAF CD 11 clone, 436–438
   HPAF CD 18 clone, 436–438
peripheral nervous system
   Ewing's sarcoma/neuroepithelioma
      cell lines, 66
   IMR-32, 48
   SK-N-SH, 48, 50
   SK-N-SH SY5Y subclone, 57
   SMS-KCNR, 58
   TC-268, 67
prostatic cell lines
   DU 145, 392, 398–399, 406
   epithelial cells, 393–395
   LNCaP, 392, 398–399, 406
   PC-3, 392, 398–399, 406
sarcoma cell lines
   A-204, 263
   A-673, 263
   HOS, 258
   HT-1080, 264
   MG-63, 258
   RD, 262
   Saos-2, 260
   U2-OS, 257
small-cell lung carcinomas
   classic type, NCI series, 130
   Dartmouth group, 131
   variant type NCI-H82, 133
urinary bladder tumor cell lines,
   344–347
   J82, 347
   RT4, 344
   SCaBER, 346

T24, 345
TCC SUP, 347
uterine cancer cell lines
   SK-UT-1, 374
   SK-UT-1B, 374
Monoclonal antibodies, OC 125,
      382–383
Mycoplasma infection
   bladder tumor cell lines, 249
   cell culture systems, 4–5

N

Neck cancer, see Head and neck cancer
Neuroblastoma
   clinical features, 45
   genetic features, 46
   pathological features, 45–46
   treatment, 47
Neuroblastoma-derived cell lines
   biochemical/biological
         characteristics, 49–59
   biological response, 61
   Class I MHC antigen expression, 55
   genetic characteristics, 59–69
   morphology, 47–49
   novel therapeutic strategies, 60–61
   proto-oncogene expression, 52
Neuroectodermal tumors, primitive,
      derived cell lines, see Primitive
      neuroectodermal tumor-derived cell
      lines
Neuroepithelioma
   clinical features, 61–62
   genetic features, 63
   pathological features, 62–63
   treatment, 63–64
Neuroepithelioma-derived cell lines
   biochemical/biological
         characteristics, 51, 65–68
   genetic characteristics, 68–69
   morphology, 64–65
   novel therapies with, 69–70
   proto-oncogene expression, 52
Non-small-cell lung carcinoma cell
      lines, see Lung carcinoma cell
      lines, non-small-cell

Non-small-cell lung
   carcinoma–neuroendocrine tumor
   (NSCLC–NE) cell lines, 134, 137

O

Oligodendroglioma-derived cell
   lines, 31
Oncogenes, see also Proto-oncogenes
   esophageal tumor cell lines, 279–280
   gastric tumor cell lines, 313
   glioma cell lines, 27–29
   prostatic cancer cell lines, 401
Osteosarcoma-derived cell lines
   HOS, 258
   MG-63, 258–259
   Saos-2, 259–260
   U2-OS, 257–258
Ovarian carcinoma cell lines
   CA 125 marker, 382–383
   establishment and maintenance,
      360–362
   growth, 377–382
   list, 361
   morphology, 364–371

P

Pancreatic tumor cell lines
   biochemical properties, 436, 439–440
   establishment and maintenance, 432
   induction of differentiation, 436–439
   list, 430–431
   morphology, 432–435
   prospects, 441
Photomicrographs, see Micrographs
Platelet-derived growth factor, autocrine
      effects in glioma cell lines, 29–30
Polymerase chain reaction, for cell line
      verification, 12
Primitive neuroectodermal tumor-derived
      cell lines, 32–33
Prostaglandins, inhibitory effects on
      gastric tumor cell lines, 312
Prostate-specific antigen, expression,
      397–398
Prostatic acid phosphatase, expression,
      398

Prostatic cancer cell lines
  antigenic determinants, 396–399
  DU 145, 389
  DuPro-1, 390
  growth factor expression/response, 402
  HONDA, 390
  immortal lines, 388–390
  invasive properties *in vitro*, 402
  karyotype analysis, 400–401
  list, 388
  LNCaP, 389
  morphology
    monolayer cultures, 391–395
    3D cultures, 396
  oncogenes, 401
  PC-3, 389
  PC 82, 390
  PC EG, 390
  PC EW, 390
  PPC-1, 389
  prospects, 405–407
  response to inhibitory factors, 402–404
  short-term strains, 390–391
  TSU-Pr1, 389
  tumor suppressor genes, 401
Proto-oncogenes, *see also* Oncogenes
  in colorectal tumor cell lines, 335–337
  in neuroblastoma- and neuroepithelioma-derived cell lines, 52

R

Rhabdomyosarcoma-derived cell lines
  A-204, 262
  A-673, 262–264
  RD, 261–262

S

Salivary gland tumor cell lines
  list, 82
  micrograph, 118
  morphology, 106
Sarcoma tumor cell lines
  establishment and maintenance, 252

fibrosarcoma HT-1080, 264
list, 256
osteosarcomas, 256–260
  HOS, 258
  MG-63, 258–259
  Saos-2, 259–260
  U2-OS, 257–258
rhabdomyosarcomas, 260–264
  A-204, 262
  A-673, 262–264
  RD, 261–262
xenografts, 255
Seed stock, development, 2–3
Small-cell lung carcinoma cell lines, *see*
  Lung carcinoma cell lines, small-cell
Spheroid cultures, glioma cells, 31

T

Testis tumors, 466–469; *see also*
  Germ-cell tumor cell lines
Thyroid tumor cell lines
  list, 85
  micrograph, 116
  morphology, 106
Transforming growth factor β
  autocrine effects in glioma cell lines, 30
  prostatic cell response, 404
Tumorigenicity
  esophageal tumor cell lines, 278
  gastric tumor cell lines, 309
  prostatic cell lines, 405
Tumor necrosis factor, prostatic cell response, 404
Tumor suppressor genes
  esophageal tumor cell lines, 279–280
  glioma cell lines, 26–27
  prostatic cancer cell lines, 401

U

Ultrastructure
  esophageal tumor cell lines, 277
  gastric tumor cell lines, 293
  urinary bladder tumor cell lines, 350

Urinary bladder tumor cell lines
  antibodies, 352–354
  chromosome analysis, 354–355
  cryopreservation, 349–350
  establishment and maintenance, 349
  HLA Class II antigen expression, 351
  HLA phenotyping, 350
  isoenzyme pattern, 354
  J82
    developed culture, 346–349
    histology, 346–349
    original tumor, 343
  list, 348
  lymphocyte cytotoxicity assays,
    351–352
  mycoplasma testing, 349
  RT4
    developed culture, 344
    histology, 344
    original tumor, 342, 344
  SCaBER
    developed culture, 345
    histology, 345
    original tumor, 342–343
  T24
    developed culture, 344
    histology, 344
    original tumor, 342

TCC SUP
  developed culture, 345–346
  histology, 345–346
  original tumor, 343
Uterine cancer cell lines
  growth, 377–382
  list, 361
  morphology, 371–377

V

Viral contamination
  cell culture systems, 5–7
  representative viruses, 6
Vitamin A, prostatic cell response, 404

X

Xenografts
  germ-cell tumor cell lines, histology,
    457
  sarcoma cells, 255

Y

Yolk sac carcinomas
  establishment and maintenance, 448
  secreted products, 462